Hydrogen Peroxide

Hydrogen peroxide (H_2O_2) is recognized as a crucial signaling molecule that mediates physiological and biochemical processes in plants, regulating various development and stress responses. *Hydrogen Peroxide: Signaling Mechanisms and Crosstalk in Plant Development and Stress Responses* presents a comprehensive overview of hydrogen peroxide's modes of action in plants, demonstrating the important role played in plant stress signaling and communication. It introduces key topics in H_2O_2 research such as plant signaling, molecular responses, and interaction with other hormones.

Features

- Discusses experiments interrelated to H_2O_2 signaling pathway in plants under various environmental conditions.
- Addresses important concerns in H_2O_2 research from a wide range of organisms, including plants and prokaryotes such as bacteria and archaea.
- Collects, summarizes, and presents developments in plant signaling and communication.
- Aids scientists and breeders in developing strategies to enhance plant growth and stress tolerance.

Environmental stress is destructively disturbing plant growth and efficiency, resulting in concerns to improve food crop yield, and H_2O_2 has immense field implications as it is vital in regulating plant growth and stress responses. *Hydrogen Peroxide: Signaling Mechanisms and Crosstalk in Plant Development and Stress Responses* is an invaluable resource for researchers and scientists to use as a guide to conduct studies on environmental conditions of the plant hydrogen peroxide signaling systems.

Hydrogen Peroxide

Signaling Mechanisms and Crosstalk in Plant Development and Stress Responses

Edited by
Mohd Tanveer Alam Khan, Taiba Saeed,
Aqeel Ahmad, Qazi Fariduddin, and Mohammad Yusuf

CRC Press
Taylor & Francis Group
Boca Raton London New York

CRC Press is an imprint of the
Taylor & Francis Group, an **informa** business

Designed cover image: Shutterstock

First edition published 2025
by CRC Press
2385 NW Executive Center Drive, Suite 320, Boca Raton FL 33431

and by CRC Press
4 Park Square, Milton Park, Abingdon, Oxon, OX14 4RN

CRC Press is an imprint of Taylor & Francis Group, LLC

Library of Congress Cataloging-in-Publication Data
Names: Khan, Tanveer Alam, editor. | Saeed, Taiba, editor. | Ahmad, Aqeel, editor. |
Fariduddin, Qazi, editor. | Yusuf, Mohammad (Lecturer in biology) editor.
Title: Hydrogen peroxide : signalling mechanisms and crosstalk in plant
development and stress responses / edited by Tanveer Alam Khan, Taiba
Saeed, Aqeel Ahmad, Qazi Fariduddin, and Mohammad Yusuf.
Description: First edition. | Boca Raton, FL : CRC Press, 2025. |
Includes bibliographical references and index. |
Identifiers: LCCN 2024043153 (print) | LCCN 2024043154 (ebook) |
ISBN 9781032610443 (hardback) | ISBN 9781032647678 (paperback) |
ISBN 9781032647692 (epub)
Subjects: LCSH: Plant cellular signal transduction. | Hydrogen peroxide. |
Plants—Effect of stress on. | Plant defenses.
Classification: LCC QK754 .H94 2025 (print) | LCC QK754 (ebook) |
DDC 571.7/42—dc23/eng/20250111
LC record available at https://lccn.loc.gov/2024043153
LC ebook record available at https://lccn.loc.gov/2024043154

ISBN: 9781032610443 (hbk)
ISBN: 9781032647678 (pbk)
ISBN: 9781032647692 (ebk)

DOI: 10.1201/9781032647692

Typeset in Times
by codeMantra

Dedication

Dedicated to my son

(Tahaam Alam Khan)

Contents

Contributors

Faheem Adil
Department of Agronomy
Zhejiang University
Hangzhou, People's Republic of China

Zill e Huma Aftab
Department of Plant Pathology
University of the Punjab
Lahore, Pakistan

Bushra Ahmad
Department of Bioengineering
Integral University
Lucknow, India

Sagheer Ahmad
School of Landscape Architecture and Art
Fujian Agriculture and Forestry University
Fuzhou, China

Waheed Akram
Department of Plant Pathology
University of the Punjab
Lahore, Pakistan

Tehmina Anjum
Department of Plant Pathology
University of the Punjab
Lahore, Pakistan

Taruba Ansari
Consultant at National Raw Drug Repository
Ghaziabad, India

Harpreet Bhatia
Department of Botany
Cluster University
Jammu, India

Adil Dilawar
Faculty of Geographical Science
Beijing Normal University
Beijing, China

Mohammad Faizan
School of Sciences
Maulana Azad National Urdu University
Hyderabad, India

Aiman Fatima
Institute of Biochemistry and Biotechnology
University of Veterinary and Animal Sciences
Lahore, Pakistan

Jamila Fatima
Faculty of Agricultural Sciences
University of the Punjab
Lahore, Pakistan

Mahnoor Fatima
Institute of Biochemistry and Biotechnology
University of Veterinary and Animal Sciences
Lahore, Pakistan

Tasneem Fatma
Department of Biosciences
Jamia Millia Islamia
New Delhi, India

Farhan Goher
College of Plant Protection
Northwest A&F University
Shaanxi, People's Republic of China

Mayank Anand Gururani
Department of Biology, College of Science
United Arab Emirates University
Al Ain, UAE

Bisma Hilal
Department of Botany
Aligarh Muslim University
Aligarh, India

Barket Hussain
Faculty of Life Sciences, Department of Botany
Aligarh Muslim University
Aligarh, India

Syed Khalida Izhar
Department of Bioengineering
Integral University
Lucknow, India

Sajeesh Kappachery
College of Science
United Arab Emirates University
Al Ain, United Arab Emirates

Sameera Karumannil
Department of Biology
United Arab Emirates University
Al Ain, United Arab Emirates

Ahamad Faiz Khan
Department of Bioengineering
Integral University
Lucknow, India

Faiza Shafique Khan
College of Horticulture and Forestry Science
Huazhong Agricultural University
Wuhan, People's Republic of China

Mohd Sajid Khan
Department of Biochemistry
Aligarh Muslim University
Aligarh, India

Tanveer Ahmad Khan
Department of Botany
Aligarh Muslim University
Aligarh, India

Zhongjian Liu
School of Landscape Architecture and Art
Fujian Agriculture and Forestry University
Fuzhou, China

Sugandha Mahajan
Department of Botany
Cluster University
Jammu, India

Alina Mansoor
Institute of Biochemistry and Biotechnology
University of Veterinary and Animal Sciences
Lahore, Pakistan

Abdul Mazeed
Department of Agriculture (IIAST)
Integral University
Lucknow, India

Farina Mujeeb
Department of Biosciences
Integral University
Lucknow, India

Nilofer
Department of Biosciences
Integral University
Lucknow, India

Vinay Kumar Pandey
Biotechnology Department
Manav Rachna International Institute of Research Studies (Deemed to be University) Faridabad,
India

Donghui Peng
School of Landscape Architecture and Art
Fujian Agriculture and Forestry University
Fuzhou, China

Fasiha Qurashi
Department of Biological Sciences
University of Veterinary and Animal Sciences
Lahore, Pakistan

Hamda Hussain Qureshi
Department of Biological Sciences
University of Veterinary and Animal Sciences
Lahore, Pakistan

Hamza Rafiq
State Key Laboratory of Crop Stress Biology for Arid Areas
Northwest A&F University
Shaanxi, China

Areeba Rehman
College of Earth and Environmental Sciences
University of the Punjab
Lahore, Pakistan

Shafaq Sahar
Department of Agronomy
Zhejiang University
Hangzhou, People's Republic of China

Mohd Saleem
Department of Botany
Aligarh Muslim University
Aligarh, India

Neha Sami
Department of Biosciences
Jamia Millia Islamia
New Delhi, India

Hamid Shahida
Department of Biosciences
Integral University
Lucknow, India

Anwar Shahzad
Department of Botany
Aligarh Muslim University
Aligarh, India

Iqra Shahzadi
School of Resource and Environmental Sciences
Wuhan University
Hubei, China

Rafeeya Shams
Department of Food Technology and Nutrition
Lovely Professional University
Phagwara, India

Shivangi Srivastava
Department of Food Technology
Harcourt Butler Technical University
Uttar Pradesh, India

Durdana Yasin
Department of Biosciences
Integral University
Lucknow, India

Nasim Ahmad Yasin
Department of Horticulture
University of the Punjab
Lahore, Pakistan

About the Editors

Dr. Mohd Tanveer Alam Khan is a Leibniz-DAAD Postdoctoral Fellow at the Leibniz Institute of Plant Genetics and Crop Plant Research (IPK) in Gatersleben, Germany. His research focuses on understanding the integrative analysis of low-temperature stress defense responses in *Arabidopsis thaliana*, particularly in relation to brassinosteroids signaling and metabolite patterns. He earned his BSc, MSc, and PhD from the Department of Botany at Aligarh Muslim University in Aligarh, India. Prior to joining the Leibniz Institute, Dr. Khan worked as a postdoctoral fellow at the National Key Laboratory of Crop Genetic Improvement at Huazhong Agricultural University in Wuhan, P.R. China. His research aims to dissect the mechanisms of abiotic stress tolerance in plants through engineered signaling, proteomics, metabolomics, and biochemical traits, both in the presence and absence of phytohormones. Over his 12 years as a researcher, Dr. Khan has published more than 34 research articles in internationally reputed journals, with a total impact factor exceeding 50 and 1374 citations. He holds an h-index of 19 and has contributed six book chapters to editions published by Springer. He has received various research fellowships during his PhD and postdoctoral tenure, including the CST-UP-RA, SERB-NPDF, and international PDF at Huazhong Agricultural University, Wuhan, China, and the Leibniz Institute of Plant Genetics and Crop Plant Research (IPK) in Gatersleben, Germany.

Dr. Taiba Saeed is an Assistant Professor in the Department of Biosciences at Integral University, Lucknow, India. Dr. Saeed is a young and active academician and researcher working in the fields of biotechnology, plant physiology, and molecular sciences. She earned her PhD degree from the Aligarh Muslim University, Aligarh, India. She extensively studied plant tissues culture, medicinal plants, plant signaling pathways, proteomics, and metabolomics. After obtaining her PhD degree, Dr. Saeed joined the Research Associate position at Biotech Park, Lucknow, under a DBT-sponsored partnership project with IBSD, Imphal, for a period of one year. Later, she moved to Germany to work as a visiting scientist at Leibniz Institute of Plant Genetics and Crop Plant Research (IPK), Germany. She has published more than 18 research articles in world-renowned journals. She has been conferred several research fellowships during her PhD and post-PhD tenure such as CST-UP-RA, DBT-RA, and international PDF in Leibniz Institute of Plant Genetics and Crop Plant Research (IPK), Gatersleben, Germany.

Dr. Aqeel Ahmad is a young and energetic researcher, working in the field of stress physiology. He earned his doctoral degree (PhD) from the University of the Punjab, Lahore, Pakistan, and performed his research work at the University of Florida, USA. Then, he joined Huazhong Agricultural University as a Postdoctoral Fellow, and now he is serving Guangdong Academy of Agricultural Sciences as a Postdoctoral Fellow. He extensively studied plant defense responses (i.e., defense genes, hormones, PR proteins, and biochemicals) against the plant pathogens. His core skills are proteomics (profiling, characterization, and supramolecular kinetics) and metabolomics, by using which he elucidated cell signaling to understand plant defenses, based on oxidative enzymes of the phenylpropanoid pathway. Transcriptomics further strengthened his scientific findings concerning cell responses toward pathogenic invasions and abiotic stresses. He has devised bioactive metabolites (e.g., benzimidazole and benzenedicarboxylic acid) to control plant pathogens and to augment the nutritional quality of our plant-based foods. He has developed techniques to make edible plants tolerant against environmental stressors by reharmonizing their osmoregulatory systems, oxidative machinery, and physiological responses. He has published 59 research manuscripts in world-renowned journals and won three research grants and one research honor award at a young age of 32 years. A wide spectrum of publication platforms is evident from his scientific articles, including the leading journals of *Food Chemistry*, *Chemosphere*, etc. His editorial activities in multiple impact factor journals have made him a distinctive and progressive figure in the researchers' pool.

Qazi Fariduddin is Professor of Botany at Aligarh Muslim University, Aligarh, India, where he has been serving as faculty since 2006. He has been extensively working in the field of agricultural biotechnology to explore the abiotic stress tolerance mechanism in plants through physiological and molecular approaches. The findings of his work have revealed that brassinosteroids (BRs) and salicylic acid improved the yield and quality of plants under low temperatures, salt, water, and heavy metal stress and could be exploited as a farmer-friendly tool to overcome the menace of crop losses due to various abiotic stresses. Moreover, his findings have also revealed the potential role of hydrogen peroxide and polyamines in conferring tolerance to abiotic stresses in crop plants. His lab is extensively using proteomic approaches to reveal the novel pathway protein expressed under various abiotic stresses in plants. He had visited Göttingen University, Göttingen, Germany, for six months under BOYSCAST Fellowship and conducted experiments related to the topic "Molecular studies of salt tolerance in *Arabidopsis thaliana*." He has visited Michigan State University, Michigan, USA, on an International Research Project with a specific objective to generate information on "Host target modification as a strategy to counter pathogen hijacking of the jasmonate hormone receptor" (published in PNAS, 2015). He has published more than 80 research papers in the international journals of high impact factor such as *Proceedings of the National Academy of Sciences*, USA, *Food Chemistry*, *Plant Physiology and Biochemistry*, *Chemosphere*, *Journal of Integrative Plant Biology*, *Environmental and Experimental Botany*, and *Ecotoxicology and Environmental Safety* with a total citation of 4925 and an h-index of 32. He had presented his findings in various conferences held in USA, Germany, China, Malaysia, etc. He has successfully completed various funded research projects from reputed funding agencies. He has also supervised six doctoral students, three MPhil students, and a number of Master's students, and presently five students are enrolled under his supervision for PhD degree.

Dr. Mohammad Yusuf is a Lecturer in the Department of Biology at United Arab Emirates University, Al Ain, UAE. He earned his PhD in Plant Physiology from the Department of Botany at Aligarh Muslim University in India. Dr. Yusuf has received several prestigious awards and fellowships, including the Dr. D.S. Kothari Postdoctoral Fellowship from the Government of India and the CST-UP Young Scientist Award from the State Government of Uttar Pradesh, India. He worked as a SERB-DST Young Scientist on a Government of India–funded project. He has also presented his work at Huazhong Agricultural University in Wuhan, China, and the Gordon Research Conference in Switzerland. In addition to editing three books with international publishers, including Springer Nature, Dr. Yusuf has co-authored several book chapters and has published over 60 research papers in peer-reviewed international journals. His work has garnered more than 5000 Google Scholar citations with a 37 h-index. Dr. Yusuf's research focuses on exploring phytohormones-mediated abiotic stress tolerance mechanisms in plants through physiological, proteomic, and molecular studies, specifically on proline metabolism.

Acknowledgments

First and foremost, I express my gratitude to Almighty God, the most merciful, for His gracious blessings upon us all, guiding all networks to work consistently and in coordination from the inception of the idea to the progress and successful completion of the final version of this book, *Hydrogen Peroxide: Signaling Mechanisms and Crosstalk in Plant Development and Stress Responses.*

We extend our heartfelt appreciation to all the contributors for their timely responses, outstanding and current contributions, and consistent support and collaboration. We are also grateful to Prof. Mayank Gururani, Department of Biology, United Arab Emirates University, Al Ain, for his inspiration. Our sincere thanks go to CRC Press/Taylor and Francis Group for their speedy acceptance of our proposal and the efficient completion of the review process, as well as for the cooperation and understanding of their staff.

Finally, we express our sincere gratitude to the members of our family for all the support they provided and the sacrifices they made during the preparation of this book.

Mohd Tanveer Alam Khan

Taiba Saeed

Aqeel Ahmad

Qazi Fariduddin

Mohammad Yusuf

1 Occurrence, Chemical Structure, and Biosynthesis of Hydrogen Peroxide in Plants

Durdana Yasin, Taiba Saeed, Neha Sami, Tasneem Fatma, and Shahida Hamid

INTRODUCTION

Hydrogen peroxide (H_2O_2) has emerged as a critical molecule in plant biology, garnering significant attention for its diverse roles in cellular function and stress responses. As a reactive oxygen species (ROS), H_2O_2 stands out due to its relatively stable nature, allowing it to act effectively as both a signaling molecule and a mediator of oxidative stress. Understanding the intricacies of H_2O_2 in plant systems is essential, not only for advancing fundamental plant science but also for its practical applications in agriculture and biotechnology.

H_2O_2 is a simple molecule composed of two hydrogen atoms and two oxygen atoms, but its simplicity belies its multifaceted roles in biological systems. In plants, H_2O_2 is involved in various physiological processes, including growth, development, and defense mechanisms (Quan et al., 2008). Unlike other ROS, which are often highly reactive and short-lived, H_2O_2 has a longer half-life and can diffuse across membranes, making it an ideal candidate for intracellular and intercellular signaling. It is involved in the regulation of numerous pathways that govern plant responses to both biotic and abiotic stresses. For instance, during pathogen attack, H_2O_2 levels can increase, leading to the activation of defense genes and the reinforcement of cell walls to prevent pathogen entry. Similarly, in response to environmental stresses such as drought, salinity, and extreme temperatures, H_2O_2 acts as a signal to trigger protective mechanisms, helping plants to cope with adverse conditions (Rajput et al., 2021).

Moreover, H_2O_2 plays a crucial role in cellular homeostasis and metabolism. It is produced as a by-product of various metabolic processes, including photosynthesis and respiration, primarily in chloroplasts and mitochondria. The controlled production and scavenging of H_2O_2 are vital for maintaining cellular redox balance. Enzymes such as catalases and peroxidases are responsible for regulating H_2O_2 levels, ensuring that its concentration remains within a range that is beneficial rather than harmful to the cell (Bhattacharjee 2005). The importance of H_2O_2 extends to its role in programmed cell death (PCD), a process crucial for plant development and defense. During PCD, precise regulation of H_2O_2 levels is necessary to ensure the orderly dismantling of cellular components without causing undue damage to surrounding tissues. This controlled cell death is essential for processes such as leaf senescence, root hair development, and the hypersensitive response to pathogens (Mondal et al., 2021).

Given the central role of H_2O_2 in plant physiology, understanding its biosynthetic pathway, metabolism, signaling pathways, and mechanisms of action is imperative. Advances in this field can lead to the development of crops that are more resilient to environmental stresses and have enhanced growth and productivity. For instance, by manipulating H_2O_2 signaling pathways, it may be possible to improve plant resistance to diseases or to enhance their tolerance to drought and salinity. In this chapter, we delve into the structure, localization, and a comprehensive overview of the mechanisms underlying H_2O_2 biosynthesis in plants and highlight its functional significance.

DOI: 10.1201/9781032647692-1

OCCURRENCE OF HYDROGEN PEROXIDE IN PLANTS

Hydrogen peroxide is ubiquitously present in plant cells and is produced as a by-product of various metabolic processes. Its occurrence spans several cellular compartments, each contributing to its synthesis and regulation. These cellular compartments are chloroplasts, mitochondria, peroxisomes, cell walls, apoplast, cytoplasm, and nuclei. In the chloroplasts, hydrogen peroxide is primarily produced during photosynthesis. The water-splitting reaction in photosystem II (PSII) generates H_2O_2 as part of the Mehler reaction, where electrons are transferred to oxygen, forming superoxide radicals. These radicals are then converted to hydrogen peroxide by superoxide dismutase (SOD) enzymes. This process is crucial for managing the redox state within chloroplasts and protecting the photosynthetic machinery from oxidative damage (Mubarakshina et al., 2010). The mitochondria, the powerhouse of the cell, also contribute to H_2O_2 production. During cellular respiration, the electron transport chain (ETC) facilitates the transfer of electrons, but occasionally, electrons prematurely reduce oxygen to form superoxide radicals. These radicals are subsequently converted to hydrogen peroxide, which is then detoxified by mitochondrial antioxidant enzymes like catalase and glutathione peroxidase (GPX) (Igamberdiev & Bykova, 2023).

Peroxisomes are central to the metabolism of hydrogen peroxide in plant cells. These organelles are involved in processes such as fatty acid β-oxidation, photorespiration, and the detoxification of harmful compounds (Pan et al., 2020). The oxidation reactions within peroxisomes generate significant amounts of H_2O_2, which is promptly decomposed by catalase to prevent oxidative stress. The apoplastic space, which includes the cell walls and the extracellular matrix, is another site of hydrogen peroxide production and localization. During pathogen attack or mechanical injury, plants produce H_2O_2 as part of the oxidative burst—a rapid and localized production of ROS (Podgórska et al., 2021). This response strengthens the cell walls and triggers defense mechanisms, such as the synthesis of antimicrobial compounds and the activation of systemic acquired resistance (SAR). Although less abundant, hydrogen peroxide is also found in the cytoplasm and nuclei of plant cells. It can diffuse into these compartments, where it participates in redox signaling, modulating the activity of transcription factors and influencing gene expression patterns. The detailed mechanism of synthesis in these compartments is discussed in the subsequent sections.

Structure of Hydrogen Peroxide

Hydrogen peroxide is a simple molecule with the chemical formula H_2O_2. Despite its simplicity, its structure has significant implications for its reactivity and function (Figure 1.1).

Bond Lengths and Angles: The O-O bond length in hydrogen peroxide is approximately 1.48 Å, while each O-H bond measures about 0.98 Å. The molecule has a bent structure with an H-O-O bond angle of about 94.8°, which contributes to its polar nature.

Electron Configuration: The oxygen atoms in H_2O_2 carry partial negative charges, and the hydrogen atoms have partial positive charges. This distribution of charge creates a dipole moment, making hydrogen peroxide a reactive and relatively unstable molecule.

Molecular Geometry: The overall geometry of hydrogen peroxide resembles an open book, with the two planes formed by the O-H bonds creating a dihedral angle of about 111.5°. This non-linear configuration is due to the repulsive forces between the lone pairs of electrons on the oxygen atoms, further influencing the molecule's reactivity.

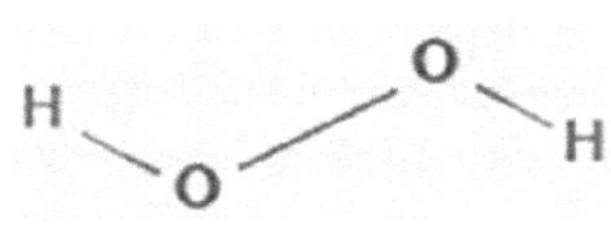

FIGURE 1.1 Structure of hydrogen peroxide.

HYDROGEN PEROXIDE BIOSYNTHESIS

The combined action of light, organic matter, and water produces H_2O_2 on its own. As a result, the environment contains small amounts of H_2O_2 (Li & Imlay, 2018). Proteins that contain flavon may be able to reduce oxygen in a way that is dependent on blue light. *Arabidopsis* and *Drosophila* ROS (H_2O_2) is generated through the blue light photoreceptor cryptochrome (Arthaut et al., 2017).

It has been demonstrated that a yeast mutant in peroxisomal acyl-CoA oxidase is deficient in light-induced H_2O_2 generation and downstream signaling, indicating that other flavin-containing enzymes may also produce superoxide ion ($O_2^{\bullet-}$) and subsequently H_2O_2 in the presence of light (Bodvard et al., 2017). As yeast lacks traditional photoreceptors, another flavin-dependent protein must give light–dark or circadian cues. In plants, there are various paths for the H_2O_2 production. A number of enzymes are involved in it that leads to the generation of hydrogen peroxides in different cellular compartments. These pathways and compartments are depicted in Figure 1.2.

OXIDASES AND H_2O_2 PRODUCTION IN PEROXISOMES AND GLYOXYSOMES

Peroxisomes and glyoxysomes are single-membrane-bound organelles specialized for having oxidase enzymes that generate superoxide ($O_2^{\bullet-}$) and H_2O_2 as by-products in various reactions (del Rio & Lopez-Huertas, 2016). The amounts of $O_2^{\bullet-}$ and H_2O_2 produced by flavin-containing oxidases and dehydrogenases during catalysis vary significantly based on the redox potential and solvent

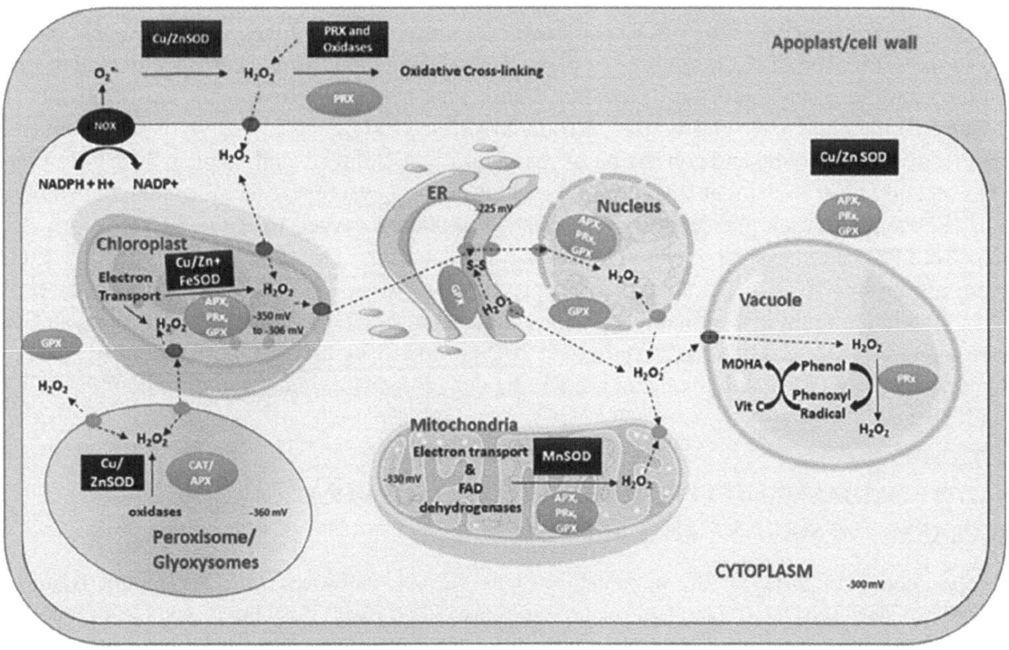

FIGURE 1.2 The diagram illustrates the primary sites of H_2O_2 production and scavenging within a typical plant cell, along with the transport pathway from the chloroplast to the nucleus, potentially via the endoplasmic reticulum (ER). It also shows the normal glutathione redox potential (mV) in various compartments (Schwarzlander et al., 2008). Light coloured oval/square boxes represent H_2O_2-removing enzymes, dark-coloured boxes indicate H_2O_2-producing enzymes, darker circles show H_2O_2-transporting aquaporins, and lighter ones are unconfirmed H_2O_2 transporters. Solid lines depict reactions, while dashed lines represent transport processes. Key elements include Vit C (ascorbate), APX (ascorbate peroxidase), CAT (catalase), FAD (flavin adenine dinucleotide), GPX (glutathione peroxidase), MDHA (monodehydroascorbate radical), NOX (NADPH oxidase), PRX (type III peroxidase), Prx (peroxiredoxin), and SOD (superoxide dismutase).

exposure level of the active site (Messner & Imlay, 2002). In plants, these oxidases, which produce $O_2^{\bullet-}$ and H_2O_2, play roles in processes such as photorespiration (via glycolate oxidase), fatty acid oxidation (via acyl-CoA oxidase), polyamine and purine breakdown, and the synthesis of certain hormones. Significant H_2O_2 fluxes are especially likely during photorespiration (Queval et al., 2007) and the utilization of fatty acids during the germination of oilseeds in glyoxysomes, a specific type of peroxisome. Additionally, peroxisomes seem to be a source of nitric oxide (NO), which opens up the possibility for the formation of peroxynitrite.

ROLE OF ETC AND H_2O_2 PRODUCTION IN MITOCHONDRIA AND CHLOROPLASTS

Strong electron transport activity is found in chloroplasts and mitochondria, where it is combined with the creation of significant electrochemical gradients across membranes that transfer energy. Superoxide dismutation presents an opportunity for the creation of H_2O_2. There are several possible locations for the synthesis of $O_2^{\bullet-}$ in chloroplasts, including ETC and photosystem I (PSI) and PSII (Foyer & Harbinson, 2019). The number of electrons that reduce oxygen to generate $O_2^{\bullet-}$ at PSI (Mehler reaction) is estimated to be between 1% and 5% (Mullineaux et al., 2018). It has been shown that oxygen photoreduction generates $O_2^{\bullet-}$ and H_2O_2 in thylakoid membranes, isolated chloroplasts, and *in situ* (Mubarakshina Borisova et al., 2012).

DCMU (3-(3,4-dichlorophenyl)-1,1-dimethylurea), the inhibitor of the photosynthetic electron transport, has been shown to block most H_2O_2 production when used with HyPer that targeted stroma. This indicates that H_2O_2 is formed downstream of PSII, most likely in the Mehler reaction (Exposito-Rodriguez et al., 2017). SOD is present in the stromal side of the thylakoid membrane and aids in the conversion of H_2O_2 to water by peroxidases. There are other oxygen photoreduction pathways that have been suggested; these pathways could serve as "valves" to reduce the formation of H_2O_2.

The enzyme known as "plastoquinol terminal oxidase (PTOX)" is in charge of taking up electrons from the compound and converting oxygen into water (Johnson and Stepien, 2016). However, it has recently been found that bryophytes, green algae, and cyanobacteria all use flavodiiron proteins (FDPs) for oxygen photoreduction. Instead of reducing oxygen to H_2O_2, FDPs absorb electrons from PSI and convert it to water. Since angiosperms lack FDPs, it is hypothesized that cyclic electron transfer (CET) serves a comparable protective purpose in these organisms (Chaux et al., 2017). In relation to this, Nicotinamide adenine dinucleotide phosphate (NADPH)-dependent CET is activated by H_2O_2 generation (Strand et al., 2015). Superoxide and H_2O_2 are produced by mitochondria via the ETC, mainly by flavin-containing dehydrogenases and respiratory complexes I, II, and II (Brand, 2016).

NADPH OXIDASE AND PEROXIDASES IN H_2O_2 GENERATION IN PLASMA MEMBRANE AND APOPLAST

The production of H_2O_2 in the apoplast involves various pathways. One significant source is NADPH oxidase (NOX). In plants, these enzymes are called respiratory burst oxidase homologues (RBOHs), named after their mammalian counterparts involved in neutrophil-mediated microbial defense (Segal, 2016). These enzymes are located in the plasma membrane and catalyze the transfer of electrons from cytosolic NADPH to molecular oxygen, producing superoxide ($O_2^{\bullet-}$) in the apoplast, which is subsequently dismutated to H_2O_2 either spontaneously or by SOD.

These enzymes are predicted to have six transmembrane domains with both the N- and C-termini located in the cytosol. The C-terminus contains binding sites for NADPH and flavin adenine dinucleotide (FAD), facilitating electron transfer to cytochromes within the transmembrane channel. The superoxide produced is rapidly converted into H_2O_2 through a dismutation reaction, which is proton-dependent and favored by the acidic conditions of the apoplast. There is also evidence suggesting the presence of SOD activity in the apoplast, possibly involving germin-like proteins.

The proton consumption during the dismutation of superoxide can increase the pH of the apoplast. Additionally, the electron transport mediated by RBOH to the extracellular space causes membrane depolarization, suggesting that NOX functions are not limited to just superoxide production (Segal, 2016). Although mutant studies are crucial for understanding NOX functions, many experiments have used the inhibitor diphenyleneiodonium (DPI), which is a general flavoprotein inhibitor rather than being specific to NOX (Riganti et al., 2004). Therefore, caution and corroborative evidence are necessary to fully establish the mechanism.

Due to the short-lived nature and poor membrane permeability of superoxide, the signaling roles of NOXs might involve the transport of H_2O_2 into the cytosol via aquaporins or interactions with plasma membrane sensor kinases. The localization of NOX in lipid rafts could enhance interactions with regulatory proteins (Nagano et al., 2016; Hao et al., 2014). Under certain conditions, such as salt stress, the production of superoxide and H_2O_2 can occur in cytoplasmic vesicles originating from the plasma membrane or the endoplasmic reticulum (ER), and RBOHD can be internalized into plasma membrane vesicles (Liu and He, 2016; Hao et al., 2014).

The N-terminus of NOX contains Ca^{2+}-binding EF-hand motifs, and its activation by Ca^{2+} has been confirmed *in vivo* (García et al., 2023). NOX-dependent ROS production in the apoplast is triggered by various stimuli, including extracellular adenosine triphosphate (ATP), hormones like abscisic acid (ABA) and ethylene, as well as pathogen- and damage-associated molecular patterns (PAMPs and DAMPs). The molecular mechanisms that activate NOX through its regulatory N-terminal domain have been thoroughly reviewed (Qi et al., 2017), involving receptor-like kinases (RLKs), phosphatidic acid, calcium influx-activated calcium-dependent protein kinases (CPKs), protein kinases (e.g., BIK1), and small GTPases (RAC/ROP). For example, RLKs and CPKs activate NOX in response to external stimuli such as pathogen attack, abiotic stress, and hormonal signals. This regulation ensures a rapid and localized production of H_2O_2, which acts as a signaling molecule to trigger downstream defense responses (Table 1.1 provides examples of the NOX interacting factors and their effect on its activity.)

In *Arabidopsis*, there are ten NOX isoforms with distinct expression patterns, each specialized for particular functions across various processes. For instance, RBOHC and RBOHH/J are involved in the polar growth of root hairs and pollen tubes, respectively (Foreman et al., 2003; Boisson-Dernier et al., 2013; Kaya et al., 2014), while RBOHD plays a role in wound responses and

TABLE 1.1
Examples of How Various Interacting Factors Modulate NADPH Oxidase Activity

Types of NOX	Interacting Factor	Effect
AtRBOHC/RHD2	Calcium	Activation via EF-hand binding
AtRBOHD	BIK1	Activation through phosphorylation
	CPK5	Activation through phosphorylation
	XLG2	Activation
	Nitric oxide	FAD binding inhibition via S-nitrosylation
	DORN1	Activation via phosphorylation
	Phosphatidic acid	Activation
AtRBOHD, AtRBOHF	Calcium	Activation via EF-hand binding
AtRBOHF	CIPK26	Activation through phosphorylation
	OST1/SnRK2.6	Phosphorylation
AtRBOHH, AtRBOHJ	Calcium	Activation via EF-hand binding
OsRbohB	OsRac1	Activation via EF-hand binding
	OsRACK1	Activation
StRBOHB	StCDPK5	Activation through phosphorylation

systemic signaling (Miller et al., 2009). Both RBOHD and RBOHF are involved in stomatal function and pathogen responses (Kadota et al., 2015).

The role of NOX in polar growth is conserved across brown seaweeds, fungi, and mammals. However, despite the foundational research on NOX's role in root hair growth (Foreman et al., 2003), the exact biochemical events it mediates remain unclear. Potential roles include cell wall remodeling or signaling (Rentel et al., 2004). Given these varied functions, it is not surprising that NOX mutants are more sensitive to environmental changes. For example, antisense suppression of the tomato RBOH family resulted in plants with multiple developmental defects and increased pathogen sensitivity (Sagi et al., 2004). Recent research continues to uncover its influences on root development (Orman-Ligeza et al., 2016).

While NOX has garnered significant attention, there are other sources of apoplastic H_2O_2. The cell wall contains numerous type III peroxidase isoenzymes that play roles in the cross-linking of cell wall polymers. Depending on the chemical environment, peroxidases have been reported to generate superoxide ($O_2^{\bullet-}$), H_2O_2, and hydroxyl radicals (Daudi et al., 2012; Kimura et al., 2014). The reaction that generates H_2O_2 is not completely understood but is favored by a high pH and requires an unidentified reductant.

TYPE III PEROXIDASE-DERIVED H_2O_2 PRODUCTION IN APOPLAST

Type III peroxidases are heme-containing enzymes that catalyze the reduction of H_2O_2 using various organic and inorganic substrates. While peroxidases are primarily involved in the scavenging of H_2O_2, certain peroxidase isoforms can also produce H_2O_2 under specific conditions, such as high pH environments. This dual functionality allows peroxidases to play a central role in regulating H_2O_2 levels within the apoplast. Peroxidases participate in lignification, suberization, and the oxidative cross-linking of cell wall components, processes that are essential for strengthening the cell wall during pathogen attack and abiotic stress. Mutant studies indicate that peroxidases are involved in the generation of H_2O_2 or other ROS during immune responses (Daudi et al., 2012), stomatal defenses, and root hair growth (Khokon et al., 2010; Mangano et al., 2017). The reaction for H_2O_2 production described by Bolwell's group is favored by high pH, aligning with the pH increases observed during elicitor/PAMP responses. In potatoes, a comparison of various PAMPs and DAMPs shows that only those causing a pH increase also induce ROS production (Daudi et al., 2012; Moroz et al., 2017). It is possible that NOX, along with the suppression of H+-ATPase activity, could contribute to a higher pH environment that favors type III peroxidase-derived H_2O_2 production.

ROLE OF OTHER OXIDASES OF APOPLAST IN H_2O_2 PRODUCTION

Several oxidases are present in the apoplast, including copper amine oxidases (CuAOs), FAD-dependent polyamine oxidases (PAOs), and oxalate oxidase that have a role in H_2O_2 generation.

Copper Amine and Polyamine Oxidases

CuAOs and PAOs are enzymes that oxidize primary amines and polyamines, respectively, producing H_2O_2 as a by-product. CuAOs, with approximately 10 genes in *Arabidopsis*, are mainly apoplastic, whereas PAOs, with five genes in *Arabidopsis*, are primarily peroxisomal but also have an apoplastic presence (notably PAO1). These enzymes have a broad substrate specificity for polyamines (Tavladoraki et al., 2016). The oxidation of amines and polyamines by these oxidases produces H_2O_2 and either an aldehyde or an amino aldehyde. Studies involving mutants or overexpression of various apoplastic amine oxidases demonstrate that these changes correlate with alterations in reactive species levels. Research using mutants and polyamine feeding has shown that these enzymes contribute to ABA and H_2O_2-dependent stomatal closure (An et al., 2008) and H_2O_2 production in pollen tubes (Wu et al., 2010). The potential signaling roles of reactive aldehydes and their further products have not been extensively explored. These enzymes have broad substrate specificity and play significant roles in plant growth, development, and stress responses. For instance, the oxidation of polyamines by PAOs contributes to H_2O_2 production during pollen tube growth, stomatal closure, and responses to abiotic stressors such as drought and salinity.

Oxalate and Ascorbate Oxidase

Apoplastic oxalate oxidase (a germin), commonly found in cereals, catalyzes the oxidation of oxalate to carbon dioxide (CO_2) and H_2O_2 (Le Deunff et al., 2004). Although *Arabidopsis* has a family of germin-like proteins that lack oxalate oxidase activity, these proteins may exhibit SOD activity. Oxalate, a product of ascorbate degradation in the cell wall, and the oxidation products of ascorbate (dehydroascorbate (DHA) and 2,3-diketogulonate) can degrade under apoplastic conditions, leading to non-enzymatic H_2O_2 production (Membré et al., 2000, Kärkönen et al., 2017). Interestingly, an unidentified oxidation product of ascorbate has been found to inhibit peroxidase activity (Kärkönen et al., 2017). Thus, oxalate oxidase is particularly important in plant defense, as the H_2O_2 generated during oxalate oxidation can act as a signaling molecule to activate defense genes and fortify cell walls against pathogen invasion.

Ascorbate (vitamin C) is a key antioxidant in plant cells, and its oxidation can lead to the production of H_2O_2. Ascorbate oxidase, an enzyme located in the cell wall, catalyzes the oxidation of ascorbate, maintaining it in an oxidized state. The oxidation of ascorbate results in the formation of DHA, which can non-enzymatically produce H_2O_2 under certain conditions.

This pathway is particularly relevant in the ER, where DHA may facilitate the formation of disulfide bonds during protein folding, indirectly influencing H_2O_2 levels (to be discussed in next section). Cell wall ascorbate oxidase maintains ascorbate in a relatively oxidized state, potentially influencing H_2O_2 production.

Overall, a complex interplay of H_2O_2 production (via NOX, peroxidase, PAO, CuAO, and oxidized ascorbate) and removal (via peroxidase) is emerging, with significant implications for growth, lignification, pathogen defense, and signaling. Extracellular H_2O_2 production by nectaries and trichomes (Peiffer et al., 2009) may also play a role in defense against insects and microbes. Nectar H_2O_2 (approximately 30 µM) is produced through glucose oxidase and a germin-like protein, which are secreted into the nectar, and by a nectary NOX (Bezzi et al., 2010; Harper et al., 2010).

ER-MEDIATED BIOSYNTHESIS OF H_2O_2

The ER provides an oxidizing environment essential for the formation of disulfide bonds in proteins, which is crucial for proper protein folding. This oxidative cross-linking is facilitated by protein disulfide isomerases (PDIs) and involves H_2O_2 in two ways (Bulleid, 2012).

In the first mechanism, a FAD-containing ER oxidase (ERO) generates an internal disulfide by reducing oxygen to H_2O_2. The oxidized ERO then oxidizes PDI, which subsequently transfers the disulfide bond to target proteins, aiding in their correct folding. This process thus results in the production of H_2O_2.

In the second mechanism, H_2O_2 is used to produce the disulfide form of a 2-cysteine peroxiredoxin (Prx) or the sulfenic acid variants of a GPX, which in turn leads the oxidation of PDI. However, it remains unclear whether both pathways are active in plants. Although the *Arabidopsis* genome have been reported to include two ERO and 13 PDI proteins, there is no report for the presence of Prx being localized in ER (Lennicke and Cochemé, 2021; Aller & Meyer, 2013). However, there are findings that back sufficient evidence to support the ERO pathway in soybean seedlings during developmental stage, where extensive disulfide bond formation occurs during storage protein synthesis (Matsusaki et al., 2016).

The potential existence of a GPX-like (GPXL) pathway in plants is suggested by findings that rice mutants in OsGPXL5 (localized in the ER/chloroplast) exhibit significantly reduced grain filling, indicating a role in synthesis of storage protein (Matsusaki et al., 2016). *Arabidopsis* GPXL3 is associated with the ER membrane (Attacha et al., 2017). GPXLs or Prxs might also help eliminate H_2O_2 produced by the ERO system. Additionally, the presence of DHA, the oxidized form of ascorbate, in the ER could potentially aid in disulfide bond formation, although this has not been thoroughly investigated (Rajput et al., 2021). More research is needed to fully understand the oxidative processes involved in protein folding in plants.

TRANSPORT OF H_2O_2 FROM ONE CELLULAR COMPARTMENT TO THE OTHER

H_2O_2 has limited ability to permeate membranes, and its transport is facilitated by aquaporin-type channel proteins. In plants, aquaporins are found on the plasma membrane intrinsic proteins (PIPs), the tonoplast intrinsic proteins (TIPs), and possibly on the inner membrane of the chloroplasts (Mubarakshina Borisova et al., 2012; Singh et al., 2020). The presence of aquaporins on other membranes remains less understood. Research on a mouse mutant of aquaporin 3 shows impaired NOX_2/H_2O_2-mediated signaling, which activates the redox-regulated transcription factor nuclear factor (NF)-κB (Hara-Chikuma et al., 2015), suggesting that aquaporins are essential for H_2O_2 signaling systems. It's believed that aquaporins involved in water transport also facilitate H_2O_2 transport, although some isoforms may have greater specificity for H_2O_2 or might not transport H_2O_2 at all (Almasalmeh et al., 2014).

In *Arabidopsis*, the aquaporin AtPIP1;4 assists in moving H_2O_2 from the apoplast to the cytosol during PAMP-triggered immunity. Mutants with reduced H_2O_2 transport show increased bacterial growth, likely due to disrupted intracellular H_2O_2 signaling (Tian et al., 2016). The AtPIP2;1 mutants show impaired stomatal closure in response to ABA and PAMP, with decreased intracellular ROS, indicating a role in both water and H_2O_2 transport (Grondin et al., 2015; Rodrigues et al., 2017). Notably, AtPIP2;1 is phosphorylated by OST1 in response to ABA (Grondin et al., 2015).

Differences in H_2O_2 permeability among maize plasma membrane aquaporins were demonstrated through yeast expression studies (Bienert and Chaumont, 2014). High concentrations of H_2O_2 (around 30 mM) reduce water permeability in maize roots, suggesting that oxidative modification of aquaporins affects their water transport capability (Ye & Steudle, 2006). Conversely, a lower H_2O_2 concentration (0.5 mM) causes AtPIP2;1 to be internalized into endosomes (Wudick et al., 2015). This indicates that aquaporin distribution and permeability are dynamic and may influence H_2O_2 signaling. For instance, the egress of H_2O_2 from photosynthetic electron transport in isolated chloroplasts is inhibited by acetazolamide, a membrane-impermeable aquaporin blocker (Mubarakshina Borisova et al., 2012). Since H_2O_2 from chloroplasts affects nuclear H_2O_2 levels and gene expression, specific chloroplast envelope aquaporins could potentially influence light signaling (Ugalde et al., 2021).

While plasma membrane aquaporins have been studied in this context, their roles in other membranes are less understood. In mammals, aquaporins are believed to facilitate H_2O_2 transport between the ER and cytosol (Appenzeller-Herzog et al., 2016), but this has not been investigated in plants. Additionally, in mammals, NOX2 and aquaporin AP3 are physically associated, facilitating H_2O_2 movement into the cytosol for signaling (Hara-Chikuma et al., 2015). Further studies using mutants and inhibitors are necessary to confirm the role of aquaporins in H_2O_2 transport in plants.

CONCLUSION

Hydrogen peroxide is a multifaceted molecule in plants, with its production tightly regulated through various enzymatic and non-enzymatic pathways. Its roles as a signaling molecule, growth regulator, and defense agent underscore its importance in plant biology. Understanding the mechanisms of H_2O_2 biosynthesis and its functional roles provides valuable insights into plant resilience and adaptation, with potential applications in agriculture and horticulture to improve crop performance and stress tolerance. H_2O_2 transport is facilitated by aquaporin-type channel proteins. These proteins, located on various membranes such as the plasma membrane, tonoplast, and potentially chloroplast inner membranes, are essential for H_2O_2 signaling pathways. Studies in *Arabidopsis* have shown that aquaporins like AtPIP1;4 and AtPIP2;1 are crucial for pathogen-triggered immunity and stomatal responses, respectively. The dynamic nature of aquaporin functionality, influenced by oxidative modifications and internalization, highlights their importance in maintaining cellular homeostasis. The understanding of aquaporins in H_2O_2 transport provides significant insights into plant stress responses and signaling mechanisms. By elucidating these pathways, researchers can develop strategies to enhance plant resistance to pathogens and environmental stresses, ultimately leading to improved crop resilience and productivity. Further studies are needed to explore the roles of aquaporins across different plant membranes and their potential applications in agricultural biotechnology. As research progresses, further elucidation of H_2O_2 dynamics in plants will continue to uncover new aspects of its biological significance.

REFERENCES

Almasalmeh A., Krenc D., Wu B., Beitz E. 2014. Structural determinants of the hydrogen peroxide permeability of aquaporins. *FEBS Journal* 281: 647–656.Aller I., Meyer A.J. 2013. The oxidative protein folding machinery in plant cells. *Protoplasma* 250: 799–816.

An Z., Jing W., Liu Y., Zhang W. 2008. Hydrogen peroxide generated by copper amine oxidase is involved in abscisic acid-induced stomatal closure in Vicia faba. *Journal of Experimental Botany* 59: 815–825.

Appenzeller-Herzog C., Banhegyi G., Bogeski I., Davies K.J., Delaunay-Moisan A., Forman H.J., Gorlach A., Kietzmann T., Laurindo F., Margittai E., et al. 2016. Transit of H_2O_2 across the endoplasmic reticulum membrane is not sluggish. *Free Radical Biology and Medicine* 94: 157–160.

Arthaut L-D, Jourdan N, Mteyrek A, Procopio M, El-Esawi M, D'Harlingue A, Bouchet P-E, Witczak J, Ritz T, Klarsfeld A., et al. 2017. Blue-light induced accumulation of reactive oxygen species is a consequence of the Drosophila cryptochrome photocycle. *PLoS One* 12: e0171836.

Attacha S, Solbach D, Bela K, Moseler A, Wagner S, Schwarzlander M, Aller I, Muller SJ, Meyer AJ. 2017. Glutathione peroxidase-like enzymes cover five distinct cell compartments and membrane surfaces in Arabidopsis thaliana. *Plant Cell & Environment* 40: 1281–1295.

Bezzi S, Kessler D, Diezel C, Muck A, Anssour S, Baldwin IT. 2010. Silencing NaTPI expression increases nectar germin, nectarins, and hydrogen peroxide levels and inhibits nectar removal from plants in nature. *Plant Physiology* 152: 2232–2242.

Bhattacharjee S. 2005. Reactive oxygen species and oxidative burst: Roles in stress, senescence and signal transduction in plants. *Current Science* 89: 1113–1121.

Bienert GP, Chaumont F. 2014. Aquaporin-facilitated transmembrane diffusion of hydrogen peroxide. *Biochimica et Biophysica Acta - General Subjects* 1840: 1596–1604.

Bodvard K, Peeters K, Roger F, Romanov N, Igbaria A, Welkenhuysen N, Palais G, Reiter W, Toledano MB, Kall M., et al. 2017. Light-sensing via hydrogen peroxide and a peroxiredoxin. *Nature Communications* 8: 14791.

Boisson-Dernier A, Lituiev DS, Nestorova A, Franck CM, Thirugnanarajah S, Grossniklaus U. 2013. ANXUR receptor-like kinases coordinate cell wall integrity with growth at the pollen tube tip via NADPH oxidases. *PLoS Biology* 11: e1001719.

Brand MD. 2016. Mitochondrial generation of superoxide and hydrogen peroxide as the source of mitochondrial redox signaling. *Free Radical Biology and Medicine* 100:14–31.

Bulleid NJ. 2012. Disulfide bond formation in the mammalian endoplasmic reticulum. *Cold Spring Harbor Perspectives in Biology* 4: a013219.

Chaux F, Burlacot A, Mekhalfi M, Auroy P, Blangy S, Richaud P, Peltier G. 2017. Flavodiiron proteins promote fast and transient O_2 photoreduction in Chlamydomonas. *Plant Physiology* 174: 1825–1836.

Daudi A, Cheng Z, O'Brien JA, Mammarella N, Khan S, Ausubel FM, Bolwell GP. 2012. The apoplastic oxidative burst peroxidase in Arabidopsis is a major component of pattern-triggered immunity. *Plant Cell* 24: 275–287

Del Rio LA, Lopez-Huertas E. 2016. ROS generation in peroxisomes and its role in cell signaling. *Plant and Cell Physiology* 57: 1364–1376.

Exposito-Rodriguez M, Laissue PP, Yvon-Durocher G, Smirnoff N, Mullineaux PM. 2017. Photosynthesis-dependent H_2O_2 transfer from chloroplasts to nuclei provides a high-light signalling mechanism. *Nature Communications* 8: 49.

Foreman J, Demidchik V, Bothwell JHF, Mylona P, Miedema H, Torres MA, Linstead P, Costa S, Brownlee C, Jones JDG., et al. 2003. Reactive oxygen species produced by NADPH oxidase regulate plant cell growth. *Nature* 422: 442–446.

Foyer, CH, Harbinson J. 2019. Oxygen metabolism and the regulation of photosynthetic electron transport. In C.H. Foyer, & P.M. Mullineaux (Eds.), *Causes of Photooxidative Stress and Amelioration of Defense Systems in Plants* (pp. 1–42). CRC Press, Boca Raton, FL.

García JG, Ansorena E, Izal I, Zalba G, De Miguel C, Milagro FI. 2023. Structure, regulation, and physiological functions of NADPH oxidase 5 (NOX_5). *Journal of Physiology and Biochemistry* 79(2): 383–395.

Grondin A, Rodrigues O, Verdoucq L, Merlot S, Leonhardt N, Maurel C. 2015. Aquaporins contribute to ABA-triggered stomatal closure through OST1-mediated phosphorylation. *Plant Cell* 27: 1945–1954.

Hao H, Fan L, Chen T, Li R, Li X, He Q, Botella MA, Lin J. 2014. Clathrin and membrane microdomains cooperatively regulate rbohD dynamics and activity in Arabidopsis. *Plant Cell* 26: 1729–1745.

Hara-Chikuma M, Satooka H, Watanabe S, Honda T, Miyachi Y, Watanabe T, Verkman AS. 2015. Aquaporin-3-mediated hydrogen peroxide transport is required for NF-jB signalling in keratinocytes and development of psoriasis. *Nature Communications* 6: 7454.

Harper AD, Stalnaker SH, Wells L, Darvill A, Thornburg R, York WS. 2010. Interaction of Nectarin 4 with a fungal protein triggers a microbial surveillance and defense mechanism in nectar. *Phytochemistry* 71: 1963–1969.

Igamberdiev AU, Bykova NV. 2023. Mitochondria in photosynthetic cells: Coordinating redox control and energy balance. *Plant Physiology* 191(4): 2104–2119.

Johnson, GN, Stepien P. 2016. Plastid terminal oxidase as a route to improving plant stress tolerance: known knowns and known unknowns. *Plant and Cell Physiology* 57(7): 1387–1396.

Kadota Y, Shirasu K, Zipfel C. 2015. Regulation of the NADPH oxidase RBOHD during plant immunity. *Plant and Cell Physiology* 56: 1472–1480.

Kärkönen A, Dewhirst RA, Mackay CL, Fry SC. 2017. Metabolites of 2,3-diketogulonate delay peroxidase action and induce non-enzymic H_2O_2 generation: Potential roles in the plant cell wall. *Archives of Biochemistry and Biophysics* 620: 12–22.

Kaya H, Nakajima R, Iwano M, Kanaoka MM, Kimura S, Takeda S, Kawarazaki . T, Senzaki E, Hamamura Y, Higashiyama T., et al. 2014. Ca^{2+}-activated reactive oxygen species production by Arabidopsis RbohH and RbohJ is essential for proper pollen tube tip growth. *Plant Cell* 26: 1069–1080.

Khokon MAR, Uraji M, Munemasa S, Okuma E, Nakamura Y, Mori IC, Murata Y. 2010. Chitosan-induced stomatal closure accompanied by peroxidase mediated reactive oxygen species production in Arabidopsis. *Bioscience, Biotechnology, and Biochemistry* 74: 2313–2315.

Kimura M, Umemoto Y, Kawano T. 2014. Hydrogen peroxide-independent generation of superoxide by plant peroxidase: Hypotheses and supportive data employing ferrous ion as a model stimulus. *Frontiers in Plant Science* 5: 285.

Le Deunff E, Davoine C, Le Dantec C, Billard JP, Huault C. 2004. Oxidative burst and expression of germin/oxo genes during wounding of ryegrass leaf blades: Comparison with senescence of leaf sheaths. *Plant Journal* 38: 421–431.

Lennicke C, Cochemé HM. 2021. Redox metabolism: ROS as specific molecular regulators of cell signaling and function. *Molecular Cell* 81(18): 3691–3707.

Li X, Imlay JA. 2018. Improved measurements of scant hydrogen peroxide enable experiments that define its threshold of toxicity for Escherichia coli. *Free Radical Biology and Medicine* 120: 217–227.

Liu Y, He C. 2016. Regulation of plant reactive oxygen species (ROS) in stress responses: learning from AtRBOHD. *Plant Cell Reports* 35: 995–1007.

Mangano S, Denita-Juarez SP, Choi H-S, Marzol E, Hwang Y, Ranocha P, Velasquez SM, Borassi C, Barberini ML, Aptekmann AA, et al. 2017. Molecular link between auxin and ROS-mediated polar growth. In *Proceedings of the National Academy of Sciences, USA* 114: 5289–5294.

Matsusaki M, Okuda A, Masuda T, Koishihara K, Mita R, Iwasaki K, Hara K, Naruo Y, Hirose A, Tsuchi Y, et al. 2016. Cooperative protein folding by two protein thiol disulfide oxidoreductases and ERO1 in soybean. *Plant Physiology* 170: 774–789.

Messner KR, Imlay JA. 2002. Mechanism of superoxide and hydrogen peroxide formation by fumarate reductase, succinate dehydrogenase, and aspartate oxidase. *Journal of Biological Chemistry* 277: 42563–42571.

Miller G, Schlauch K, Tam R, Cortes D, Torres MA, Shulaev V, Dangl JL, Mittler R. 2009. The plant NADPH oxidase RBOHD mediates rapid systemic signaling in response to diverse stimuli. *Science Signaling* 2: ra45.

Mondal R, Antony S, Roy S, Chattopadhyay SK. 2021. Programmed cell death (PCD) in plant: Molecular mechanism, regulation, and cellular dysfunction in response to development and stress. *Regulation and Dysfunction of Apoptosis* 2: 1–20.

Moroz N, Fritch KR, Marcec MJ, Tripathi D, Smertenko A, Tanaka K. 2017. Extracellular alkalinization as a defense response in potato cells. *Frontiers in Plant Science* 8: 1–11.

Mubarakshina Borisova MM, Kozuleva MA, Rudenko NN, Naydov IA, Klenina IB, Ivanov BN. 2012. Photosynthetic electron flow to oxygen and diffusion of hydrogen peroxide through the chloroplast envelope via aquaporins. *Biochimica et Biophysica Acta* 1817: 1314–1321.

Mubarakshina MM, Ivanov BN, Naydov IA, Hillier W, Badger MR, Krieger-Liszkay A. 2010. Production and diffusion of chloroplastic H_2O_2 and its implication to signalling. *Journal of Experimental Botany* 61(13): 3577–3587.

Mullineaux PM, Exposito-Rodriguez M, Laissue PP, Smirnoff N. 2018. ROS dependent signalling pathways in plants and algae exposed to high light: Comparisons with other eukaryotes. *Free Radical Biology and Medicine* 122: 52–64.

Nagano M, Ishikawa T, Fujiwara M, Fukao Y, Kawano Y, Kawai-Yamada M, Shimamoto K. 2016. Plasma membrane microdomains are essential for Rac1-RbohB/H-mediated immunity in rice. *Plant Cell* 28: 1966–1983.

Orman-Ligeza B, Parizot B, de Rycke R, Fernandez A, Himschoot EA. 2016. RBOH-mediated ROS production facilitates lateral root emergence in Arabidopsis. *Development* 143: 3328–3339.

Pan R, Liu J, Wang S, Hu J. 2020. Peroxisomes: Versatile organelles with diverse roles in plants. *New Phytologist* 225(4): 1410–1427.

Peiffer M, Tooker JF, Luthe DS, Felton GW. 2009. Plants on early alert: glandular trichomes as sensors for insect herbivores. *New Phytologist* 184: 644–656.

Podgórska A, Burian M, Dobrzyńska K, Rasmusson AG, Szal B. 2021. Respiratory burst oxidases and apoplastic peroxidases facilitate ammonium syndrome development in Arabidopsis. *Environmental and Experimental Botany* 181: 104279.

Qi J, Wang J, Gong Z, Zhou JM. 2017. Apoplastic ROS signaling in plant immunity. *Current Opinion in Plant Biology* 38: 92–100.

Quan LJ, Zhang B, Shi WW, Li HY. 2008. Hydrogen peroxide in plants: A versatile molecule of the reactive oxygen species network. *Journal of Integrative Plant Biology*, 50(1), 2–18.

Queval G, Issakidis-Bourguet E, Hoeberichts FA, Vandorpe M, Gakiere B, Vanacker H, Miginiac-Maslow M, Van Breusegem F, Noctor G. 2007. Conditional oxidative stress responses in the Arabidopsis photorespiratory mutant cat2 demonstrate that redox state is a key modulator of daylength dependent gene expression, and define photoperiod as a crucial factor in the regulation of H_2O_2-induced cell death. *Plant Journal* 52: 640–657.

Rajput VD, Harish, Singh RK, Verma KK, Sharma L, Quiroz-Figueroa FR, Meena M, Gour VS, Minkina T, Sushkova S, Mandzhieva S. 2021. Recent developments in enzymatic antioxidant defence mechanism in plants with special reference to abiotic stress. *Biology* 10(4): 267.

Rentel MC, Lecourieux D, Ouaked F, Usher SL, Petersen L, Okamoto H, Knight H, Peck SC, Grierson CS, Hirt H, et al. 2004. OXI1 kinase is necessary for oxidative burst-mediated signalling in Arabidopsis. *Nature* 427: 858–861.

Riganti C, Gazzano E, Polimeni M, Costamagna C, Bosia A. 2004. Diphenyleneiodonium inhibits the cell redox metabolism and induces oxidative stress. *Biochemistry* 279: 47726–47731.

Rodrigues O, Reshetnyak G, Grondin A, Saijo Y, Leonhardt N, Maurel C, Verdoucq L. 2017. Aquaporins facilitate hydrogen peroxide entry into guard cells to mediate ABA- and pathogen-triggered stomatal closure. *Proceedings of the National Academy of Sciences, USA* 114: 9200–9205.

Sagi M, Davydov O, Orazova S, Yesbergenova Z, Ophir R, Stratmann JW, Fluhr R. 2004. Plant Respiratory Burst Oxidase Homologs impinge on wound responsiveness and development in *Lycopersicon esculentum*. *Plant Cell* 16: 616–628.

Schwarzländer M, Fricker, MD, Müller C, Marty L, Brach T, Novak J, Sweetlove LJ, Hell R, Meyer AJ. 2008. Confocal imaging of glutathione redox potential in living plant cells. *Journal of Microscopy 231*(2): 299–316.

Segal AW. 2016. NADPH oxidases as electrochemical generators to produce ion fluxes and turgor in fungi, plants and humans. *Open Biology* 6: 160028.

Singh RK, Deshmukh R, Muthamilarasan M, Rani R, Prasad M. 2020. Versatile roles of aquaporin in physiological processes and stress tolerance in plants. *Plant Physiology and Biochemistry*, 149: 178–189.

Strand DD, Livingston AK, Satoh-Cruz M, Froehlich JE, Maurino VG, Kramer DM. 2015. Activation of cyclic electron flow by hydrogen peroxide in vivo. *Proceedings of the National Academy of Sciences, USA* 112: 5539–5544.

Tavladoraki P, Cona A, Angelini R. 2016. Copper-containing amine oxidases and FAD-dependent polyamine oxidases are key players in plant tissue differentiation and organ development. *Frontiers in Plant Science* 7: 824.

Tian S, Wang X, Li P, Wang H, Ji H, Xie J, Qiu Q, Shen D, Dong H. 2016. Plant aquaporin AtPIP1;4 links apoplastic H_2O_2 induction to disease immunity pathways. *Plant Physiology* 171(3):1635–1650.
Ugalde JM, Fuchs P, Nietzel T, Cutolo EA, Homagk M, Vothknecht UC, Meyer AJ. 2021. Chloroplast-derived photo-oxidative stress causes changes in H_2O_2 and E GSH in other subcellular compartments. *Plant Physiology* 186(1): 125–141.
Wu J, Shang Z, Wu J, Jiang X, Moschou PN, Sun W, Roubelakis-Angelakis KA, Zhang S. 2010. Spermidine oxidase-derived H_2O_2 regulates pollen plasma membrane hyperpolarization-activated Ca^{2+}-permeable channels and pollen tube growth. *Plant Journal* 63: 1042–1053.
Wudick MM, Li X, Valentini V, Geldner N, Chory J, Lin J, Maurel C, Luu DT. 2015. Subcellular redistribution of root aquaporins induced by hydrogen peroxide. *Molecular Plant* 8: 1103–1114.
Ye Q, Steudle E. 2006. Oxidative gating of water channels (aquaporins) in corn roots. *Plant, Cell & Environment* 29: 459–470.

2 Ongoing Story of Hydrogen Peroxide During the Lifespan of Plants
From Development to Stress

Faiza Shafique Khan and Farhan Goher

INTRODUCTION

Exposure of plants to changing environmental conditions induces the accumulation of reactive oxygen species (ROS), including hydrogen peroxide (H_2O_2), superoxide radical O_2^-, and hydroxyl radical OH^-. Abiotic stresses such as drought, cold, heat, and salinity increase ROS production and disturb scavenging, which stimulates oxidative damage (Kapoor et al. 2019). Redox reactions and ROS production in cells can be conceded by enzymatic and non-enzymatic antioxidants that sustain ROS to a specific lethal level. This synergistic effect maintains the balance of ROS generation and scavenging in plant cells (Mittler et al. 2011; Mittler 2017; Mittler and Van Breusegem 2023). ROS production that could enhance the scavenging facility of the antioxidative system in plant cells is stated as oxidative stress, which leads toward cell death (Vandepoele et al. 2011; Tripathy and Oelmüller 2012).

H_2O_2 is a reactive signaling molecule, actively involved in various physiological and developmental mechanisms in plants and acts as a signaling molecule during oxidative stress (Figure 2.1) (Jahan et al. 2023; ul Islam et al. 2023). Firstly, H_2O_2 was considered toxic and harmful for plants and widely linked with senescence, necrosis, and cell death (Saxena et al. 2016; Mittler and Van Breusegem 2023). Secondly, in the past few years, the constructive role of H_2O_2 in plant growth and development has widely been explored. H_2O_2 is involved in stomatal development (Shi et al. 2022), seed germination (Cheng et al. 2022), axillary bud outgrowth and dormancy release (Chen et al. 2016; Porcher et al. 2020; Hubmann et al. 2022), root shoot development (Passaia et al. 2013), lateral root development (Chen et al. 2018), fiber cell development (Li et al. 2007), somatic embryogenesis (Cheng et al. 2015), and shoot organogenesis (Tian et al. 2003). H_2O_2 also associates with plant growth regulators such as salicylic acid (SA), auxin (Soares et al. 2020), ethylene, gibberellic acid (GA), and abscisic acid (ABA) and induces abiotic stress resistance and plant development (Tian et al. 2018; Gong et al. 2022; Guo et al. 2023). H_2O_2 promotes seed dormancy released in *Arabidopsis* by mediating ABA catabolism and GA biosynthesis (Liu et al. 2010). A decrease in H_2O_2 production in the early stage of endo-dormancy is linked with an increase in ABA production in apples (Sapkota et al. 2021). Exogenous application of optimum concentration of indole-3-acetic acid (IAA) increases endogenous H_2O_2 content in wild-type root tips of tomatoes, and H_2O_2 is IAA controlled in the growing part of the root (Ivanchenko et al. 2013). Moreover, ABA also plays an important role in drought stress response as well as interacts with H_2O_2 for signaling via MAPK cascade. In the transduction of ABA, H_2O_2 serves as a secondary signal transmitter (Miao et al. 2006). H_2O_2 accumulation triggered plasma membrane calcium Ca^{2+} channels for ABA-mediated stomatal movement in *Vicia faba* (Zhang et al. 2001). Some other developmental mechanism, such as endodormancy and bud burst, is generally escorted by high production of H_2O_2 and O^{-2} in buds (Pérez et al. 2008). Chilling conditions enhance H_2O_2 production of the flower bud with the onset of dormancy release (Kuroda et al. 2002; Hubmann et al. 2022). However, it might be difficult to

DOI: 10.1201/9781032647692-2

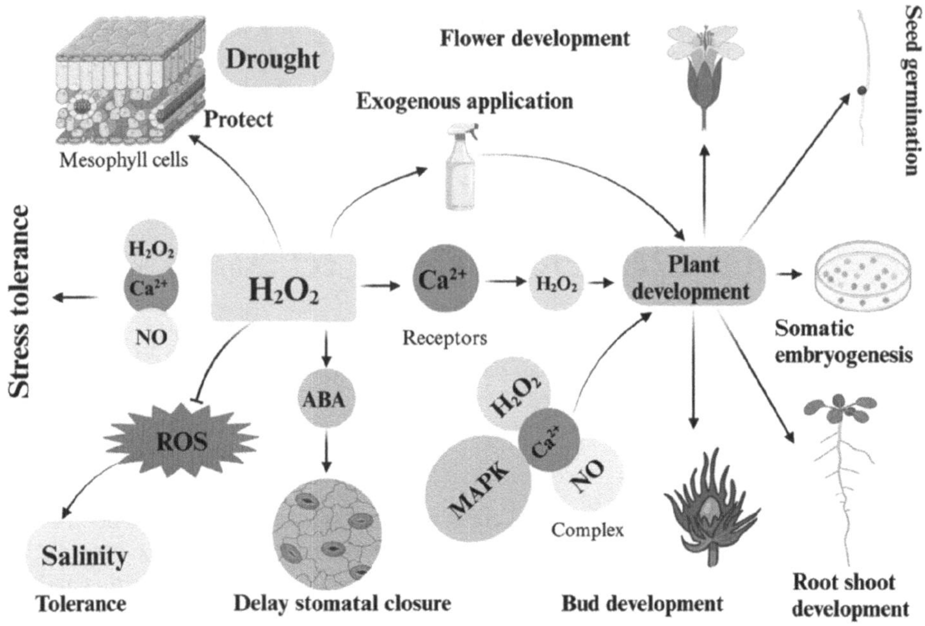

FIGURE 2.1 Graphic representation of H_2O_2 in response to plant development and abiotic stress tolerance in plants

understand the underlying molecular mechanism in the H_2O_2 signaling pathway regarding damage and development.

Many studies showed that H_2O_2 could respond to abiotic stresses such as heavy metal, drought, high temperature, cold and salinity (Wen et al. 2013; Orabi et al. 2015; Wu et al. 2015; Ashraf et al. 2015; Dikilitas et al. 2020; Jira-anunkul and Pattanagul 2021; Ma et al. 2022; Niu and Liao 2016; Khan et al. 2018;). Further research findings on plants have shown that exogenous treatment of H_2O_2 under optimal range can increase abiotic stress tolerance by modulating physiological and biochemical mechanisms such as stress-inducing pathways and photosynthetic activity (Bagheri et al. 2019; Ellouzi et al. 2021; Özdamar et al. 2022; Zhou et al. 2023; Barzotto et al. 2023). These results suggest the possible positive role of H_2O_2 signaling, signal transduction, and regulation in gene expression patterns. H_2O_2 interacts synergistically or antagonistically with Ca^{2+}, nitric oxide (NO), plant growth regulators, enzymatic and non-enzymatic antioxidants to mediate plant development and abiotic stress tolerance in plants (Liu et al. 2020; Mukherjee and Corpas 2023; Wang and Yao 2023). It is now clear that H_2O_2 acts as a reactive signaling molecule. This study specifies the H_2O_2 role in plant development and abiotic stress response and its interrelationship with Ca^{2+} and plant growth regulators.

ROLE OF H_2O_2 AND SIGNALING PATHWAY DURING ROOT DEVELOPMENT

H_2O_2 regulates plant growth and development and plays an important role in root elongation in many plants (Liao et al. 2012a; Han et al. 2015; Gong et al. 2021), including Arabidopsis, tomato, rice, cucumber, wheat, and soybean (Hameed et al. 2004; Su et al. 2006; Passaia et al. 2013; Xiong et al. 2015; Sun et al. 2016; Mei et al. 2017; Gong et al. 2021). Root development and root apex signaling pathway is a cyclic process by the continuous production of cells in the maturation region. ROS generation varies on different developmental stages of the cell cycle, which regulate the expression of cyclin-dependent protein kinase (CDK) (Tognetti et al. 2017). H_2O_2 production varies among meristematic and elongation zones of root apex where external factors such as pH and nutrient stress regulate cell expansion. ROS regulates the expression of CDK proteins, which play an important role in cell division and cell maintenance (Tognetti et al. 2017). The CsCYCD3 (cyclin-dependent protein kinase) protein interacts with *CsWUS (WUSCHEL)* in citrus and is involved in cell maintenance (Shafique Khan et al. 2021). H_2O_2 is also linked with other growth regulators during root development, such as NO and H_2S in plants during stress (Li et al. 2014, 2018, 2020; Ma et al. 2020). NO and H_2S do not function independently but play a significant role in root development with ROS signaling. NO interacts with H_2O_2 and other oxidants and stimulates protein nitration or loss of function of target proteins (Nicolas-Francès et al. 2022). NO and H_2O_2 interplay at the level of their biosynthesis in roots and other plant organs during physiological and abiotic stress changes (Suhel et al. 2020). MAPK signaling pathway regulates the interrelations between NO and H_2O_2 generation (Lin et al. 2019).

Specifically, MAPK6 synthesis is mandatory for H_2O_2 and NO production during *Arabidopsis* root development (Wang et al. 2010). H_2O_2 activates MAPK4, MPK6, MAPK1/2, and MAPK12 (Vandepoele et al. 2011; Zhou et al. 2014). Some previous studies reported that H_2O_2 stimulated the expression of genes encoding proteins related to cell expansion, cell cycle, and cell elasticity, such as 14-3-3-like protein (Balazadeh et al. 2010). H_2O_2 supports a transiently dose-dependent increase in Ca^{2+} influx in the root epidermis of *Arabidopsis*. H_2O_2 signaling induces root elongation by mediating Ca^{2+} influx in the plasma membrane of root cells in *Arabidopsis* signaling (Han et al. 2015). Two Ca^{2+} channels regulate H_2O_2 content in the root elongation zone (Demidchik et al. 2007). Ca^{2+} signaling is potentially involved in H_2O_2-induced adventitious root formation in marigolds because Ca^{2+} removal could inhibit H_2O_2-mediated adventitious root development (Liao et al. 2012a). H_2O_2 may activate the Ca^{2+} receptors and regulate proteins to increase Ca^{2+} level and Ca^{2+} content may increase H_2O_2 production during plant development. Exogenous treatment of H_2O_2 in optimized concentration (0–100 μM) increased root diameter and cell expansion, whereas high concentration of H_2O_2 (100uM-500uM) inhibited root elongation in rice (Xiong et al. 2015). A 0.1 mM concentration of H_2O_2 potentially improved plant shoot and root, fresh and dry mass in tomatoes (Nazir et al. 2019). Exogenous application of H_2O_2 on cucumber seedlings increases adventitious root formation (Sun et al. 2016). Similarly, in pistachio tree seedlings, exogenous treatment of H_2O_2 (1–10 mM) increases root and shoot thickness and fresh and dry weight (Bagheri et al. 2019). H_2O_2 plays critical role in the plant development mechanisms are describe in (Table 2.1).

TABLE 2.1
Role of H_2O_2 in Plant Development

Plant	Function	Mechanism	Reference
Tomato	Stomatal closure	H_2O_2 disrupts the cellular redox status of glutathione, which is essential for BR-mediated stomatal opening	Xia et al. (2014)
Arabidopsis	Stomatal development	H_2O_2 activates *SnRK1* by inducing *KIN10*, which stabilizes the *SPCH* to promote stomatal development	Shi et al. (2022)
Arabidopsis		Induced stomatal closure	Ge et al. (2015)
Peas	Seed germination	Involved in cellular detox signaling that influences plant development	Barba-Espín et al. (2012)
Melon	Seed germination	H_2O_2 and Ca^{2+} counteracted ABA to induce seed germination	Cheng et al. (2022)
Arabidopsis	Leaf senescence	H_2O_2 as a signaling molecule involved in leaf senescence	Zentgraf et al. (2022)
Capsicum	Leaf abscission	H_2O_2 acts downstream from ethylene and plays a role in abscission in vitro and in planta	Sakamoto et al. (2008)
Mung bean	Photosynthetic rate	Exogenous application of H_2O_2 photosynthetic activity	Khan et al. (2015)
Tomato	Axillary bud outgrowth	H_2O_2 promotes IAA biosynthesis in the apex, inhibits CK biosynthesis and bud growth	Chen et al. (2016)
Rosebush	Axillary bud outgrowth	H_2O_2 metabolism contributes to axillary bud outgrowth	Porcher et al. (2020)

(Continued)

TABLE 2.1 (*Continued*)
Role of H_2O_2 in Plant Development

Plant	Function	Mechanism	Reference
Rice	Root shoot development	OsGPX3 gene silencing impairs development and leads to stress-induced accumulation of H_2O_2	Passaia et al. (2013)
Marigold	Root development	H_2O_2 and NO are involved in IBA-mediated adventitious root formation	Liao et al. (2011)
Soybean	Lateral root development	H_2O_2 plays an important role in lateral root development	Su et al. (2006)
Tomato	Lateral root development	RBOH1-mediated H_2O_2 plays a role in H_2S-elicited lateral root development	Mei et al. (2017)
Alfalfa	Lateral root development	H_2O_2 is involved in lateral root development triggered by melatonin	Chen et al. (2018)
Cucumber	Adventitious root formation	H_2O_2 involved in salicylic acid-induced adventitious root formation	Gong et al. (2022)
Soybean	Nodule development and senescence	Cysteine protease and high accumulation of H_2O_2 involved in nodule development and PCD	Alesandrini et al. (2003)
Cotton	Fiber cell development	*GhAPX1* expression is upregulated in response to cellular H_2O_2 and ethylene. *GhAPX1* encodes functional enzymes that are involved in H_2O_2 hemostasis during fiber development	Li et al. (2007)
Potato	Induction and formation of tubers	Accumulation in stolons plays a role in initiating tuber induction and regulating tuberization gene expression	Lei et al. (2023)
Rice	Aerenchyma development in stem	H_2O_2 promotes aerenchyma development in a dose-dependent manner	Steffens et al. (2011)
Arabidopsis	Embryogenic root meristem development	H_2O_2 response gene (*HRG1, HRG2*) regulate embryogenic meristem activity	Gong et al. (2021)
Cotton	Somatic embryogenesis	Enhance embryogenic callus growth and embryo formation	Cheng et al. (2015)
Strawberry	Shoot organogenesis in callus	H_2O_2 promotes the regeneration of shoot bud	Tian et al. (2003)

ROLE OF H_2O_2 SIGNALING DURING FLORAL DEVELOPMENT

Vegetative to reproductive phase transition is a critical developmental mechanism linked with external environmental factors; for example, exogenous application of SA and drought stress induced early flowering in sweet orange trees by increasing expression pattern of flowering-related genes (*Citrus sinensis*) (Khan et al. 2022). Similarly, the exogenous application of H_2O_2 (600 µM) increases flower density and improves the average number of flowers opening in lily flowers (Liao et al. 2012b). H_2O_2 significantly promotes the reproductive development in Litchi (*Litchi chinensis*) by increasing the expression pattern of flowering-related genes such as *LcLFY* (Zhou et al. 2012). H_2O_2, with other ROS, plays a central role in gametophyte and pollen's possible functioning (Černý et al. 2018). In *Paeonia suffruticosa*, exogenous application of optimum concentration of H_2O_2 reduced pollen germination, whereas application of H_2O_2 scavengers such as catalase and ascorbic acid promotes pollen germination and pollen tube development (He et al. 2006). Spermidine oxidase-derived H_2O_2 signals calcium Ca^{2+} influx and regulates pollen tube growth (Wu et al. 2010). Spermidine induces an increase in H_2O_2 content in the pollen tube that causes pollen tube hyperpolarization-activated Ca^{2+}-permeable channel activation to induce and enhance Ca^{2+}. Single channel activation and whole cell clamp data suggest that spermidine-derived H_2O_2 is responsible for hyperpolarization-activated Ca^{2+} current (Wu et al. 2010).

ROLE OF H_2O_2 AND SIGNALING DURING BUD DEVELOPMENT

Bud development or axillary bud developmental mechanism is regulated by plant growth regulators, redox, and sugar signaling (Hubmann et al. 2022). H_2O_2 mediates response to various stimuli, such as plant growth regulators in plant development (Kuroda et al. 2002, Chen et al. 2016). Exogenous application of H_2O_2 on quiescent axillary rose buds inhibited bud outgrowth (Porcher et al. 2020). Further, a study on enzyme activity assays for H_2O_2 converting enzyme in apple bud revealed that antioxidative enzyme CAT, POX, and cwPOX activities were reported between the phase transition of dormancy release and the beginning of ontogenetic development (Hubmann et al. 2022). One previous study confirmed that *WHITEFLY INDUCED 1 (WFI1)* and *RESPIRATORY BURST OXIDASE HOMOLOG 1 (RBOH1)* genes were involved in H_2O_2 production in apoplast and enhanced bud outgrowth in tomatoes. Furthermore, *WFI1* and *RBOH1* genes decrease the *FZY* expression level that is responsible for IAA biosynthesis and IAA accumulation in the apex, whereas they increase the expression pattern of *IPT2* involved in cytokinin biosynthesis and accumulation in the stem node (Chen et al. 2016). The highest concentration of H_2O_2 was detected during endodormancy release in the floral bud of pear (*Pyrus pyrifolia Nakai* cv. Cuiguan), whereas it decreased during endodormancy in the wonhwang cultivar during bud break stage (Hussain et al. 2015).

ROLE OF H_2O_2 SIGNALING DURING ABIOTIC STRESS TOLERANCE AND PLANT DEVELOPMENT

Excessive accumulation of ROS induced by abiotic stresses caused severe damage to chloroplast and mitochondria ultrastructure (Huang et al. 2015; Khan et al. 2022). The modification of chloroplast or mitochondrial ultrastructure was correlated with higher ROS (H_2O_2) and malondialdehyde accumulation, as well as lower antioxidant enzymes under stress conditions (Mittler and Van Breusegem 2023). H_2O_2 is a key signaling molecule associated with abiotic stress signaling pathways (Huang et al. 2015; Dikilitas et al. 2020). Various studies proved that H_2O_2 could induce abiotic stress tolerance by protecting organelle structure (Niu and Liao 2016; Hou et al. 2018). For example, H_2O_2 protects chloroplast internal structure to protect photosynthesis. The interaction between H_2O_2 and NO protects mesophyll cells and improves overall photosynthetic activity under drought conditions in marigolds during adventitious root formation. NO may acts on upstream or downstream signaling molecules of H_2O_2. Endogenous H_2O_2 and NO are also involved in the regulation of antioxidant enzyme activities by ABA-induced drought tolerance in bermudagrass (Lu et al. 2009). ABA can induce the synthesis of ROS in plant cells and H_2O_2 production in guard cells and plays an important role in ABA-mediated stomatal closer (Kwak et al. 2003). NO production acts downstream of H_2O_2 in ABA-mediated stomatal closer (Bright et al. 2006). Additionally, H_2O_2 also regulates the expression of abiotic stress-related genes and antioxidant enzyme activity during abiotic stress conditions (Table 2.2) (Niu and Liao 2016).

The interaction between H_2O_2 and NO also alleviates the drastic effect of drought stress by increasing antioxidant enzyme defense system to protect cell membrane and ion homeostasis (Ma et al. 2021). It has been reported that H_2O_2 and Ca^{2+} signaling were involved in ABA-mediated response to drought stress in *Arabidopsis* via calcium-dependent protein kinase 8 (CPK8), which regulates CAT3 (catalase 3) activity mediating stomatal movement (Zou et al. 2015). H_2O_2-mediated Ca^{2+} channels and MAPK activation play an important role in abiotic stress tolerance (Li et al. 2019). During abiotic stress conditions, ROS accumulation in cells causes severe oxidative damage. H_2O_2 also interplays with ethylene and brassinosteroids (BRs) in response to salt tolerance in tomatoes. BRs regulate ethylene biosynthesis and signaling in tomato seedlings during salt stress. Exogenous treatment of BR activates ethylene synthesis enzyme 1-aminocyclopropane-1-carboxylate synthase (ACS) and ethylene signaling pathway (Zhu et al. 2016). Silencing of ethylene signaling components inhibited the BR-induced salt tolerance and restrained BR-induced activities. The high content of H_2O_2 and ethylene is observed during BR treatment in tomato seedlings. H_2O_2 is involved in BR-induced ethylene production and ethylene also promotes H_2O_2 production

TABLE 2.2
Role of H_2O_2 in Response to Abiotic Stresses

Plant	Stress	Function	Reference
Cucumber	Chilling tolerance	H_2O_2 regulates ABA regulation in grafting-induced chilling tolerance	Lv et al. (2022)
	Salt tolerance	H_2O_2 and Ca^{2+} are involved in GR24-induced salt tolerance	Zhang et al. (2022)
	Salt tolerance	H_2O_2 and ABA delay stomatal closure during salt stress	Ma et al. (2022)
	Salt stress	H_2O_2 mediates spermidine-induced autophagy to improve tolerance	Zhang et al. (2021)
Grapevine	Cold tolerance	H_2O_2 plays an important role in BRs inducing cold tolerance	Wang et al. (2023)
	Salt tolerance	Exogenous treatment of H_2O_2 improves antioxidant mechanism against sat stress	Özdamar et al. (2022)
Tomato	Saline, alkaline tolerance	Polyamines triggered H_2O_2-mediated spermidine to induce salt tolerance	Yang et al. (2022)
	Cold	H_2O_2 and NO are involved in trehalose-regulated oxidative stress tolerance	Liu et al. (2020)
	Chilling tolerance	SlMAPK1/2/3 associated with H_2O_2 induces chilling tolerance	Wang et al. (2017)
Tobacco	Salt tolerance	H_2O_2 and Ca^{2+} signal transduction pathways and antioxidant activities improve salt tolerance	Che et al. (2022)
Watermelon	Cold tolerance	Exogenous treatment of H_2O_2 on *Cl2-CP* silence watermelon plant improves antioxidant activity.	Xu et al. (2022)
Maize	Drought	H_2O_2 modulates the antioxidant system	Ashraf et al. (2015)
Brassica napus	Drought	Combine the application of H_2O_2 and Ca^{2+} to recover water content reduction and chlorosis	Khan et al. (2017)
Rice	Drought	H_2O_2 exogenous application on seeds increases yield and reproductive stage in rice	Jira-anunkul and Pattanagul (2021)
	Drought	Rice seed priming with H_2O_2 reduces the salt stress effect	Hemalatha et al. (2017)
	Heat	OsMHAR4 mediates H_2O_2 and enhances heat tolerance	Liu et al. (2018)
	Drought	OsASR5 enhances drought tolerance via H_2O_2 and ABA pathway	Li et al. (2017)
	Drought	SNAC1-targeted gene OsSRO1c regulates H_2O_2 and modulates oxidative stress response	You et al. (2013)
Wheat	Drought	H_2O_2 pretreatment of wheat seed improves tolerance	He and Gao (2009)
Cauliflower	Salt stress	H_2O_2 regulates the drastic effect of salt stress by preventing ROS production	Ellouzi et al. (2021)
Carrot		*DcSOD* gene binds and interacts with H_2O_2 and may be involved in abiotic stress response	Zameer et al. (2022)

(Vandenabeele et al. 2003; Xia et al. 2014; Zhu et al. 2016). This study revealed that BR could promote ethylene production or H_2O_2 synthesis. BR also induced ROS production and activates MAPK, which could increase ACS stabilization and ACC biosynthesis and generate impact on ethylene production. H_2O_2-responsive MPK6, MPK4, MPK1, and MAPKKK are highly active under oxidative stress and changing climatic conditions. Exogenous treatment of H_2O_2 leads to MAPK signaling pathways (Ruffels et al. 2004). MEKKK1 pathway is highly active under oxidative stress conditions and acts as an activator of two MAPK (MKK1 and MKK2) (Mittler et al. 2011).

H_2O_2 also regulates BR signaling via oxidation of the *BRASSINAZOLE-RESISTANT1(BZR1)* transcription factor (Tian et al. 2018). Oxidative changes increase BZR1 activity by enhancing its interaction with prominent regulators of auxin signaling and light signaling pathways such as *AUXIN RESPONSE FACTOR6 (ARF6)* and *PHYTOCHROME INTERACTING FACTOR4*

(PIF4). H_2O_2 induces BZR1 oxidation and increases the transcriptional activity of ARF6 and PIF4. Oxidized BZR1 is inhibited by a thioredoxin protein (TRXh5). Overexpression of TRXh5 and mutation in BZR1 disrupts BZR1 function in cell elongation and QC cell division. BZR1 activity is regulated by the opposing redox actions of H_2O_2 and TRXh5 (Tian et al. 2018).

Auxin is significantly involved in stress-related hormonal signaling networks. Auxin receptor *TRANSPORT INHIBITOR RESPONSE 1 (TIR1)/AUXIN SIGNALING F-BOX PROTEINS (AFB)* is involved in oxidative stress tolerance by regulating H_2O_2 and antioxidant enzyme activities (Iglesias et al. 2010). H_2O_2 cross talk with auxins and regulate cell cycle under cadmium stress condition in rice. H_2O_2 influences the expression of auxin signaling-related genes such as *OsPIN*, *OsIAA*, and *OsARF* under Cd stress conditions (Zhao et al. 2012). H_2O_2 also modifies the expression pattern of cell cycle-related genes under Cd-treated roots. The effect of the H_2O_2 root growth system is linked with auxin signaling and modifies the expression of cell cycle genes as well as auxin signaling-related genes under cadmium treatment (Zhao et al. 2012).

The exogenous application of epigallocatechin-3-gallate (EGCG) stimulates *RBOH1* and triggers membrane-bound NADPH oxidase. NADPH leads superoxidase synthesis followed by its dismutation to H_2O_2 in apoplastic. Further *RBOH1*-mediated apoplastic H_2O_2 is involved in ROS signaling toward activation of antioxidant enzymes such as POD, SOD, CAT, and APX as well as the synthesis of non-enzymatic antioxidants (flavonoids, glutathione, ascorbic acid), which minimized the drastic effect of oxidative damage caused by abiotic stress (Mittler 2017; Li et al. 2019). RtGL1 is a chloroplast-localized plasma membrane encoding protein regulated by H_2O_2, abiotic stresses, and ABA. The overexpression of RtGL1 in *Arabidopsis* significantly improves chlorophyll accumulation, primary root development, and seed germination during drought and salt stress conditions.

Moreover, relative expression patterns of antioxidant and proline biosynthesis-related genes increase and influence glutathione biosynthesis (Ma et al. 2021). Glutathione biosynthesis leads to glutathione glutathione-dependent detoxification of ROS and proline content under abiotic stress conditions. RtGL1 also affected H_2O_2 concentration and glutathione/ascorbate ratios. RtGL1 positively regulates GSH biosynthesis-related genes and decreases Na^+ content in the leaf and balances the lower ratio of Na^+/K^+ ions in transgenic plants in comparison with wild-type *Arabidopsis* plants (Ma et al. 2021). In mangrove tree species exogenous application of H_2O_2 and Ca^{2+} regulates the ion fluxes under NaCl stress. Under salt stress conditions, H_2O_2 interacts with Ca^{2+} and mediates antioxidant enzyme activities and root ion balance and upregulates the expression pattern of antioxidant biosynthesis-related genes (Lu et al. 2013).

Signaling crosstalk between H_2O_2 and NO is involved in inducing the thermotolerance in maize seedlings (Li et al. 2015). H_2O_2 acts as upstream signaling of NO in the heat shock pathway in *Arabidopsis* seedlings (Wang et al. 2014). Ca^{2+} binding protein OsANN1 increases heat tolerance by regulating H_2O_2 production. Excessive production of H_2O_2 upregulates the antioxidant (CAT, SOD) production and upregulates the expression pattern of OsANN1 under heat stress conditions (Qiao et al. 2015). H_2O_2 and Ca^{2+} are also involved in various aspects of plant growth and development and abiotic stress response. H_2O_2 interrelationship with Ca^{2+} and NO indicates a feedback mechanism during abiotic stress conditions.

Plant growth and development. During the signaling pathway, Ca^{2+} acts as a point of convergence between H_2O_2 and NO in plants (Niu and Liao 2016). Moreover, Ca^{2+} sensors, including calcium-dependent protein kinases (CDPKs), calmodulins (CaMs), CaM-like proteins (CMLs), and calcineurin B-like proteins (CBLs), are involved in jasmonate (Soares et al. 2020) signaling and biosynthesis (Khan et al. 2023). Jasmonates are effective inducers of H_2O_2 accumulation in plant cell cultures (Wang et al. 2012). Methyl jasmonate (MeJA) plays an essential role in the signal transduction pathway in grape cells and regulates NO and H_2O_2 by enhancing the activities of enzymes involved in phytoalexin synthesis (Wang et al. 2012). MeJA can also induce stomatal closer mediated by ROS-dependent signaling pathway (Suhita et al. 2004). Exogenous treatment of SA induced drought tolerance by increasing antioxidant enzyme activities and promoted early flowering in sweet oranges (Khan et al. 2022). Under abiotic stress conditions, endogenous SA and H_2O_2

increase in plant cells, whereas H_2O_2 degrades in enzymes, including CAT and APX (Yuan and Lin 2008). Individual H_2O_2 is not sufficient to stimulate defense response and cell death. SA-deficient transgenic *Arabidopsis* line overexpressing salicylate hydroxylase (*NahG*) gene exposed to ROS (O^{-3}) accumulated more H_2O_2 as compared with wild-type plants but are unable to induce expression of defense-related genes such as GPX and GSTI (Rao et al. 2002). There is a need to uncover more about underlying molecular mechanisms with ABA-induced stress tolerance and plant development for a better understanding of H_2O_2's role in plant development and abiotic stress tolerance.

CONCLUSION

Plants have specific mechanisms to utilize ROS (H_2O_2) for signaling under abiotic stress conditions that confer drastic effects of abiotic stresses and induce tolerance (Figure 2.1). H_2O_2 acts as a signaling molecule and performs a significant role in plant development and abiotic stress tolerance. Many researchers have suggested the pivotal role of H_2O_2 in intracellular and systematic signaling pathways that induce tolerance and also regulate some developmental mechanisms in plants. Recent findings have shown that effective ROS signaling may require antioxidant enzyme invasions such as induction of MAPK, Ca^{2+} regulations, plant hormonal signal transduction, redox-sensitive transcription factors, and NO. There is a complex interaction between H_2O_2, Ca^{2+}, plant growth regulators, MAPK, and NO in plants. The crosstalks among them have a functional role in regulating plant development and physiological mechanisms, which may increase signal transduction and signal reception in plants. Future work will need to focus on the H_2O_2 signaling network and ROS scavenging system with respect to changes under stress conditions, and H_2O_2 interplay with other factors is necessary and is of great importance if we are improving the plants' tolerance to oxidative stress using biotechnological approaches.

REFERENCES

Alesandrini F, Mathis R, Van de Sype G, Hérouart D, Puppo A (2003) Possible roles for a cysteine protease and hydrogen peroxide in soybean nodule development and senescence. *New Phytologist* 158(1):131–138. https://doi.org/10.1046/j.1469-8137.2003.00720.x.

Ashraf MA, Rasheed R, Hussain I, Iqbal M, Haider MZ, Parveen S, Sajid MA (2015) Hydrogen peroxide modulates antioxidant system and nutrient relation in maize (Zea mays L.) under water-deficit conditions. *Archives of Agronomy Soil Science* 61(4):507–523.

Bagheri M, Gholami M, Baninasab B (2019) Hydrogen peroxide-induced salt tolerance in relation to antioxidant systems in pistachio seedlings. *Scientia Horticulturae* 243:207–213.

Balazadeh S, Wu A, Mueller-Roeber B (2010) Salt-triggered expression of the ANAC092-dependent senescence regulon in Arabidopsis thaliana. *Plant Signaling Behavior* 5(6):733–735.

Barba-Espín G, Hernández JA, Diaz-Vivancos P (2012) Role of H_2O_2 in pea seed germination. *Plant Signaling Behavior* 7(2):193–195.

Barzotto GR, Cardoso CP, Jorge LG, Campos FG, Boaro CSF (2023) Hydrogen peroxide signal photosynthetic acclimation of Solanum lycopersicum L. cv Micro-Tom under water deficit. *Scientific Reports* 13(1):13059.

Bright J, Desikan R, Hancock JT, Weir IS, Neill SJ (2006) ABA-induced NO generation and stomatal closure in Arabidopsis are dependent on H_2O_2 synthesis. *The Plant Journal* 45 (1):113–122.

Černý M, Habánová H, Berka M, Luklová M, Brzobohatý B (2018) Hydrogen peroxide: its role in plant biology and crosstalk with signalling networks. *International Journal of Molecular Sciences* 19(9):2812.

Che Y, Yao T, Wang H, Wang Z, Zhang H, Sun G, Zhang H (2022) Potassium ion regulates hormone, Ca^{2+} and H_2O_2 signal transduction and antioxidant activities to improve salt stress resistance in tobacco. *Plant Physiology Biochemistry* 186:40–51.

Chen XJ, Xia XJ, Guo X, Zhou YH, Shi K, Zhou J, Yu JQ (2016) Apoplastic H_2O_2 plays a critical role in axillary bud outgrowth by altering auxin and cytokinin homeostasis in tomato plants. *New Phytologist* 211 (4):1266–1278.

Chen Z, Gu Q, Yu X, Huang L, Xu S, Wang R, Shen W, Shen W (2018) Hydrogen peroxide acts downstream of melatonin to induce lateral root formation. *Annals of Botany* 121(6):1127–1136.

Cheng M, Guo Y, Liu Q, Nan S, Xue Y, Wei C, Zhang Y, Luan F, Zhang X, Li H (2022) H_2O_2 and Ca^{2+} signaling crosstalk counteracts ABA to induce seed germination. *Antioxidants* 11(8):1594.

Cheng W-H, Wang F-L, Cheng X-Q, Zhu Q-H, Sun Y-Q, Zhu H-G, Sun J (2015) Polyamine and its metabolite H_2O_2 play a key role in the conversion of embryogenic callus into somatic embryos in upland cotton (Gossypium hirsutum L.). *Frontiers in Plant Science* 6:1063.

Demidchik V, Shabala SN, Davies JM (2007) Spatial variation in H_2O_2 response of Arabidopsis thaliana root epidermal Ca^{2+} flux and plasma membrane Ca^{2+} channels. *The Plant Journal* 49 (3):377–386

Dikilitas M, Simsek E, Roychoudhury A (2020) Modulation of abiotic stress tolerance through hydrogen peroxide. In *Protective Chemical Agents in the Amelioration of Plant Abiotic Stress: Biochemical Molecular Perspectives*, pp. 147–173. https://doi.org/10.1002/9781119552154.ch7.

Ellouzi H, Oueslati S, Hessini K, Rabhi M, Abdelly C (2021) Seed-priming with H_2O_2 alleviates subsequent salt stress by preventing ROS production and amplifying antioxidant defense in cauliflower seeds and seedlings. *Scientia Horticulturae* 288:110360.

Ge XM, Cai HL, Lei X, Zhou X, Yue M, He JM (2015) Heterotrimeric G protein mediates ethylene-induced stomatal closure via hydrogen peroxide synthesis in Arabidopsis. *The Plant Journal* 82(1):138–150.

Gong F, Yao Z, Liu Y, Sun M, Peng X (2021) H_2O_2 response gene 1/2 are novel sensors or responders of H_2O_2 and involve in maintaining embryonic root meristem activity in Arabidopsis thaliana. *Plant Science* 310:110981.

Gong W, Niu L, Wang C, Wei L, Pan Y, Liao W (2022) Hydrogen peroxide is involved in salicylic acid-induced adventitious rooting in cucumber under cadmium stress. *Journal of Plant Biology* 65:1–10.

Guo Y, Zhu J, Liu J, Xue Y, Chang J, Zhang Y, Ahammed GJ, Wei C, Ma J, Li P (2023) Melatonin delays ABA-induced leaf senescence via H_2O_2-dependent calcium signalling. *Plant, Cell Environment* 46(1):171–184.

Hameed A, Farooq S, Iqbal N, Arshad R (2004) Influence of exogenous application of hydrogen peroxide on root and seedling growth on wheat (Triticum aestivum L.). *International Journal of Agriculture and Biology* 6(2):366–369.

Han S, Fang L, Ren X, Wang W, Jiang J (2015) MPK 6 controls H_2O_2-induced root elongation by mediating Ca^{2+} influx across the plasma membrane of root cells in Arabidopsis seedlings. *New Phytologist* 205(2):695–706.

He J-M, Liu Z-H, Xu H, She X-P, Huang C (2006) The involvement of hydrogen peroxide in UV-B-inhibited pollen germination and tube growth of Paeonia suffruticosa and Paulownia tomentosa in vitro. *Plant Growth Regulation* 49:199–208.

He L, Gao Z (2009) Pretreatment of seed with H_2O_2 enhances drought tolerance of wheat (Triticum aestivum L.) seedlings. *African Journal of Biotechnology* 8(22):6151.

Hemalatha G, Renugadevi J, Eevera T (2017) Studies on seed priming with hydrogen peroxide for mitigating salt stress in rice. *International Journal of Current Microbiology and Applied Sciences* 6(6):691–695.

Hou J, Zhang Q, Zhou Y, Ahammed GJ, Zhou Y, Yu J, Fang H, Xia X (2018) Glutaredoxin GRXS16 mediates brassinosteroid-induced apoplastic H_2O_2 production to promote pesticide metabolism in tomato. *Environmental Pollution* 240:227–234.

Huang Y-W, Zhou Z-Q, Yang H-X, Wei C-X, Wan Y-Y, Wang X-J, Bai J-G (2015) Glucose application protects chloroplast ultrastructure in heat-stressed cucumber leaves through modifying antioxidant enzyme activity. *Biologia Plantarum* 59:131–138.

Hubmann AM, Jammer A, Monschein S (2022) Activities of H_2O_2-converting enzymes in apple leaf buds during dormancy release in consideration of the ratio change between bud scales and physiologically active tissues. *Horticulturae* 8(11):982.

Hussain S, Liu G, Liu D, Ahmed M, Hussain N, Teng Y (2015) Study on the expression of dehydrin genes and activities of antioxidativeenzymes in floral buds of two sand pear (Pyrus pyrifolia Nakai) cultivars requiring different chilling hours for bud break. *Turkish Journal of Agriculture Forestry* 39(6):930–939.

Iglesias MJ, Terrile MC, Bartoli CG, D'Ippólito S, Casalongué CA (2010) Auxin signaling participates in the adaptative response against oxidative stress and salinity by interacting with redox metabolism in Arabidopsis. *Plant Molecular Biology* 74:215–222.

Ivanchenko MG, Den Os D, Monshausen GB, Dubrovsky JG, Bednářová A, Krishnan N (2013) Auxin increases the hydrogen peroxide (H_2O_2) concentration in tomato (Solanum lycopersicum) root tips while inhibiting root growth. *Annals of Botany* 112(6):1107–1116.

Jahan A, Khan MMA, Ahmad B, Ahmed KBM, Pandey RP, Gulfishan M (2023) Hydrogen peroxide: Regulator of plant development and abiotic stress response. In: *Reactive Oxygen Species: Prospects in Plant Metabolism*. Springer, pp. 213–228. https://doi.org/10.1007/978-981-19-9794-5_12.

Jira-anunkul W, Pattanagul W (2021) Effects of hydrogen peroxide application on agronomic traits of rice (Oryza sativa L.) under drought stress. *Plant, Soil, Environment* 67(4):221–229.

Kapoor D, Singh S, Kumar V, Romero R, Prasad R, Singh J (2019) Antioxidant enzymes regulation in plants in reference to reactive oxygen species (ROS) and reactive nitrogen species (RNS). *Plant Gene* 19:100182.

Khan A, Anwar Y, Hasan MM, Iqbal A, Ali M, Alharby HF, Hakeem KR, Hasanuzzaman M (2017) Attenuation of drought stress in Brassica seedlings with exogenous application of Ca^{2+} and H_2O_2. *Plants* 6(2):20.

Khan F, Goher F, Paulsmeyer M, Hu CG, Zhang JZ (2023) Calcium (Ca^{2+}) sensors and MYC2 are crucial players during jasmonates-mediated abiotic stress tolerance in plants. *Plant Biology* 25:1025.

Khan FS, Gan Z-M, Li E-Q, Ren M-K, Hu C-G, Zhang J-Z (2022) Transcriptomic and physiological analysis reveals interplay between salicylic acid and drought stress in citrus tree floral initiation. *Planta* 255:1–22.

Khan T, Yusuf M, Fariduddin Q (2018) Hydrogen peroxide in regulation of plant metabolism: Signalling and its effect under abiotic stress. *Photosynthetica* 56(4):1237–1248.

Khan TA, Yusuf M, Fariduddin Q (2015) Seed treatment with H_2O_2 modifies net photosynthetic rate and antioxidant system in mung bean (Vigna radiata L. Wilczek) plants. *Israel Journal of Plant Sciences* 62 3):167–175.

Kuroda H, Sugiura T, Ito D (2002) Changes in hydrogen peroxide content in flower buds of Japanese pear (Pyrus pyrifolia Nakai) in relation to breaking of endodormancy. *Journal of the Japanese Society for Horticultural Science* 71(5):610–616.

Kwak JM, Mori IC, Pei Z-M, Leonhardt N, Torres MA, Dangl JL, Bloom RE, Bodde S, Jones JD, Schroeder JI (2003) NADPH oxidase AtrbohD and AtrbohF genes function in ROS-dependent ABA signaling in Arabidopsis. *The EMBO Journal* 22(11):2623–2633.

Lei C, Ye M, Li C, Gong M (2023) H_2O_2 participates in the induction and formation of potato tubers by activating tuberization-related signal transduction pathways. *Agronomy* 13(5):1398.

Li HB, Qin YM, Pang Y, Song WQ, Mei WQ, Zhu YX (2007) A cotton ascorbate peroxidase is involved in hydrogen peroxide homeostasis during fibre cell development. *New Phytologist* 175 (3):462–471

Li J, Chen S, Wang X, Shi C, Liu H, Yang J, Shi W, Guo J, Jia H (2018) Hydrogen sulfide disturbs actin polymerization via S-sulfhydration resulting in stunted root hair growth. *Plant Physiology* 178 (2):936–949

Li J, Jia H, Wang J, Cao Q, Wen Z (2014) Hydrogen sulfide is involved in maintaining ion homeostasis via regulating plasma membrane Na^+/H^+ antiporter system in the hydrogen peroxide-dependent manner in salt-stress Arabidopsis thaliana root. *Protoplasma* 251:899–912.

Li J, Li Y, Yin Z, Jiang J, Zhang M, Guo X, Ye Z, Zhao Y, Xiong H, Zhang Z (2017) Os ASR 5 enhances drought tolerance through a stomatal closure pathway associated with ABA and H_2O_2 signalling in rice. *Plant Biotechnology Journal* 15(2):183–196.

Li X, Li Y, Ahammed GJ, Zhang X-N, Ying L, Zhang L, Yan P, Zhang L-P, Li Q-Y, Han W-Y (2019) RBOH1-dependent apoplastic H_2O_2 mediates epigallocatechin-3-gallate-induced abiotic stress tolerance in Solanum lycopersicum L. *Environmental Experimental Botany* 161:357–366.

Li Y, Wu Y, Liao W, Hu L, Dawuda MM, Jin X, Tang Z, Yang J, Yu J (2020) Nitric oxide is involved in the brassinolide-induced adventitious root development in cucumber. *BMC Plant Biology* 20(1):1–12.

Li Z, Luo L, Sun Y (2015) Signal crosstalk between nitric oxide and hydrogen sulfide may be involved in hydrogen peroxide-induced thermotolerance in maize seedlings. *Russian Journal of Plant Physiology* 62:507–514.

Liao W-B, Zhang M-L, Huang G-B, Yu J-H (2012a) Ca^{2+} and CaM are involved in NO-and H_2O_2-induced adventitious root development in marigold. *Journal of Plant Growth Regulation* 31:253–264.

Liao W-B, Zhang M-L, Huang G-B, Yu J-H (2012b) Hydrogen peroxide in the vase solution increases vase life and keeping quality of cut Orientalx Trumpet hybrid lily ‘Manissa’. *Scientia Horticulturae* 139:32–38.

Liao W, Huang G, Yu J, Zhang M, Shi X (2011) Nitric oxide and hydrogen peroxide are involved in indole-3-butyric acid-induced adventitious root development in marigold. *The Journal of Horticultural Science Biotechnology* 86(2):159–165.

Lin H-H, King Y-C, Li Y-C, Lin C-C, Chen Y-C, Lin J-S, Jeng S-T (2019) The p38-like MAP kinase modulated H_2O_2 accumulation in wounding signaling pathways of sweet potato. *Plant Science* 280:305–313.

Liu J, Sun X, Xu F, Zhang Y, Zhang Q, Miao R, Zhang J, Liang J, Xu W (2018) Suppression of OsMDHAR4 enhances heat tolerance by mediating H_2O_2-induced stomatal closure in rice plants. *Rice* 11 (1):1–12.

Liu T, Ye X, Li M, Li J, Qi H, Hu X (2020) H_2O_2 and NO are involved in trehalose-regulated oxidative stress tolerance in cold-stressed tomato plants. *Environmental Experimental Botany* 171:103961.

Liu Y, Ye N, Liu R, Chen M, Zhang J (2010) H_2O_2 mediates the regulation of ABA catabolism and GA biosynthesis in Arabidopsis seed dormancy and germination. *Journal of Experimental Botany* 61 (11):2979–2990

Lu S, Su W, Li H, Guo Z (2009) Abscisic acid improves drought tolerance of triploid bermudagrass and involves H_2O_2-and NO-induced antioxidant enzyme activities. *Plant Physiology Biochemistry* 47(2):132–138.

Lu Y, Li N, Sun J, Hou P, Jing X, Zhu H, Deng S, Han Y, Huang X, Ma X (2013) Exogenous hydrogen peroxide, nitric oxide and calcium mediate root ion fluxes in two non-secretor mangrove species subjected to NaCl stress. *Tree Physiology* 33(1):81–95.

Lv C, Li F, Ai X, Bi H (2022) H_2O_2 participates in ABA regulation of grafting-induced chilling tolerance in cucumber. *Plant Cell Reports* 41 (4):1115–1130.

Ma B, Suo Y, Zhang J, Xing N, Gao Z, Lin X, Zheng L, Wang Y (2021) Glutaredoxin like protein (RtGRL1) regulates H_2O_2 and Na^+ accumulation by maintaining the glutathione pool during abiotic stress. *Plant Physiology Biochemistry* 159:135–147.

Ma M, Wendehenne D, Philippot L, Hänsch R, Flemetakis E, Hu B, Rennenberg H (2020) Physiological significance of pedospheric nitric oxide for root growth, development and organismic interactions. *Plant, Cell Environmental* 43(10):2336–2354.

Ma S, Zhou X, Jahan MS, Guo S, Tian M, Zhou R, Liu H, Feng B, Shu S (2022) Putrescine regulates stomatal opening of cucumber leaves under salt stress via the H_2O_2-mediated signaling pathway. *Plant Physiology Biochemistry* 170:87–97.

Mei Y, Chen H, Shen W, Shen W, Huang L (2017) Hydrogen peroxide is involved in hydrogen sulfide-induced lateral root formation in tomato seedlings. *BMC Plant Biology* 17(1):162. https://doi.org/10.1186/s12870-017-1110-7.

Miao Y, Lv D, Wang P, Wang X-C, Chen J, Miao C, Song C-P (2006) An Arabidopsis glutathione peroxidase functions as both a redox transducer and a scavenger in abscisic acid and drought stress responses. *The Plant Cell* 18(10):2749–2766.

Mittler R (2017) ROS are good. *Trends in Plant Science* 22(1):11–19.

Mittler R, Van Breusegem F (2023) *Oxidative Stress Response in Plants*, vol. 105. Academic Press, Cambridge, MA.

Mittler R, Vanderauwera S, Suzuki N, Miller G, Tognetti VB, Vandepoele K, Gollery M, Shulaev V, Van Breusegem F (2011) ROS signaling: The new wave? *Trends in Plant Science* 16 (6):300–309.

Mukherjee S, Corpas FJ (2023) H_2O_2, NO, and H_2S networks during root development and signalling under physiological and challenging environments: Beneficial or toxic? *Plant, Cell Environment* 46(3):688–717.

Nazir F, Hussain A, Fariduddin Q (2019) Hydrogen peroxide modulate photosynthesis and antioxidant systems in tomato (Solanum lycopersicum L.) plants under copper stress. *Chemosphere* 230:544–558.

Nicolas-Francès V, Rossi J, Rosnoblet C, Pichereaux C, Hichami S, Astier J, Klinguer A, Wendehenne D, Besson-Bard A (2022) S-nitrosation of Arabidopsis thaliana protein tyrosine phosphatase 1 prevents its irreversible oxidation by hydrogen peroxide. *Frontiers in Plant Science* 13:807249.

Niu L, Liao W (2016) Hydrogen peroxide signaling in plant development and abiotic responses: crosstalk with nitric oxide and calcium. *Frontiers in Plant Science* 7:230.

Orabi SA, Dawood MG, Salman S (2015) Comparative study between the physiological role of hydrogen peroxide and salicylic acid in alleviating the harmful effect of low temperature on tomato plants grown under sand-ponic culture. *Science of Agricultural* 9(1):49–59.

Özdamar FÖ, Furtana GB, Duman H, Tipirdamaz R (2022) Improvement of the antioxidant mechanism by exogenous H_2O_2 in two genotypes of eggplant (Solanum melongena L.) under salt stress. *Journal of Anatolian Environmental Animal Sciences* 7(4):465–471.

Passaia G, Fonini LS, Caverzan A, Jardim-Messeder D, Christoff AP, Gaeta ML, de Araujo Mariath JE, Margis R, Margis-Pinheiro M (2013) The mitochondrial glutathione peroxidase GPX3 is essential for H_2O_2 homeostasis and root and shoot development in rice. *Plant Science* 208:93–101.

Pérez FJ, Vergara R, Rubio S (2008) H_2O_2 is involved in the dormancy-breaking effect of hydrogen cyanamide in grapevine buds. *Plant Growth Regulation* 55:149–155.

Porcher A, Guérin V, Montrichard F, Lebrec A, Lothier J, Vian A (2020) Ascorbate glutathione-dependent H_2O_2 scavenging is an important process in axillary bud outgrowth in rosebush. *Annals of Botany* 126(6):1049–1062.

Qiao B, Zhang Q, Liu D, Wang H, Yin J, Wang R, He M, Cui M, Shang Z, Wang D (2015) A calcium-binding protein, rice annexin OsANN1, enhances heat stress tolerance by modulating the production of H_2O_2. *Journal of Experimental Botany* 66(19):5853–5866.

Rao MV, Lee H, Davis KR (2002) Ozone-induced ethylene production is dependent on salicylic acid, and both salicylic acid and ethylene act in concert to regulate ozone-induced cell death. *The Plant Journal* 32(4):447–456.

Ruffels J, Griffin M, Dickenson JM (2004) Activation of ERK1/2, JNK and PKB by hydrogen peroxide in human SH-SY5Y neuroblastoma cells: Role of ERK1/2 in H_2O_2-induced cell death. *European Journal of Pharmacology* 483(2–3):163–173.

Sakamoto M, Munemura I, Tomita R, Kobayashi K (2008) Involvement of hydrogen peroxide in leaf abscission signaling, revealed by analysis with an in vitro abscission system in Capsicum plants. *The Plant Journal* 56(1):13–27.

Sapkota S, Liu J, Islam MT, Sherif SM (2021) Changes in reactive oxygen species, antioxidants and carbohydrate metabolism in relation to dormancy transition and bud break in apple (Malus× domestica Borkh) Cultivars. *Antioxidants* 10(10):1549.

Saxena I, Srikanth S, Chen Z (2016) Cross talk between H_2O_2 and interacting signal molecules under plant stress response. *Frontiers in Plant Science* 7:570.

Shafique Khan F, Zeng R-F, Gan Z-M, Zhang J-Z, Hu C-G (2021) Genome-wide identification and expression profiling of the WOX gene family in Citrus sinensis and functional analysis of a CsWUS member. *International Journal of Molecular Sciences* 22(9):4919.

Shi W, Wang L, Yao L, Hao W, Han C, Fan M, Wang W, Bai M-Y (2022) Spatially patterned hydrogen peroxide orchestrates stomatal development in Arabidopsis. *Nature Communications* 13(1):5040.

Soares JM, Weber KC, Qiu W, Stanton D, Mahmoud LM, Wu H, Huyck P, Zale J, Al Jasim K, Grosser JW (2020) The vascular targeted citrus FLOWERING LOCUS T3 gene promotes non-inductive early flowering in transgenic Carrizo rootstocks and grafted juvenile scions. *Scientific Reports* 10(1):21404.

Steffens B, Geske T, Sauter M (2011) Aerenchyma formation in the rice stem and its promotion by H_2O_2. *New Phytologist* 190(2):369–378.

Su GX, Zhang WH, Liu YL (2006) Involvement of hydrogen peroxide generated by polyamine oxidative degradation in the development of lateral roots in soybean. *Journal of Integrative Plant Biology* 48(4):426–432.

Suhel M, Husain T, Singh S, Pandey A, Singh SK, Fatima A, Singh VP, Prasad SM (2020) Interplay of nitric oxide and hydrogen peroxide in root development. *Reactive Oxygen Species* 9(25):1–12.

Suhita D, Raghavendra AS, Kwak JM, Vavasseur A (2004) Cytoplasmic alkalization precedes reactive oxygen species production during methyl jasmonate-and abscisic acid-induced stomatal closure. *Plant Physiology* 134(4):1536–1545.

Sun Y, Wang H, Liu S, Peng X (2016) Exogenous application of hydrogen peroxide alleviates drought stress in cucumber seedlings. *South African Journal of Botany* 106:23–28.

Tian M, Gu Q, Zhu M (2003) The involvement of hydrogen peroxide and antioxidant enzymes in the process of shoot organogenesis of strawberry callus. *Plant Science* 165(4):701–707.

Tian Y, Fan M, Qin Z, Lv H, Wang M, Zhang Z, Zhou W, Zhao N, Li X, Han C (2018) Hydrogen peroxide positively regulates brassinosteroid signaling through oxidation of the BRASSINAZOLE-RESISTANT1 transcription factor. *Nature Communications* 9(1):1063.

Tognetti VB, Bielach A, Hrtyan M (2017) Redox regulation at the site of primary growth: auxin, cytokinin and ROS crosstalk. *Plant, Cell Environmental* 40(11):2586–2605.

Tripathy BC, Oelmüller R (2012) Reactive oxygen species generation and signaling in plants. *Plant Signaling Behavior* 7(12):1621–1633.

ul Islam SN, Asgher M, Khan NA (2023) Hydrogen peroxide and its role in abiotic stress tolerance in plants. In: Fatma, M., Sehar, Z., Khan, N.A. (eds.) *Gasotransmitters Signaling in Plant Abiotic Stress: Gasotransmitters in Adaptation of Plants to Abiotic Stress*. Springer, Cham, pp 167–195.

Vandenabeele S, Van Der Kelen K, Dat J, Gadjev I, Boonefaes T, Morsa S, Rottiers P, Slooten L, Van Montagu M, Zabeau M (2003) A comprehensive analysis of hydrogen peroxide-induced gene expression in tobacco. *Proceedings of the National Academy of Sciences* 100(26):16113–16118.

Vandepoele K, Gollery M, Shulaev V, Van Breusegem F (2011) ROS signaling: The new wave. *Trends in Plant Science* 16(6):300309.

Wang K, Zheng Y, Tang W, Li T, Zhang Q, Shang H (2012) Effects of methyl jasmonate treatment on levels of nitric oxide and hydrogen peroxide and phytoalexin synthesis in postharvest grape berries. *Acta Horticulturae Sinica* 39(8):1559–1566.

Wang L, Guo Y, Jia L, Chu H, Zhou S, Chen K, Wu D, Zhao L (2014) Hydrogen peroxide acts upstream of nitric oxide in the heat shock pathway in Arabidopsis seedlings. *Plant Physiology* 164(4):2184–2196.

Wang L, Zhao R, Zheng Y, Chen L, Li R, Ma J, Hong X, Ma P, Sheng J, Shen L (2017) SlMAPK1/2/3 and antioxidant enzymes are associated with H_2O_2-induced chilling tolerance in tomato plants. *Journal of Agricultural Food Chemistry* 65(32):6812–6820.

Wang P, Du Y, Li Y, Ren D, Song C-P (2010) Hydrogen peroxide–mediated activation of MAP kinase 6 modulates nitric oxide biosynthesis and signal transduction in Arabidopsis. *The Plant Cell* 22(9):2981–2998.

Wang Y, Jiang Q, Wang X, Xi Z (2023) Brassinosteroid stimulates hydrogen peroxide biosynthesis and reduces the effect of cold stress. *Journal of Plant Growth Regulation* 42(6):3757–3769.

Wang Y, Yao R (2023) H_2O_2 and Ca^{2+} are involved in the embryogenic potential loss of cells during long-term proliferation culture in Pinus massoniana. *Plant Cell, Tissue Organ Culture* 154:1–16.

Wen J-F, Gong M, Liu Y, Hu J-L, Deng M (2013) Effect of hydrogen peroxide on growth and activity of some enzymes involved in proline metabolism of sweet corn seedlings under copper stress. *Scientia Horticulturae* 164:366–371.

Wu D, Chu H, Jia L, Chen K, Zhao L (2015) A feedback inhibition between nitric oxide and hydrogen peroxide in the heat shock pathway in Arabidopsis seedlings. *Plant Growth Regulation* 75:503–509.

Wu J, Shang Z, Wu J, Jiang X, Moschou PN, Sun W, Roubelakis-Angelakis KA, Zhang S (2010) Spermidine oxidase-derived H_2O_2 regulates pollen plasma membrane hyperpolarization-activated Ca^{2+}-permeable channels and pollen tube growth. *The Plant Journal* 63(6):1042–1053.

Xia XJ, Gao CJ, Song LX, Zhou YH, Shi K, Yu JQ (2014) Role of H_2O_2 dynamics in brassinosteroid-induced stomatal closure and opening in Solanum lycopersicum. *Plant, Cell Environment* 37(9):2036–2050.

Xiong J, Yang Y, Fu G, Tao L (2015) Novel roles of hydrogen peroxide (H_2O_2) in regulating pectin synthesis and demethyl esterification in the cell wall of rice (Oryza sativa) root tips. *New Phytologist* 206(1):118–126.

Xu A, Cheng F, Zhou S, Hu H, Bie Z (2022) Chilling-induced H_2O_2 signaling activates the antioxidant enzymes in alleviating the photooxidative damage caused by loss of function of 2-Cys peroxiredoxin in watermelon. *Plant Stress* 6:100108.

Yang J, Wang P, Li S, Liu T, Hu X (2022) Polyamine oxidase triggers H_2O_2-mediated spermidine improved oxidative stress tolerance of tomato seedlings subjected to saline-alkaline stress. *International Journal of Molecular Sciences* 23(3):1625.

You J, Zong W, Li X, Ning J, Hu H, Li X, Xiao J, Xiong L (2013) The SNAC1-targeted gene OsSRO1c modulates stomatal closure and oxidative stress tolerance by regulating hydrogen peroxide in rice. *Journal of Experimental Botany* 64(2):569–583.

Yuan S, Lin H-H (2008) Minireview: Role of salicylic acid in plant abiotic stress. *Zeitschrift für Naturforschung C* 63(5–6):313–320.

Zameer R, Fatima K, Azeem F, ALgwaiz HI, Sadaqat M, Rasheed A, Batool R, Shah AN, Zaynab M, Shah AA (2022) Genome-wide characterization of superoxide dismutase (SOD) genes in daucus carota: Novel insights into structure, expression, and binding interaction with hydrogen peroxide (H_2O_2) under abiotic stress condition. *Frontiers in Plant Science* 13:870241.

Zentgraf U, Andrade-Galan AG, Bieker S (2022) Specificity of H_2O_2 signaling in leaf senescence: Is the ratio of H_2O_2 contents in different cellular compartments sensed in Arabidopsis plants? *Cellular Molecular Biology Letters* 27(1):1–19.

Zhang X-H, Ma C, Zhang L, Su M, Wang J, Zheng S, Zhang T-G (2022) GR24-mediated enhancement of salt tolerance and roles of H_2O_2 and Ca_{2+} in regulating this enhancement in cucumber. *Journal of Plant Physiology* 270:153640.

Zhang X, Zhang L, Dong F, Gao J, Galbraith DW, Song C-P (2001) Hydrogen peroxide is involved in abscisic acid-induced stomatal closure in Vicia faba. *Plant Physiology* 126(4):1438–1448.

Zhang Y, Wang Y, Wen W, Shi Z, Gu Q, Ahammed GJ, Cao K, Shah Jahan M, Shu S, Wang J (2021) Hydrogen peroxide mediates spermidine-induced autophagy to alleviate salt stress in cucumber. *Autophagy* 17(10):2876–2890.

Zhao FY, Han MM, Zhang SY, Wang K, Zhang CR, Liu T, Liu W (2012) Hydrogen peroxide-mediated growth of the root system occurs via auxin signaling modification and variations in the expression of cell-cycle genes in rice seedlings exposed to cadmium stress. *Journal of Integrative Plant Biology* 54(12):991–1006.

Zhou B, Li N, Zhang Z, Huang X, Chen H, Hu Z, Pang X, Liu W, Lu Y (2012) Hydrogen peroxide and nitric oxide promote reproductive growth in Litchi chinensis. *Biologia Plantarum* 56:321–329.

Zhou H, Ashworth K, Dodd IC (2023) Exogenous monoterpenes mitigate H_2O_2-induced lipid damage but do not attenuate photosynthetic decline during water deficit in tomato. *Journal of Experimental Botany* 74:5327.

Zhou J, Xia X-J, Zhou Y-H, Shi K, Chen Z, Yu J-Q (2014) RBOH1-dependent H_2O_2 production and subsequent activation of MPK1/2 play an important role in acclimation-induced cross-tolerance in tomato. *Journal of Experimental Botany* 65(2):595–607.

Zhu T, Deng X, Zhou X, Zhu L, Zou L, Li P, Zhang D, Lin H (2016) Ethylene and hydrogen peroxide are involved in brassinosteroid-induced salt tolerance in tomato. *Scientific Reports* 6 (1):35392.

Zou J-J, Li X-D, Ratnasekera D, Wang C, Liu W-X, Song L-F, Zhang W-Z, Wu W-H (2015) Arabidopsis CALCIUM-DEPENDENT PROTEIN KINASE8 and CATALASE3 function in abscisic acid-mediated signaling and H_2O_2 homeostasis in stomatal guard cells under drought stress. *The Plant Cell* 27(5):1445–1460.

3 Controlling Mechanisms of Hydrogen Peroxide Signaling in Plants

Sagheer Ahmad, Donghui Peng, and Zhongjian Liu

INTRODUCTION

Hydrogen peroxide (H_2O_2) is a crucial reactive oxygen species (ROS) in plant cells, functioning both as a potentially damaging byproduct of metabolic processes and as a vital signaling molecule. From the last two decades, H_2O_2 has emerged as a key player regulating pollen growth, photosynthesis, leaf senescence, flower and fruit development, and stomatal movement. Moreover, it facilitates seed germination, stimulates oxidative burst to organic molecules, and works in association with plant hormones (auxin, cytokinin, gibberellin, jasmonic acid (JA), ethylene, abscisic acid (ABA), salicylic acid (SA), Ca^{2+}, and brassinosteroids), thereby regulating various stresses (Figure 3.1). Understanding the balance between its roles in signaling and oxidative stress is essential for comprehending plant biology, especially under conditions of environmental stress. This chapter explores the controlling mechanisms of H_2O_2 signaling in plants, focusing on its production, perception, downstream effects, and integration with other signaling pathways. Additionally, we discuss the physiological and developmental roles of H_2O_2, emphasizing the intricate balance plants maintain to leverage their signaling capabilities while minimizing potential damage.

PRODUCTION AND LOCALIZATION OF HYDROGEN PEROXIDE

Enzymatic Sources

The production of H_2O_2 in plants is highly regulated and primarily occurs through enzymatic reactions. Key enzymes involved in H_2O_2 generation include:

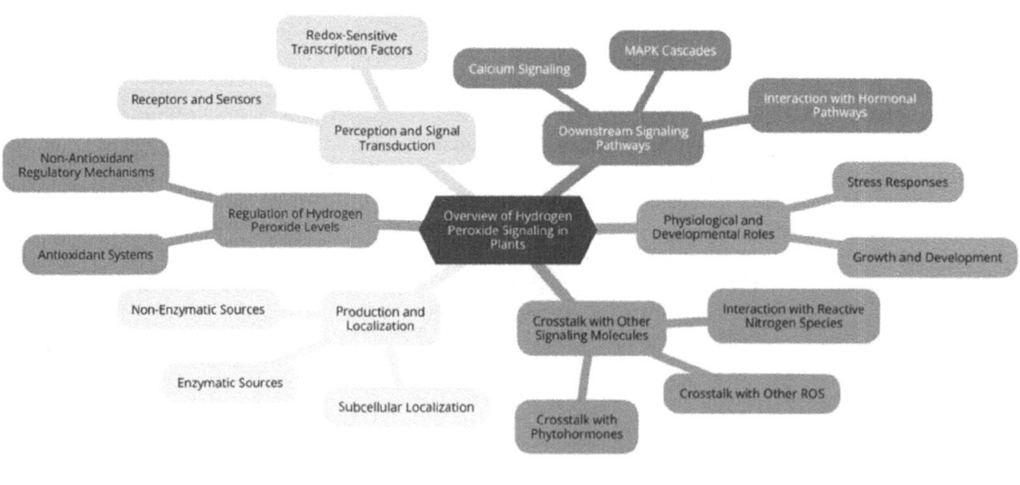

FIGURE 3.1 H_2O_2 signaling in plants.

DOI: 10.1201/9781032647692-3

- **NADPH Oxidases (RBOHs):** These membrane-bound enzymes produce superoxide ($O_2^{\bullet -}$) by transferring electrons from NADPH to oxygen. The superoxide is subsequently dismutated to H_2O_2 either spontaneously or through the action of superoxide dismutase (SOD). The RBOH family members play a critical role in mediating H_2O_2 production in response to environmental stimuli, such as pathogen attack and abiotic stress conditions. The activity of RBOHs is regulated by calcium binding, phosphorylation, and interaction with other proteins, ensuring precise control over H_2O_2 production (Chen & Yang, 2020).
- **Superoxide Dismutases (SODs):** SODs catalyze the dismutation of superoxide radicals to H_2O_2 and oxygen. SODs are present in various cellular compartments, including the cytosol, chloroplasts, mitochondria, and peroxisomes, linking superoxide production to H_2O_2 signaling across different cellular locations. Different isoforms of SOD, such as Cu/Zn-SOD, Mn-SOD, and Fe-SOD, are tailored to specific cellular environments, ensuring efficient conversion of superoxide to H_2O_2 (Tyagi, Shumayla, Singh, & Upadhyay, 2019).
- **Peroxisomal Enzymes:** Peroxisomes are key sites for H_2O_2 metabolism. Enzymes such as glycolate oxidase and urate oxidase generate H_2O_2 as a byproduct of their metabolic reactions. Catalase, another peroxisomal enzyme, decomposes H_2O_2 into water and oxygen, thus playing a crucial role in maintaining H_2O_2 homeostasis within the cell. The dynamic regulation of peroxisomal enzymes ensures that H_2O_2 levels remain balanced, preventing oxidative damage while enabling signaling (Walker, Pomatto, Tripathi, & Davies, 2018).

Non-enzymatic Sources

H_2O_2 can also be produced through non-enzymatic processes. These include the auto-oxidation of various metabolites and redox reactions involving oxygen. While non-enzymatic production is less controlled, it can contribute to overall cellular H_2O_2 levels, particularly under stress conditions where metabolic imbalances occur. For instance, the spontaneous oxidation of ascorbate in the apoplast can lead to localized H_2O_2 production, influencing cell wall modifications and defense responses (Sharova, Medvedev, & Demidchik, 2020).

Subcellular Localization

The subcellular localization of H_2O_2 production is critical for its signaling specificity. H2O2 is produced in various cellular compartments, including:

- **Chloroplasts:** Photosynthetic electron transport in chloroplasts can lead to the formation of superoxide, which is converted to H_2O_2 by chloroplastic SODs. H_2O_2 produced in chloroplasts is involved in signaling processes related to photosynthesis, oxidative stress responses, and programmed cell death. Chloroplastic H_2O_2 can also modulate the expression of nuclear genes through retrograde signaling, linking chloroplast function with overall cellular responses (Breeze & Mullineaux, 2022).
- **Mitochondria:** Mitochondrial electron transport chains are another major source of superoxide, which is subsequently converted to H_2O_2. Mitochondrial H_2O_2 signaling is implicated in metabolic regulation, stress responses, and cellular energy homeostasis. The interplay between mitochondrial ROS production and cellular redox status is crucial for maintaining metabolic balance and adapting to changing environmental conditions (Rigoulet, Yoboue, & Devin, 2011).
- **Peroxisomes:** As mentioned earlier, peroxisomes are central to H_2O_2 metabolism. They generate and detoxify H_2O_2, playing a vital role in regulating cellular redox balance and signaling. Peroxisomal H_2O_2 production is tightly linked to various metabolic pathways, including photorespiration and fatty acid oxidation, highlighting the importance of peroxisomes in cellular homeostasis (Bonekamp, Völkl, Fahimi, & Schrader, 2009).

- **Apoplast:** The extracellular space or apoplast can also accumulate H_2O_2, particularly in response to pathogen attack. Apoplastic H_2O_2 acts as a signaling molecule in plant defense responses, triggering localized cell death and activating systemic acquired resistance (SAR). The spatial and temporal regulation of apoplastic H_2O_2 production is critical for effective defense responses, preventing the spread of pathogens and coordinating immune signaling (Saxena, Srikanth, & Chen, 2016).

PERCEPTION AND SIGNAL TRANSDUCTION

Receptors and Sensors

Plants have evolved sophisticated mechanisms to perceive and respond to H_2O_2:

- **Receptor-Like Kinases (RLKs):** Some RLKs have been implicated in the perception of ROS, including H_2O_2. These receptors can initiate downstream signaling cascades that regulate gene expression, stress responses, and developmental processes. For example, the *Arabidopsis* RLK known as HPCA1 has been shown to act as an H_2O_2 sensor, modulating stomatal closure in response to oxidative stress (Wu et al., 2020). The identification of specific RLKs that respond to H_2O_2 highlights the complexity and specificity of ROS signaling in plants.
- **Peroxiredoxins and Thiol Peroxidases:** These proteins can sense H_2O_2 levels through the oxidation of cysteine residues. The oxidation induces conformational changes that affect their enzymatic activity and interaction with other proteins. For instance, the *Arabidopsis* peroxiredoxin PRX1 is involved in redox signaling by sensing H_2O_2 and transducing the signal to other cellular components. The redox-sensitive nature of these proteins allows them to act as both sensors and regulators of H_2O_2 signaling (Liebthal, Maynard, & Dietz, 2018).
- **Aquaporins:** Some aquaporins, specifically PIPs (plasma membrane intrinsic proteins), facilitate the transport of H_2O_2 across membranes (Jang, Rhee, Chung, & Kang, 2012), influencing its intracellular distribution and signaling. The regulated expression and gating of aquaporins determine the spatiotemporal dynamics of H_2O_2 and impact various physiological processes.

Redox-Sensitive Transcription Factors

H_2O_2 modulates the activity of several transcription factors through redox modifications, thereby influencing gene expression patterns:

- **AP2/ERF Family:** These transcription factors play significant roles in ethylene signaling and stress responses. H_2O_2 can modify cysteine residues in AP2/ERF proteins, altering their DNA-binding affinity and transcriptional activity (He, Van Breusegem, & Mhamdi, 2018). This redox regulation integrates H_2O_2 signaling with ethylene-mediated stress responses. For example, ERF6 is activated by H_2O_2 and regulates genes involved in oxidative stress tolerance and cell death.
- **NAC and WRKY Families:** Redox modifications of cysteine residues in NAC and WRKY transcription factors modulate their activity, impacting various biological processes, including stress responses, development, and senescence. For example, the *Arabidopsis* NAC transcription factor ANAC092 is activated by H_2O_2 and regulates the expression of genes involved in oxidative stress responses and leaf senescence (Balazadeh, Wu, & Mueller-Roeber, 2010). The redox regulation of these transcription factors enables precise control over stress-responsive gene expression.

- **bZIP and MYB Families:** These transcription factors are also regulated by H_2O_2 through redox modifications. bZIP28 and MYB30, for example, are activated by H_2O_2 and play crucial roles in ER stress responses and pathogen defense, respectively (Wu, Wang, Fang, & Xu, 2018). The involvement of multiple transcription factor families in H_2O_2 signaling underscores the complexity and versatility of ROS-regulated gene expression.

DOWNSTREAM SIGNALING PATHWAYS

MAPK Cascades

Mitogen-activated protein kinase (MAPK) cascades are major conduits for H_2O_2 signaling, translating the oxidative signal into appropriate cellular responses:

- **MAPK Activation:** H_2O_2 can activate MAPKs, such as MPK3, MPK4, and MPK6, through a series of phosphorylation events (Lee & Esselman, 2002). These MAPKs regulate gene expression, cell cycle progression, and stress responses by phosphorylating target proteins, including transcription factors, enzymes, and other signaling molecules. The activation of MAPK cascades by H_2O_2 highlights the integration of ROS signaling with kinase-mediated phosphorylation networks.
- **Gene Expression Regulation:** H_2O_2-induced MAPK activation leads to the transcriptional upregulation of genes involved in antioxidant defense, stress tolerance, and programmed cell death. For example, H_2O_2 activates MPK6 (Zhang et al., 2017), which phosphorylates and activates the transcription factor WRKY33, promoting the expression of defense-related genes. The transcriptional responses mediated by MAPK cascades are crucial for mounting effective stress responses.

Calcium Signaling

H_2O_2 can trigger changes in cytosolic calcium levels, serving as a secondary messenger in various signaling pathways:

- **Calcium Influx:** H_2O_2 can induce the opening of calcium channels in the plasma membrane, leading to an influx of calcium ions into the cytosol. This calcium signal can activate calcium-dependent protein kinases (CDPKs) and other calcium-binding proteins, which transduce the signal to downstream targets. The interplay between H_2O_2 and calcium signaling is essential for coordinating cellular responses to oxidative stress and environmental cues (Yan, Wei, Zhang, Cheng, & Liu, 2006).
- **Calcium-Dependent Kinases:** CDPKs activated by H_2O_2-induced calcium influx phosphorylate target proteins, modulating their activity and function. For instance, CDPK5 is activated by H_2O_2 and regulates stomatal closure and pathogen defense (Al Mamun, Lee, Bae, & Kim, 2023). The integration of calcium signaling with H_2O_2 signaling ensures precise regulation of cellular responses to environmental stimuli.

Interaction with Hormonal Pathways

H_2O_2 signaling interacts with various plant hormone pathways, including:

- **Abscisic Acid (ABA):** H_2O_2 is a key mediator of ABA signaling, particularly in the regulation of stomatal closure during drought stress (Wang & Song, 2008). ABA induces H_2O_2 production, which in turn activates calcium channels and MAPKs, leading to stomatal closure and reduced water loss. The crosstalk between H_2O_2 and ABA signaling is crucial for plant adaptation to water deficit conditions.

- **Salicylic Acid (SA):** H_2O_2 and SA act synergistically in plant defense responses against pathogens. SA can enhance H_2O_2 production, while H_2O_2 can amplify SA signaling, leading to the activation of defense genes and SAR (Rao, Paliyath, Ormrod, Murr, & Watkins, 1997). The positive feedback loop between H_2O_2 and SA signaling strengthens plant immune responses and enhances pathogen resistance.
- **Ethylene:** H_2O_2 modulates ethylene signaling by influencing the stability and activity of ethylene-responsive transcription factors. Ethylene can also regulate H_2O_2 production, integrating ROS and hormonal signaling in stress responses and developmental processes. The interplay between H_2O_2 and ethylene signaling is essential for coordinating plant responses to biotic and abiotic stresses (Wang, Huang, Liang, & Bi, 2012).
- **Auxin:** H_2O_2 influences auxin signaling by modulating the activity of auxin transporters and response regulators. This interaction affects various developmental processes, including root growth, lateral root formation, and tropic responses. For example, H_2O_2 can inhibit the activity of auxin transporters, leading to altered auxin distribution and root growth patterns (Su et al., 2016).

REGULATION OF HYDROGEN PEROXIDE LEVELS

Antioxidant Systems

Plants possess an array of antioxidant systems to regulate H_2O_2 levels and prevent oxidative damage:

- **Ascorbate-Glutathione Cycle:** This cycle involves the reduction of H_2O_2 to water by ascorbate peroxidase (APX) using ascorbate as an electron donor (Pang & Wang, 2010). The resulting dehydroascorbate is then reduced back to ascorbate by dehydroascorbate reductase (DHAR) using glutathione (GSH) as an electron donor. GSH is regenerated by glutathione reductase (GR) using NADPH, ensuring a continuous supply of reducing equivalents. The ascorbate-glutathione cycle is a critical component of the cellular antioxidant defense, maintaining redox homeostasis and regulating H_2O_2 signaling.
- **Peroxiredoxins and Glutathione Peroxidases:** These enzymes reduce H_2O_2 and organic hydroperoxides, protecting cells from oxidative stress. Peroxiredoxins (PRXs) reduce H_2O_2 to water using thioredoxin (Trx) as an electron donor (Liebthal et al., 2018). Glutathione peroxidases (GPXs) reduce H_2O_2 and lipid hydroperoxides using GSH as an electron donor, generating oxidized glutathione (GSSG), which is subsequently reduced back to GSH by GR. The coordinated activity of PRXs and GPXs ensures efficient detoxification of H_2O_2 and maintenance of redox balance (Rhee, Woo, & Kang, 2018).

Non-antioxidant Regulatory Mechanisms

In addition to antioxidants, non-antioxidant proteins modulate H_2O_2 levels and its signaling outcomes:

- **Heat Shock Proteins (HSPs):** HSPs act as molecular chaperones, ensuring proper protein folding and preventing aggregation (Cox, Carver, & Ecroyd, 2014). Some HSPs also have redox-regulatory functions, modulating the activity of H_2O_2-responsive proteins and pathways. For example, the *Arabidopsis* HSP17.6A interacts with catalase, enhancing its stability and activity under oxidative stress. The involvement of HSPs in H_2O_2 signaling underscores the importance of protein homeostasis in cellular redox regulation.

- **Kinases and Phosphatases:** Protein kinases and phosphatases regulate H_2O_2 signaling by modulating the phosphorylation status of target proteins (Cho et al., 2004). For instance, the protein kinase OXI1 is activated by H_2O_2 (Rentel et al., 2004) and regulates the phosphorylation of downstream signaling components, including MAPKs and transcription factors. Similarly, protein phosphatases, such as PP2A, modulate H_2O_2 signaling by dephosphorylating and inactivating target proteins. The dynamic regulation of protein phosphorylation and dephosphorylation is essential for fine-tuning H_2O_2 signaling and ensuring appropriate cellular responses.

PHYSIOLOGICAL AND DEVELOPMENTAL ROLES

Stress Responses

H_2O_2 signaling is pivotal in plant responses to various stresses, including:

- **Abiotic Stress:** Drought, salinity, and extreme temperatures often lead to increased H_2O_2 production, triggering stress adaptation mechanisms. H_2O_2 mediates the activation of antioxidant defense, osmolyte accumulation, and protective proteins, enhancing stress tolerance. For instance, H_2O_2-induced activation of MAPKs and transcription factors leads to the expression of genes involved in osmoprotection, such as proline and trehalose biosynthesis, as well as the synthesis of HSPs and late embryogenesis abundant (LEA) proteins (Azar, 2007).
- **Biotic Stress:** H_2O_2 is a crucial mediator of plant defense responses against pathogens and herbivores (Sood et al., 2021). H_2O_2 accumulation at infection sites leads to the reinforcement of cell walls, production of antimicrobial compounds, and activation of defense-related genes. H_2O_2 also acts as a signaling molecule in SAR, enhancing the overall immune response of the plant. The interplay between H_2O_2 and other defense-related signaling molecules, such as SA and JA, ensures a coordinated and effective defense response.

Growth and Development

Beyond stress responses, H_2O_2 influences several developmental processes:

- **Root Growth and Development:** H_2O_2 regulates root hair formation, lateral root development, and overall root architecture. H_2O_2 modulates the activity of auxin transporters and signaling components, influencing root growth patterns and responses to environmental cues. For example, localized H_2O_2 production at root hair initiation sites promotes cell wall loosening and extension, facilitating root hair elongation (Bahmani, Kim, Modareszadeh, & Hwang, 2022). Additionally, H_2O_2 interacts with ethylene and auxin signaling pathways to regulate lateral root formation and root branching (Chen et al., 2016).
- **Seed Germination:** H_2O_2 levels modulate the breakdown of seed dormancy and promote germination (Bailly, 2019). H_2O_2 signaling interacts with hormonal pathways, including gibberellins and ABA, to regulate the expression of genes involved in germination and early seedling growth. The interplay between H_2O_2 and gibberellins promotes the degradation of ABA, facilitating seed germination under favorable conditions (Liu, Ye, Liu, Chen, & Zhang, 2010). Furthermore, H_2O_2-mediated activation of MAPKs and transcription factors modulates the expression of cell wall-modifying enzymes, such as endo-β-mannanases, promoting seed coat weakening and radicle emergence (Sen, Hamouzová, Košnarová, & Soukup, 2024).
- **Leaf Senescence:** H_2O_2 accumulation is associated with the regulation of senescence-related gene expression and leaf aging (Shi et al., 2019). H_2O_2 signaling coordinates the breakdown of cellular components, nutrient remobilization, and activation of senescence-associated pathways, ensuring an orderly progression of leaf senescence. The

interaction between H_2O_2 and hormonal signals, such as ethylene, JA, and SA, regulates the expression of senescence-associated genes, including those involved in chlorophyll degradation, proteolysis, and nutrient recycling. H_2O_2-induced activation of transcription factors, such as WRKY53 (Miao, Laun, Zimmermann, & Zentgraf, 2004) and ORE1 (La et al., 2022), further modulates the expression of senescence-related genes, promoting the orderly dismantling of cellular structures and nutrient mobilization to developing tissues.

CROSSTALK WITH OTHER SIGNALING MOLECULES

Interaction with Reactive Nitrogen Species

H_2O_2 signaling interacts with reactive nitrogen species (RNS), such as nitric oxide (NO), to regulate various physiological processes:

- **S-nitrosylation:** NO can react with thiol groups in proteins to form S-nitrosothiols (Stamler, 1995), a process known as S-nitrosylation. This post-translational modification can modulate the activity of H_2O_2-responsive proteins, such as PRXs and transcription factors, integrating ROS and RNS signaling. S-nitrosylation can enhance or inhibit the activity of target proteins, fine-tuning their function in response to changing cellular redox states.
- **Peroxynitrite Formation:** H_2O_2 can react with NO to form peroxynitrite ($ONOO^-$), a potent oxidant that can modify proteins, lipids, and nucleic acids. Peroxynitrite formation can act as a signaling mechanism, modulating the activity of target proteins and triggering stress responses. The balance between H_2O_2 and NO levels is crucial for maintaining cellular homeostasis and preventing oxidative and nitrosative stress (Choi, Kim, & Moon, 2020).

CROSSTALK WITH OTHER ROS

H_2O_2 signaling is intricately linked with other ROS, such as superoxide ($O_2^{\bullet-}$) and singlet oxygen ($^1O_2$):

- **Redox Balance:** The interplay between different ROS determines the overall redox state of the cell. For example, SODs convert superoxide to H_2O_2, linking the signaling pathways of these two ROS. The coordinated regulation of ROS production and detoxification ensures that cells can effectively respond to environmental cues without succumbing to oxidative damage (Liochev & Fridovich, 2007).
- **ROS Wave:** The concept of a "ROS wave" refers to the propagation of ROS signals from one part of the plant to another, facilitating systemic responses to local stress. H_2O_2 can amplify and propagate ROS signals (Rigoulet et al., 2011), coordinating stress responses at the whole plant level. The ROS wave involves the interplay between different ROS and antioxidants, creating a dynamic redox signaling network that integrates local and systemic responses.

CROSSTALK WITH PHYTOHORMONES

H_2O_2 signaling interacts with various phytohormones to regulate growth, development, and stress responses:

- **Auxin:** H_2O_2 modulates auxin signaling by influencing the activity of auxin transporters and response regulators. This interaction affects various developmental processes, including root growth, lateral root formation, and tropic responses. For example, H_2O_2 can inhibit the activity of auxin transporters, leading to altered auxin distribution and root growth patterns.

- **Gibberellins:** H_2O_2 signaling interacts with gibberellins to regulate seed germination, stem elongation, and flowering. The balance between H_2O_2 and gibberellin levels determines the timing and extent of these developmental processes (Dinler, Cetinkaya, Akgun, & Korkmaz, 2021). H_2O_2-mediated activation of MAPKs and transcription factors modulates the expression of gibberellin-responsive genes, integrating ROS and hormonal signaling.
- **Cytokinins:** H_2O_2 signaling can influence cytokinin signaling, affecting cell division, differentiation, and shoot development. The crosstalk between H_2O_2 and cytokinins modulates the expression of genes involved in cell cycle regulation and meristem activity (Lewandowska, Messens, & Van). H_2O_2-induced changes in cytokinin signaling can impact tissue patterning, organogenesis, and stress responses.

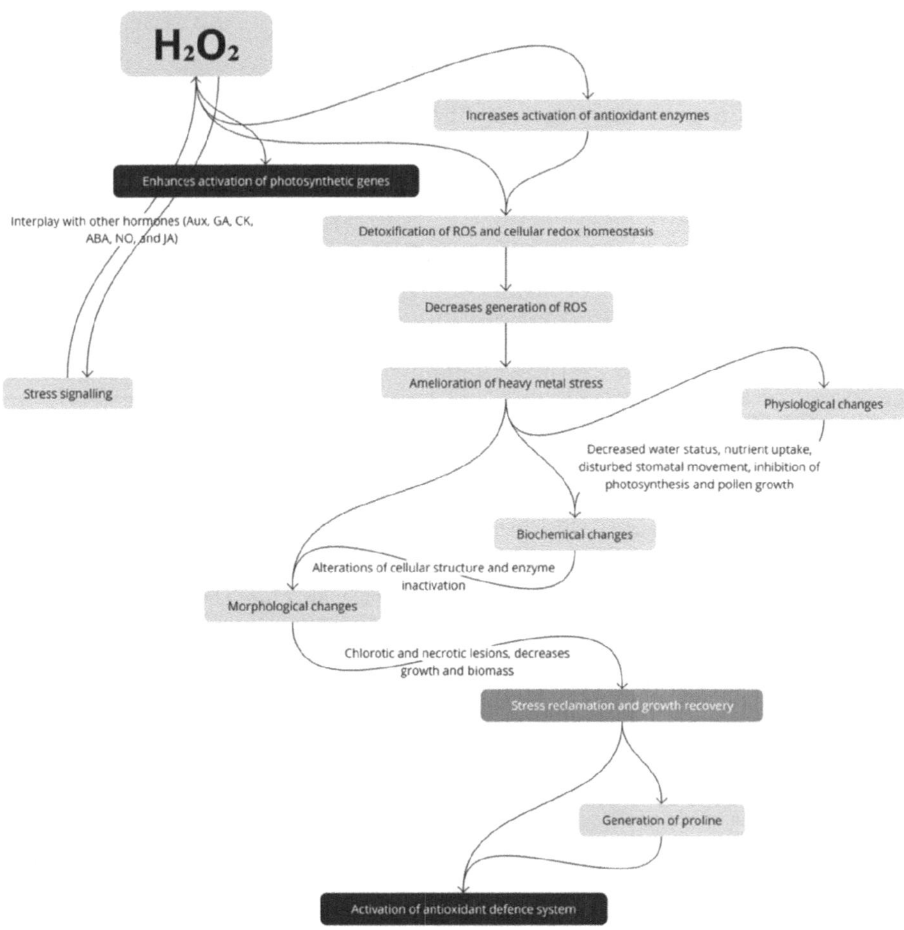

FIGURE 3.2 Overview of H_2O_2 signaling targets in plants.

CONCLUSION

Hydrogen peroxide (H_2O_2) serves as a multifaceted signaling molecule in plants, orchestrating a wide array of physiological and developmental processes (Figure 3.2). Its production, perception, and downstream signaling mechanisms are finely tuned to maintain a balance between beneficial signaling and detrimental oxidative stress. Through complex interactions with other signaling molecules and pathways, H_2O_2 integrates environmental cues and internal signals, enabling plants to adapt to changing conditions and optimize their growth and development. Future research will continue to uncover the intricate details of H_2O_2 signaling networks, revealing new insights into plant biology and offering potential applications in agriculture and environmental management. Understanding the controlling mechanisms of H_2O_2 signaling will be crucial for developing strategies to enhance plant resilience, productivity, and sustainability in the face of global challenges such as climate change and food security.

REFERENCES

Al Mamun, M., Lee, B.-R., Bae, D.-W., & Kim, T.-H. (2023). Pathogen Xanthomonas campestris-induced immune-related genes integrated with phytohormones are involved in stomatal closure by interactively regulating ROS and Ca^{2+} signaling in the Xcc–Brassica napus pathosystem. *Physiological and Molecular Plant Pathology, 127*, 102127.

Azar, Z. (2007). Upstream mechanisms responsible for H_2O_2-induced activation of MAPK and PKB in vascular smooth muscle cells. https://hdl.handle.net/1866/15375.

Bahmani, R., Kim, D., Modareszadeh, M., & Hwang, S. (2022). Cadmium enhances root hair elongation through reactive oxygen species in Arabidopsis. *Environmental and Experimental Botany, 196*, 104813.

Bailly, C. (2019). The signalling role of ROS in the regulation of seed germination and dormancy. *Biochemical Journal, 476*(20), 3019–3032.

Balazadeh, S., Wu, A., & Mueller-Roeber, B. (2010). Salt-triggered expression of the ANAC092-dependent senescence regulon in Arabidopsis thaliana. *Plant Signaling & Behavior, 5*(6), 733–735.

Bonekamp, N. A., Völkl, A., Fahimi, H. D., & Schrader, M. (2009). Reactive oxygen species and peroxisomes: Struggling for balance. *Biofactors, 35*(4), 346–355.

Breeze, E., & Mullineaux, P. M. (2022). The passage of H_2O_2 from chloroplasts to their associated nucleus during retrograde signalling: Reflections on the role of the nuclear envelope. *Plants, 11*(4), 552.

Chen, Q., & Yang, G. (2020). Signal function studies of ROS, especially RBOH-dependent ROS, in plant growth, development and environmental stress. *Journal of Plant Growth Regulation, 39*(1), 157–171.

Chen, X. J., Xia, X. J., Guo, X., Zhou, Y. H., Shi, K., Zhou, J., & Yu, J. Q. (2016). Apoplastic H_2O_2 plays a critical role in axillary bud outgrowth by altering auxin and cytokinin homeostasis in tomato plants. *New Phytologist, 211*(4), 1266–1278.

Cho, S.-H., Lee, C.-H., Ahn, Y., Kim, H., Kim, H., Ahn, C.-Y., ... Lee, S.-R. (2004). Redox regulation of PTEN and protein tyrosine phosphatases in H_2O_2-mediated cell signaling. *FEBS Letters, 560*(1–3), 7–13.

Choi, Y., Kim, J., & Moon, I. (2020). Simulation and economic assessment of using H_2O_2 solution in wet scrubber for large marine vessels. *Energy, 194*, 116907.

Cox, D., Carver, J. A., & Ecroyd, H. (2014). Preventing α-synuclein aggregation: The role of the small heat-shock molecular chaperone proteins. *Biochimica et Biophysica Acta (BBA)-Molecular Basis of Disease, 1842*(9), 1830–1843.

Dinler, B., Cetinkaya, H., Akgun, M., & Korkmaz, K. (2021). Simultaneous treatment of different gibberellic acid doses induces ion accumulation and response mechanisms to salt damage in maize roots. *Journal of Plant Biochemistry and Physiology, 9*(3), 258.

He, H., Van Breusegem, F., & Mhamdi, A. (2018). Redox-dependent control of nuclear transcription in plants. *Journal of Experimental Botany, 69*(14), 3359–3372.

Jang, J. Y., Rhee, J. Y., Chung, G. C., & Kang, H. (2012). Aquaporin as a membrane transporter of hydrogen peroxide in plant response to stresses. *Plant Signaling & Behavior, 7*(9), 1180–1181.

La, V. H., Nguyen, T. H. A., Ngo, X. B., Tran, V. D., Khuat, H. T., Bui, T. T., . . . Nguyen, T. D. (2022). At-ore1 gene induces distinct novel H_2O_2-NACs signaling in regulating the leaf senescence in soybeans (Glycine max L.). *Agronomy, 12*(9), 2110.

Lee, K., & Esselman, W. J. (2002). Inhibition of PTPs by H_2O_2 regulates the activation of distinct MAPK pathways. *Free Radical Biology and Medicine, 33*(8), 1121–1132.

Lewandowska, A., Messens, J., & Van, F. (2021). Plant stress signalling: role of H_2O_2 and phytohormones. *Forward and reverse: looking for components of oxidative stress signalling pathways* (Doctoral dissertation, Ghent University).
Liebthal, M., Maynard, D., & Dietz, K.-J. (2018). Peroxiredoxins and redox signaling in plants. *Antioxidants & Redox Signaling, 28*(7), 609–624.
Liochev, S. I., & Fridovich, I. (2007). The effects of superoxide dismutase on H_2O_2 formation. *Free Radical Biology and Medicine, 42*(10), 1465–1469.
Liu, Y., Ye, N., Liu, R., Chen, M., & Zhang, J. (2010). H_2O_2 mediates the regulation of ABA catabolism and GA biosynthesis in Arabidopsis seed dormancy and germination. *Journal of Experimental Botany, 61*(11), 2979–2990.
Miao, Y., Laun, T., Zimmermann, P., & Zentgraf, U. (2004). Targets of the WRKY53 transcription factor and its role during leaf senescence in Arabidopsis. *Plant Molecular Biology, 55*, 853–867.
Pang, C.-H., & Wang, B.-S. (2010). Role of ascorbate peroxidase and glutathione reductase in ascorbate–glutathione cycle and stress tolerance in plants. In: *Ascorbate-Glutathione Pathway and Stress Tolerance in Plants*, 91–113. https://doi.org/10.1007/978-90-481-9404-9_3.
Rao, M. V., Paliyath, G., Ormrod, D. P., Murr, D. P., & Watkins, C. B. (1997). Influence of salicylic acid on H_2O_2 production, oxidative stress, and H_2O_2-metabolizing enzymes (salicylic acid-mediated oxidative damage requires H_2O_2). *Plant Physiology, 115*(1), 137–149.
Rentel, M. C., Lecourieux, D., Ouaked, F., Usher, S. L., Petersen, L., Okamoto, H., … Hirt, H. (2004). OXI1 kinase is necessary for oxidative burst-mediated signalling in Arabidopsis. *Nature, 427*(6977), 858–861.
Rhee, S. G., Woo, H. A., & Kang, D. (2018). The role of peroxiredoxins in the transduction of H_2O_2 signals. *Antioxidants & Redox Signaling, 28*(7), 537–557.
Rigoulet, M., Yoboue, E. D., & Devin, A. (2011). Mitochondrial ROS generation and its regulation: mechanisms involved in H_2O_2 signaling. *Antioxidants & Redox Signaling, 14*(3), 459–468.
Saxena, I., Srikanth, S., & Chen, Z. (2016). Cross talk between H_2O_2 and interacting signal molecules under plant stress response. *Frontiers in Plant Science, 7*, 570.
Sen, M. K., Hamouzová, K., Košnarová, P., & Soukup, J. (2024). H_2O_2-mediated signaling in plant stress caused by herbicides: Its role in metabolism and degradation pathways. *Plant Science, 346*, 112166.
Sharova, E., Medvedev, S., & Demidchik, V. (2020). Ascorbate in the apoplast: Metabolism and functions. *Russian Journal of Plant Physiology, 67*, 207–220.
Shi, X., Xu, S., Mu, D., Sadeghnezhad, E., Li, Q., Ma, Z., … Wang, L. (2019). Exogenous melatonin delays dark-induced grape leaf senescence by regulation of antioxidant system and senescence associated genes (SAGs). *Plants, 8*(10), 366.
Sood, M., Kapoor, D., Kumar, V., Kalia, N., Bhardwaj, R., Sidhu, G. P., & Sharma, A. (2021). Mechanisms of plant defense under pathogen stress: A review. *Current Protein and Peptide Science, 22*(5), 376–395.
Stamler, J. (1995). S-nitrosothiols and the bioregulatory actions of nitrogen oxides through reactions with thiol groups. In: *The Role of Nitric Oxide in Physiology and Pathophysiology*, 19–36. https://doi.org/10.1007/978-3-642-79130-7_4.
Su, C., Liu, L., Liu, H., Ferguson, B. J., Zou, Y., Zhao, Y., … Li, X. (2016). H_2O_2 regulates root system architecture by modulating the polar transport and redistribution of auxin. *Journal of Plant Biology, 59*, 260–270.
Tyagi, S., Shumayla, Singh, S. P., & Upadhyay, S. K. (2019). Role of superoxide dismutases (SODs) in stress tolerance in plants. In: *Molecular Approaches in Plant Biology and Environmental Challenges*, 51–77. https://doi.org/10.1007/978-981-15-0690-1_3.
Walker, C. L., Pomatto, L. C., Tripathi, D. N., & Davies, K. J. (2018). Redox regulation of homeostasis and proteostasis in peroxisomes. *Physiological Reviews, 98*(1), 89–115.
Wang, H., Huang, J., Liang, X., & Bi, Y. (2012). Involvement of hydrogen peroxide, calcium, and ethylene in the induction of the alternative pathway in chilling-stressed Arabidopsis callus. *Planta, 235*, 53–67.
Wang, P., & Song, C.-P. (2008). Guard-cell signalling for hydrogen peroxide and abscisic acid. *New Phytologist, 178*(4), 703–718.
Wu, F., Chi, Y., Jiang, Z., Xu, Y., Xie, L., Huang, F., … Wu, X. (2020). Hydrogen peroxide sensor HPCA1 is an LRR receptor kinase in Arabidopsis. *Nature, 578*(7796), 577–581.
Wu, Z., Wang, H., Fang, S., & Xu, C. (2018). Roles of endoplasmic reticulum stress and autophagy on H_2O_2-induced oxidative stress injury in HepG2 cells. *Molecular Medicine Reports, 18*(5), 4163–4174.
Yan, Y., Wei, C., Zhang, W., Cheng, H. & Liu, J. (2006). Cross-talk between calcium and reactive oxygen species signaling. *Acta Pharmacologica Sinica, 27*(7), 821–826.
Zhang, T.-Y., Li, F.-C., Fan, C.-M., Li, X., Zhang, F.-F., & He, J.-M. (2017). Role and interrelationship of MEK1-MPK6 cascade, hydrogen peroxide and nitric oxide in darkness-induced stomatal closure. *Plant Science, 262*, 190–199.

4 Signal Transduction of Hydrogen Peroxide under Abiotic Stresses

Bisma Hilal and Qazi Fariduddin

INTRODUCTION

Plant growth, development, and survival are constantly shaped and influenced by genetic and environmental factors. Owing to their sessile nature, plants have developed mechanisms that enable them to harness their metabolism and, therefore, grow under adverse environmental conditions, such as by incorporating primary metabolic products into decisive processes. One example of such products of metabolic reactions occurring at multiple sites within a plant cell that regulate plant growth and development is reactive oxygen species (ROS) (Das and Roychoudhury 2014). Production of ROS is one of the earliest responses shown by plant cells subjected to different environmental stresses and include superoxide ($O_2^{\bullet-}$), hydroxyl (OH^-), hydroperoxyl, and peroxy radicals and nonradical molecules such as hydrogen peroxide (H_2O_2) and singlet oxygen (^1O) (Dikilitas et al. 2018; Khan et al. 2022). Such molecules are produced in different cell organelles like peroxisomes, chloroplasts, mitochondria, plasma membrane, endoplasmic reticulum, apoplast, and cell wall. Increasing concentrations of ROS result in the oxidation of lipids, proteins, and carbohydrates resulting in pigment breakdown, along with single and double strand breaks in DNA, thereby impairing enzyme activities and affecting normal cellular functions which can eventually lead to dysfunction in metabolic activities and cell death (Bose et al. 2014; Banerjee and Roychoudhury 2017; Khan et al. 2020, 2023a, b). Concurrently, plants have also developed sophisticated antioxidant system on the alert for mounting H_2O_2 levels which provides its effective scavenging (Apel and Hirt 2004; Gechev et al. 2006; Miller et al. 2010; Khan et al., 2023a, b; Hilal et al. 2023). The antioxidant system is comprised of various antioxidant enzymes including catalase, ascorbate peroxidases, glutathione reductases, and peroxiredoxins along with non-enzymatic antioxidants such as ascorbic acid, tocopherols, and flavonoids (Willekens et al. 1995; Miller et al. 2010). However, below the threshold levels, ROS may be essential in stress responses for triggering defense-related mechanisms, stress-responsive pathways, and could act as signaling components for cellular metabolism (You and Chan 2015; Choudhury et al. 2017). Among the ROS compounds, H_2O_2 has received most attention of the scientific community over the past few decades because H_2O_2 generation is indispensable during plant responses and has high stability and long half-life, which allows it to have great ability to buffer other ROS molecules and also serves as a key signaling molecule. Therefore, H_2O_2 has emerged as a vital player in the signaling nexus, orchestrating cellular processes in response to abiotic stresses. H_2O_2 acts as a crucial secondary messenger in abiotic stress signal transduction by modulating various signaling pathways such as H_2O_2-mediated activation of mitogen-activated protein kinases (MAPKs) contributing to stress-responsive gene expressionthrough phosphorylation of transcription factors (TFs), and H_2O_2-mediated rapid changes in cytosolic Ca^{2+} levels. Additionally, it has also been proposed that crosstalk of H_2O_2 with other plant growth regulators is involved in signal transduction of stress resilience responses, and growth and development in the plant (Dempsey and Klessig 1995; Wendehenne et al. 2004; Quan et al. 2008). Therefore, this chapter deals with H_2O_2 as a versatile signaling molecule that plays a complex role in signaling networks for deciphering plant

DOI: 10.1201/9781032647692-4

responses to adverse environmental conditions and also various aspects of H_2O_2 such as its sensing on the cell surface and cytoplasmic level, H_2O_2-mediated gene regulation through crosstalk with crucial signaling pathway components like Ca^{2+}, MAPK and plant growth regulators such as nitric oxide (NO), abscisic acid, and salicylic acid (SA) during plant growth, development, and resistance against abiotic stresses.

SENSING OF H_2O_2 SIGNALS

For proteins to detect the presence of ROS like H_2O_2 and act as signal intermediaries, they must specifically recognize H_2O_2 or have a direct interaction with H_2O_2 that results in signal transmission. Due to the small size of H_2O_2, the presence of a specific receptor protein is unlikely; however, the direct interaction between the protein and H_2O_2 is expected as the oxidizing nature of H_2O_2 allows it to modify thiol residues present in the proteins (Foyer and Noctor 2005). The initial events in the ROS-sensing mechanism include oxidative post-translational modifications (PTMs) of sensory proteins and oxidation of metabolites. Primary targets of H_2O_2-mediated oxidative PTMs in ROS-sensitive proteins include sulfur atoms in cysteine and methionine residues (Waszczak et al. 2018). Around 100 cytosolic proteins in H_2O_2-treated plant cells were found sulfenylated as identified by Waszczak et al. (2014); therefore, it can be speculated that sulfenylation seems to be a widespread PTM. H_2O_2 perception also mediates oxidation of methionine to methionine sulfoxide which can be reduced back to methionine by a large group of methionine sulfoxide reductases (Sadanandom et al. 2000). About 400 proteins in *Arabidopsis* underwent methionine oxidation after being exposed to photorespiratory stress (Jacques et al. 2015), demonstrating that this PTM is widespread. However, signaling events triggered by oxidation of methionine are still largely unknown compared to mechanisms involving cysteine oxidative PTMs. Some proteins involved in H_2O_2 sensing have disparate roles in major metabolic pathways like glycolysis. Cytosolic glyceraldehyde-3-phosphate dehydrogenase (GAPDH) undergoes reversible redox PTMs on its thiol group after reacting with H_2O_2 (Yang and Zhai 2017). The oxidized compounds are further reduced by reducing agents like glutathione (GSH), thioredoxin (Trx), and glutaredoxin (Grx) (Yang and Zhai 2017). Further, several transcriptomic studies have revealed some proteins that might be engaged in H_2O_2 signaling and perception like ETR1, a histidine kinase receptor involved in ethylene signaling (Desikan et al. 2005). In *Arabidopsis*, ETR1 receptor is required for H_2O_2 perception leading to stomatal closure. Stomata in etrl-1 mutant with Cys-65Tyr mutation showed reduced closure in response to H_2O_2, indicating that the thiol of Cys-65 is important for H_2O_2 signaling in plants (Desikan et al. 2005). Extracellular H_2O_2, synthesized in apoplast either by nicotinamide adenine dinucleotide phosphate (NADPH) oxidases and superoxide dismutase (SOD) or induced by various abiotic stresses, enters the cytosol through aquaporin membrane proteins and oxidizes cytoplasmic proteins to regulate stress responses, signaling, and growth and development. Along with some ion channels and receptors, many of these H_2O_2-modified cysteine-containing proteins function as H_2O_2 sensors, as discussed previously. However, sensors for H_2O_2 perception on cell surface remain largely unknown until recently, Wu et al. (2020) identified HPCA1, a leucine-rich-repeat receptor kinase localized to the plasma membrane and is oxidized by H_2O_2 on the thiol groups of Cys residues located on the extracellular domain. This modification leads to autophosphorylation of HPCA1 and mediates H_2O_2-induced activation of Ca^{2+} channels, followed by stomatal closure (Wu et al. 2020).

H_2O_2 AS A SIGNALING MOLECULE UNDER NORMAL CONDITIONS IN PLANTS

ROS are inevitable byproducts of aerobic metabolism and have been primarily considered toxic derivatives; however, in contrast with previous views, ROS are necessary and important signaling molecules during normal plant metabolism and for the adjustment of the cellular machinery in response to environmental factors (Mittler et al. 2011; Wrzaczek et al. 2013). Therefore, they are also produced in a regulated manner and are involved in a wide range of functions including cell signaling and defense against environmental cues (Halliwell and Gutteridge 1999). Relative to other ROS, the specific properties of H_2O_2 such as no unpaired electrons, moderately reactive, relative stability as well as capability to diffuse through membranes make it an ideal signaling molecule for carrying out a variety of functions at lower levels (Reeves et al. 2002). NADPH oxidases are plasma membrane-localized enzymes, known as respiratory burst oxidase homologues (Rbohs) which produce H_2O_2. Rbohs reduce O_2 to $O_2^{\bullet-}$ using NADPH as an electron donor. The SOD then converts $O_2^{\bullet-}$ to H_2O_2 (Sagi and Fluhr 2001, 2006). It has been characterized that the genome of *Arabidopsis* consists of ten NADPH oxidase-encoding genes, named AtRbohA to J, that display diverse patterns of expression during plant development and in response to stress factors (Fluhr 2009; Marino et al. 2012). Mutants have been developed to reveal the involvement of ROS and their spatial distribution along the root for normal root growth (Tsukagoshi et al. 2010; Ivanchenko et al. 2013). It has been found that H_2O_2 was mostly detected in the differentiated zone along the root (Dunand et al. 2007). H_2O_2 has been reported to play an important role in the regulation of pectin synthesis, pectin methylesterase activity, and pectin demethylesterification. Exogenous H_2O_2 enhances cell expansion and root diameter by increasing pectin content and demethylesterification (Xiong et al. 2015). It has also been studied that apoplastic H_2O_2 is actively involved in the regulation of apical dominance in tomato plants by inhibiting lateral bud outgrowth via promoting auxin biosynthesis in the apical bud and reducing cytokinin biosynthesis in the node, which results in an increased ratio of auxin: cytokinin in the node. Moreover, H_2O_2 triggers changes in redox homeostasis and also induces the expression of key genes of auxin transport and signaling proteins (Chen et al. 2016). Shi et al. (2015) used the pharmacological approach to determine the signaling relationships of H_2O_2 and NO in CO_2-regulated stomatal movement and it has been found that both H_2O_2 and NO induce stomatal closure. Elevated CO_2 triggered H_2O_2 and NO accumulation in guard cells while inhibiting RBOH1 or NR, downregulated the elevated CO_2-induced SLAC1 (SLOW ANION CHANNEL-ASSOCIATED 1) transcript abundance and stomatal closure by suppressing the accumulation of H_2O_2 and NO. In another report, Lv et al. (2018) also revealed that H_2O_2 acts as a signaling intermediate in stomatal closure induced by strigolactones. Moreover, several studies have demonstrated that H_2O_2, NADPH oxidase, and Ca^{2+}-calmodulin (CaM) are involved in abscisic acid-mediated increase in activities of antioxidant enzymes, e.g., catalase, SOD, and ascorbate peroxidase, in leaf samples of maize plants (Jiang and Zhang 2003; Hu et al. 2005, 2007). Additionally, it has also been evaluated that H_2O_2 production induced by abscisic acid mediates NO generation leading to the closure of stomata (Bright et al. 2006). Zhang et al. (2006) have revealed that activation of MAPK by abscisic acid increases H_2O_2 production and thus forms a positive feedback loop, which suggested the involvement of H_2O_2 in various signaling processes. Similarly, it has also been studied that H_2O_2 production induced by abscisic acid mediates NO generation, which in turn activates MAPK and results in upregulation of the expression and activities of antioxidant enzymes (Zhang et al. 2007).

Relative to other ROS, H_2O_2 has a characteristic feature, i.e., it is a long-lived molecule (1 ms) and can permeate through cellular membranes which makes it compatible as a signaling molecule during growth and developmental processes (Bienert et al. 2006; Sauer et al. 2001; Van Breusegem and Dat 2006). The signal transduction of H_2O_2 is coordinated by receptors or sensors, MAPK, and TFs which are proposed to be evolutionarily conserved (Georgiou 2002; Liu et al. 2005). In plants, H_2O_2 signaling is governed by complex network that includes various protein kinases and TFs (Miller and Mittler 2006; Kaminaka et al. 2006). The regulated action of TGACG motif-binding factor 1, which is a redox TF and non-expressor of pathogenesis-related genes 1, is essential for defense gene expression and SAR (systemic acquired disease resistance) in *Arabidopsis thaliana* (Després et al. 2003; Mou et al. 2003). Various molecular techniques have led to comprehensive inventories of genome-wide H_2O_2-related gene expression in different organisms including *A. thaliana* (Vanderauwera et al. 2005). The different molecular technologies have demonstrated that H_2O_2 has a significant impact on the transcriptome and regulated the expression of a wide range of genes.

Therefore, H_2O_2 has a multifaceted impact on gene expression, influencing not only genes responsible for detoxifying ROS but also activating genes involved in signal transduction, transcriptional regulation, and the metabolism of proteins, carbohydrates, and lipids. This demonstrates the intricate nature of H_2O_2 toward transcriptional regulation (Vandenbroucke et al. 2008).

H_2O_2 AS A SIGNALING MOLECULE UNDER ABIOTIC STRESSES

Global crop productivity is being impacted by changes in weather patterns brought on by climate change in addition to an ever-growing demographic demand. The effects of abiotic stresses have grown more significant than they had in previous decades, particularly in light of global climate change. Abiotic stress conditions like drought, heat, salinity, and cold, as well as their various combinations, severely reduce crop productivity all over the world each year (Khan et al. 2022). To combat oxidative stress, plants have evolved an extensive system to regulate accumulation of ROS concentration in the cell (Mittler et al. 2011). Surprisingly, plants also take advantage of lower levels of ROS and use them as signaling molecules to regulate a wide range of plant developmental processes like cell elongation and differentiation, morphogenesis, proper development of root system, and regulation of stomatal aperture and responses to environmental stresses (Tsukagoshi et al. 2010; Bhattacharjee 2012). Among ROS and other oxygen-derived free radicals, H_2O_2 has received significant interest in the last decade due to its phytohormonal and signaling properties and toxic levels. Higher concentrations of H_2O_2 result in oxidative stress and damage cellular components, and in extreme cases lead to cell death through DNA damages or methylation. On the other hand, low reactivity compared to other ROS, relatively long life span and small size of H_2O_2 molecules permit it to traverse through biological membranes to different cellular compartments, enabling signaling functions particularly in the context of plant processes like antioxidative defense, stress acclimation, stomatal behavior, cell wall cross-linking, regulation of the cell cycle, phytoalexin production, and photosynthesis (Khan et al. 2018). Hence, the dual role of H_2O_2, the toxicity or danger on one hand and signaling cascades on the other, makes H_2O_2 a versatile molecule whose concentration must be carefully managed within plant cells. Numerous signaling networks, including protein kinase networks, Ca^{2+} signaling, hormonal networks, etc., have been linked to hydrogen peroxide signaling. Other signaling mechanisms, such as G-protein and various plant growth regulators, have also been reported to interact with H_2O_2 and regulate various developmental processes in plants under a variety of abiotic stresses as discussed below.

HYDROGEN PEROXIDE AND CA^{2+} CROSSTALK UNDER VARIOUS ABIOTIC STRESSES

Ca^{2+} is one of the key elements in cellular signaling and acts as a secondary messenger in numerous signaling pathways (Sanders et al. 1999). An important early signal in both plant development and response to abiotic stresses is elevation of cellular Ca^{2+}. The influx of Ca^{2+} and consequent increase in cytosolic Ca^{2+} is primarily enabled by membrane-localized Ca^{2+}-permeable ion channels, including ionotropic glutamate receptors (GLRs), cyclic nucleotide gated channels (CNGCs), mechanosensitive-like channels (MSLs), annexins, and reduced hyperosmolarity-induced [Ca^{2+}] cyt increase channels (OSCAs) (Kudla et al. 2018). The rise in intracellular Ca^{2+} concentration initiates a signaling cascade and regulates a series of cellular processes like stomatal closure (Zou et al. 2015), leaf de-etiolation (Huang et al. 2012), seed germination (Kong et al. 2015), root growth and development (Liao et al. 2012), and is also involved in regulating responses to a wide range of environmental stresses like drought (Zou et al. 2015), light (Hu et al. 2015), heavy metal (Li et al. 2015a), salt (Tepe and Aydemir 2015), and heat shock (Urao et al. 1994). H_2O_2 and Ca^{2+} have been reported to interact with each other in response to abiotic stresses in plants, in addition to their interactions in growth and developmental processes (Niu and Liao 2016). In *Arabidopsis*, H_2O_2 activates

Ca^{2+}-permeable channels in the plasma membrane, which further mediate increase in cytosolic Ca^{2+}, leading to stomatal closure. As discussed earlier, HPCA1, a cell surface receptor protein for extracellular H_2O_2, mediates activation of Ca^{2+} channels in guard cells (Wu et al. 2020). Under drought stress, spermidine-induced H_2O_2 production requires release of Ca^{2+}; however, spermidine-induced Ca^{2+} release was also necessary for H_2O_2 generation, suggesting a reciprocal influence between polyamine-induced H_2O_2 and Ca^{2+} signaling. This influence was further connected to antioxidant defense and the expression of dehydrin genes in the leaves of white clover (Li et al. 2015b). The role of alternative respiratory pathway (ARP) in plant thermogenesis and responses to environmental stresses like ozone, salinity, and low temperatures is well known (Simons and Lambers 2018; Ederli et al. 2006; Feng et al. 2008; Wang et al. 2010a). Ethylene-dependent ARP induction by chilling is mediated by H_2O_2 and Ca^{2+}, both of which are independent upstream signaling elements in this pathway (Wang et al. 2012). H_2O_2 functions as a second messenger in the induction of thermotolerance and acts upstream of NO which further increases the expression of CaM3, hence regulating the DNA-binding activity of heat shock factors (HSFs) (Wang et al. 2014). Further, the interaction between H_2O_2 and Ca^{2+}-CaM also regulates antioxidant system which alleviates stress damages in plants (Hu et al. 2007). The combined application of H_2O_2 and Ca^{2+} can also enhance the ameliorative effect on salt stress by decreasing the salt-induced H_2O_2 accumulation, counteracting saline-caused plant growth inhibition and stimulating the overall antioxidant activity in naked oat seedlings (Xu et al. 2008). In addition, NADPH oxidase-dependent H_2O_2 generation played a role in the regulation of total Ca^{2+} in wheat roots, which in turn regulates the physiological responses to salinity stress (Yang et al. 2007). Therefore, H_2O_2 has emerged as a key player in complex signaling networks and works in association with other signaling components as comprehensively illustrated in Figure 4.1 in regulating diverse plant responses.

MAPK AND H_2O_2 SIGNALING IN MEDIATING ALLEVIATION OF ABIOTIC STRESSES

Transduction of extracellular stimulus signals to cellular machinery occurs through a large network of MAPKs, which phosphorylate specific proteins and amplify the signal. It has been reported that the phosphorylation cascades, which include MAPKs, MAPK kinases (MAPKKs), MEK, and MAPKK kinases (MAPKKKs), operate across multiple signal transduction pathways in different organisms (Morris 2001). These protein kinases are either activated by H_2O_2 accumulation or themselves cause H_2O_2-induced oxidative burst, in both animals and plants, which may result in the modulation of gene expression (Pitzschke et al. 2009; Torres and Forman 2003), thereby providing the linkage between upstream H_2O_2 signal and downstream gene expression. MAPK activation is a crucial step in mediating cellular responses to various stresses. In *Arabidopsis*, H_2O_2 activates the ANP (*Arabidopsis* NPK1-like protein kinase, in which NPK is a *Nicotiana* protein kinase) class of MAPKKKs which further triggers a specific class of stress-induced MAPKs like AtMPK3 and AtMPK6 (Ichimura et al. 2000). Overexpression of ANP1 in transgenic plants has shown increased tolerance to salt stress, freezing and heat shock (Kovtun et al. 2000). Accumulation of H_2O_2 is also involved in upregulating MPK3/6 responses to heavy metals like cadmium (Liu et al. 2010). Further, H_2O_2 also initiates the expression of AtNDPK2 gene which is located upstream of AtMPK3 and AtMPK6 signaling cascade, and is involved in fortifying plants against oxidative stresses (Moon et al. 2003). Moreover, H_2O_2 in crosstalk with Ca^{2+} upregulated the expression of SlMPK7, a novel MAPK gene, thereby enhancing the antioxidant enzyme activity and overall increase in tolerance to chilling stress (Yu et al. 2016). H_2O_2 also acts a main signaling molecule in strigolactone-induced low-temperature stress tolerance by triggering NADPH oxidase genes and MAPK signaling pathway (Zhang et al. 2020). MAPK also confers tolerance to other abiotic stresses like drought and salinity via MAPK pathways (Wang et al. 2010b). Under drought stress, H_2O_2 interacts with ABA, NO, Ca^{2+}, and MAPK signaling pathways to induce stomatal closure and increase antioxidant gene expression (Neill 2007).

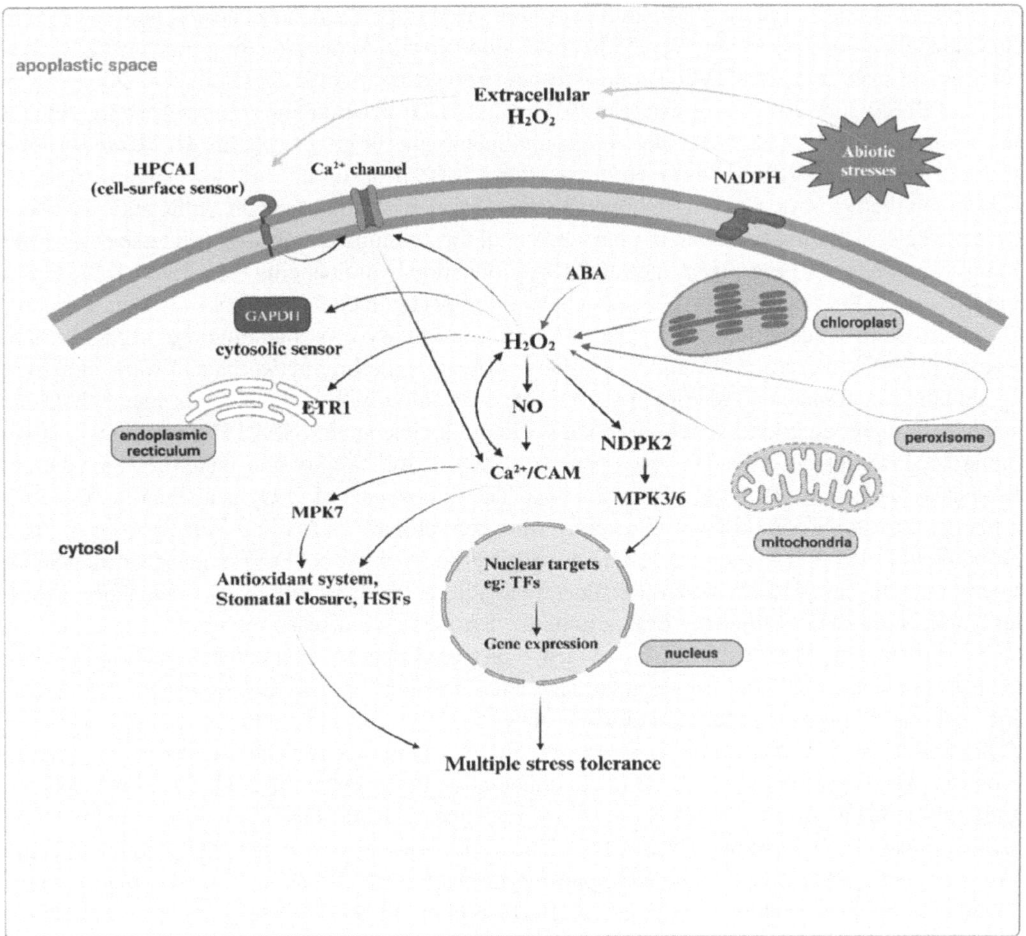

FIGURE 4.1 A schematic representation of sensing of H_2O_2 at the cell surface and cytoplasmic level, its generation via different cell organelles and interaction with different signaling pathways involved in plant growth processes and abiotic stress tolerance. Extracellular H_2O_2 generated via NADPH oxidases or different oxidative stresses is sensed by HPCA1, which upon autophosphorylation mediates activation of Ca^{2+} channels. At the cytoplasmic level, H_2O_2 perception is mediated by GAPDH and ETR1. Further H_2O_2 in crosstalk with different signaling molecules like NO, Ca^{2+}, ABA, and MAPK mediates the activation of different TFs involved in regulating numerous plant developmental processes contributing to abiotic stress tolerance in plants.

H_2O_2 IN SIGNALING NETWORKS AND ITS CROSSTALK WITH PGRS

Increased production of ROS in plant cells is a common physiological event that takes place in response to abiotic and biotic stresses. This increase in ROS production in different cellular compartments such as plasma membrane, chloroplast, mitochondria, and peroxisome or as by-product of various metabolic processes disrupts cellular homeostasis (Suzuki et al. 2012; Groß et al. 2013). For instance, in response to heat, cold, wounding, and pathogen infection, H_2O_2 generation in plants is facilitated by NADPH oxidase-dependent oxidative burst (Baxter et al. 2014). H_2O_2 acts as a second messenger, facilitating the acquisition of resilience to environmental perturbations (Desikan et al. 2003). Therefore, H_2O_2 is a signaling molecule because plants can sense, transport and mediate H_2O_2-induced cellular responses and regulate the adaptive responses in plants such as against pathogen (Neill et al. 2002), plant senescence (Bhattacharjee 2005), regulation of cell expansion (Suzuki et al. 1999), and programmed cell death (Mittler 2002). H_2O_2 is also involved in regulating the activities of many signaling players like protein kinases, protein phosphatases, and TFs (Cheng and Chunpeng 2006). Moreover, H_2O_2 also interconnects with other signaling transducers such as NO, Ca^{2+}, abscisic acid, and SA that control response downstream of H_2O_2 (Neill et al. 2002; Petrov and Van Breusegem 2012). Abscisic acid-stimulated H_2O_2 accumulation is involved in stomatal closure by activation of Ca^{2+} channels in plasma membrane of guard cells in *Arabidopsis* (Pel et al. 2000; Neill et al. 2002), where H_2O_2 functions as both upstream and downstream signaling molecule in the cytosolic Ca^{2+} signaling pathways (Sun et al. 2010). Upon the pathogenic exposure, SA is a crucial signaling molecule in plants essential for development of SAR (Pieterse et al. 2009; Dempsey and Klessig 1994). H_2O_2 facilitates SA production by influencing the activation of benzoic acid 2-hydroxylase, which converts benzoic acid into SA (León et al. 1995). SA also has a capability to increase the endogenous H_2O_2 concentration, particularly by stimulating the activity of SOD (Rao et al. 1997); therefore, SA and H_2O_2 share different signaling networks and work in activating signaling components. Additionally, ROS are associated with cellular redox metabolism, for instance through thioredoxins, peroxiredoxins, glutaredoxins, and/or NADPH (Suzuki et al. 2012). Therefore, it is implicated that the stress-generated ROS disturbs redox state balance, which thereby stimulates downstream ROS signaling cascades. The importance of H_2O_2 as a signaling molecule in response to stress conditions was recently well exemplified (Petrov and Van Breusegem 2012; Strand et al. 2015). It has been well studied that respiratory burst NADPH oxidases mediate ROS generation in the apoplast for signal transduction to adjacent cells as well as for intracellular signaling (Quan et al. 2008). Enhanced ROS stimulated SA-mediated disease resistance (Chaouch et al. 2012). Furthermore, there is increasing evidence proposed that NO is a potent signaling molecule involved in plant growth and development and defense responses such as root growth, leaf expansion, photomorphogenesis, and phytoalexin production (Noritake et al. 1996; Beligni and Lamattina 2000), and it has been found that H_2O_2 powers the NO synthesis and then works in combination in promoting plant disease resistance (Delledonne et al. 2001; Hancock et al. 2002; Wendehenne et al. 2004). NO is regarded as essential to SA-mediated SAR response (Durner et al. 1998; Song and Goodman 2007). Moreover, NO in association with H_2O_2 also activates SAR (Małolepsza and Rózalska 2005). Recent studies on NO synthase in plant peroxisomes also suggest that it acts as a source of signal molecules such as H_2O_2 (Barroso et al. 1999). Therefore, H_2O_2 as well as NO are endogenous signaling molecules; however, their higher levels can be toxic (Delledonne 2005; Bailey-Serres and Mittler 2006). At their optimal levels they are involved in signaling networks such as during pathogen infection both H_2O_2 and NO work concurrently and induce abscisic acid-mediated stomatal closure (Desikan et al. 2004). Interestingly, all the above-discussed signaling molecules also activate antioxidant system and, therefore, alleviate oxidative damage (Uchida et al. 2002; Zhou et al. 2005; Zhang et al. 2006). In *Zea mays*, abscisic acid, H_2O_2 along with NO form the signaling networks under dehydration conditions (drought stress) in which endogenous production of abscisic acid induces H_2O_2 generation which thereby stimulates NO production and works in coordination to improve the activities of antioxidant enzymes (Zhang et al. 2006, 2007). Additionally, upregulation of activity and gene expression of antioxidant enzymes need MAPK signaling which is regulated by H_2O_2 (Zhang et al. 2006, 2007). H_2O_2 and NO also work in coordination and act as key signaling molecules in acclimation to salt stress. Proteomic studies implicated that 85 proteins differ quantitatively under salt stress in citrus plants. However, when plants treated with H_2O_2 or NO prevented the alteration in the accumulation levels of 45 proteins and mitigated the negative impact of salt stress (Tanou et al. 2009). Therefore, H_2O_2, in association with various plant growth regulators, is involved in signaling networks to regulate diverse plant response under environmental perturbations.

CONCLUSION

Plants have evolved mechanisms to use potentially harmful ROS for signaling over the course of time. This transforms ROS, particularly H_2O_2, into signaling agents that have a significant effect on plant growth and development. From the last few decades, the perception of H_2O_2 as a toxic product has changed as it is involved in diverse plant responses such as stomatal closure, photosynthesis, and resilience to abiotic stresses. More importantly, it acts as a key signaling molecule in diverse signaling networks in association with other signaling components such as MAPKKK, Ca^{2+}, NO, ABA, among others. Production of H_2O_2 increases in response to environmental perturbations and helps in stress tolerance through activation of antioxidant system. Therefore, signal transduction of H_2O_2 under abiotic stresses represents a dynamic and intricate interplay between redox signaling, MAPK, and calcium-mediated pathways as well as crosstalk with plant growth regulator signaling cascades. Since H_2O_2 is generated in different cell compartments, the spatial and temporal generation of H_2O_2 as a signaling molecule and its feedback control seem very important to uncover along with its precise functions in plants.

REFERENCES

Apel K, Hirt H (2004) Reactive oxygen species: Metabolism, oxidative stress, and signal transduction. *Ann Rev Plant Biol* 55:373–399. https://doi.org/10.1146/annurev.arplant.55.031903.141701

Bailey-Serres J, Mittler R (2006) The roles of reactive oxygen species in plant cells. *Plant Physiol* 141:311–311. https://doi.org/10.1104/PP.104.900191.

Banerjee A, Roychoudhury A (2017) Abiotic stress, generation of reactive oxygen species, and their consequences. In: *React Oxyg Species Plants Boon Or Bane - Revisiting Role ROS*, 23–50. https://doi.org/10.1002/9781119324928.CH2.

Barroso JB, Corpas FJ, Carreras A, Sandalio LM, Valderrama R, Palma JM, Lupiánez JA, Del Río LA (1999) Localization of nitric-oxide synthase in plant peroxisomes. *J Biol Chem* 274:36729–36733. https://doi.org/10.1074/jbc.274.51.36729.

Baxter A, Mittler R, Suzuki N (2014) ROS as key players in plant stress signalling. *J Exp Bot* 65:1229–1240. https://doi.org/10.1093/JXB/ERT375.

Beligni MV, Lamattina L (2000) Nitric oxide stimulates seed germination and de-etiolation, and inhibits hypocotyl elongation, three light-inducible responses in plants. *Planta* 210:215–221. https://doi.org/10.1007/PL00008128.

Bhattacharjee S (2012) An inductive pulse of hydrogen peroxide pretreatment restores redox-homeostasis and oxidative membrane damage under extremes of temperature in two rice cultivars. *Plant Growth Regul* 68:395–410. https://doi.org/10.1007/S10725-012-9728-9.

Bhattachrjee S (2005). Reactive oxygen species and oxidative burst: Roles in stress, senescence and signal transduction in plant. *Curr Sci* 89:1113–1121.

Bienert GP, Schjoerring JK, Jahn TP (2006) Membrane transport of hydrogen peroxide. *Biochim Biophys Acta - Biomembr* 1758:994–1003. https://doi.org/10.1016/J.BBAMEM.2006.02.015.

Bose J, Rodrigo-Moreno A, Shabala S (2014) ROS homeostasis in halophytes in the context of salinity stress tolerance. *J Exp Bot* 65:1241–1257. https://doi.org/10.1093/JXB/ERT430.

Bright J, Desikan R, Hancock JT, Weir IS, Neill SJ (2006) ABA-induced NO generation and stomatal closure in *Arabidopsis* are dependent on H_2O_2 synthesis. *Plant J* 45:113–122. https://doi.org/10.1111/J.1365-313X.2005.02615.X.

Chaouch S, Queval G, Noctor G (2012) AtRbohF is a crucial modulator of defence-associated metabolism and a key actor in the interplay between intracellular oxidative stress and pathogenesis responses in *Arabidopsis*. *Plant J* 69:613–627. https://doi.org/10.1111/J.1365-313X.2011.04816.X.

Chen XJ, Xia XJ, Guo X, Zhou YH, Shi K, Zhou J, Yu JQ (2016) Apoplastic H_2O_2 plays a critical role in axillary bud outgrowth by altering auxin and cytokinin homeostasis in tomato plants. *New Phytol* 211:1266–1278. https://doi.org/10.1111/NPH.14015.

Cheng Y, Chunpeng S (2006) Hydrogen peroxide homeostasis and signaling in plant cells. *Sci China Ser C Life Sci* 49:1–12. https://doi.org/10.1007/s11427-005-0071-5

Choudhury FK, Rivero RM, Blumwald E, Mittler R (2017) Reactive oxygen species, abiotic stress and stress combination. *Plant J* 90:856–867. https://doi.org/10.1111/TPJ.13299.

Das K, Roychoudhury A (2014) Reactive oxygen species (ROS) and response of antioxidants as ROS-scavengers during environmental stress in plants. *Front Environ Sci* 2:121942. https://doi.org/10.3389/fenvs.2014.00053

Delledonne M (2005) NO news is good news for plants. *Curr Opin Plant Biol* 8:390–396. https://doi.org/10.1016/J.PBI.2005.05.002.

Delledonne M, Zeier J, Marocco A, Lamb C (2001) Signal interactions between nitric oxide and reactive oxygen intermediates in the plant hypersensitive disease resistance response. *Proc Natl Acad Sci USA* 98:13454–13459. https://doi.org/10.1073/PNAS.231178298.

Dempsey DA, Klessig DF (1994) Salicylic acid, active oxygen species and systemic acquired resistance in plants. *Trends Cell Biol* 4:334–338. https://doi.org/10.1016/0962-8924(94)90235-6.

Dempsey DA, Klessig DF (1995) Signals in plant disease resistance. *Bull Inst Pasteur* 93:167–186. https://doi.org/10.1016/0020-2452(96)81488-6.

Desikan R, Cheung MK, Bright J, Henson D, Hancock JT, Neill SJ (2004) ABA, hydrogen peroxide and nitric oxide signalling in stomatal guard cells. *J Exp Bot* 55:205–212. https://doi.org/10.1093/JXB/ERH033.

Desikan R, Hancock JT, Bright J, Harrison J, Weir I, Hooley R, Neill SJ (2005) A role for ETR1 in hydrogen peroxide signaling in stomatal guard cells. *Plant Physiol* 137:831. https://doi.org/10.1104/PP.104.056994.

Desikan R, Hancock JT, Neill SJ (2003) Oxidative stress signalling. In: *Plant Responses to Abiotic Stress. Topics in Current Genetics,* 121–149. https://doi.org/10.1007/978-3-540-39402-0_6

Després C, Chubak C, Rochon A, Clark R, Bethune T, Desveaux D, Fobert PR (2003) *The Arabidopsis* NPR1 disease resistance protein is a novel cofactor that confers redox regulation of DNA binding activity to the basic domain/leucine zipper transcription factor TGA1. *Plant Cell* 15:2181–2191. https://doi.org/10.1105/TPC.012849.

Dikilitas M, Karakas S, Ahmad P (2018) Predisposition of crop plants to stress is directly related to their DNA health. In: *Plant Microbiome: Stress Response. Microorganisms for Sustainability*, 233–254. https://doi.org/10.1007/978-981-10-5514-0_11.

Dunand C, Crèvecoeur M, Penel C (2007) Distribution of superoxide and hydrogen peroxide in *Arabidopsis* root and their influence on root development: possible interaction with peroxidases. *New Phytol* 174:332–341. https://doi.org/10.1111/J.1469-8137.2007.01995.X.

Durner J, Wendehenne D, Klessig DF (1998) Defense gene induction in tobacco by nitric oxide, cyclic GMP, and cyclic ADP-ribose. *Proc Natl Acad Sci* 95:10328–10333. https://doi.org/10.1073/PNAS.95.17.10328.

Ederli L, Morettini R, Borgogni A, Wasternack C, Miersch O, Reale L, Ferranti F, Tosti N, Pasqualini S (2006) Interaction between nitric oxide and ethylene in the induction of alternative oxidase in ozone-treated tobacco plants. *Plant Physiol* 142:595–608. https://doi.org/10.1104/PP.106.085472.

Feng H, Li X, Duan J, Li H, Liang H (2008) Chilling tolerance of wheat seedlings is related to an enhanced alternative respiratory pathway. *Crop Sci* 48:2381–2388. https://doi.org/10.2135/CROPSCI2007.04.0232.

Fluhr R (2009) Reactive oxygen-generating NADPH oxidases in plants. 1–23. https://doi.org/10.1007/978-3-642-00390-5_1.

Foyer CH, Noctor G (2005) Redox homeostasis and antioxidant signaling: A metabolic interface between stress perception and physiological responses. *Plant Cell* 17:1866–1875. https://doi.org/10.1105/TPC.105.033589.

Gechev TS, Van Breusegem F, Stone JM, Denev I, Laloi C (2006) Reactive oxygen species as signals that modulate plant stress responses and programmed cell death. *BioEssays* 28:1091–1101. https://doi.org/10.1002/BIES.20493.

Georgiou G (2002) How to flip the (redox) switch. *Cell* 111:607–610. https://doi.org/10.1016/S0092-8674(02)01165-0.

Groß F, Durner J, Gaupels F (2013) Nitric oxide, antioxidants and prooxidants in plant defence responses. *Front Plant Sci* 4. https://doi.org/10.3389/FPLS.2013.00419.

Halliwell B, Gutteridge JMC (1999). *Free Radicals in Biology and Medicine*. Oxford: Oxford University Press.

Hancock JT, Desikan R, Clarke A, Hurst RD, Neill SJ (2002) Cell signalling following plant/pathogen interactions involves the generation of reactive oxygen and reactive nitrogen species. *Plant Physiol Biochem* 40:611–617. https://doi.org/10.1016/S0981-9428(02)01403-1.

Hilal B, Khan TA, Fariduddin Q. (2023) Recent advances and mechanistic interactions of hydrogen sulfide with plant growth regulators in relation to abiotic stress tolerance in plants. *Plant Physiol Biochem* 196:1065–1083.

Hu X, Jiang M, Zhang A, Lu J (2005) Abscisic acid-induced apoplastic H_2O_2 accumulation up-regulates the activities of chloroplastic and cytosolic antioxidant enzymes in maize leaves. *Planta* 223:57–68. https://doi.org/10.1007/s00425-005-0068-0

Hu X, Jiang M, Zhang J, Zhang A, Lin F, Tan M (2007) Calcium–calmodulin is required for abscisic acid-induced antioxidant defense and functions both upstream and downstream of H_2O_2 production in leaves of maize (*Zea mays*) plants. *New Phytol* 173:27–38. https://doi.org/10.1111/J.1469-8137.2006.01888.X.

Hu Z, Li T, Zheng J, Yang K, He X, Leng P (2015) Ca^{2+} signal contributing to the synthesis and emission of monoterpenes regulated by light intensity in *Lilium 'siberia'*. *Plant Physiol Biochem* 91:1–9. https://doi.org/10.1016/J.PLAPHY.2015.03.005.

Huang SS, Chen J, Dong XJ, Patton J, Pei ZM, Zheng HL (2012) Calcium and calcium receptor CAS promote *Arabidopsis thaliana* de-etiolation. *Physiol Plant* 144:73–82. https://doi.org/10.1111/J.1399-3054.2011.01523.X.

Ichimura K, Mizoguchi T, Yoshida R, Yuasa T, Shinozaki K (2000) Various abiotic stresses rapidly activate *Arabidopsis* MAP kinases ATMPK4 and ATMPK6. *Plant J* 24:655–665. https://doi.org/10.1046/J.1365-313X.2000.00913.X.

Ivanchenko MG, Den Os D, Monshausen GB, Dubrovsky JG, Bednářová A, Krishnan N (2013) Auxin increases the hydrogen peroxide (H_2O_2) concentration in tomato (*Solanum lycopersicum*) root tips while inhibiting root growth. *Ann Bot* 112:1107–1116. https://doi.org/10.1093/AOB/MCT181.

Jacques S, Ghesquière B, De Bock PJ, Demol H, Wahni K, Willems P, Messens J, Van Breusegem F, Gevaert K (2015) Protein methionine sulfoxide dynamics in *Arabidopsis thaliana* under oxidative stress. *Mol Cell Proteomics* 14:1217–1229. https://doi.org/10.1074/MCP.M114.043729.

Jiang M, Zhang J (2003) Cross-talk between calcium and reactive oxygen species originated from NADPH oxidase in abscisic acid-induced antioxidant defence in leaves of maize seedlings. *Plant Cell Environ* 26:929–939. https://doi.org/10.1046/J.1365-3040.2003.01025.X.

Kaminaka H, Näke C, Epple P, Dittgen J, Schütze K, Chaban C, Holt BF, Merkle T, Schäfer E, Harter K, Dangl JL (2006) bZIP10-LSD1 antagonism modulates basal defense and cell death in *Arabidopsis* following infection. *EMBO J* 25:4400–4411. https://doi.org/10.1038/SJ.EMBOJ.7601312.

Khan TA, Fariduddin Q, Nazir F, Saleem M (2020) Melatonin in business with abiotic stresses in plants. *Physiol Mol Biol Plants* 26, 1931–1944.

Khan TA, Saleem M, Fariduddin Q (2022) Recent advances and mechanistic insights on melatonin-mediated salt stress signaling in plants. *Plant Physiol Biochem* 188, 97–107.

Khan TA, Saleem M, Fariduddin Q. (2023a). Melatonin influences stomatal behavior, root morphology, cell viability, photosynthetic responses, fruit yield, and fruit quality of tomato plants exposed to salt stress. *J Plant Growth Regul* 42(4):2408–32.

Khan TA, Saleem M, Fariduddin Q. (2023b) Differential drought stress tolerance in five varieties of tomato (*Solanum lycopersicum* L.): An evaluation of photosynthesis and antioxidant system. *J Soil Sci Plant Nutr* 23(2):2810–31.

Khan TA, Yusuf M, Fariduddin Q (2018) Hydrogen peroxide in regulation of plant metabolism: Signalling and its effect under abiotic stress. *Photosynth* 56:1237–1248. https://doi.org/10.1007/S11099-018-0830-8.

Kong D, Ju C, Parihar A, Kim S, Cho D, Kwak JM (2015) *Arabidopsis* glutamate receptor homolog3.5 modulates cytosolic Ca^{2+} level to counteract effect of abscisic acid in seed germination. *Plant Physiol* 167:1630–1642. https://doi.org/10.1104/PP.114.251298.

Kovtun Y, Chiu WL, Tena G, Sheen J (2000) Functional analysis of oxidative stress-activated mitogen-activated protein kinase cascade in plants. *Proc Natl Acad Sci USA* 97:2940–2945. https://doi.org/10.1073/PNAS.97.6.2940.

Kudla J, Becker D, Grill E, Hedrich R, Hippler M, Kummer U, Parniske M, Romeis T, Schumacher K (2018) Advances and current challenges in calcium signaling. *New Phytol* 218:414–431. https://doi.org/10.1111/NPH.14966.

Li P, Zhao C, Zhang Y, Wang X, Wang X, Wang J, Wang F, Bi Y (2015a) Calcium alleviates cadmium-induced inhibition on root growth by maintaining auxin homeostasis in *Arabidopsis* seedlings. *Protoplasma* 253:185–200. https://doi.org/10.1007/S00709-015-0810-9.

Li Z, Zhang Y, Peng D, Wang X, Peng Y, He X, Zhang X, Ma X, Huang L, Yan Y (2015b) Polyamine regulates tolerance to water stress in leaves of white clover associated with antioxidant defense and dehydrin genes via involvement in calcium messenger system and hydrogen peroxide signaling. *Front Physiol* 6:162311. https://doi.org/10.3389/fphys.2015.00280

Liao WB, Zhang ML, Huang GB, Yu JH (2012) Ca^{2+} and CaM are involved in NO- and H_2O_2-induced adventitious root development in marigold. *J Plant Growth Regul* 31:253–264. http://dx.doi.org/10.1007/s00344-011-9235-7

Liu H, Colavitti R, Rovira II, Finkel T (2005) Redox-dependent transcriptional regulation. *Circ Res* 97: 967–974. https://doi.org/10.1161/01.RES.0000188210.72062.10.

Liu XM, Kim KE, Kim KC, Nguyen XC, Han HJ, Jung MS, Kim HS, Kim SH, Park HC, Yun DJ, Chung WS (2010) Cadmium activates *Arabidopsis* MPK3 and MPK6 via accumulation of reactive oxygen species. *Phytochemistry* 71:614–618. https://doi.org/10.1016/J.PHYTOCHEM.2010.01.005.

Lv S, Zhang Y, Li C, Liu Z, Yang N, Pan L, Wu J, Wang J, Yang J, Lv Y, Zhang Y, Jiang W, She X, Wang G (2018) Strigolactone-triggered stomatal closure requires hydrogen peroxide synthesis and nitric oxide production in an abscisic acid-independent manner. *New Phytol* 217:290–304. https://doi.org/10.1111/NPH.14813.

Małolepsza U, Rózalska S (2005) Nitric oxide and hydrogen peroxide in tomato resistance: Nitric oxide modulates hydrogen peroxide level in o-hydroxyethylorutin-induced resistance to *Botrytis cinerea* in tomato. *Plant Physiol Biochem* 43:623–635. https://doi.org/10.1016/J.PLAPHY.2005.04.002.

Marino D, Dunand C, Puppo A, Pauly N (2012) A burst of plant NADPH oxidases. *Trends Plant Sci* 17:9–15. https://doi.org/10.1016/j.tplants.2011.10.001.

Miller G, Mittler R (2006) Could heat shock transcription factors function as hydrogen peroxide sensors in plants? *Ann Bot* 98:279–288. https://doi.org/10.1093/AOB/MCL107.

Miller G, Suzuki N, Ciftci-Yilmaz S, Mittler R (2010) Reactive oxygen species homeostasis and signalling during drought and salinity stresses. *Plant Cell Environ* 33:453–467. https://doi.org/10.1111/J.1365-3040.2009.02041.X.

Mittler R (2002) Oxidative stress, antioxidants and stress tolerance. *Trends Plant Sci* 7:405–410. https://doi.org/10.1016/S1360-1385(02)02312-9.

Mittler R, Vanderauwera S, Suzuki N, Miller G, Tognetti VB, Vandepoele K, Gollery M, Shulaev V, Van Breusegem F (2011) ROS signaling: The new wave? *Trends Plant Sci* 16:300–309. https://doi.org/10.1016/j.tplants.2011.03.007.

Moon H, Lee B, Choi G, et al (2003) NDP kinase 2 interacts with two oxidative stress-activated MAPKs to regulate cellular redox state and enhances multiple stress tolerance in transgenic plants. *Proc Natl Acad Sci USA* 100:358–363. https://doi.org/10.1073/PNAS.252641899.

Morris PC (2001) MAP kinase signal transduction pathways in plants. *New Phytol* 151:67–89. https://doi.org/10.1046/J.1469-8137.2001.00167.X.

Mou Z, Fan W, Dong X (2003) Inducers of plant systemic acquired resistance Regulate NPR1 function through redox changes. *Cell* 113:935–944. https://doi.org/10.1016/S0092-8674(03)00429-X.

Neill S (2007) Interactions between abscisic acid, hydrogen peroxide and nitric oxide mediate survival responses during water stress. *New Phytol* 175:4–6. https://doi.org/10.1111/J.1469-8137.2007.02112.X.

Neill SJ, Desikan R, Clarke A, Hancock JT (2002) Nitric oxide is a novel component of abscisic acid signaling in stomatal guard cells. *Plant Physiol* 128:13–16. https://doi.org/10.1104/PP.010707.

Niu L, Liao W (2016) Hydrogen peroxide signaling in plant development and abiotic responses: Crosstalk with nitric oxide and calcium. *Front Plant Sci* 7:177967. https://doi.org/10.3389/fpls.2016.00230

Noritake T, Kawakita K, Doke N (1996) Nitric oxide induces phytoalexin accumulation in potato tuber tissues. *Plant Cell Physiol* 37:113–116. https://doi.org/10.1093/OXFORDJOURNALS.PCP.A028908.

Pel ZM, Murata Y, Benning G, Thomine S, Klüsener B, Allen GJ, Grill E, Schroeder JI (2000) Calcium channels activated by hydrogen peroxide mediate abscisic acid signalling in guard cells. *Nature* 406:731–734. https://doi.org/10.1038/35021067.

Petrov VD, Van Breusegem F (2012) Hydrogen peroxide—a central hub for information flow in plant cells. *AoB Plants* 2012:14. https://doi.org/10.1093/AOBPLA/PLS014.

Pieterse CMJ, Leon-Reyes A, Van Der Ent S, Van Wees SCM (2009) Networking by small-molecule hormones in plant immunity. *Nat Chem Biol* 5:308–316. https://doi.org/10.1038/nchembio.164.

Pitzschke A, Djamei A, Dé Rique Bitton F, Hirt H (2009) A major role of the MEKK1–MKK1/2–MPK4 pathway in ROS signalling. *Mol Plant* 2:120–137. https://doi.org/10.1093/mp/ssn079.

Quan LJ, Zhang B, Shi WW, Li HY (2008) Hydrogen peroxide in plants: A versatile molecule of the reactive oxygen species network. *J Integr Plant Biol* 50:2–18. https://doi.org/10.1111/J.1744-7909.2007.00599.X.

Rao MV, Paliyath G, Ormrod DP, Murr DP, Watkins CB (1997) Influence of salicylic acid on H_2O_2 production, oxidative stress, and H_2O_2-metabolizing enzymes (salicylic acid-mediated oxidative damage requires H_2O_2). *Plant Physiol* 115:137–149. https://doi.org/10.1104/PP.115.1.137.

Reeves EP, Lu H, Jacobs HL, Messina CGM, Bolsover S, Gabellall G, Potma EO, Warley A, Roes J, Segal AW (2002) Killing activity of neutrophils is mediated through activation of proteases by K^+ flux. *Nature* 416:291–297. https://doi.org/10.1038/416291a.

Sadanandom A, Poghosyan Z, Fairbairn DJ, Murphy DJ (2000) Differential regulation of plastidial and cytosolic isoforms of peptide methionine sulfoxide reductase in *Arabidopsis*. *Plant Physiol* 123:255–263. https://doi.org/10.1104/PP.123.1.255.

Sagi M, Fluhr R (2001) Superoxide production by plant homologues of the gp91phox NADPH oxidase. Modulation of activity by calcium and by tobacco mosaic virus infection. *Plant Physiol* 126:1281–1290. https://doi.org/10.1104/PP.126.3.1281.

Sagi M, Fluhr R (2006) Production of reactive oxygen species by plant NADPH oxidases. *Plant Physiol* 141:336–340. https://doi.org/10.1104/PP.106.078089.

Sanders D, Brownlee C, Harper JF (1999) Communicating with calcium. *Plant Cell* 11:691. https://doi.org/10.1105/TPC.11.4.691.

Sauer H, Wartenberg M, Hescheler J (2001) Reactive oxygen species as intracellular messengers during cell growth and differentiation. *Cell Physiol Biochem* 11:173–186. https://doi.org/10.1159/000047804.

Shi K, Li X, Zhang H, Zhang G, Liu Y, Zhou Y, Xia X, Chen Z, Yu J (2015) Guard cell hydrogen peroxide and nitric oxide mediate elevated CO_2-induced stomatal movement in tomato. *New Phytol* 208:342–353. https://doi.org/10.1111/NPH.13621.

Simons BH, Lambers H (2018) The alternative oxidase: Is it a respiratory pathway allowing a plant to cope with stress? In: *Plant Responses to Environ Stress*, 265–286. https://doi.org/10.1201/9780203743157-12.

Song F, Goodman RM (2007) Activity of nitric oxide is dependent on, but is partially required for function of, salicylic acid in the signaling pathway in tobacco systemic acquired resistance. 14:1458–1462. https://doi.org/10.1094/MPMI.2001.14.12.1458.

Strand DD, Livingston AK, Satoh-Cruz M, Froehlich JE, Maurino VG, Kramer DM (2015) Activation of cyclic electron flow by hydrogen peroxide in vivo. *Proc Natl Acad Sci USA* 112:5539–5544. https://doi.org/10.1073/PNAS.1418223112.

Sun J, Wang MJ, Ding MQ, Deng SR, Liu MQ, Lu CF, Zhou XY, Shen X, Zheng XJ, Zhang ZK, Song J, Hu ZM, Xu Y, Chen SL (2010) H_2O_2 and cytosolic Ca^{2+} signals triggered by the PM H^+-coupled transport system mediate K^+/Na^+ homeostasis in NaCl-stressed *Populus euphratica* cells. *Plant Cell Environ* 33:943–958. https://doi.org/10.1111/J.1365-3040.2010.02118.X.

Suzuki K, Yano A, Shinshi H (1999). Slow and prolonged activation of the p47 protein kinase during hypersensitive cell death in cultured tobacco cells. *Plant Physiol* 119, 1465–1472.

Suzuki N, Koussevitzky S, Mittler R, Miller G (2012) ROS and redox signalling in the response of plants to abiotic stress. *Plant Cell Environ* 35:259–270. https://doi.org/10.1111/J.1365-3040.2011.02336.X.

Tanou G, Job C, Rajjou L, Arc E, Belghazi M, Diamantidis G, Molassiotis A, Job D (2009) Proteomics reveals the overlapping roles of hydrogen peroxide and nitric oxide in the acclimation of citrus plants to salinity. *Plant J* 60:795–804. https://doi.org/10.1111/J.1365-313X.2009.04000.X.

Tepe HD, Aydemir T (2015) Protective effects of Ca^{2+} against NaCl induced salt stress in two lentil (*Lens culinaris*) cultivars. *African J Agric Res* 10:2389–2398. https://doi.org/10.5897/AJAR2014.9479

Torres M, Forman HJ (2003) Redox signaling and the MAP kinase pathways. *Biofactors* 17:287–296. https://doi.org/10.1002/BIOF.5520170128.

Tsukagoshi H, Busch W, Benfey PN (2010) Transcriptional regulation of ROS controls transition from proliferation to differentiation in the root. *Cell* 143:606–616. https://doi.org/10.1016/j.cell.2010.10.020.

Uchida A, Jagendorf AT, Hibino T, Takabe T, Takabe T (2002) Effects of hydrogen peroxide and nitric oxide on both salt and heat stress tolerance in rice. *Plant Sci* 163:515–523. https://doi.org/10.1016/S0168-9452(02)00159-0.

Urao T, Katagiri T, Mizoguchi T, Yamaguchi-Shinozaki K, Hayashida N, Shinozaki K (1994) Two genes that encode Ca^{2+}-dependent protein kinases are induced by drought and high-salt stresses in *Arabidopsis thaliana. MGG Mol Gen Genet* 244:331–340. https://doi.org/10.1007/BF00286684.

Van Breusegem F, Dat JF (2006) Reactive oxygen species in plant cell death. *Plant Physiol* 141:384–390. https://doi.org/10.1104/PP.106.078295.

Vandenbroucke K, Robbens S, Vandepoele K, Inzé D, Van De Peer Y, Van Breusegem F (2008) Hydrogen peroxide–induced gene expression across kingdoms: A comparative analysis. *Mol Biol E* 25:507–516. https://doi.org/10.1093/MOLBEV/MSM276.

Vanderauwera S, Zimmermann P, Rombauts S, Vandenabeele S, Langebartels C, Gruissem W, Inzé D, Van Breusegem F (2005) Genome-wide analysis of hydrogen peroxide-regulated gene expression in *Arabidopsis* reveals a high light-induced transcriptional cluster involved in anthocyanin biosynthesis. *Plant Physiol* 139:806–821. https://doi.org/10.1104/PP.105.065896.

Wang H, Huang J, Liang X, Bi Y (2012) Involvement of hydrogen peroxide, calcium, and ethylene in the induction of the alternative pathway in chilling-stressed *Arabidopsis* callus. *Planta* 235:53–67. https://doi.org/10.1007/S00425-011-1488-7.

Wang H, Liang X, Huang J, Zhang D, Lu H, Liu Z, Bi Y (2010a) Involvement of ethylene and hydrogen peroxide in induction of alternative respiratory pathway in salt-treated *Arabidopsis* calluses. *Plant Cell Physiol* 51:1754–1765. https://doi.org/10.1093/PCP/PCQ134.

Wang J, Ding H, Zhang A, Ma F, Cao J, Jiang M (2010b) A novel mitogen-activated protein kinase gene in maize (*Zea mays*), ZmMPK3, is involved in response to diverse environmental cues. *J Integr Plant Biol* 52:442–452. https://doi.org/10.1111/J.1744-7909.2010.00906.X.

Wang L, Guo Y, Jia L, Chu H, Zhou S, Chen K, Wu D, Zhao L (2014) Hydrogen peroxide acts upstream of nitric oxide in the heat shock pathway in *Arabidopsis* seedlings. *Plant Physiol* 164:2184–2196. https://doi.org/10.1104/PP.113.229369.

Waszczak C, Akter S, Eeckhout D, Persiau G, Wahni K, Bodra N, Van Molle I, De Smet B, Vertommen D, Gevaert K, De Jaeger G, Van Montagu M, Messens J, Van Breusegem F (2014) Sulfenome mining in *Arabidopsis thaliana. Proc Natl Acad Sci USA* 111:11545–11550. https://doi.org/10.1073/PNAS.1411607111.

Waszczak C, Carmody M, Kangasjärvi J (2018) Reactive oxygen species in plant signaling. *Ann Rev Plant Biol* 69:209–236. https://doi.org/10.1146/ANNUREV-ARPLANT-042817-040322.

Wendehenne D, Durner J, Klessig DF (2004) Nitric oxide: A new player in plant signalling and defence responses. *Curr Opin Plant Biol* 7:449–455. https://doi.org/10.1016/J.PBI.2004.04.002.

Willekens H, Inzé D, Van Montagu M, van Camp W (1995) Catalases in plants. *Mol Breed New Strateg Plant Improv* 1:207–228. https://doi.org/10.1007/BF02277422.

Wrzaczek M, Brosché M, Kangasjärvi J (2013) ROS signaling loops — production, perception, regulation. *Curr Opin Plant Biol* 16:575–582. https://doi.org/10.1016/J.PBI.2013.07.002.

Wu F, Chi Y, Jiang Z, et al (2020) Hydrogen peroxide sensor HPCA1 is an LRR receptor kinase in *Arabidopsis. Nature* 578:577–581. https://doi.org/10.1038/s41586-020-2032-3.

Xiong J, Yang Y, Fu G, Tao L (2015) Novel roles of hydrogen peroxide (H_2O_2) in regulating pectin synthesis and demethylesterification in the cell wall of rice (*Oryza sativa*) root tips. *New Phytol* 206:118–126. https://doi.org/10.1111/NPH.13285.

Xu Q, Xu X, Zhao Y, Jiao K, Herbert SJ, Hao L (2008) Salicylic acid, hydrogen peroxide and calcium-induced saline tolerance associated with endogenous hydrogen peroxide homeostasis in naked oat seedlings. *Plant Growth Regul* 54:249–259. http://dx.doi.org/10.1007/s10725-007-9247-2

Yang SS, Zhai QH (2017) Cytosolic GAPDH: A key mediator in redox signal transduction in plants. *Biol Plant* 61:417–426. https://doi.org/10.1007/s10535-017-0706-y

Yang Y, Xu S, An L, Chen N (2007) NADPH oxidase-dependent hydrogen peroxide production, induced by salinity stress, may be involved in the regulation of total calcium in roots of wheat. *J Plant Physiol* 164:1429–1435. https://doi.org/10.1016/J.JPLPH.2006.08.009.

You J, Chan Z (2015) Ros regulation during abiotic stress responses in crop plants. *Front Plant Sci* 6:165102. https://doi.org/10.3389/fpls.2015.01092

Yu L, Yan J, Yang Y, He L, Zhu W (2016) Enhanced tolerance to chilling stress in tomato by overexpression of a mitogen-activated protein kinase, SlMPK7. *Plant Mol Biol Report* 34:76–88. http://dx.doi.org/10.1007/s11105-015-0897-3

Zhang A, Jiang M, Zhang J, Ding H, Xu S, Hu X, Tan M (2007) Nitric oxide induced by hydrogen peroxide mediates abscisic acid-induced activation of the mitogen-activated protein kinase cascade involved in antioxidant defense in maize leaves. *New Phytol* 175:36–50. https://doi.org/10.1111/J.1469-8137.2007.02071.X.

Zhang A, Jiang M, Zhang J, Tan M, Hu X (2006) Mitogen-activated protein kinase is involved in abscisic acid-induced antioxidant defense and acts downstream of reactive oxygen species production in leaves of maize plants. *Plant Physiol* 141:475–487. https://doi.org/10.1104/PP.105.075416.

Zhang X, Zhang L, Sun Y, Zheng S, Wang J, Zhang T (2020) Hydrogen peroxide is involved in strigolactone induced low temperature stress tolerance in rape seedlings (*Brassica rapa* L.). *Plant Physiol Biochem* 157:402–415. https://doi.org/10.1016/J.PLAPHY.2020.11.006.

Zhou B, Guo Z, Xing J, Huang B (2005) Nitric oxide is involved in abscisic acid-induced antioxidant activities in *Stylosanthes guianensis. J Exp Bot* 56:3223–3228. https://doi.org/10.1093/JXB/ERI319.

Zou JJ, Li XD, Ratnasekera D, Wang C, Liu WX, Song LF, Zhang WZ, Wu WH (2015) *Arabidopsis* CALCIUM-DEPENDENT PROTEIN KINASE8 and CATALASE3 function in abscisic acid-mediated signaling and H_2O_2 homeostasis in stomatal guard cells under drought stress. *Plant Cell* 27:1445–1460. https://doi.org/10.1105/TPC.15.00144.

5 Role of H_2O_2 in Seed Germination

Taiba Saeed, Durdana Yasin, and Anwar Shahzad

INTRODUCTION

Seed germination is a fundamental process in the life cycle of plants. It marks the transition from a dormant seed to an actively growing seedling, laying the foundation for a plant's life. However, the successful germination of seeds is not guaranteed, as many seeds exhibit dormancy—a state of arrested development and reduced metabolic activity. Seed dormancy allows seeds to survive adverse environmental conditions and synchronize germination with favorable circumstances. Understanding the mechanisms of seed dormancy and its regulation is of paramount importance in agriculture, ecology, and conservation. Seed dormancy is a complex and highly regulated process involving multiple genetic, physiological, and environmental factors. It serves as a survival mechanism for seeds, allowing them to delay germination until conditions are favorable for the seedling's survival. Two primary types of seed dormancy exist: exogenous dormancy and endogenous dormancy. Exogenous dormancy is primarily driven by external factors that prevent the germination of seeds (Geneve, 2003). These factors include hard seed coats and the presence of chemical inhibitors. These inhibitors may include allelopathic chemicals produced by other plants in the vicinity or compounds naturally occurring in the seed itself. Breaking exogenous dormancy typically involves the techniques of scarification and stratification (Górnik et al., 2023). Scarification involves mechanical treatments such as abrasion, sandpaper rubbing, or acid treatment that can be used to break down hard seed coats, allowing water penetration and germination to occur. Cold stratification involves exposing seeds to cold, moist conditions for a specific period. This treatment is particularly effective for seeds with morphological dormancy (MD) and can simulate the natural winter chilling that some seeds require for germination (Schütz and Rave,1999).

Endogenous dormancy, on the other hand, is governed by internal factors within the seed, including hormonal regulation and genetic predisposition (Nautiyal et al., 2023). It is further categorized into three main types: physiological dormancy (PD), MD, and combined dormancy (CD). PD is the most common form of endogenous dormancy and is primarily controlled by hormonal interactions, particularly the balance between abscisic acid (ABA) and gibberellic acid (GA). ABA acts as an inhibitory hormone, while GAs promote germination (Baskin and Baskin, 2021). MD primarily involves the development of a dormant embryo within a seed that requires a specific period of cold stratification to mature fully. This type of dormancy is often observed in temperate plant species. CD is a more complex form of dormancy that combines elements of both PD and MD. Seeds exhibiting CD have hormonal regulation mechanisms, as in PD, but also require specific environmental cues or treatments, as in MD, to initiate germination (Baskin and Baskin, 2021).

The transition from dormancy to germination begins when the dry seed comes into contact with water and ends when the radicle has emerged through all the coats of the developing embryo (Wang et al., 2020). This encounter activates the internal metabolic process, involving the careful equilibrium of phytohormones (Oracz and Karpiński, 2016) in the presence of optimum light and temperature to overcome the seed's dormancy (Shu et al., 2016). ABA and GA play a key role in several physiological processes during seed germination, and increased biosynthesis of GA and ABA degradation, resulting in a low ABA:GA ratio that favors the release of dormancy and triggers germination (Sano and Marion-Poll, 2021). Light, temperature, and soil conditions are major signals

DOI: 10.1201/9781032647692-5

that can be perceived by seeds to regulate the timing of germination. The regulatory effect of light on seed germination depends on its spectrum (Springthorpe and Penfield, 2015). Blue light activates ABA and delays seed germination, whereas red or far-red light plays a key role in the activation of seed germination via the activation of GA biosynthesis and restricting the production of ABA (Jiang et al., 2021). Seed germination is dependent on the surrounding temperature, which can delay or expedite the germination process after sowing. Most species germinate in the presence of temperatures between 15°C and 30°C (Yang et al., 2020). Weakening of the endosperm is a prerequisite for the initiation of seed germination and is driven by various internal and external factors. The decay of the endosperm is directly linked to the production of ROS in response to the availability of external environmental signals (Zhang et al., 2019).

H_2O_2: DUAL ROLE AS SIGNALING MOLECULE AND STRESSOR

Hydrogen peroxide (H_2O_2) is a versatile signaling molecule in plants, participating in various physiological processes, including growth, development, and stress responses. While H_2O_2 primarily serves as a signaling molecule in stress adaptation, excessive accumulation can lead to oxidative damage and stress amplification (Hossain et al., 2015). Plants are constantly exposed to a wide array of environmental stresses, including drought, salinity, extreme temperatures, pathogens, and pollutants. To survive and thrive under these adverse conditions, plants have developed intricate defense mechanisms. Among the key players in plant stress responses is H_2O_2, a reactive oxygen species (ROS) that serves a dual role in stress adaptation (Møller et al., 2007).

H_2O_2 is not only a toxic byproduct of metabolic processes but also a pivotal signaling molecule involved in the regulation of various physiological and stress-related processes (Reczek and Chandel, 2015).

In plant cells, H_2O_2 is primarily generated through the activity of several enzymes, including nicotinamide adenine dinucleotide phosphate (NADPH) oxidase, xanthine oxidase, and peroxidases (PODs) (Sutherland, 1991; Corpas et al., 2001; Bhattacharjee, 2005). H_2O_2 can also be produced non-enzymatically through reactions involving superoxide radicals or by the breakdown of other ROS (Kärkönen et al., 2017). Plants tightly regulate H_2O_2 levels through a balance of production and scavenging mechanisms. Antioxidant enzymes such as catalase (CAT), PODs, and superoxide dismutase (SOD) help neutralize excess H_2O_2, maintaining cellular homeostasis (Sharma et al., 2019).

ROLE OF H_2O_2 IN ABIOTIC STRESS RESPONSES

Drought Stress: H_2O_2 acts as a secondary messenger in the response to drought stress. It activates various stress-responsive genes and pathways, helping plants adapt to limited water availability (Zou et al., 2015).

Salinity Stress: H_2O_2 plays a crucial role in mitigating salt stress by regulating ion transport, enhancing antioxidant defense mechanisms, and promoting osmotic adjustment (Li et al., 2011).

Temperature Stress: During temperature stress, H_2O_2 mediates the activation of heat shock proteins (HSPs) and antioxidant enzymes, aiding in thermal tolerance (Cavusoglu and Kabar, 2010).

Heavy Metal Stress: Plants exposed to heavy metal stress accumulate H_2O_2, which triggers the expression of metal detoxification genes and antioxidants (Nazir et al., 2020).

ROLE IN BIOTIC STRESS RESPONSES

Pathogen Defense: H_2O_2 is involved in plant defense against pathogens. It activates hypersensitive response (HR) and systemic acquired resistance (SAR), leading to the synthesis of pathogenesis-related (PR) proteins (Orozco-Cárdenas et al., 2001).

Herbivore Resistance: Plants release H_2O_2 as a defense response to herbivore attack. This signaling molecule can trigger the production of secondary metabolites and the recruitment of beneficial predators (Maffei et al., 2007).

ROLE OF H_2O_2 AS A STRESSOR: OXIDATIVE DAMAGE AND STRESS AMPLIFICATION

While H_2O_2 serves as a signaling molecule in stress adaptation, excessive accumulation can have detrimental effects on plant cells. Uncontrolled H_2O_2 levels can lead to oxidative damage to cellular components, including lipids, proteins, and DNA. This oxidative stress can further exacerbate the impact of abiotic and biotic stresses, creating a feedback loop of stress amplification (Ransy et al., 2020).

OXIDATIVE STRESS AND CELLULAR DAMAGE

Lipid Peroxidation: Excessive H_2O_2 can initiate lipid peroxidation, leading to membrane damage and increased permeability (Smirnoff and Amaud, 2019).

Protein Oxidation: Oxidized proteins lose their functionality, disrupting various cellular processes and enzyme activities (Smirnoff and Amaud, 2019).

DNA Damage: H_2O_2-induced DNA damage can result in mutations and genome instability (Gechev et al., 2013).

AMPLIFICATION OF STRESS RESPONSES

ROS Burst: Oxidative stress induced by H_2O_2 can trigger the rapid production of other ROS, such as superoxide radicals ($O_2^{\bullet-}$) and hydroxyl radicals ($^{\bullet}OH$), intensifying the cellular damage (Mandal et al., 2022).

Activation of Cell Death Pathways: Prolonged exposure to high H_2O_2 levels can activate programmed cell death (PCD) pathways, contributing to tissue damage and plant susceptibility to stress (Gadjev et al., 2008).

EFFECT OF PRE-SOWING SEED PRIMING WITH H_2O_2 ON GERMINATION RATES

Seed priming represents a crucial pre-sowing approach accomplished through the application of diverse chemical, physical, and biological agents, leading to enhanced germination rates (Srivastava et al., 2021). The priming process involves inducing "stress memory" in plants, triggering heightened metabolic activities and the development of defense mechanisms, including osmotic adjustments and antioxidant defenses. This, in turn, results in improved seed germination and overall plant performance, particularly in stressful conditions (Srivastava et al., 2021). The various strategies employed for seed treatment are categorized as either conventional/invasive (such as hydro-, osmo-, hormonal, halo-, solid matrix, nutrient, bio-, or nano-priming) or physical/non-invasive (including magneto, UV irradiation, γ-radiation, cold plasma, electron, and laser priming). Invasive seed priming methods are considered cost-effective, as the implementation of non-invasive techniques is financially demanding, and skilled personnel are required to operate the associated instruments (Thakur et al., 2022).

Hydrogen peroxide has been documented as beneficial for enhancing the germination of various seeds, such as zinnia (Ogawa and Iwabuchi, 2001), barley and wheat (Smilanick et al., 1994), eastern gamagrass (Klein et al., 2008), rice (Sasaki et al., 2005), maize (Wahid et al., 2008), watermelon (Duval and NeSmith, 2000; Jaskani et al., 2006), camphor tree (Chien and Lin, 1994), and muscadine (Conner, 2008). Ogawa and Iwabuchi (2001) observed that the germination of zinnia seeds improved due to the oxidation of germination inhibitors in the pericarp by H_2O_2. Klein et al. (2008) demonstrated that 15% H_2O_2 treatment for 18 hours could serve as a substitute for a six-week stratification at 4°C in breaking the dormancy of eastern gamagrass. Duval and NeSmith (2000) proposed that watermelon germination barriers related to the seed coat could be overcome through seed coat alteration and the application of H_2O_2. Narimanov (2000) noticed that brief soaking of barley, maize, haricot, melon, vegetable marrow, garden radish, and carrot seeds in an H_2O_2 solution resulted in enhanced germination, early sprout appearance, and accelerated plant development. Szopińska (2014) suggested that treating zinnia seeds with H_2O_2 effectively improves germination, especially in heavily infected seeds, with germination rates primarily influenced by the concentration of hydrogen peroxide rather than the treatment duration.

Rajashekar and Baek (2014) discovered that immersing bean (*Phaseolus vulgaris*) seeds in a solution containing H_2O_2 effectively counteracted the adverse hypoxic effects on hypocotyl elongation and stem growth. Jafariyan and Zarea (2016) found that soaking wheat seeds in an 80% H_2O_2 solution for 8 hours, followed by inoculation with *Azospirillum*, led to the highest grain yields. Nandi et al. (2017) demonstrated that a 1% H_2O_2 solution was effective in increasing seed germination percentage, vigor index, and inhibiting the mycelial growth of fungus in chili seeds. Karthikeyan (2018) reported that immersing mung bean seeds in a solution of H_2O_2 and urea (NH_2-CO-NH_2) enhanced both germination and seedling growth, invigorating the seeds. Hajihashemi et al. (2020) found that pre-treating *Chenopodium quinoa* (quinoa) seeds with $CaCl_2$, H_2O_2, and sodium nitroprusside (SNP) mitigated the adverse effects of salt stress on seed germination. The pre-treated seeds exhibited a significant increase in germination rate, relative germination rate, and germination index (GI), while the mean germination time (MGT) was significantly reduced under both optimal and stress conditions. Barba-Espin et al. (2022) reported that immersing peach seeds without endocarp in hydrogen peroxide after eight weeks of stratification increased the germination rate and resulted in seedlings displaying robust vegetative growth. Table 5.1 represents the effect of chemo-priming of seeds on seed germination and other attributes.

MECHANISM OF H_2O_2 ACTION DURING SEED GERMINATION

Under the light of various studies discussed above, exogenous application of H_2O_2 has resulted in increased seed germination and improved growth in different plant species. A varied number of reports explained different mechanisms related to H_2O_2-mediated stimulation of seed germination. The most suggested mechanism involves breakdown of H_2O_2 resulting in the production of O_2 which is utilized in mitochondrial metabolism and respiration, oxidation of germination inhibitors, activation of redox-sensitive proteins, and modifications of mRNA transcripts, proteins as well as hormones (Bentsik and Koornneef, 2008; Nonogaki et al., 2010; Bewley et al., 2013). It has been proposed that H_2O_2 and ROS bring about oxidation of stored proteins, resulting in the modification of enzymatic and binding properties of proteins (Barba-Espín et al., 2010). Such modified proteins act as signals, stimulating the mobilization of nutrients from storage organs, thus playing a significant role in releasing dormancy and inducing germination (Diaz-Vivancos et al., 2013). Another suggested mode of action of H_2O_2 on seed germination is through the regulation of phytohormone levels which are known to have key roles in maintaining and releasing seed dormancy (Wojtyla et al., 2016; Carrera-Castaño et al., 2020). H_2O_2 acts as a signal triggering the genes for ABA catabolism and GA_3 biosynthesis. It has also been reported that the expression of genes encoding enzymes that hydrolyze the testa and endosperm is regulated by H_2O_2 (Lariguet et al., 2013). During the germination process, H_2O_2 serves as a protective factor against pathogens. The production of superoxide radicals ($O2^-$), H_2O_2, and hydroxyl radicals ($^{\bullet}OH$) in radish (*Raphanus sativus*) seeds has been identified as indicative of active and developmentally controlled physiological processes. These processes are presumed to play a crucial role in safeguarding emerging seedlings from potential damage caused by pathogens (Schopfer et al., 2001). Further evidence supporting the protective role of H_2O_2 against pathogens during germination and early seedling development

TABLE 5.1
Effect of Chemo-priming of Seeds on Germination and Other Attributes

S.No.	Crop	Compounds Used for Chemo-priming (H_2O_2 Alone or in Combination with Other Compounds)	Effect	References
1	Barley, maize, haricot, melon, vegetable marrow, garden radish, and carrot	Short-time soaking of seeds in 6×10^{-2}–6×10^{-5} M H_2O_2 solution	Enhances their germination, promotes early appearance of sprouts, and accelerates the development of plants	Narimanov (2000)
2	Rice	100 mM H_2O_2	Accelerated the germination percentage of seeds and induced acclimation of the plants to salinity	Gondim et al. (2010)
3	Pea	20 mM H_2O_2	Increased the germination as well as the seedling growth	Barba-Espín et al. (2012)
4	Wheat	50–200 µM H_2O_2 treatment	Stimulated the germination of seeds but inhibited root and shoot growth	Lu et al. (2013)
5	Maize	0.5 mM salicylic acid (SA) + 50 mM H_2O_2	Shortened seed germination time and enhanced seed vigor and seedling growth	Li et al. (2017)
6	Barley	50–150 mM H_2O_2	Increased germination of seeds	Delis-Hechavarría et al. (2021)
7	Wheat	1 mM SA and 0.1 mM H_2O_2	Improve germination and seedling growth under salt stress	Tania et al. (2021)
8	Cauliflower	120 µM H_2O_2	Improved germination percentages and plant performance under salt stress conditions	Ellouzi et al. (2021)
9	Sugarbeet	1% H_2O_2 solution	Increased germination of seeds	Chu et al. (2022)
10	*Cannabis sativa*	1% H_2O_2 solution	Enhance germination and survival rates while decreasing microbial contamination in antibiotic-free tissue culture media	Ahsan et al. (2022)

comes from studies on isolated lupine (*Lupinus luteus*) embryonic axes inoculated with *Fusarium oxysporum*. This inoculation leads to the accumulation of H_2O_2 and free radicals (Morkunas et al., 2004). Szopińska (2014) also suggested that the antimicrobial activity of H_2O_2 contributes to the improved germination rates, particularly in seeds that are infested with pathogens.

ASSOCIATION OF H_2O_2 SIGNALING WITH PHYTOHORMONES

Seed germination is regulated by the variations in the redox condition of the seeds and the interaction of ROS with phytohormones for varied plant responses has been well described in literature (Hernández Cortés, 2022). Interaction of ROS with phytohormone systems results in the stimulation of signaling pathways involved in different plant physiological processes, including seed germination and establishment of plantlets (Diaz-Vivancos et al., 2013). H_2O_2, being an important ROS, shares a complex interaction with hormonal networks in regulating seed dormancy and germination. Such interactions can be positive or negative mainly depending on the organs and biological processes. Different phytohormones play different roles during germination and the balance between ABA and GA_3 is known to be the most important factor governing dormancy release and the onset of germination (Weiss and Ori, 2007). GA_3 promotes seed germination in cereals by upregulating the transcription of α-amylase enzyme in the aleurone layer of cereal seeds while ABA has been known to suppress the transcription of the same (Liu et al., 2010). Sneideris et al. (2015) showed that hormonal priming of seeds with auxin, cytokinin, GA_3, and ethylene resulted in increased germination rates under control and salinity (cadmium) stress. Also, low concentration of ABA was found to trigger germination, while at higher concentrations it proved to be inhibitory. Seed priming with cytokinins, particularly kinetin, has been related to improved germination and vigor in many plant species (Araujo et al., 2016; Sharma et al., 2022). Cytokinin application has been proved to be an important regulatory mechanism in coordinating early seedling establishment (Wang et al., 2012). In *Arabidopsis*, Guan et al. (2014) showed that cytokinin induces a genetic cascade leading to the degradation of ABI5 protein specifically, thereby antagonizing the ABA-induced inhibition of post-germinative growth. Auxins, such as indole-3-acetic acid (IAA), are involved in various aspects of plant growth and development, including embryogenesis and germination. They play a role in cell elongation, seed coat rupture, and the development of the embryonic axis. Recent findings, however, show that auxin suppresses seed germination in an ABA-dependent manner. It was observed that seeds from transgenic plants overproducing auxin (iaaM-OX) exhibit arrested germination. Conversely, mutants with reduced auxin levels (yucl/yuc6) demonstrate an increased rate of seed germination (Liu et al., 2013). Similarly, mutants in auxin signaling, which have a compromised ability to degrade Aux/IAA repressor proteins, exhibit enhanced germination (Kepinski and Leyser, 2005; Liu et al., 2013). Gain-of-function mutants of Aux/IAA repressors, such as IAA7/AXR2 and IAA17/AXR3, also show an elevated germination rate (Liu et al., 2013). Genetic and biochemical evidence suggests that ABSCISSIC ACID INSENSITIVE3 (ABI3) serves as the downstream regulatory component in auxin-mediated seed dormancy (Belin et al., 2009; Liu et al., 2013). These molecular observations imply that inhibiting auxin signaling through Aux/IAA may be accountable for promoting seed germination. Moreover, recent research indicates that increased seed germination rate in *Arabidopsis* is facilitated by the downregulation of ABI3 transcription through the auxin signaling repressor IAA8 (Hussain et al., 2020).

Regulation of seed dormancy and germination is not only governed by interplay between various hormones, but also determined by the interaction between ROS and phytohormone signaling networks which help in the modulation of gene expression and cellular redox reactions (Xia et al., 2015). The concentration of ROS within the seed undergoes strict regulation through a multifactorial network of ROS-removing enzymes, including SOD, CAT, and POD. These enzymes enable ROS to function as cellular messengers by maintaining their levels within a critical range known as the "oxidative window." The progression toward germination is inhibited when the ROS concentration exceeds or falls below the defined oxidative window for germination (Morohashi, 2002). Of the various ROS generated, H_2O_2 has a prolonged lifespan and the ability to diffuse across membranes, reaching targets distant from its production sites. Several enzymes, including oxalate oxidase, xanthine oxidase, membrane-linked NADPH oxidase, amine oxidase (AO), and extracellular PODs, contribute to the production of H_2O_2. These enzymes can generate superoxide radicals, which are subsequently transformed into hydrogen peroxide (Bolwell and Wojtaszek, 1997). The precise regulation of H_2O_2 accumulation by the cell's antioxidant machinery is essential for maintaining a delicate balance between oxidative signaling, which promotes germination, and oxidative damage, which hinders or delays germination (Barba-Espin et al., 2011). H_2O_2 plays a regulatory role in the expression of various genes associated with the positive control of germination. This regulation occurs through processes such as protein carbonylation, activation, and modulation of kinase transduction cascades, coupled with alterations in cellular redox states (El-Maarouf-Bouteau et al., 2013). Bailly et al. (2008) observed a higher accumulation of H_2O_2 in non-dormant imbibed seeds compared to dormant imbibed seeds.

Research has demonstrated that H_2O_2 plays a regulatory role in seed germination by simultaneously inducing the catabolism of ABA and the biosynthesis of GA_3. H_2O_2 promotes GA biosynthesis by upregulating the expression of genes involved in the GA biosynthetic pathway, such as GA_{3ox} and GA_{20ox} genes (Liu et al., 2010). Several studies support these findings, where exogenous H_2O_2 increased the expression of CYP707A genes responsible for encoding ABA 8′-hydroxylases and GA_3 synthesis in dormant *Arabidopsis* seeds (Liu et al., 2010). Contrastingly, it has been demonstrated that externally applied ABA inhibits the accumulation of ROS in various plant species, including barley (Ishibashi et al., 2012), rice (Ye et al., 2012), lettuce (Zhang et al., 2014), and sunflower (El-Maarouf-Bouteau et al., 2015). On the other hand, the addition of GA enhances ROS production, particularly through superoxide and H_2O_2, in radish plants (Schopfer et al., 2001), *Arabidopsis* (Liu et al., 2010; Lariguet et al., 2013), and barley (Bahin et al., 2011). Bahin et al. (2011) proposed that H_2O_2 contributes to alleviating barley seed dormancy by activating GA signaling and/or biosynthesis rather than inhibiting ABA signaling. Their study revealed that exogenously applied H_2O_2 does not affect ABA biosynthesis and signaling significantly but exerts a more pronounced effect on GA signaling, influencing hormonal balance and subsequently initiating germination. The modulation of phytohormone balance during germination due to exogenously applied H_2O_2 is influenced by changes in H_2O_2 levels in seeds treated with GA and ABA. In this context, increased superoxide and H_2O_2 production have been observed in *Arabidopsis* seeds treated with GA, while decreases in ROS have been identified in seeds treated with ABA (Lariguet et al., 2013).

Anand et al. (2019) reported that the magnetoprimed tomato seeds exhibited increased seed germination rates due to the increased levels of hydrogen peroxide. The relative transcript profiles of genes responsible for H_2O_2 synthesis, such as copper-AO, receptor for activated C kinase 1 (RACK1) homologue (ArcA2), and SOD1 and SOD9, showed an increase in magnetoprimed tomato seeds compared to unprimed ones. Among the genes associated with H_2O_2 scavenging, namely metallothionein (MT1, MT3, and MT4), CAT1, and ascorbate POD (APX1 and APX2), MT1 and MT4 exhibited a fourteen to fifteen-fold increase, respectively, in transcript abundance in primed seeds compared to the control.

ASSOCIATION OF H_2O_2 SIGNALING WITH OTHER SIGNALING MOLECULES

Although there has been extensive research on the roles of H_2O_2 and nitric oxide (NO) in seed biology, information about other signaling molecules and their interactions is still limited. NO and H_2O_2 play parallel roles in disrupting germination dormancy through their interactions with ABA. The interactive role played by ABA, NO, and H_2O_2 during seed germination is quite different from their known regulation in stomatal guard cell behavior (Arc et al., 2013a,b). The NO has been found to break seed dormancy by eliciting the H_2O_2-responsive expression of ABA catabolic genes. Exogenous application of H_2O_2 during imbibition resulted in increased NO levels, suggesting that H_2O_2 also regulates NO production (Liu et al., 2010). Like H_2O_2, the exogenous application of NO also reduces seed dormancy and diminishes the inhibitory effects of ABA on seed germination (Bethke et al., 2004, 2006). NO applications have been found to increase seed germination under stress environments (Kopyra and Gwóźdź, 2003; Zheng et al., 2009).

Liu et al. (2010) proposed a hypothetical model that explains the interrelationships between H_2O_2 and NO in the regulation of seed germination by joint actions of ABA and GA. According to this model, H_2O_2 can interrupt the dormancy of *Arabidopsis* seeds through two pathways. The first pathway relies on the enhancement of ABA catabolism and GA biosynthesis. The signaling molecule (NO) does not regulate GA biosynthesis directly but instead acts as a temporary signaling molecule involved in the H_2O_2 regulation of ABA catabolism. The second pathway assumes the negative regulation of GA biosynthesis by ABA. Gniazdowska et al. (2010) proposed a function for H_2O_2 in apple seed germination and its role in the downstream signaling of NO and HCN in the activation of

ET biosynthesis during early seedling growth. They also found that the activities of crucial enzymes involved in ET metabolism are modified by HCN and NO treatments. Oracz et al. (2009) presented a comprehensive scheme on the mechanism of HCN-dependent dormancy alleviation and on its crosstalk to ROS as a decisive signaling element involved in seed germination. The dominant role of ROS and reactive nitrogen species (RNS) in the regulation of seed dormancy and germination is also discussed. The authors postulate that NO may play a key role in germination vigor, which may result from crosstalk between NO and ROS (Arc et al., 2013a). Krasuska and Gniazdowska (2012) stated that ROS, NO, and HCN can simultaneously affect embryo dormancy release processes and that their accurate levels are essential to seed germination and development regulation. Based on the "oxidative window" model, a model proposed by Bailly et al. (2008) that describes regulating functions of ROS in seed dormancy/germination switch, Krasuska and Gniazdowska (2012) presented the "nitrosative door" hypothesis, which focuses on the concentration-dependent role of RNS (mainly in terms of NO in seed physiology). They also proposed that RNS and ROS levels are strictly regulated by ROS scavenging enzymes.

Wang et al. (2015a) presented a mechanism of NO suppression on the inhibitory effects of ABA on seed germination. Based on studies related to interactions between ABA, NO, and ROS in stomatal guard cells and based on their own results, Wang et al. (2015a) suggested that NO negatively regulates ABA signaling through *S*-nitrosylation of SnRK2s proteins not only in terms of stomatal closure but also in terms of the inhibition of seed germination and seedling growth. They proposed that *S*-nitrosylation of SnRK2s proteins serves as a key component of signaling crosstalk between ABA and NO that regulates *Arabidopsis* seed germination. They described a mechanism for NO involvement in dormancy release and germination promotion. Based on their findings, endogenous and exogenously applied NO exerts inhibitory effects on the kinase activities of SnRK2.2 and SnRK2.3 via *S*-nitrosylation and thus blocks ABA signaling (Wang et al., 2015b). Interest in the H_2S molecule has grown in plant biology research. This is due to its signaling functions and interactions with H_2O_2 and NO during plant development and stress responses (Calderwood and Kopriva, 2014; Hancock and Whiteman, 2014, 2015; Jin and Pei, 2015) and during seed germination (Li, 2013). Improved germination and decrease in germination time periods have been observed in common bean (*P. vulgaris*), maize (*Zea mays*), wheat (*Triticum aestivum*), and pea seeds subjected to H_2S treatments (Dooley et al., 2013). These results suggest that H_2S plays an important role as a signaling molecule that can accelerate growth rates of numerous plant species. Positive effects of H_2S and H_2O_2 treatments on the promotion of mung bean seed germination have been observed by Li and He (2015). These authors suggest that H_2O_2 and H_2S may promote the germination of mung bean seeds by mobilizing reserve proteins and that H_2O_2 may serve as a downstream signaling molecule of H_2S. Li et al. (2012) proposed the existence of crosstalk between H_2O_2 and H_2S during seed germination. The authors found improved germination percentages for *Jatropha curcas* seeds soaked in H_2O_2 accompanied by an increase in L-cysteine desulfhydrase activity that induces H_2S accumulation. Moreover, Li et al. (2012) observed better germination performance after adding H_2S to a soaking solution and postulated that this improvement is mediated by H_2S. In the signaling process mediated by H_2S during seed germination, both H_2O_2 and NO play important roles. In NaCl-stressed alfalfa (*Medicago sativa*) seeds, both H_2S (sodium hydrosulfide) and NO donors (SNP) can significantly attenuate seed germination and seedling growth inhibition by protecting against oxidative damage (Wang et al., 2012). The authors also showed that the application of H_2S donor enhances NO accumulation while the addition of 2-phenyl-4,4,5,5-tetramethylimidazoline-1-oxyl 3-oxide (PTIO), a specific NO scavenger, diminishes the positive impacts of H_2S on germination and NaCl stress tolerance, suggesting the presence of interactions between H_2S and NO in germinating seeds. The pretreatment of *C. quinoa* (quinoa) seeds with $CaCl_2$, H_2O_2, and SNP limited the adverse effect of salt stress on seed germination. The pre-treated seeds showed a significant increase in germination rate, relative germination rate, and GI while the MGT was significantly reduced under both optimal and stress conditions (Figure 5.1).

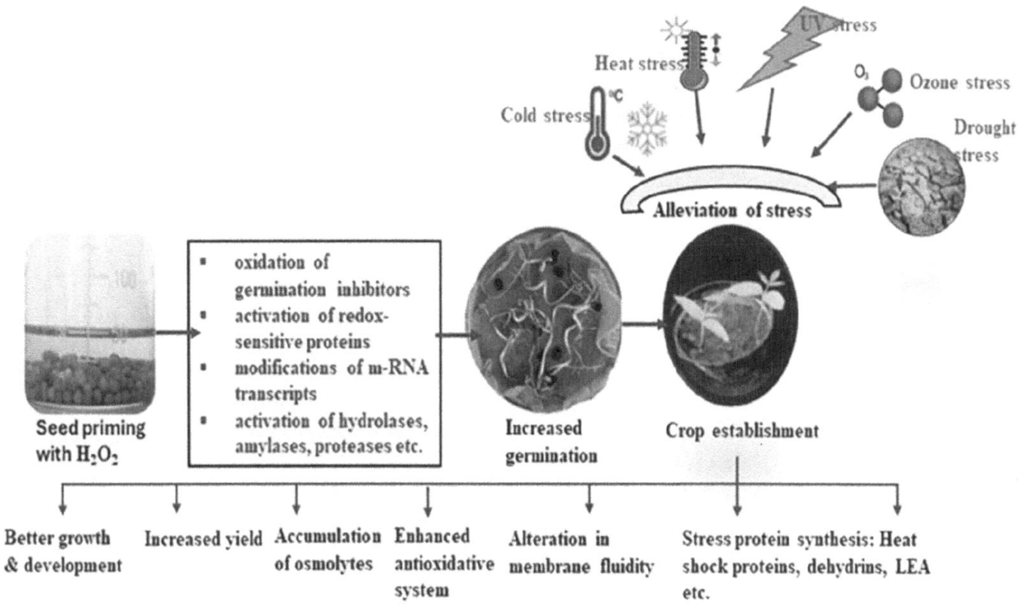

FIGURE 5.1 Effect of seed priming with hydrogen peroxide on seed germination and seedling growth.

ASSOCIATION OF H_2O_2 IN MITIGATING ABIOTIC STRESSES

Hydrogen peroxide plays a crucial role in mitigating abiotic stresses during seed germination by acting as a signaling molecule, enhancing antioxidant defenses, interacting with plant hormones, modifying cell walls, and inducing metabolic adjustments. Seed priming enhances the stress tolerance of germinating seeds and the plants that grow from them by promoting quicker transitions to the germinated state and activating stress-responsive systems. These strategies form what is known as "priming memory" or stress imprinting mechanisms. This involves genetic or biochemical changes induced by priming, which improve the plant's ability to withstand future stress exposures (Chen and Arora, 2013). Studies have shown that soaking seeds in H_2O_2 or applying H_2O_2 as a foliar spray induces low levels of oxidative stress. This oxidative stress, particularly through H_2O_2-dependent signaling networks, triggers the accumulation of latent defense proteins. This process creates primed states in the plants, enhancing their stress responses (Hossain et al., 2015).

Primed seeds of *Brassica napus* showed salinity stress tolerance during post-priming germination after initial exposure to salt stress (Kubala et al., 2015). Likewise, there are several reports where exogenous application of H_2O_2 in the form of seed soaking or foliar spray results in increased tolerance to salinity, drought, chilling, high temperatures, and heavy metal stress. Seed priming with H_2O_2 has demonstrated beneficial effects in improving salt and high-temperature tolerance in *Hordeum vulgare*, drought tolerance in *T. aestivum* and *P. vulgaris*, as well as soil moisture stress tolerance in *Gossypium hirsutum* (He et al., 2009; Cavusoglu and Kabar, 2010; Abass and Mohamed, 2011; Santhy et al., 2014). Moreover, it has been seen that treating seeds with exogenous H_2O_2 and allowing them to germinate without drying to their original moisture content has been found to enhance the salt tolerance of wheat seedlings. This process alleviates oxidative damage and boosts the expression of stress proteins (Wahid et al., 2007). Similarly, such treatment has known to mitigate aluminum-induced oxidative stress in wheat seedlings (Xu et al., 2011) and modulates the antioxidant system and nutrient relations in maize plants under water-deficit conditions (Ashraf et al., 2015). The efficacy of H_2O_2 priming in facilitating faster germination under salt or drought stress was also found in pepper seeds. This acceleration in germination was found to be characterized by a reduction in MGT. Additionally, there is an observed increase in the GI associated with H_2O_2 priming under these stress conditions (Gammoudi et al., 2020). Seed priming with hydrogen peroxide in combination with other compounds has also known to alleviate abiotic stress during seed germination. Research indicates that using H_2O_2 in conjunction with salicylic acid (SA) as priming agents can alleviate the negative effects of salinity on the germination and growth of kidney beans (Tania et al., 2022). Additionally, the physiological conditioning of cotton seeds with hydrogen peroxide and SNP significantly enhances seed germination under salinity stress, leading to improved seedling growth performance (dos Santos Guaraldo et al., 2023).

ASSOCIATION OF H_2O_2 IN SEED AGING

Seed aging involves the gradual accumulation of damage to cellular components, which in turn results in a loss of seed viability and vigor. This process occurs during prolonged seed storage and escalates when seeds are stored in improper conditions (especially in high-temperature and moisture conditions) (Arc et al., 2011). As stored seeds lose longevity over time, it is critically important to understand the mechanisms of the aging process that are related to agronomic and ecological (*ex situ* seed conservation) factors. It is worth emphasizing that rates of aging and seed longevity vary between species. The lifespan of stored seeds depends not only on storage periods and conditions but also on genetic, physiological, and morphological factors (Ventura et al., 2012). According to the "free radical theory of aging," the driving force behind most alterations that occur during the aging of living organisms is ROS activity. This assumption also refers to aged seeds and is supported by numerous reports (Ratajczak et al., 2015). ROS production in dry stored seeds ensues as a result of non-enzymatic processes, e.g., Amadori and Maillard reactions and lipid peroxidation (El-Maarouf-Bouteau and Bailly, 2008).

H_2O_2, a stable ROS, can traverse membranes over considerable distances, contributing to the aging process (Ratajczak et al., 2015; Kurek et al., 2019). Studies have demonstrated a negative correlation between seed viability/germination capacity and H_2O_2 accumulation during aging in various seeds, including artificially aged sunflower (Kibinza et al., 2011), *Fagus sylvatica* (Pukacka and Ratajczak, 2007), and wheat (Lehner et al., 2008), as well as naturally aged cotton (Goel and Sheoran, 2003) and cucumis seeds (Ahmad et al., 2021). Ahmad et al. (2021) noted that H_2O_2 within an oxidative window can be efficiently neutralized, but beyond this threshold, it becomes toxic, impacting seed longevity as enzymes are inactivated during storage. It is speculated that seed aging in storage is influenced by the interplay between hydrogen peroxide and ABA. In rice, H_2O_2 accumulation in the radicle is associated with reduced seed vigor and asynchronous seedling emergence (Zheng et al., 2024). Overexpression of the ABA catabolism gene, ABA 8'-hydroxylase 2 (OsABA8ox2), in rice seeds significantly decreases seed vigor, while knockout of OsABA8ox2 enhances it. Additionally, seeds pre-treated with ABA or the H_2O_2 inhibitor diphenyleneiodonium chloride (DPI) exhibit increased resistance to aging and more robust early seedling establishment. Thus, ABA inhibits rice seed aging by reducing H_2O_2 accumulation in the radicle (Zheng et al., 2024).

CONCLUSIONS

Seeds are fundamentally important to plants as a means of propagation, making germination a critical phase as seeds change from a dormant state to metabolically active state. Understanding the intricate combination of environmental and endogenous signals that regulate germination is crucial. Research has shown that the interplay between H_2O_2 signaling pathways, phytohormones such as ABA and GA, and other signaling molecules plays an integrative role in the switch between dormancy and germination. While the accumulation of H_2O_2 and other ROS during storage can facilitate germination, it also has detrimental effects on seed viability. Pre-sowing seed priming has been demonstrated to improve seed quality, enhancing germination performance and vigor while mitigating the effects of seed aging. Future research should focus on understanding how seed priming influences H_2O_2-mediated signaling networks. The use of various mutants and the development of new techniques will provide fresh perspectives and facilitate a more comprehensive understanding and validation of these processes.

REFERENCES

Abass SM, Mohamed HI (2011) Alleviation of adverse effects of drought stress on common bean (*Phaseolus vulgaris* L.) by exogenous application of hydrogen peroxide. *Bangladesh J. Bot.* 41:75–83.

Ahmad D, Jain SK, Joshi MA, Anand A, Tomar BS, Kumar S, Hasan M (2021) H_2O_2 as a better index of seed quality and mechanism of cucumber (*Cucumis sativus*) seed deterioration. *Ind. J. Agricult. Sci.* 91(10):1500–1504.

Ahsan SM, Shin JH, Choi HW (2022) Availability of hydrogen peroxide solutions as a germination liquid medium for contamination-free in vitro seedling development of *Cannabis sativa*. *Hortic. Sci. Technol.* 40(6):605–613.

Anand A, Kumari A, Thakur M, Koul A (2019) Hydrogen peroxide signaling integrates with phytohormones during the germination of magnetoprimed tomato seeds. *Sci Rep.* 9:8814.

Araujo SS, Paparella S, Dondi D, Bentivoglio A, Carbonera D, Balestrazzi A (2016) Physical methods for seed invigoration: Advantages and challenges in seed technology. *Front. Plant Sci.* 7:646.

Arc E, Galland M, Godin B, Cueff G, Rajjou L (2013a). Nitric oxide implication in the control of seed dormancy and germination. *Front. Plant Sci.* 4:346

Arc E, Ogé L, Grappin P, Rajjou L (2011) Plant seed: A relevant model to study aging processes. *Field of Biological Aging: Past, Present and Future*, Ed. Olgun A. (Kerala: Transworld Research Network) pp. 87–102.

Arc E, Sechet J, Corbineau F, Rajjou L, Marion-Poll A (2013b) ABA crosstalk with ethylene and nitric oxide in seed dormancy and germination. *Front. Plant Sci.* 4:63

Ashraf MA, Rasheed R, Hussain I, Iqbal M, Haider MZ, Parveen S, Sajid MA (2015). Hydrogen peroxide modulates antioxidant system and nutrient relation in maize (*Zea mays* L.) under water-deficit conditions. *Arch. Agron. Soil Sci.* 61: 507–523.

Bahin E, Bailly C, Sotta B, Kranner I, Corbineau F, Leymarie J (2011) Crosstalk between reactive oxygen species and hormonal signalling pathways regulates grain dormancy in barley. *Plant Cell Environ.* 34: 980–993.

Bailly C, El-Maarouf-Bouteau H, Corbineau F (2008) From intracellular signaling networks to cell death: The dual role of reactive oxygen species in seed physiology. *C. R. Biol.* 331:806–814.

Barba-Espin G, Diaz-Vivancos P, Clemente-Moreno MJ, Albacete A, Faize L, Faize M., Pé rez-Alfocea, F, Hernández JA (2010) Interaction between hydrogen peroxide and plant hormones during germination and the early growth of pea seedlings. *Plant Cell Environ.* 33:981–994.

Barba-Espín G, Diaz-Vivancos P, Job D, Belghazi M, Job C, Hernández JA (2011). Understanding the role of H_2O_2 during pea seed germination: A combined proteomic and hormone profiling approach. *Plant Cell Environ.* 34:1907–1919.

Barba-Espín G, Hernández JA, Diaz-Vivancos P (2012) Role of H_2O_2 in pea seed germination. *Plant Signal Behav.* 7(2):193–195.

Barba-Espín G, Hernández JA, Martínez-Andújar C, Díaz-Vivancos P (2022) Hydrogen peroxide imbibition following cold stratification promotes seed germination rate and uniformity in peach cv. GF305. *Seeds* 1(1):28–35.

Baskin JM, Baskin CC (2021). The great diversity in kinds of seed dormancy: a revision of the Nikolaeva–Baskin classification system for primary seed dormancy. *Seed Sci. Res.* 31(4):249–277.

Belin C, Megies C, Hauserova E, Lopez-Molina L (2009) Abscisic acid represses growth of the Arabidopsis embryonic axis after germination by enhancing auxin signaling. *Plant Cell* 21:2253–2268.

Bentsik L, Koornneef M (2008) Seed dormancy and germination. *Arabidopsis Book* 6:e0119.

Bethke PC, Gubler F, Jacobsen JV, Jones RL (2004) Dormancy of *Arabidopsis* seeds barley grains can be broken by nitric oxide. *Planta* 219:847–855.

Bethke PC, Libourel IGL, Jones RL (2006) Nitric oxide reduces seed dormancy in *Arabidopsis*. *J. Exp. Bot.* 57:517–526.

Bewley JD, Bradford KJ, Hilhorst HMW, Nonogaki H (2013) *Seeds: Physiology of Development, Germination and Dormancy.* New York: Springer.

Bhattacharjee, S. (2005). Reactive oxygen species and oxidative burst: Roles in stress, senescence and signal transducation in plants. *Curr. Sci.* 89:1113–1121.
Bolwell GP, Wojtaszek P (1997) Mechanisms for the generation of reactive oxygen species in plant defence – a broad perspective. *Physiol. Mol. Plant Pathol.* 51:347–366.
Calderwood A, Kopriva S (2014) Hydrogen sulfide in plants: From dissipation of excess sulfur to signaling molecule. *Nitric Oxide* 41:72–78.
Carrera-Castaño G, Calleja-Cabrera J, Pernas M, Gómez L, Oñate-Sánchez L (2020) An updated overview on the regulation of seed germination. *Plants (Basel)* 9(6):703.
Cavusoglu K, Kabar K (2010) Effects of hydrogen peroxide on the germination and early seedling growth of barley under NaCl and high temperature stresses. *Eurasia J. Biosci.* 4:70–79.
Chen K, Arora R (2013) Priming memory invokes seed stress-tolerance. *Environ. Exp. Bot.* 94:33–45.
Chien CT, Lin TP (1994) Mechanism of hydrogen peroxide in improving the germination of Cinnamonum camphora seed. *Seed Sci. Technol.* 22:231–236.
Chu C, Poore RC, Bolton MD, Fugate KK (2022) Mechanism of sugarbeet seed germination enhanced by hydrogen peroxide. *Front Plant Sci.* 13: 888519.
Conner PJ (2008) Effects of stratification, germination temperature and pre-treatment with gibberellic acid and hydrogen peroxide on germination of 'Fry' muscadine (*Vitis rotundifolia*) seed. *HortScience* 43(3): 853–856.
Corpas FJ, Barroso JB, del Río LA (2001). Peroxisomes as a source of reactive oxygen species and nitric oxide signal molecules in plant cells. *Trends Plant Sci.* 6(4):145–150.
Delis-Hechavarría EA, Guevara-González RG, Ocampo-Velazquez RV, Gómez-Soto JG, Tovar-Pérez EG, García-Trejo JF, Torres-Pacheco I (2021). Pre-germination treatment with hydrogen peroxide as a controlled elicitation strategy to improve chemical properties of hydroponic barley fodder. *Crop Pasture Sci.72(10)*: 815-822.
Diaz-Vivancos P, Barba-Espín G, Hernández JA (2013) Elucidating hormonal/ROS networks during seed germination: insights and perspectives. *Plant Cell Rep.* 32:1491–1502.
Dooley FD, Nair SP, Ward PD (2013) Increased growth and germination success in plants following hydrogen sulfide administration. *PLoS One* 8:e62048.
dos Santos Guaraldo MM, Pereira TM, dos Santos HO, de Oliveira TL, Pereira WVS, de Resende Von Pinho EV (2023) Priming with sodium nitroprusside and hydrogen peroxide increases cotton seed tolerance to salinity and water deficit during seed germination and seedling development. *Environ. Exper. Botany* 209:105294.
Duval JR, NeSmith DS (2000) Treatment with hydrogen peroxide and seedcoat removal or clipping improve germination of 'Genesis' triploid watermelon. *HortScience* 35:85–86.
Ellouzi H, Oueslati S, Hessini K, Rabhi M, Abdelly C (2021) Seed-priming with H_2O_2 alleviates subsequent salt stress by preventing ROS production and amplifying antioxidant defense in cauliflower seeds and seedlings. *Sci. Hortic.* 288:110360.
El-Maarouf-Bouteau H, Bailly C (2008) Oxidative signaling in seed germination and dormancy. *Plant Signal. Behav.* 3:175–182.
El-Maarouf-Bouteau H, Meimoun P, Job C, Job D, Bailly C (2013) Role of protein and mRNA oxidation in seed dormancy and germination. *Front. Plant Sci.* 4:77
El-Maarouf-Bouteau H, Sajjad Y, Bazin J, Langlade N, Cristescu SM, Balzergue S, Baudouin E, Bailly C (2015) Reactive oxygen species, abscisic acid and ethylene interact to regulate sunflower seed germination. *Plant Cell Environ.* 38:364–374.
Gadjev I, Stone JM, Gechev TS (2008). Programmed cell death in plants: New insights into redox regulation and the role of hydrogen peroxide. *Int. Rev. Cell Mol. Biol.* 270:87–144.
Gammoudi N, Karmous I, Zerria K, Loumerem M, Ferchichi A, Nagaz K (2020) Efficiency of pepper seed invigoration through hydrogen peroxide priming to improve in vitro salt and drought stress tolerance. *Horticul. Environ. Biotechnol.* 61:703–714.
Gechev T, Mehterov N, Denev I, Hille J (2013). A simple and powerful approach for isolation of Arabidopsis mutants with increased tolerance to H_2O_2-induced cell death. *Method Enzymol.* 527: 203–220.
Geneve RL (2003). Impact of temperature on seed dormancy. *HortScience* 38(3):336–341.
Gniazdowska A, Krasuska U, Czajkowska K, Bogatek R (2010) Nitric oxide, hydrogen cyanide and ethylene are required in the control of germination and undisturbed development of young apple seedlings. *Plant Growth Regul.* 61:75–84.
Goel A, Sheoran IS (2003) Lipid peroxidation and peroxide-scavenging enzymes in cotton seeds under natural ageing. *Biol. Plant.* 46:429–434.

Gondim FA, Gomes-Filho E, Lacerda CF, Prisco JT, Neto ADA, Marques EC (2010) Pretreatment with H_2O_2 in maize seeds: Effects on germination and seedling acclimation to salt stress Braz. *J. Plant Physiol.* 22(2):103–112.

Górnik K, Sas-Paszt L, Seliga Ł, Pluta S, Derkowska E, Głuszek S, Mosa WF (2023). The effect of different stratification and scarification treatments on breaking the dormancy of Saskatoon berry seeds. *Agronomy* 13(2):520.

Guan C, Wang X, Feng J, Hong S, Liang Y, Ren B, Zuo J (2014) Cytokinin antagonizes abscisic acid-mediated inhibition of cotyledon greening by promoting the degradation of abscisic acid insensitive protein in Arabidopsis. *Plant Physiol.* 164:1515–1526.

Hajihashemi S, Skalicky M, Brestic M, Pavla V (2020) Cross-talk between nitric oxide, hydrogen peroxide and calcium in salt-stressed *Chenopodium quinoa* Willd. At seed germination stage. *Plant Physiol. Biochem.* 154: 657–664.

Hancock JT, Whiteman M (2014) Hydrogen sulfide and cell signaling: Team player or referee? *Plant Physiol. Biochem.* 78:37–42.

Hancock JT, Whiteman M (2015) Hydrogen sulfide signaling: Interactions with nitric oxide and reactive oxygen species. *Ann. N. Y. Acad. Sci.* 1365:5–14.

He L, Gao Z, Li R (2009) Pretreatment of seed with H_2O_2 enhances drought tolerance of wheat seedlings. *Afr. J. Biotechnol.* 8:6151–6157.

Hernández Cortés JA (2022) Seed science research: Global trends in seed biology and technology. *Seeds* 1(1):1–4.

Hossain MA, Bhattacharjee S, Armin SM, Qian P, Xin W, Li HY, Burritt DJ, Fujita M, Tran Lam-Son P (2015) Hydrogen peroxide priming modulates abiotic oxidative stress tolerance: Insights from ROS detoxification and scavenging. *Front. Plant Sci.* 6:420.

Hossain MA, Bhattacharjee S, Armin SM, Qian P, Xin W, Li HY, Tran LSP (2015). Hydrogen peroxide priming modulates abiotic oxidative stress tolerance: Insights from ROS detoxification and scavenging. *Front. Plant Sci.* 6:141857.

Hussain S, Kim SH, Bahk S, Ali A, Nguyen XC, Yun DJ, Chung WS (2020) The auxin signaling repressor IAA8 promotes seed germination through down-regulation of *ABI3* transcription in *Arabidopsis. Front. Plant Sci.* 11:111.

Ishibashi Y, Tawaratsumida T, Kondo K, Kasa S, Sakamoto M, Aoki N, Zheng S-H, Yuasa T, Iwaya-Inoue M (2012) Reactive oxygen species are involved in gibberellin/abscisic acid signaling in barley aleurone cells. *Plant Physiol.* 158:1705–1714.

Jafariyan T, Zarea MJ (2016) Hydrogen peroxide affects plant growth promoting effects of *Azospirillum. J. Crop Sci. Biotechnol.* 19:167–175.

Jaskani MJ, Kwon SW, Kim DH, Abbas H (2006) Seed treatments and orientation affects germination and seedling emergence in tetraploid watermelon. *Pakistan J. Bot.* 38(1): 89–98.

Jiang A, Guo Z, Pan J, Yang Y, Zhuang Y, Zuo D, Hao C, Gao Z, Xin P, Chu J, Zhong S, Li L (2021) The PIF1-miR408-PLANTACYANIN repression cascade regulates light-dependent seed germination. *Plant Cell.* 33:1506–1529.

Jin Z, Pei Y (2015) Physiological implications of hydrogen sulfide in plants: pleasant exploration behind its unpleasant odour. *Oxid. Med. Cell. Longev.* 2015: 397502.

Kärkönen A, Dewhirst RA, Mackay CL, Fry SC (2017) Metabolites of 2, 3-diketogulonate delay peroxidase action and induce non-enzymic H_2O_2 generation: Potential roles in the plant cell wall. *Arch. Biochem. Biophys.* 620:12–22.

Karthikeyan S. (2018) Role of hydrogen peroxide (H_2O_2) and urea (NH_2-CO-NH_2) in mung bean (*Vigna radiata*) seed germination process. *IJSER* 9(7):116–119.

Kepinski S, Leyser O (2005) The *Arabidopsis* F-box protein TIR1 is an auxin receptor. *Nature* 435:446–451.

Kibinza S, Bazin J, Bailly C, Farrant JM, Corbineau F, El-Maarouf-Bouteau H (2011). Catalase is a key enzyme in seed recovery from ageing during priming. *Plant Sci.* 181:309–315.

Klein JD, Wood LA, Geneve RL (2008) Hydrogen peroxide and color sorting improves germination and vigor of eastern gamagrass (*Tripsacum dactyloides*) seeds. *Acta Hortic.* 782:93–97.

Kopyra M, Gwóźdź EA (2003) Nitric oxide stimulates seed germination and counteracts the inhibitory effect of heavy metals and salinity on root growth of *Lupinus luteus. Plant Physiol. Biochem.* 41:1011–1017.

Krasuska U, Gniazdowska A (2012). Nitric oxide and hydrogen cyanide as regulating factors of enzymatic antioxidant system in germinating apple embryos. *Acta Physiol. Plant.* 34:683–692.

Kubala S, Wojtyla Ł, Quinet M, Lechowska K, Lutts S, Garnczarska M (2015). Enhanced expression of the proline synthesis gene P5CSA in relations to seed osmopriming improvement of Brassica *napus* germination under salinity stress. *J. Plant Physiol.* 183:1–12.

Kurek K, Plitta-Michalak B, Ratajczak E (2019) Reactive oxygen species as potential drivers of the seed aging process. *Plants* 8(6):174.

Lariguet P, Ranocha P, De Meyer M, Barbier O, Penel C, Dunand C (2013) Identification of a hydrogen peroxide signalling pathway in the control of light-dependent germination in *Arabidopsis*. *Planta* 238:381–395.

Lehner A, Mamadou N, Poels P, Côme D, Bailly C, Corbineau F (2008) Changes in soluble carbohydrates, lipid peroxidation and antioxidant enzyme activities in the embryo during ageing in wheat grains. *J. Cereal Sci.* 47:555–565.

Li ZG (2013) Hydrogen sulfide: A multifunctional gaseous molecule in plants. *Rus. J. Plant Physiol.* 60:733–740.

Li ZG, He QQ (2015) Hydrogen peroxide might be a downstream signal molecule of hydrogen sulfide in seed germination of mung bean (*Vigna radiata*). *Biologia* 70:753–759.

Li ZG, Gong M, Liu P (2012) Hydrogen sulfide is a mediator in H_2O_2-induced seed germination in *Jatropha curcas*. *Acta Physiol. Plant.* 34:2207–2213.

Li JT, Qiu ZB, Zhang XW, Wang LS (2011) Exogenous hydrogen peroxide can enhance tolerance of wheat seedlings to salt stress. *Acta Physiol. Plant.*33:835–842.

Li Z, Xu J, Gao Y, Wang C, Guo G, Luo Y, Huang Y, Hu W, Sheteiwy MS, Guan Y, Hu J (2017) The synergistic priming effect of exogenous salicylic acid and H_2O_2 on chilling tolerance enhancement during maize (*Zea mays* L.) seed germination. *Front Plant Sci.* 8:1153.

Liu WZ, Kong DD, Gu XX, Gao HB, Wang JZ, Xia M, Gao Q, Tian LL, Xu ZH, Bao F, Hu Y, Ye NS, Pei ZM, He YK (2013) Cytokinins can act as suppressors of nitric oxide in *Arabidopsis*. *Proc. Natl. Acad. Sci. USA* 110:1548–1553.

Liu Y, Ye N, Liu R, Chen M, Zhang J (2010) H_2O_2 mediates the regulation of ABA catabolism and GA biosynthesis in *Arabidopsis* seed dormancy and germination. *J. Exp. Bot.* 61:2979–2990.

Lu J, Li XN, Yang YL, Jia LY, You J, Wang WR (2013) Effect of hydrogen peroxide on seedling growth and antioxidants in two wheat cultivars. *Biol. Plant.* 57(3):487–494.

Maffei ME, Mithöfer A, Boland W (2007). Insects feeding on plants: rapid signals and responses preceding the induction of phytochemical release. *Phytochemistry* 68(22–24):2946–2959.

Mandal M, Sarkar M, Khan A, Biswas M, Masi A, Rakwal R, Sarkar A (2022). Reactive Oxygen Species (ROS) and Reactive Nitrogen Species (RNS) in plants–maintenance of structural individuality and functional blend. *Adv. Redox Res.* 5:100039.

Møller IM, Jensen PE, Hansson A (2007). Oxidative modifications to cellular components in plants. *Annu. Rev. Plant Biol.* 58:459–481.

Morkunas I, Bednarski W, Kozłowska M (2004) Response of embryo axes of germinating seeds of yellow lupine to *Fusarium oxysporum. Plant Physiol. Biochem.* 42:493–499.

Morohashi Y (2002) Peroxidase activity develops in the micropylar endosperm of tomato seeds prior to radicle protrusion. *J. Exp. Bot.* 53:1643–1650.

Nandi M, Pervaz Z, Alam AS, Islam MS, Mahmud MR (2017) Effect of hydrogen peroxide treatment on health and quality of chilli seed. *Int. J. Plant Pathol.* 8:8–13.

Narimanov AA (2000) Presowing treatment of seeds with hydrogen peroxide promotes germination and development of plants. *Biológia (Bratislava)* 55(4):425–428.

Nautiyal PC, Sivasubramaniam K, Dadlani M (2023). Seed dormancy and regulation of germination. *Seed Science and Technology*, 39–66. https://doi.org/10.1007/978-981-19-5888-5_3.

Nazir F, Fariduddin Q, Khan TA (2020). Hydrogen peroxide as a signalling molecule in plants and its crosstalk with other plant growth regulators under heavy metal stress. *Chemosphere* 252:126486.

Nonogaki H, Bassel GW, Bewley JD (2010) Germination – still a mystery. *Plant Sci.* 179:574–581.

Ogawa K, Iwabuchi M (2001) A mechanism for promoting the germination of Zinnia elegans seeds by hydrogen peroxide. *Plant Cell Physiol.* 42(3):286–291.

Oracz K, El-Maarouf-Bouteau H, Kranner I, Bogatek R, Corbineau F, Bailly C (2009) The mechanisms involved in seed dormancy alleviation by hydrogen cyanide unravel the role of reactive oxygen species as key factors of cellular signaling during germination. *Plant Physiol.* 150:494–505.

Oracz K, Karpiński S (2016) Phytohormones signaling pathways and ROS involvement in seed germination. *Front. Plant Sci.* 7:864.

Orozco-Cárdenas ML, Narváez-Vásquez J, Ryan CA (2001) Hydrogen peroxide acts as a second messenger for the induction of defense genes in tomato plants in response to wounding, systemin, and methyl jasmonate. *The Plant Cell* 13(1):179–191.

Pukacka S, Ratajczak E (2007) Age-related biochemical changes during storage of beech (*Fagus sylvatica* L.) seeds. *Seed Sci. Res.* 17:45–53.

Rajashekar C, Baek K-H (2014) Hydrogen peroxide alleviates hypoxia during imbibition and germination of bean seeds (*Phaseolus vulgaris* L.). *Am. J. Plant Sci.* 5:3572–3584.

Ransy C, Vaz C, Lombès A, Bouillaud F (2020) Use of H_2O_2 to cause oxidative stress, the catalase issue. *Int. J. Mol. Sci.* 21(23):9149.

Ratajczak E, Małecka A, Bagniewska-Zadworna A, Kalemba EM (2015) The production, localization and spreading of reactive oxygen species contributes to the low vitality of long-term stored common beech (*Fagus sylvatica* L.) seeds. *J. Plant Physiol.* 174:147–156.

Reczek CR, Chandel NS (2015) ROS-dependent signal transduction. *Curr. Opin. Cell Biol.* 33:8–13.

Sano N, Marion-Poll A (2021) ABA metabolism and homeostasis in seed dormancy and germination. *Int. J. Mol. Sci.* 22(10):5069.

Santhy V, Meshram M, Wakde R, Vijaja Kumari PR (2014) Hydrogen peroxide pre-treatment for seed enhancement in cotton (*Gossypium hirstum* L.). *Afr. J. Agric. Res.* 9:1982–1989.

Sasaki K, Kishtami S, Fumitaka A, Sato T (2005) Promotion of seedling growth of seeds of rice (*Oryza sativa* L. cv. Hitomebore) by treatment with H_2O_2 before sowing. *Plant Prod. Sci.* 8(5):509–514.

Schopfer P, Plachy C, Frahry G (2001) Release of reactive oxygen intermediates (superoxide radicals, hydrogen peroxide, and hydroxyl radicals) and peroxidase in germinating radish seeds controlled by light, gibberellin, and abscisic acid. *Plant Physiol.* 125:1591–1602.

Schütz W, Rave G. (1999). The effect of cold stratification and light on the seed germination of temperate sedges (Carex) from various habitats and implications for regenerative strategies. *Plant Ecol.* 144:215–230.

Sharma P, Jha AB, Dubey RS (2019) Oxidative stress and antioxidative defense system in plants growing under abiotic stresses. In *Handbook of Plant and Crop Stress, Fourth Edition* (CRC Press) pp. 93–136.

Sharma S, Kaur P, Gaikwad K (2022) Role of cytokinins in seed development in pulses and oilseed crops: Current status and future perspective. *Front Genet.* 13: 940660.

Shu K, Liu XD, Xie Q, He Z-H (2016) Two faces of one seed: Hormonal regulation of dormancy and germination. *Mol. Plant.* 9:34–45.

Smilanick JL, Goates BJ, Denis-Arrue R, Simmons GF, Peterson GL, Henson DJ, Riji RE (1994) Germinability of Tilletia spp. after hydrogen peroxide treatment. *Plant Dis.* 78:861–865.

Smirnoff N, Arnaud D (2019) Hydrogen peroxide metabolism and functions in plants. *New Phytol.* 221(3):1197–1214.

Sneideris LC, Gavassi MA, Campos ML, D'Amico-Damião V, Carvalho RF (2015) Effects of hormonal priming on seed germination of pigeon pea under cadmium stress. *An. Acad. Bras. Cienc.* 87:1847–1852.

Springthorpe V, Penfield S (2015) Flowering time and seed dormancy control use external coincidence to generate life history strategy. *Elife* 4:e05557.

Srivastava AK, Kumar JS, Suprasanna P (2021) Seed 'primeomics': plants memorize their germination under stress. *Biol. Rev.* 96:1723–1743.

Sutherland, M. W. (1991). The generation of oxygen radicals during host plant responses to infection. *Physiol. Mol. Plant Pathol.* 39(2):79–93.

Szopińska D (2014) Effects of hydrogen peroxide treatment on the germination, vigour and health of *Zinnia elegans* seeds. *Folia Hort.* 26/1:19–29.

Tania SS, Rahaman MM, Rauf F, Suborna MA, Kabir MH, Hoque MA, Rahman MS (2021) Seed priming with salicylic acid (SA) and hydrogen peroxide (H_2O_2) improve germination and seedling growth of wheat (*Triticum aestivum*) under salt stress. *Asian J. Res. Crop Sci.* 6:60–69.

Tania SS, Rhaman MS, Rauf F, Rahaman MM, Kabir MH, Hoque MA, Murata Y (2022) Alleviation of salt-inhibited germination and seedling growth of kidney bean by seed priming and exogenous application of salicylic acid (SA) and hydrogen peroxide (H_2O_2). *Seeds* 1(2):87–98.

Thakur M, Tiwari S, Kataria S, Anand A (2022) Recent advances in seed priming strategies for enhancing planting value of vegetable seeds. *Sci. Hortic.* 305:111355.

Ventura L, Donà M, Macovei A, Carbonera D, Buttafava A, Mondoni A, Rossi G, Balestrazzi A (2012) Understanding the molecular pathways associated with seed vigor. *Plant Physiol. Biochem.* 60:196–206.

Wahid A, Parveen M, Gelani S, Basra MSA (2007) Pretreatment of seeds with H_2O_2 improves salt tolerance of wheat seedling by alleviation of oxidative damage and expression of stress proteins. *J. Plant Physiol.* 164:283–294.

Wahid A, Sehar S, Perveen M, Gelani S, Basra SMA, Farooq M (2008) Seed pre-treatment with hydrogen peroxide improves heat tolerance in maize at germination and seedling growth stages. *Seed Sci. Technol.* 36(3):633–645.

Wang H, Zhao K, Li X, Chen X, Liu W, Wang J (2020) Factors affecting seed germination and emergence of *Aegilops tauschii*. *Weed Res.* 60:171–181.

Wang HL, Wang L, Tian CY, Huang ZY (2012) Germination dimorphism in *Suaeda acuminata*: A new combination of dormancy types for heteromorphic seeds. *S. Afr. J. Bot.* 78:270–275.

Wang P, Zhu JK, Lang Z (2015a) Nitric oxide suppresses the inhibitory effect of abscisic acid on seed germination by S-nitrosylation of SnRK2 proteins. *Plant Signal. Behav.* 10:e1031939.

Wang Y, Li Y, Xue H, Pritchard HW, Wang X (2015b). Reactive oxygen species (ROS)-provoked mitochondria-dependent cell death during ageing of elm (*Ulmus pumila* L.) seeds. *Plant J.* 81:438–452.

Weiss D, Ori N (2007) Mechanisms of cross talk between gibberellin and other hormones. *Plant Physiol.* 144:1240–1246.

Wojtyla Ł, Lechowska K, Kubala S, Garnczarska M. (2016). Different modes of hydrogen peroxide action during seed germination. *Front Plant Sci.* 4;7:66.

Xia X-J, Zhou Y-H, Shi K, Zhou J, Foyer CH, Yu J-Q (2015) Interplay between reactive oxygen species and hormones in the control of plant development and stress tolerance. *J. Exp. Bot.* 66:2839–2856.

Xu FJ, Jin CW, Liu WJ, Zhang YS, Lin XY (2011) Pretreatment with H_2O_2 alleviates aluminum-induced oxidative stress in wheat seedlings. *J. Integr. Plant Biol.* 53:44–53.

Yang L-W, Liu S, Lin R (2020) The role of light in regulating seed dormancy and germination. *J. Integr. Plant Biol.* 62:1310–1326.

Ye N, Zhu G, Liu Y, Zhang A, Li Y, Liu R, Shi L, Jia L, Zhang J (2012) Ascorbic acid and reactive oxygen species are involved in the inhibition of seed germination by abscisic acid in rice seeds. *J. Exp. Bot.* 63:1809–1822.

Zhang K, Yao L, Zhang Y, Baskin JM, Baskin CC, Xiong Z, Tao J (2019) A review of the seed biology of *Paeonia* species (Paeoniaceae), with particular reference to dormancy and germination. *Planta.* 249:291–303.

Zhang Y, Chen B, Xu Z, Shi Z, Chen S, Huang X, Chen J, Wang X (2014) Involvement of reactive oxygen species in endosperm cap weakening and embryo elongation growth during lettuce seed germination. *J. Exp. Bot.* 65(12):3189–200.

Zheng C, Jiang D, Liu F, Dai T, Liu W, Jing Q, Cao W (2009) Exogenous nitric oxide improves seed germination in wheat against mitochondrial oxidative damage induced by high salinity. *Environ. Exp. Bot.* 67:222–227.

Zheng Q, Teng Z, Zhang J, Ye N (2024) ABA inhibits rice seed aging by reducing H_2O_2 accumulation in the radicle of seeds. *Plants* 13:809.

Zou JJ., Li XD, Ratnasekera D, Wang C, Liu WX, Song LF, Wu WH (2015) Arabidopsis Calcium-dependent protein kinase8 and Catalase3 function in abscisic acid-mediated signaling and H_2O_2 homeostasis in stomatal guard cells under drought stress. *Plant Cell* 27(5):1445–1460.

6 Physiological and Molecular Inferences of Plant Hydrogen Peroxide Metabolism During Stress Conditions

Sameera Karumannil, Taiba Saeed,
Mohd Tanveer Alam Khan, Mohammad Yusuf,
Sajeesh Kappachery, and Mayank Anand Gururani

INTRODUCTION

Hydrogen peroxide (H_2O_2) stands as a pivotal player in the intricate tapestry of plant biology, contributing to diverse physiological processes essential for growth, development, and responses to environmental challenges. This small, reactive oxygen species (ROS) molecule serves as both a signaling messenger and a double-edged sword, participating in fundamental cellular events and orchestrating defense strategies against various stressors. Arising primarily as a byproduct of aerobic metabolism within cellular organelles such as chloroplasts, peroxisomes, and mitochondria, H_2O_2 levels are tightly regulated to maintain redox homeostasis in plants. Its production, however, is not limited to routine metabolic activities. Environmental stimuli, ranging from biotic factors like pathogen attacks to abiotic stresses such as drought and high light intensity, can stimulate a surge in H_2O_2 levels. This surge often acts as a molecular herald, prompting plants to activate intricate defense mechanisms. H_2O_2 operates as a versatile secondary messenger in cellular signaling pathways. It plays a crucial role in modulating gene expression, influencing the activation of stress-responsive genes, and orchestrating adaptive responses. Under pathogenic assaults, H_2O_2 induces programmed cell death and hypersensitive responses, limiting the spread of pathogens within plant tissues. Moreover, H_2O_2 contributes to abiotic stress tolerance, facilitating the activation of genes that enhance resilience against environmental challenges. Its involvement in redox signaling extends to the modification of proteins, particularly redox-sensitive enzymes, and transcription factors (TFs), further shaping the molecular landscape in response to changing conditions. The delicate balance between H_2O_2 production and scavenging through antioxidant systems is paramount. Plants deploy a sophisticated arsenal of enzymes, including catalase, peroxidase (POX), and superoxide dismutase (SOD), to detoxify excess hydrogen peroxide and mitigate potential cellular damage. As research advances, unraveling the nuances of H_2O_2 involvement in plant processes becomes imperative. The knowledge gleaned holds significant promise for deciphering stress response mechanisms, enabling the development of strategies for crop improvement and fostering agricultural sustainability in the face of a changing climate (Figure 6.1).

OVERVIEW OF ABIOTIC STRESS AND ITS IMPACT ON PLANT GROWTH AND DEVELOPMENT

In contrast to animals, plants exhibit a sessile nature and continually face disruptions from unfavorable abiotic factors such as salinity, cold, heat, drought, and heavy metal stresses. These stressors

DOI: 10.1201/9781032647692-6

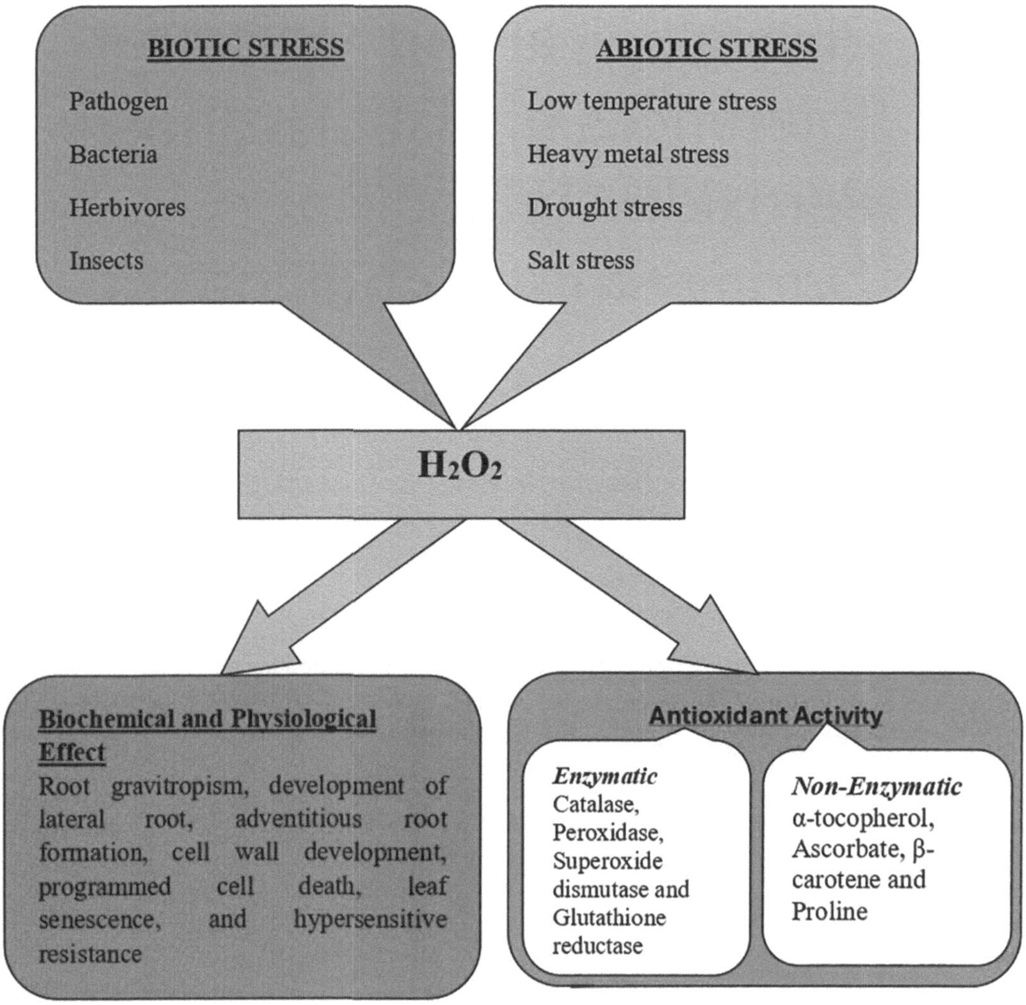

FIGURE 6.1 Central role of hydrogen peroxide in plants under biotic and abiotic stress conditions.

have profound effects on basic physiological functions, impairing photosynthesis and the absorption of nutrients and water. They also hinder general metabolic processes. Furthermore, these stresses negatively impact crucial phases of plant growth, such as germination, flowering, and seed production; all these result in reduced crop yields, compromised quality, and increased susceptibility to diseases (Radha et al., 2023; Zhang et al., 2022). Abiotic stress can induce the generation of ROS, including singlet oxygen ($^1O_2$), hydroxyl radicals ($OH^{\cdot}$), H_2O_2, and superoxide radicals ($O_2^{\cdot-}$), each with a characteristic oxidizing potential and half-life. Both positive and negative consequences can be seen when ROS generation occurs in plant cells (Noctor and Foyer, 2016; Sachdev et al., 2021). Even in standard conditions, higher plants naturally generate ROS as part of their metabolic processes. Nevertheless, excessive levels of ROS can cause apoptotic cell death due to lipid peroxidation, protein denaturation, DNA mutations, disturbance of cellular homeostasis, and various forms of oxidative harm to cells (Jira-anunkul and Pattanagul, 2021; Pandey and Shukla, 2015). To avoid the potentially harmful effects of ROS, cells constantly remove ROS through cellular antioxidant mechanisms (Mignolet-Spruyt et al., 2016). Various functions in normal plant physiological processes are fulfilled by ROS, particularly H_2O_2, contributing to the plant's capacity to withstand different stresses.

H_2O_2 AND THEIR ROLE AS PLANT HORMONES

Among the diverse forms ROS, H_2O_2 stands out as the most prevalent and comparatively long-lived molecule. Research indicates that the generation of ROS can result from both normal metabolism and stress reactions. Plants consistently produce H_2O_2 through various enzymatic and non-enzymatic processes. Usually, SOD catalyzes the reduction of superoxide radicals, which results in the formation of H_2O_2 (Petrov and Van Breusegem, 2012). Referred to as the "master hormone," H_2O_2 regulates the balance of ROS and hormones, playing a pivotal role in facilitating plant development and growth. Furthermore, it coordinates adaptive responses to enable plants to thrive amidst environmental changes (Khan et al., 2018). This potent molecule is produced in different cellular compartments: mitochondria during respiration, chloroplasts during photosynthesis, and by enzymes like apoplastic SODs and nicotinamide adenine dinucleotide phosphate (NADPH) oxidases, such as RBOHs (respiratory burst oxidase homologs). Peroxisomes also contribute significantly to H_2O_2 production, especially during photorespiration. The spatial separation of ROS generation allows precise control over H_2O_2 signaling, facilitating communication between the nucleus and organelles (Exposito-Rodriguez et al., 2017; Mhamdi, 2022; Noctor and Foyer, 2016).

In an active plant cell, the photoactive chloroplast and peroxisome system contribute to around 90% of the total H_2O_2 production. At low nanomolar concentrations, H_2O_2 acts as a signaling molecule, sharing similarities with phytohormones. Research by Slesak et al. (2007) suggests that H_2O_2 plays a crucial role in diverse physiological processes, including photosynthesis, respiration, translocation, and transpiration, owing to its exceptional stability as a ROS. This signaling role of H_2O_2 has practical implications for agriculture, as it contributes to increased crop yield and overall productivity (Ismail et al., 2015). Understanding the intricate role of H_2O_2 in plant physiology provides insights that can be harnessed for optimizing agricultural practices and crop management strategies. Currently, there is widespread acknowledgment that H_2O_2 is vital for many physiological and developmental processes in plants, including seed germination (Barba-Espín et al., 2011), senescence (Liao et al., 2012), regulation of stomatal aperture (Ge et al., 2015), root system development (Hernández-Barrera et al., 2015; Liao et al., 2009; Ma et al., 2014; Niu and Liao, 2016), flower development (Liu et al., 2013), senescence (Liao et al., 2012), and numerous other processes. Common methods for H_2O_2 application include seed priming and foliar spray due to their simplicity and cost-effectiveness.

The responses of several signaling components such as TFs, protein phosphatases, and protein kinases can be controlled by H_2O_2. Cheng and Song (2006) and Neill et al. (2002) also reported that H_2O_2 communicates with other signaling molecules, contributing to the formation of a pathway within the signaling network that governs downstream responses triggered by H_2O_2. Nevertheless, an excessive production of H_2O_2 can lead to cellular damage and eventual plant death. Plants have developed inherent antioxidant systems that utilize an array of antioxidative enzymes and antioxidants. These systems work to eliminate or detoxify these molecules, maintaining them at a baseline non-toxic level. The enzymes responsible for eliminating H_2O_2 include catalase (CAT; Willekens et al., 1997), POX (Fan and Huang, 2012), ascorbate peroxidase (APX), and glutathione reductase (GR) (Jahan and Anis, 2017). In the non-enzymatic pathway, the reduction of H_2O_2 levels is attributed to ascorbic acid (AsA) and glutathione (GSH) (Kapoor et al., 2015; Khan et al., 2018). Within plant cells engaged in photosynthesis, CAT serves as the primary enzymatic scavenger of H_2O_2, transforming it into water (H_2O) and oxygen (O_2).

ROLE OF H_2O_2 IN PLANT GROWTH AND DEVELOPMENT

In the last decade, H_2O_2 has attracted considerable attention from the scientific community among various ROS. H_2O_2 links the signaling pathways of different phytohormones and functions as a central signaling molecule. It also acts as a second messenger, responding to different stimuli that affect the growth and development of plants. It has been shown that this versatile molecule regulates a number of physiological processes, including the development of resistance, the strengthening of cell walls, the induction of senescence, the stimulation of phytoalexin production, the regulation of photosynthesis, the influence on stomatal opening, and the regulation of the cell cycle (Petrov and Van Breusegem, 2012). The concentration of H_2O_2 determines its biological effects, as do the plant's developmental stage, the production site, and prior exposure to various stresses. In plants, H_2O_2 usually has dual roles: at low concentrations, it functions as a signaling molecule, and at greater concentrations, it causes cell death (Gechev and Hille, 2005). Aquaporins play a crucial role in aiding the intracellular transport of H_2O_2 by allowing its diffusion across membranes. This ability is essential for H_2O_2 to carry out its intermediary functions within cells (Bienert et al., 2007). H_2O_2-induced cell death plays a crucial role in some developmental processes and environmental responses, including allelopathic plant–plant interactions, leaf senescence, aleurone cell death, and hypersensitive reactions to pathogenic infections.

Generally, H_2O_2 has been employed as a stimulant for initiating germination. The immersion of seeds in H_2O_2 has demonstrated enhanced germination in various plants, including zinnia, barley, almond, camphor, spinach, *Panicum virgatum*, *Sorghastrum nutans*, and *Andropogon gerardii* (Barba-Espín et al., 2011; Chien & Lin, 1994; Ogawa & Iwabuchi, 2001; Sarath et al., 2007; Zeinalabedini et al., 2009). It has been observed that several plant hormones interact with H_2O_2 during germination and the early stages of plant growth (Barba-Espin et al., 2010). The positive impact of H_2O_2 on seed germination in spinach may be attributed to its scavenging ability. This ability results in increased oxygen (O_2) production, which is essential for mitochondrial respiration and metabolic activities (Katzman et al., 2001). Another rationale is that H_2O_2 facilitates the softening of hard seeds, enhancing their interaction with water (Chien and Lin, 1994).

In plants, elevated levels of H_2O_2 not only induce the expression of defense genes but also trigger the activation of resistance genes (Mittler et al., 2004; Niu and Liao, 2016). Subsequent research has presented evidence that H_2O_2 itself acts as a central signaling molecule, coordinating a series of responses and activating various other crucial signal molecules in plants, including Ca^{2+}, abscisic acid (ABA), salicylic acid (SA), ethylene, jasmonic acid (JA), and nitric oxide (NO) (Khan et al., 2018; Nazir et al., 2021). Together, these signal molecules serve a complex function in both the growth and development of the plant and the signal transmission of resistance responses. Despite its multifunctionality, the potential danger associated with elevated concentrations necessitates stringent control over H_2O_2 levels in plant cells (Petrov and Van Breusegem, 2012).

ROLE OF H_2O_2 IN PROMOTING CELL ELONGATION AND DIVISION

H_2O_2 is actively involved in resistance mechanisms, including the strengthening of plant cell walls through processes such as lignification and the cross-linking of cell wall structural proteins, as well as the production and enhancement of resistance through phytoalexins (Dempsey and Klessig, 1995). H_2O_2 in grapes can influence cell wall loosening during phase II to regulate cell growth or contribute to cellular development during the first cell division in phase I (Geros et al., 2012).

The use of a luminescence probe allows the observation of H_2O_2 accumulation in germinating radish seeds. Moreover, H_2O_2 has the ability to regulate the plant cell cycle. In the case of tobacco treated with a fungal elicitor, the production of H_2O_2 and activation of mitogen-activated protein kinase (MAPK) protein were observed (Suzuki et al., 1999). This key signal protein regulates the cell cycle, and its orthologous proteins, MAPKKK (mitogen-activated protein kinase kinase kinase), ANP1 (*Arabidopsis* NPK1-like protein kinase 1), and NPK1 (nucleus- and phragmoplast-localized protein kinase 1), are linked to the processes that occur between H_2O_2 and the cell cycle (Suzuki et al., 1999).

H_2O_2 AND ROOT DEVELOPMENT

Among the array of signaling molecules, notably H_2O_2 exhibits diverse functions in the regulation of various aspects of root development. This includes the maintenance of meristem, elongation of roots, branching of lateral roots, differentiation of vascular tissue, formation of endodermis and adventitious rooting. According to Xiong et al. (2015), the application of H_2O_2 promotes expansion of cells and an increase in root diameter; however, it hinders the elongation of roots. Moreover, the administration of H_2O_2 extends the area where pectin synthesis occurs in the tips of roots, elevates pectin levels, and boosts the activity of pectin methyl esterase (PME), thereby facilitating the process of pectin demethylesterification.

The assumption that complex interactions among different signaling molecules lead to changes in the regulation of root morphology and root architecture is highly supported by current experimental evidence. Furthermore, the regulation of these signaling molecules is complex under both normal and stressful conditions (Eljebbawi et al., 2021). Precise control of H_2O_2 levels is especially vital within the meristematic and elongation zones of developing root tips. In these areas, environmental factors like nutrient stress, alterations in pH, and various challenges exert an impact on cell expansion and division in the vicinity of the quiescent center (Mukherjee and Corpas, 2023).

Joo et al. (2001) reported that auxin-induced ROS could function as a downstream element in auxin-related signal transmission, playing a role in the control of root gravitropism. Several investigations suggest the presence of a signaling pathway involving ROS that impacts the development of root hairs in *Arabidopsis*. This pathway includes different enzymes like OXI1 kinase and NADPH oxidase, as indicated by studies conducted by Foreman et al. (2003) and Rentel et al. (2004). The accumulation of H_2O_2 at the apex of developing *Arabidopsis* roots is associated with constraining growth and the production of root hairs (Dunand et al., 2007). Furthermore, H_2O_2 is presumed to have a notable impact on the development of lateral roots in soybean (Su et al., 2016), and on the initiation and growth of adventitious roots in *Cucumis sativus* (Li et al., 2007). Moreover, several lines of evidence suggest that the application of H_2O_2 and NO can trigger the formation of adventitious roots in plants. Independent and synergistic treatments with H_2O_2 and NO were found to boost the development of adventitious roots in marigold (Liao et al., 2009). This enhancement was achieved by promoting the activities of enzymes such as PPO (polyphenol oxidase) and IAAO (indole-3-acetic acid oxidase), increasing carbohydrate and nitrogen content, and concomitantly suppressing polyphenol production in chrysanthemum (Liao et al., 2010). Auxin, a key phytohormone in regulating root growth, is widely acknowledged for its role in modulating H_2O_2 levels. This modulation, in turn, facilitates cell growth and the formation of lateral roots (Qu et al., 2017; Su et al., 2016; Takáč et al., 2016). The characterization of the gene *OsNOX3*, responsible for encoding NADPH oxidase and producing apoplastic superoxide ions (O_2^-), unveiled its involvement in the regulation of both the initiation and elongation of root hairs in rice (Wang et al., 2018a). Nevertheless, a recent study revealed that H_2O_2 facilitates the initiation of lateral roots downstream of an auxin-like indoleamine molecule, known as melatonin (Chen et al., 2018). Additionally, H_2O_2 has been identified as influencing root tip growth. Within cells, superoxide ions (O_2^-) quickly transform into H_2O_2, either spontaneously or through the action of SOD. In rice, the POX gene *PRX102* is crucial for the growth of root hair tips, as it regulates the balance of H_2O_2 (Moon et al., 2023).

H_2O_2 AND PHOTOSYNTHESIS

Plants primarily utilize photosynthesis as the key process for assimilating carbon. Enhancing the efficiency or overall capacity of photosynthesis, which postpones leaf senescence, results in an increased production of photoassimilates (Van Camp, 2005). Significant research has been conducted to showcase the impact of H_2O_2 on the net photosynthetic rate and related characteristics in plants (Fariduddin et al., 2014; Hasan et al., 2016; Khan et al., 2015). Additionally, under field conditions, Khandaker et al. (2012) observed increased photosynthetic rates and higher dry matter content in wax apple leaves following the external application of H_2O_2. Several studies have explored the impact of H_2O_2 on photosynthesis, yielding diverse observations. The enhanced photosynthetic efficiency attributed to H_2O_2 treatment may be linked to an increase in the maximum carboxylation rate of Rubisco (V_{cmax}) and initial Rubisco activity. Remarkably, H_2O_2 primarily controls the activation state of Rubisco, potentially through direct interaction with the Rubisco enzyme, as demonstrated by Jiang et al. (2012). Moreover, H_2O_2 treatments noticeably enhance the photochemical efficiency of PSII (ΦPSII), mainly by increasing the photochemical quenching coefficient (qP) rather than altering the F_v/F_m ratio, the maximum quantum yield of PSII. The need for ATP and NADPH in the Calvin cycle is thought to be the cause of rise in qP that results from H_2O_2 inducing PSII (Nogués and Baker, 2000). Consequently, the observed rise in qP might be explained by the increased activity of the Calvin cycle brought on by H_2O_2.

H_2O_2 AND REPRODUCTIVE DEVELOPMENT

The process of pollen tube germination, growth, and guidance within pistil tissues is highly regulated and involves a continuous exchange of both physical and chemical signals. The process of pollen tube germination, growth, and guidance within pistil tissues is highly regulated and involves a continuous exchange of both physical and chemical signals (Chae and Lord, 2011). H_2O_2 and other ROS play a crucial role in guiding pollen and facilitating the fusion of gametophytes in generatively propagating plants. Studies have demonstrated that receptive stigmas from various species accumulate elevated levels of H_2O_2, which subsequently decrease as the stigmas support pollen development (Serrano et al., 2012, 2015). Various functions have been theorized for H_2O_2 in the stigma, including loosening cell-wall components to facilitate pollen tube penetration, providing defense against pathogens, and contributing to pollen-pistil recognition. It has been proposed that the decrease in H_2O_2 on the stigma is triggered by NO produced by the pollen (Serrano et al., 2012). The accumulation of ROS, particularly the hydroxyl radical primarily derived from H_2O_2, is essential for the rupture of the pollen tube and the subsequent release of sperm cells (Duan et al., 2014).

H_2O_2 AND FLOWERING TIME

Arabidopsis thaliana plants that overexpress glycolate oxidase in their chloroplasts experience the accumulation of both H_2O_2 and glyoxylate, leading to growth retardation and delayed flowering. The increased levels of H_2O_2 in these plants influence the expression of TFs that regulate both plant growth and flowering (Fahnenstich et al., 2008). In the "millet stage," litchi plants exposed to a chilling-induced blooming treatment have higher levels of NO and H_2O_2 (Zhou et al., 2012). The increased amounts of H_2O_2 and NO inhibit the growth of primitive leaves and stimulate the expression of *LcLFY*, the litchi *LEAFY* homolog, which in turn promotes reproductive growth. A recently proposed mechanism provides insight into a novel way in which H_2O_2 governs the development of the shoot apical meristem (SAM). This mechanism involves the interaction between TMF (TERMINATING FLOWER), a crucial TF which regulates the timing of flowering transition in tomatoes, and the floral identity gene *ANANTHA* (Huang et al., 2021). *TMF* accomplishes this by preventing SAM from pre-maturing. Disulfide linkages that bind many *TMF* molecules together are formed when cysteine residues in *TMF* detect the redox state of the cell. The phase separation of *TMF* is accelerated by this process, which also results in an increase in intrinsically disordered regions. To efficiently inhibit the expression of the floral identity gene *ANANTHA*, *TMF* binds to and sequesters its promoter by oxidation-triggered phase separation (Foyer, 2021; Huang et al., 2021).

H_2O_2 AND CROP PRODUCTIVITY

Utilizing H_2O_2 presents a simple, cost-efficient, and practical approach to mitigate abiotic stress in plants. This method enhances various plant attributes, including yield, photosynthetic efficiency, relative water content (RWC), membrane stability index, and overall plant growth (Hajivar and Zare-Bavani, 2019). Applying a 5 mM H_2O_2 treatment to wax apple fruit resulted in an augmentation of size, set, number, biomass, and yield compared to the control (Khandaker et al., 2012). There is compelling evidence indicating that H_2O_2 application contributes to increased yield in numerous plants. Orabi_et al. (2018) observed that spraying canola plants with H_2O_2 increased the activity of antioxidant enzymes like CAT and APX while decreasing lipid peroxidation. Onion yield was greatly increased by using 1–2 mmol/L H_2O_2 (Semida, 2016). Therefore, it is proposed that H_2O_2 induces physiological changes in plants that let them continue growing even in adverse environments, leading to higher yields. It is also important to note that applying H_2O_2 foliar spray dramatically lowers the quantity and weight of unfilled grains, an effect more pronounced than seed priming.

H_2O_2 AS REGULATORS OF ABIOTIC STRESS RESPONSES

It is commonly acknowledged that oxidative stress plays a major role in cellular damage. Plants have evolved regulatory mechanisms to adapt to a variety of environmental challenges in response to stress conditions. Stress causes a rise in ROS concentration in cells, including H_2O_2. Several studies consider that H_2O_2 is essential for acclimatization and cross-tolerance (Neill et al., 2002; Saxena et al., 2016). According to Hossain et al. (2015), plants are protected against abiotic stressors by applying H_2O_2 either concurrently or as a pretreatment for seeds or seedlings. This safeguarding mechanism operates by reinstating redox balance, relieving oxidative harm to membranes, proteins, and lipids, and adjusting stress-signaling pathways. In the regulation of plant development and stress tolerance, there is a growing focus on ROS generation and its interplay with different hormonal signaling pathways. The principal phytohormones coordinating responses of plants to various stresses include jasmonates, ABA, SA, and ethylene. H_2O_2 is used by each of these phytohormones in either an upstream or downstream signaling cascade (Xia et al., 2015). However, the crucial role of H_2O_2 as a potent regulator in signaling pathways is widely recognized whenever a plant undergoes stress (Černý et al., 2018; Luna, 2004). Consequently, concentrations of H_2O_2 are frequently employed as an indicative measure of oxidative stress occurrence in plants exposed to stressful conditions. External application of H_2O_2 has been documented to provide stress tolerance in various crops (Sohag et al., 2020). Several reports have investigated the effects of H_2O_2 on various abiotic stresses, but there is variability in the observations among these studies. It has been suggested that changes in ROS-detoxification mechanisms or the control of several stress-responsive pathways could be the source of the increased stress tolerance shown in plants that have received external H_2O_2 treatment (Hossain et al., 2015; Khan et al., 2018; Terzi et al., 2014). Previous research has demonstrated that H_2O_2 attenuates the deleterious effects of stress by controlling stomatal closure (Kolla et al., 2007; Quan et al., 2008; Wang and Song, 2008) or by stimulating the production of oligosaccharides, which preserve the water content of leaves (Ishibashi et al., 2011). In general, when plants face stress conditions such as ultraviolet radiation, salt, drought, light fluctuations, metal exposure, or extreme temperatures, the production of H_2O_2 induces resistance to multiple stresses, safeguarding plants from damage.

H_2O_2 IN DROUGHT STRESS RESPONSES

Drought represents a critical abiotic stress factor that globally restricts crop productivity, and its impact is escalating with climate change. Regardless of the plant's growth stage, drought imposes a detrimental influence, significantly reducing yield at every developmental phase (Oguz et al., 2022). However, ROS generation in cellular compartments such as peroxisomes, mitochondria, and chloroplasts is known to increase during drought stress. But during times of water scarcity, plants have developed adaptation mechanisms that allow them to survive and even thrive. Drought stress causes plants to respond physiologically and biochemically through the activation of several genes, each of which serves a different role. Drought stress affects a number of physiological functions, such as growth, photosynthesis, enzyme activity, ROS generation, and membrane integrity (Farooq et al., 2009; Li et al., 2011).

Among ROS, H_2O_2 play pivotal roles in various developmental stages and has been employed to enhance plants' tolerance against drought. Zhongjing et al. (2009) explored the impact of H_2O_2 priming on enhancing drought tolerance in cucumber growth. Drought stress altered chloroplast morphology, but H_2O_2 priming increased the activity of antioxidant enzymes (SOD, CAT, GPOX, APX, DHAR, GR), along with elevated levels of GSH and AsA. Interestingly, it had no effect on chloroplast structure. Furthermore, it resulted in reduced levels of malondialdehyde (MDA), H_2O_2, and superoxide anion ($O^{\bullet -2}$), suggesting that H_2O_2 priming mitigated the accumulation of ROS and alleviated chloroplast membrane damage induced by drought stress. In a comparable study, Ishibashi et al. (2011) discovered that treating soybean plants with H_2O_2 through spraying alleviated symptoms of drought stress. The leaves that underwent treatment showed elevated levels of RWC, photosynthetic rate, and stomatal conductance in comparison to those sprayed with distilled water. In contrast to the spray of distilled water, the application of H_2O_2 resulted in an upregulation of genes related to oligosaccharide synthesis. This implies that H_2O_2 plays a role in preserving leaf water levels in soybean plants experiencing drought stress. Hossain et al. (2013) investigated how H_2O_2 priming improves drought tolerance in mustard seedlings. The study found that pre-treating seedlings with H_2O_2 reduced MDA content, stabilized cell membranes, and lowered endogenous H_2O_2 levels. Additionally, antioxidant enzyme activities and glyoxalase pathways were better maintained in H_2O_2 pretreated seedlings during drought stress, indicating enhanced tolerance through the activation of both ROS and methylglyoxal detoxification pathways.

In wheat, the application of H_2O_2 as a priming agent has been shown to enhance growth, increase water potential, accumulate osmolytes, strengthen the antioxidant system, and improve yield in drought-stressed environments (Habib et al., 2020). Similarly, the foliar application of low concentrations of H_2O_2 has been found to mitigate yield loss in cotton by enhancing antioxidant enzyme activities (Sarwar et al., 2018). The beneficial effects of H_2O_2 have also been documented in wheat, where 1,000-grain weight, grains per spike, and certain physiological traits showed improvement under drought stress conditions (Farooq et al., 2017). Additionally, it has been demonstrated that the performance of quinoa can be enhanced through the exogenous application of H_2O_2. Exogenous H_2O_2 not only improved emergence but also conferred tolerance to impending drought stress. In addition, compared to untreated plants under drought stress, the administration of exogenous H_2O_2 resulted in increased stomatal conductance, higher levels of proline and total soluble sugars, as well as elevated transpiration and photosynthetic rates. Applying H_2O_2 to drought-stressed plants resulted in a 30% decrease in the content of ABA as well as an increase in the activities of antioxidant enzymes such as SOD, POD, CAT, and APX (Iqbal et al., 2018). Moreover, seed priming, using H_2O_2 through foliar application, proves advantageous in increasing crop yield and minimizing yield loss by mitigating the effects of drought stress in rice during its reproductive stage (Jira-anunkul and Pattanagul, 2021). The exploration of H_2O_2 pretreatment on seeds for enhancing seedling growth and fortifying defense responses in three wheat cultivars under drought stress has revealed that such pretreatment led to the increased expression of proteins associated with defense mechanisms, redox balance, and photosynthesis (Bhardwaj et al., 2021). These findings illustrate that H_2O_2 pretreatment enhances tolerance to osmotic stress by elevating levels of soluble sugars, proline, and polyamines. Reis et al. (2022) examined the role of H_2O_2 in regulating root-to-shoot communication under drought and found that elevated H_2O_2 levels in the root system serve as a long-distance signal, resulting in reduced transpiration even in the absence of water limitation in the shoot. However, its impact is limited when there is a decrease in the shoot water potential.

H_2O_2 IN SALINITY STRESS RESPONSES

The presence of salinity in soil or water poses a significant challenge to the growth and productivity of numerous crops, particularly in arid and semi-arid conditions. Plant growth and development are adversely affected by salinity for a variety of reasons, including low soil solution osmotic potential (which causes moisture stress), imbalances in nutrition, specific ion interactions (which cause salt stress), or a combination of these (Ashraf, 2004). However, currently H_2O_2 priming has emerged as a potentially effective method for reducing the negative effects of salt. Studies on maize (de Azevedo Neto et al., 2005; Gondim et al., 2013), strawberries (Christou et al., 2014), and rice (Carvalho et al., 2011) have reported that applying H_2O_2 externally improves the plants' ability to withstand salt. However, Uchida et al. (2002) investigated the impact of low-concentration pretreatments (<10 μM) of NO and H_2O_2 on oxidative stress in rice plants exposed to heat or salt. Their results revealed that, when subjected to stress, seedlings treated with H_2O_2, or NO exhibited greener leaves and higher photosynthetic activity compared to control plants. This suggests that NO and H_2O_2 serve as signaling molecules in regulating tolerance to heat and salt stress. Additionally, the pretreatments increased the expression of stress-related genes and the activities of ROS-scavenging enzymes. de Azevedo Neto et al. (2005) studied the influence of externally applied H_2O_2 on salt stress acclimation in a salt-sensitive maize genotype, specifically examining plant growth, lipid peroxidation, and the activity of antioxidative enzymes in leaves and roots. The research indicated that pre-treating with 1 μM H_2O_2 for 2 days in the hydroponic solution enhanced salt tolerance when the plants were subsequently exposed to salt stress. Wahid et al. (2007) observed that salinity tolerance in *Triticum aestivum* was enhanced by exogenous H_2O_2 when seeds were immersed in H_2O_2 solutions and then cultivated in saline conditions. Furthermore, it enhanced membrane properties by reducing ion leakage and relative membrane permeability (RMP), leading to the production of two heat-stable proteins in the pretreated seedlings.

Gondim et al. (2013) conducted three studies to investigate the influence of H_2O_2 on maize seedlings' growth and acclimation under salinity stress. The first study showed that H_2O_2 accelerated seed germination at 100 mM. In the second study, seed pretreatment with H_2O_2 upregulated APX and CAT activities, while GPX activity was lower. The third experiment revealed that seed pretreatment with H_2O_2 enhanced seedling tolerance to salinity, improving growth rates, potentially linked to variations in antioxidant enzyme activities. Subsequently, Ashfaque et al. (2014) observed a positive impact on wheat plant growth with the application of H_2O_2, both in saline and non-saline conditions. Specifically, 50 or 100 μM H_2O_2 was found to decrease the severity of salt stress. This reduction was associated with lower Na^+ and Cl^- ion levels, increased proline content, improved N absorption, better water relations, higher levels of photosynthetic pigments, and faster growth rates. H_2O_2 application also demonstrated positive effects on these parameters under non-saline conditions. Furthermore, Silva et al. (2019) assessed the impact of H_2O_2 application on salt stress tolerance and essential oil yield in basil. The findings suggest that introducing H_2O_2 into the nutrient solution in hydroponic systems with brackish water is a practical approach for cultivating basil, suitable for both fresh consumption and essential oil production in such regions. In a different study assessing the positive impact of H_2O_2 pretreatment on mitigating seawater stress in wheat cultivars, salt-sensitive Gemmiza 11 and salt-tolerant Misr 1, it was found that under normal conditions, H_2O_2 eustress significantly enhanced plant growth parameters in both cultivars. Furthermore, under salt stress, seeds subjected to H_2O_2 eustress exhibited reduced damage to plant growth and physiological parameters, accompanied by increased pigment content, proline levels, mineral uptake, and decreased Na^+ content and lipid peroxidation (Abdel Latef et al., 2019). According to Migahid et al. (2019), H_2O_2 supplementation and magnetic biostimulation efficiently scavenged and mitigated the harmful effects of salt stress on *Silybum marianum* seedlings at the early stage of development.

H_2O_2 IN EXTREME TEMPERATURE STRESS RESPONSES

High/low temperature is a major stress that affects the defensive system, chlorophyll levels, photosynthetic apparatus, antioxidant enzymes, and yield of crop plants. Heat-stressed plants exhibit a rise in endogenous H_2O_2 levels, like many abiotic stresses, and exogenous pretreatments have been found to be beneficial in boosting plant heat tolerance (Hossain et al., 2013; Mostofa et al., 2013). Plants that are subjected to heat stress accumulate ROS, including H_2O_2. These ROS can cause oxidative stress by altering the nature of membranes, breaking down proteins, and deactivating enzymes. Ultimately, this can reduce the viability of plant cells (Hasanuzzaman et al., 2013). It was shown that by reducing oxidative damage, foliar application of H_2O_2 in *Agrostis stolonifera* induced heat stress tolerance (Larkindale and Huang, 2004). According to research by Kang et al. (2009), the foliar application of H_2O_2 in cucumber and tomato seedlings induced thermotolerance and activated antioxidant enzymes, reducing electrolyte leakage and MDA content while increasing the amount of water-soluble protein and chlorophyll in heat-stressed seedlings. Superoxide anions and H_2O_2 levels were also lowered by the treatment. APX and glucose-6-phosphate dehydrogenase (G6PDH) specific activities, as well as the gel activities of several antioxidant enzymes, were significantly elevated, indicating that foliar application of H_2O_2 could be a useful strategy for mitigating heat stress in crops. Though, when comparing the salt-sensitive rice cultivar Ratna with the salt-resistant SR 26B, Ratna exhibited superior responses to H_2O_2-mediated acclamatory performances and the reestablishment of redox balance under severe temperatures. These results underscore the importance of the H_2O_2 'inductive pulse' in restoring redox homeostasis and mitigating oxidative damage to membrane proteins and lipids during the post-germination recovery period. This process is essential for adapting to unfavorable temperature stress (Bhattacharjee, 2012).

Wang et al. (2014) examined the benefits of exogenous H_2O_2 on improving heat stress tolerance in turfgrass species, particularly ryegrass and tall fescue. Before heat stress, seedlings that had received a pretreatment of H_2O_2 showed higher activity of CAT, POD, GR, APX, and GPX, greater pools of GSH and AsA, and a lower GSH/GSSG ratio. These pretreated seedlings underwent heat stress and showed reduced levels of oxidative damage and H_2O_2, as well as increased activity of APX, GR, GST, and GPX. These findings suggest that H_2O_2 may strengthen antioxidant defense systems and improve thermotolerance in turfgrass species. Applying H_2O_2 to leaves during heat stress significantly elevated SOD and CAT contents, decreased MDA contents, and induced thermotolerance by capturing ROS. Moreover, studies revealed that pre-treating cucumber leaves with H_2O_2 lessened the effects of heat stress, enhanced growth, and reduced oxidative stress by safeguarding DNA structure through heat-induced alterations in DNA methylation (Gao et al., 2010; Hossain et al., 2015).

In comparison to untreated seedlings, Shu et al. (2007) reported that pre-treating cold-sensitive *Vigna radiata* seedlings with H_2O_2 improved chilling tolerance and reduced electrolyte leakage. They observed that the H_2O_2-induced chilling tolerance was linked to GSH. Similarly, Işeri et al. (2013) reported that during brief periods of cold stress, tomato roots maintained a high RWC when pretreated with a 1 mmol/L H_2O_2 solution for an hour. Additionally, compared to tomato plants that were not treated with H_2O_2, H_2O_2-treated plants showed noticeably greater amounts of anthocyanin in their leaves. According to Orabi et al. (2015), the two tomato cultivars grown in a sand-ponic culture at low temperatures showed significantly positive responses to all treatments involving H_2O_2 and SA (0.5 mM and 1.0 mM). These responses also included GA3, IAA, and ABA growth regulators; POD and PPO antioxidant enzyme activity; and the quantity and quality of fruit yield. Nevertheless, EL levels, MDA, and CAT enzyme activity all significantly decreased because of these treatments. Notably, the 1.0 mM H_2O_2 treatment exhibited the most pronounced effects. High temperatures during cotton crop reproductive stages result in a significant yield reduction and decreased fiber quality. This is attributed to fewer sympodial branches and reduced boll weight, linked to impaired leaf photosynthesis in heat-stressed plants. Improvement in corn seed germination and seedling growth under chilling stress was notably achieved by priming seeds with a combination of SA and H_2O_2. This enhancement was significantly associated with ROS, energy supply, metabolites and the regulation of hormone metabolism (Li et al., 2017). Further, high temperatures during cotton crop reproductive stages result in a significant yield reduction and decreased fiber quality. This is attributed to fewer sympodial branches and reduced boll weight, linked to impaired leaf photosynthesis in heat-stressed plants. Growth regulators, particularly H_2O_2, mitigate heat-induced damage by enhancing SOD and CAT antioxidant enzyme activities and improving leaf chlorophyll and photosynthetic efficiency under medium and high-temperature conditions

(Sarwar et al., 2018). Exposing Phalaenopsis seedlings to low temperatures increased the activities of GR, CAT and APX with fluctuations. However, H_2O_2 pretreatments suppressed CAT and APX at a lower level. Remarkably, the GSH-to-oxidized GSH ratio was reduced by the elevated GR activity, peaking at 4–8 hours and on day 2 during chilling. This implies that exogenous H_2O_2 improves Phalaenopsis seedlings' ability to withstand cold, possibly via modifying the GSH-related oxidant defense system (Chen and Ko, 2021).

H_2O_2 IN HEAVY METAL STRESS RESPONSES

Contamination of soils, air and groundwater with toxic metals poses a significant challenge in our industrialized world. Several studies have shown that when plants are exposed to various metals, it triggers oxidative stress. This stress is marked by an imbalance between the production and neutralization of ROS, including H_2O_2 (Schützendübel and Polle, 2002; Sharma and Dietz, 2009). It is also reported that metals can impede the activity of enzymes by attaching themselves to functional groups or by relocating necessary cations in particular binding sites. This interference can disrupt the function of enzymes crucial for antioxidative defense, as well as physiological processes like respiration and photosynthesis, leading to increased ROS production like H_2O_2 (Cuypers et al., 2016). However, studies show that pre-treating plants with H_2O_2, known as H_2O_2 priming, can alleviate adverse effects from factors like salinity, drought, heat, chilling, and metals (Hossain et al., 2015). Pretreatment with H_2O_2 at concentrations between 100 and 500 µM prior to metal exposure has been shown to mitigate metal-induced effects, including chlorophyll degradation, lipid peroxidation, reduced growth and programmed cell death, in a variety of plant species (Chao et al., 2009; Xu et al., 2011; Yıldız et al., 2013). In a hydroponic experiment examining the impact of H_2O_2 on cadmium (Cd) tolerance in rice seedlings, exposure to Cd significantly inhibited plant growth. However, pretreatment with 100 µM H_2O_2 for 1 day alleviated Cd stress by enhancing the activities of antioxidative and detoxification enzymes and increasing the levels of reduced GSH and AsA. As a result, there was less restricted seedling growth due to a decrease in H_2O_2 and MDA concentration (Hu et al., 2009). Moreover, Xu et al. (2011) found that H_2O_2-induced overexpression of AsA and GSH metabolism in wheat seedlings increased tolerance to aluminum (Al). Al stress increased superoxide anion and H_2O_2 levels elevated programmed cell death, MDA levels, and impeded root growth in resistant and susceptible rice genotypes. In comparison to seedlings exposed to Al stress alone, H_2O_2-primed, Al-stressed seedlings displayed even higher activity of MDHAR, DHAR, GPX, CAT, POD, and GR, along with raised GSH and AsA levels and a more favorable redox state. Yıldız et al. (2013) examined the reducing effects of H_2O_2 on the toxicity of Cr in canola plants, taking into account the levels of metallothionein protein (BnMP1), growth, chlorophyll content, lipid peroxidation, and thiol content. Although SOD and CAT activities remained unchanged, there was an increase in POD and SOD activities under Cr stress. Furthermore, H_2O_2 pretreatment elevated APX and POD activities in response to Cr stress. Additionally, under Cu stress, Guzel and Terzi (2013) investigated the effects of exogenous H_2O_2 on maize plant development, water balance, proline concentration, mineral ion content, and the amounts of soluble protein and total sugar in leaves. In comparison to control plants, the H_2O_2-pretreated seedlings showed improved growth, higher water content, higher mineral concentration, higher proline content, and higher levels of total soluble sugar and soluble protein. According to Nazir et al. (2019), the application of H_2O_2 boosted photosynthesis, growth, and the metabolic state of the plants, which helped the tomato plants tolerate Cu stress better. Brassinosteroids and H_2O_2, applied through different mechanisms, enhanced photosynthetic efficiency, altered chloroplast ultrastructure, influenced stomatal behavior, modified root structure, increased cell viability, and stimulated the production of antioxidants and proline. These changes collectively improved the resilience of tomato plants to Cu stress (Nazir et al., 2021).

CROSSTALK BETWEEN H_2O_2 AND OTHER HORMONAL PATHWAYS

The crosstalk between H_2O_2 signaling and other hormonal pathways is a fascinating aspect of plant biology that underscores the complexity and integration of various signaling networks. Plants utilize a multitude of hormones to regulate growth, development, and responses to environmental stimuli, and H_2O_2 emerges as a key player in modulating and integrating these hormonal pathways. Here, we explore the intricate crosstalk between H_2O_2 and some major hormonal pathways in plants:

H_2O_2 AND ABSCISIC ACID

ABA plays a regulatory role in multiple plant processes, including leaf growth, bud and seed dormancy, stomatal behavior, and water transport, exerting its influence on aquaporins. It is crucial for responding to drought and salinity. ABA signaling involves interaction with H_2O_2 through MAPK cascades, where H_2O_2 serves as a significant secondary transmitter in ABA transduction (Miao et al., 2006). In *Vicia faba*, ABA-mediated stomatal behavior triggers plasma membrane calcium channels through H_2O_2 accumulation (Zhang et al., 2001). Essential catalytic subunit genes Atrboh D and Atrboh F, embedded in the plasma membrane, are crucial for ABA-triggered stomatal ROS production in plants like *Vicia faba* (Kwak et al., 2003). NADPH oxidase, implicated in ABA-triggered stomatal ROS, can also provoke Ca^{2+} and MAPK signaling routes and initiate hormone signal attenuation pathways (Overmyer et al., 2003). Abiotic stress factors modulate the ABA/gibberellic acid (GA) ratio, impacting DELLA protein output and H_2O_2 concentration (Considine and Foyer, 2014). In *Zea mays* leaves subjected to drought stress, both ABA and H_2O_2 concentrations increase. Exogenous ABA treatment enhances *OsRab16* A expression (an ABA-responsive gene) and accelerates H_2O_2 generation in rice seedlings grown in potassium-rich soils (Liu et al., 2012). NADPH oxidase is confirmed as the enzyme responsible for H_2O_2 production in rice seedlings (Tsai and Kao, 2004). In pumpkin-grafted cucumber leaves, induced ABA accumulation triggers H_2O_2 production, leading to the upregulation of antioxidant defense system genes (SOD, POD, and cAPX) (Shu et al., 2016). ABA enrichment in tobacco activates the 9-cis epoxycarotenoid dioxygenase gene (*SgNCED1*), enhancing H_2O_2 and NO in cells and leading to the activation of various antioxidant enzymes (Zhang et al., 2009). Elevated H_2O_2 influences ABA output, stimulating NADPH oxidase and contributing to prolonged stress resilience. The interplay between H_2O_2 and ABA regulates freezing tolerance in wheat plants triggered by SA. Elevated H_2O_2 and ABA independently activate protective mechanisms, enhancing cold-responsive gene expression and antioxidant enzyme activity to improve photosynthesis efficiency in wheat plants exposed to freezing stress (Wang et al., 2018b).

Ishibashi et al. (2017) explored the relationship between H_2O_2 and ABA in dormant and non-dormant barley seeds. Non-dormant seeds exhibited a higher H_2O_2 concentration after imbibition, with H_2O_2 controlling ABA content through an ABA-80-hydroxylase catabolic enzyme. NADPH oxidase activity producing ROS was smaller in dormant seeds, while catalase activity was greater, controlled by the ABA-responsive TF HvABI5. This balance between ABA and ROS in barley seeds influences dormancy and germination. In Arabidopsis, exogenous inhibitors of H_2O_2 and NADPH oxidase increased ABA catabolism and activated genes for GA synthesis in dormant seeds (Liu et al., 2010b). In *C. sativus*, H_2O_2 is implicated in ABA-modulated development of adventitious roots, influencing various physiological parameters (Li et al., 2016). In conclusion, the crosstalk between ABA and H_2O_2 signaling pathways in plants involves shared components, such as TFs and protein kinases, leading to coordinated responses to environmental stress. This intricate interplay allows plants to integrate signals from ABA and H_2O_2 to fine-tune their adaptive responses for survival under challenging conditions.

H_2O_2 AND BRASSINOSTEROIDS

The positive modulation of brassinosteroids (BRs) signaling by H_2O_2 involves the activation of the TF BRASSINAZOLE-RESISTANT1 (BZR1), a pivotal promoter of BR signaling (Ding, 2018; Tian et al., 2018). Within the root stem cell niche, externally applied BR boosts H_2O_2 production, impacting BR-mediated QC division and cell elasticity (Tian et al., 2018). Enhanced NADPH oxidase activity results in H_2O_2 accumulation in cucumber plants treated with BR, while elevated RBOH1 activity achieves a similar outcome in tomatoes (Nie et al., 2013). The silencing of RBOH in *Nicotiana benthamiana* disrupts AOX activity triggered by BR, diminishing ROS detoxification, and rendering the plant more vulnerable to abiotic factors (Deng et al., 2015). However, H_2O_2 functions as a secondary agent in carbohydrate metabolism and CO_2 assimilation induced by BRs through redox signaling in *C. sativus*, enhancing photosynthetic efficiency and productivity (Jiang et al., 2012). The solubility and concentrations of H_2O_2 play a critical role in regulating stomatal movement in response to varying BR doses in Solanum lycopersicum (Xia et al., 2014). A transient increase in H_2O_2 resulting from a low BR dose serves as a catalyst for the regeneration and/or biosynthesis of GSH, leading to a decreased redox status that selectively regulates H_2O_2 levels, adversely affecting the ABA-responsive mechanism. Conversely, a high BR concentration can induce an excessive rise in H_2O_2, potentially forming a synergistic amplification loop with ABA signaling, resulting in stomatal closure (Xia et al., 2014).

HYDROGEN PEROXIDE AND CYTOKININ

Cytokinins (CKs) are linked to a variety of physiological, cognitive, and metabolic functions in plants, encompassing cell division, leaf senescence, and the facilitation of light-regulated plant development (Chang et al., 2016). H_2O_2 plays a pivotal role in maintaining the optimal CK concentration necessary for plant development, amplifying cytokinin oxidase and gene expression (Causin et al., 2009). In wheat, the collaborative interplay between CK and H_2O_2 regulates shade-induced leaf senescence (Causin et al., 2009). A significant association exists between H_2O_2 and plant growth regulators (PGRs) like ABA and zeatin riboside (ZR), crucial for stimulating proteins involved in plant development and signaling (Barba-Espin et al., 2010). The enhancement in germination efficiency in peas may be attributed to the reduction in more active zeatin (zeatin riboside) triggered by exogenous H_2O_2. These studies suggest that H_2O_2, in conjunction with other PGRs, effectively acts as a messenger in seed germination processes, indicating potential interplay between CK and H_2O_2, possibly in an antagonistic manner. H_2O_2 suppresses CK synthesis, promotes degradation, and decreases CK concentrations in roots and shoots (Nishiyama et al., 2011).

The roots of the transgenic line (S41) of creeping bentgrass exhibited significantly higher CK content and lower H_2O_2 levels than the wild-type (WT) roots (Xu et al., 2016). Augmenting CK content enhances the growth and fibrous root systems of creeping bentgrass and monocot species, such as barley (*Hordeum vulgare*), achieved through overexpression of the isopentyltransferase (IPT) gene and RNAi silencing of HvCKX1, which encodes cytokinin oxidase catalyzing CK cleavage (Zalewski et al., 2010; Merewitz et al., 2011). In *A. thaliana*, exogenously sourced zeatin induces stomatal closure, associated with increased accumulation of H_2O_2 in guard cells and the activation of HISTIDINE KINASE3 (AHK3) and RESPONSE REGULATOR2 (ARR2) receptors (Arnaud et al., 2017). CKs play a pivotal role in cellular-level signal transduction processes, initiating transcription, regulating protein standardization and stabilization, and governing phosphorylase activity (To et al., 2007). However, it has been revealed that H_2O_2 up-regulates an influenced gene, the C_2H_2 zinc finger ZAT11, which plays a more prominent role in stress responses and triggers programmed cell death (Gechev et al., 2005).

HYDROGEN PEROXIDE AND ETHYLENE

Ethylene, a gaseous plant growth regulator, plays diverse roles in plant development, encompassing processes such as leaf, flower, and fruit senescence, as well as conferring tolerance to abiotic stress factors (Iqbal et al., 2017). The interaction between H_2O_2 and ethylene can either enhance or inhibit stress response or productivity in a synergistic or antagonistic manner (Zhu et al., 2016). The interplay between H_2O_2 and ethylene is primarily linked to *ETHYLENE INSENSITIVE3* (EIN3) and its close homolog EIN3-Likel (EIL1), both of which are ethylene-activated TFs (Zhang et al., 2016). Ethylene response factors (ERFs) play a crucial role in promoting ethylene signaling and diminishing the abiotic stress tolerance of plants. In *Nicotiana tabacum*, TERF1 is activated by externally sourced ROS and is regulated by upstream EIN3-like regulators, LeEIL3, and LeEIL4 (Zhang et al., 2016). The activation of TERF1 in tobacco seedlings influences various stress-responsive genes, enhancing stress resilience to H_2O_2 by associating with their promoters, such as GCC-box, DRE/CRT, CE1, JERE, and CT-rich elements (Muller and Munne-Bosch, 2015).

In tomatoes, it has been proposed that ethylene and H_2O_2 act synergistically, with H_2O_2 being a major contributor to ethylene production, mediating cell death caused by Cd (Liu et al., 2008). Under salt stress, H_2O_2 and ethylene may be involved in regulating the alternative respiratory pathway (AP) (Wang et al., 2010). Zhu et al. (2016) demonstrated the significant roles of ethylene and H_2O_2 in triggering BR-reliant endurance in tomato plants under salt stress. Oxygen deficiencies in plants shift energy metabolism from aerobic to anaerobic, affecting water and nutrient absorption and inducing the production of H_2O_2 and ethylene through hypoxia signaling. In Arabidopsis, H_2O_2 and ethylene interaction influences the gene expression of AtERF73/HRE1 and ADH1 during the early phases of hypoxia signaling (Yang, 2014). Hypoxia signaling stimulates ethylene enzymes like ACC synthase (ACS) and ACC oxidase (ACO) (Peng et al., 2005).

SIGNALING OF H_2O_2

H_2O_2 functions as a signaling molecule in various cellular processes, both within and between cells. This is particularly significant because intercellular signaling requires transport across at least one membrane. Many cell types spontaneously create H_2O_2, either continuously or in response to external stimuli (Burdon et al., 1995). Plant tissues create ROS in response to various stresses, including superoxide radical ($O_2^{\cdot-}$) and H_2O_2 (Desikan et al., 2003). According to Burdon et al. (1995), H_2O_2 is essential for transducing defensive signals in plants, causing gene expression and protein synthesis. According to Sanders et al. (1999), calcium (Ca^{2+}) is a crucial component of cellular signaling and functions as a secondary messenger. Calcium enters the cell through cation channels in the endoplasmic reticulum, plasma membrane, and tonoplast to create cytoplasmic Ca^{2+} ($[Ca^{2+}]_{cyt}$). This sets off biological reactions to various stimuli for development and environmental pressures (White, 2000; Sanders et al., 2002). Research indicates that brief variations in $[Ca_{2+}]$cyt are essential for signaling reactions to different kinds of stress (Monroy and Dhindsa, 1995; Sanders et al., 1999). In tobacco and Arabidopsis, cold shock causes a transient rise in $[Ca^{2+}]_{cyt}$ (Polisensky and Braam, 1996; Lewis et al., 1997). These results imply that a critical component of the cold shock response is sustaining intracellular Ca^{2+} levels through its influx. According to Pei et al. (2000), stomatal closure in *Arabidopsis* is caused by a signaling cascade involving both Ca^{2+} and H_2O_2. H_2O_2-activated Ca^{2+} channels facilitate Ca^{2+} input into protoplasts and raise $[Ca^{2+}]_{cyt}$ in guard cells. The mechanism may be autocatalytic since the formation of H_2O_2 during the oxidative burst requires a constant Ca2+ influx, which activates the NADPH oxidase located in the plasma membrane (Grant et al., 2000; Yang and Poovaiah, 2002). Furthermore, elevated $[Ca^{2+}]_{cyt}$ arouses the calmodulin calcium sensor, which in turn signals downstream targets such as CAT to catalyze the breakdown of H_2O_2 in order to lower its levels.

Additionally, H2O2 modifies a number of signaling elements, including TFs, protein phosphatases, and protein kinases (Cheng and Song, 2006). It is a component of a network that regulates reactions downstream of H_2O_2 through interactions with other signaling molecules and pathways

(Neill et al., 2002). Research suggests that H_2O_2 controls genes downstream. In plants, for instance, a vast system engaged in transmitting H_2O_2 signals is formed by MAPK cascades, which operate downstream of H_2O_2 (Zhang et al., 2006; Xing et al., 2008). In a cascade, MAPKs activate NPK1, which then converts extracellular stimuli to defense-related gene expression (Kovtun et al., 2000; Desikan et al., 2001). Even when given externally to the roots, H_2O_2 quickly activates MAPKs in shoots (Capone et al., 2004). By triggering the genes involved in ABA catabolism, H_2O_2 regulates seed dormancy and germination (Liu et al., 2010b). Zinc finger TFs often control defense genes linked to ROS. When H_2O_2 and methyl viologen (MV) are administered to Arabidopsis, oxidative stress causes an increase in the expression of the zinc finger proteins ZAT7 and ZAT12 (Rizhsky, 2004). Under conditions of salt, drought, and osmotic stress, ZAT10 functions as a repressor as well as an inducer of genes that respond to ROS (Sakamoto et al., 2004; Mittler, 2006). ZAT6 enhances resistance to bacterial infections, drought, cold, and salt stress via modifying ROS levels and SA-related gene expression (Shi et al., 2014).

A model of the H_2O_2 signaling pathway has been proposed based on the findings mentioned above (Figure 6.2). A cell surface receptor may pick up on the H_2O_2 signal, which would raise $[Ca^{2+}]_{cyt}$. The activation of signaling proteins such protein kinases or phosphatases by the increased

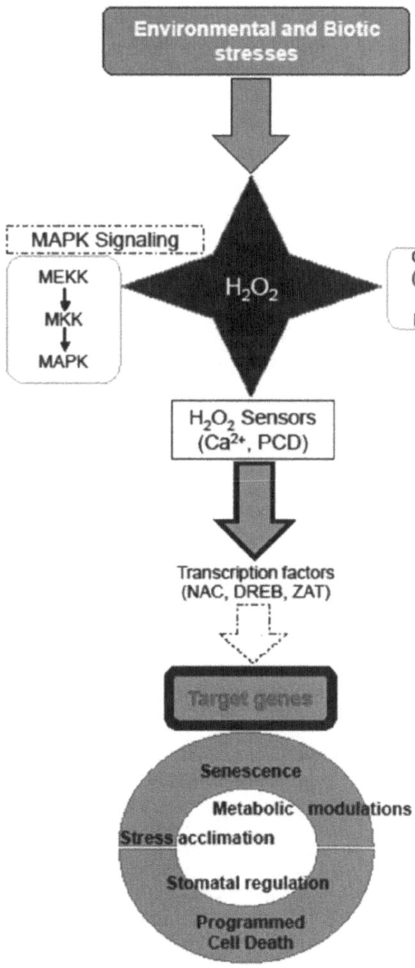

FIGURE 6.2 A schematic representation of signaling components and their interactions in the hydrogen peroxide signaling network and promising outcome in plants. DREB, dehydration-responsive element-binding protein; H_2O_2, hydrogen peroxide; PCD, programmed cell death; MAPK, mitogen-activated protein kinase.

$[Ca^{2+}]_{cyt}$ may set off a cascade that mediates the phosphorylation or dephosphorylation of TFs. Furthermore, H_2O_2 may directly oxidize H_2O_2-responsive TFs by oxidizing cysteine residue thiols. This would activate transcription. Upon entering the nucleus, the activated TF interacts with the target promoters' corresponding cis-acting region to control the expression of genes linked to cold stress.

CONCLUSION AND FUTURE PERSPECTIVES

H_2O_2 is highly promising molecule, plays a crucial role in enhancing various plant functions, yet it remains a challenging molecule to fully understand. It is increasingly evident that H_2O_2 functions as a versatile signaling molecule in plants. An endeavor has been undertaken to delve deeper into how H_2O_2 effectively alleviates stress-induced modifications in developmental processes, spanning plant growth, photosynthesis, stomatal behavior, pollen development, flowering, fruit growth, development, and leaf senescence. The interaction between H_2O_2 and other PGRs is crucial for governing diverse responses under both normal and ecological conditions. While H_2O_2 serves as a protective agent against the detrimental effects of environmental stresses, a comprehensive understanding of the intricate crosstalk is imperative to assess the regulation of numerous physiological processes guided by H_2O_2 and PGRs (such as ABA, CKs, ET, and BRs). Interestingly, recent molecular research suggests that H_2O_2 also plays a role in modulating signal transduction through MAPKs and TFs, serving as a crucial strategy to adapt to environmental conditions and enabling extracellular changes to impact intracellular processes, thereby maintaining cellular homeostasis. We posit that H_2O_2 proves to be exceptionally beneficial for the growth and metabolism of plants. The exploration of the specific functions of H_2O_2 signaling in the interaction between crops and ecological factors holds great promise for the immediate future.

REFERENCES

Abdel Latef, A. A. H., Kordrostami, M., Zakir, A., Zaki, H., & Saleh, O. M. (2019). Eustress with H_2O_2 facilitates plant growth by improving tolerance to salt stress in two wheat cultivars. *Plants*, 8(9), Article 9. https://doi.org/10.3390/plants8090303.

Arnaud, D., Lee, S., Takebayashi, Y., Choi, D., Choi, J., Sakakibara, H., & Hwang, I. (2017). Cytokinin-mediated regulation of reactive oxygen species homeostasis modulates stomatal immunity in Arabidopsis. *Plant Cell*, 29 (3), 543–559.

Ashfaque, F., Khan, M., & Khan, N. (2014). Exogenously applied H_2O_2 promotes proline accumulation, water relations, photosynthetic efficiency and growth of wheat (triticum aestivum l.) under salt stress. *Ann Res Rev Biol*, 4, 105–120.

Ashraf, M. (2004). Some important physiological selection criteria for salt tolerance in plants. Flora - Morphology, Distribution. *Funct Ecol Plants*, 199(5), 361–376. https://doi.org/10.1078/0367-2530-00165.

Barba-Espin, G., Diaz-Vivancos, P., Clemente-Moreno, M. J., Albacete, A., Faize, L., Faize, M., Pérez-Alfocea, F., & Hernández, J. A. (2010). Interaction between hydrogen peroxide and plant hormones during germination and the early growth of pea seedlings. *Plant Cell Environ*, 33(6), 981–994. https://doi.org/10.1111/j.1365-3040.2010.02120.x.

Barba-Espín, G., Diaz-Vivancos, P., Job, D., Belghazi, M., Job, C., & Hernández, J. A. (2011). Understanding the role of H(2)O(2) during pea seed germination: A combined proteomic and hormone profiling approach. *Plant Cell Environ*, 34(11), 1907–1919. https://doi.org/10.1111/j.1365-3040.2011.02386.x.

Bhardwaj, R. D., Singh, N., Sharma, A., Joshi, R., & Srivastava, P. (2021). Hydrogen peroxide regulates antioxidant responses and redox related proteins in drought stressed wheat seedlings. *Physiol Mol Biol Plants*, 27(1), 151–163. https://doi.org/10.1007/s12298-021-00937-z.

Bhattacharjee, S. (2012). An inductive pulse of hydrogen peroxide pretreatment restores redox-homeostasis and oxidative membrane damage under extremes of temperature in two rice cultivars. *Plant Growth Regul*, 68. https://doi.org/10.1007/s10725-012-9728-9.

Bienert, G. P., Møller, A. L. B., Kristiansen, K. A., Schulz, A., Møller, I. M., Schjoerring, J. K., & Jahn, T. P. (2007). Specific aquaporins facilitate the diffusion of hydrogen peroxide across membranes. *J Biol Chem*, 282(2), 1183–1192. https://doi.org/10.1074/jbc.M603761200.

Burdon, R. H., Alliangana, D., & Gill, V. (1995). Hydrogen peroxide and the proliferation of BHK-21 cell. *Free Radic Res* 23, 471–486.

Capone, R., Tiwari, B.S., & Levine, A. (2004). Rapid transmission of oxidative and nitrosative stress signals from roots to shoots in Arabidopsis. *Plant Physiol Biochem*, 42, 425–428.

Carvalho, R. F., Piotto, F. A., Schmidt, D., Peters, L. P., Monteiro, C. C., & de Azevedo, R. A. (2011). Seed priming with hormones does not alleviate induced oxidative stress in maize seedlings subjected to salt stress. *Sci Agric*, 68(5), 598–602.

Causin, H.F., Roberts, I.N., Criado, M.V., Gallego, S.M., Pena, L.B., del Carmen Ríos, M., & Barneix, A.J. (2009). Changes in hydrogen peroxide homeostasis and cytokinin levels contribute to the regulation of shade-induced senescence in wheat leaves. *Plant Sci.* 177(6), 698–704.

Černý, M., Habánová, H., Berka, M., Luklová, M., & Brzobohatý, B. (2018). Hydrogen peroxide: Its role in plant biology and crosstalk with signalling networks. *Int J Mol Sci*, 19(9), 2812. https://doi.org/10.3390/ijms19092812.

Chae, K., & Lord, E. M. (2011). Pollen tube growth and guidance: Roles of small, secreted proteins. *Ann Bot*, 108(4), 627–636. https://doi.org/10.1093/aob/mcr015.

Chang, Z., Liu, Y., Dong, H., Teng, K., Han, L., & Zhang, X. (2016). Effects of cytokinin and nitrogen on drought tolerance of creeping bentgrass. *PLoS One* 11 (4), e0154005.

Chao, Y.-Y., Hsu, Y. T., & Kao, C. H. (2009). Involvement of glutathione in heat shock– and hydrogen peroxide–induced cadmium tolerance of rice (Oryza sativa L.) seedlings. *Plant Soil*, 318(1), 37–45. https://doi.org/10.1007/s11104-008-9815-x.

Chen, W.-L., & Ko, Y.-T. (2021). Exogenous hydrogen peroxide induces chilling tolerance in Phalaenopsis seedlings through glutathione-related antioxidant system. *Sci Hortic*, 289, 110421. https://doi.org/10.1016/j.scienta.2021.110421.

Chen, Z., Gu, Q., Yu, X., Huang, L., Xu, S., Wang, R., Shen, W., & Shen, W. (2018). Hydrogen peroxide acts downstream of melatonin to induce lateral root formation. *Ann Bot*, 121(6), 1127–1136. https://doi.org/10.1093/aob/mcx207.

Cheng, Y., & Song, C. (2006). Hydrogen peroxide homeostasis and signaling in plant cells. *Sci China C Life Sci*, 49, 1–11.

Chien, C., & Lin, T. (1994). Mechanism of hydrogen peroxide in improving the germination of Cinnamomum camphora seed. Seed Science and Technology.

Christou, A., Manganaris, G. A., & Fotopoulos, V. (2014). Systemic mitigation of salt stress by hydrogen peroxide and sodium nitroprusside in strawberry plants via transcriptional regulation of enzymatic and non-enzymatic antioxidants. *Environ Exper Bot*, 107, 46–54. https://doi.org/10.1016/j.envexpbot.2014.05.009.

Considine, M.J., & Foyer, C.H. (2014). Redox regulation of plant development. *Antioxid Redox Signal*, 21(9), 1305–1326.

Cuypers, A., Hendrix, S., Amaral Dos Reis, R., De Smet, S., Deckers, J., Gielen, H., Jozefczak, M., Loix, C., Vercampt, H., Vangronsveld, J., & Keunen, E. (2016). Hydrogen peroxide, signaling in disguise during metal phytotoxicity. *Front Plant Sci*, 7, 470. https://doi.org/10.3389/fpls.2016.00470.

de Azevedo Neto, A. D., Prisco, J. T., Enéas-Filho, J., Medeiros, J.-V. R., & Gomes-Filho, E. (2005). Hydrogen peroxide pre-treatment induces salt-stress acclimation in maize plants. *J Plant Physiol*, 162(10), 1114–1122. https://doi.org/10.1016/j.jplph.2005.01.007.

Dempsey, D. A., & Klessig, D. F. (1995). Signals in plant disease resistance. *Bulletin de l'Institut Pasteur*, 93(3), 167–186. https://doi.org/10.1016/0020-2452(96)81488-6.

Deng, X.G., Zhu, T., Zhang, D.W., & Lin, H.H. (2015). The alternative respiratory pathway is involved in brassinosteroid-induced environmental stress tolerance in Nicotiana benthamiana. *J Exp Bot*, 66, 6219–6232.

Desikan, R., Hancock, J.T., & Neill, S.J. (2003). Oxidative stress signalling. *Top Curr Genet*, 4, 129–149.

Desikan, R., Mackerness, A.H., Hancock, J.T., & Neill, S.J. (2001). Regulation of the Arabidopsis transcriptome by oxidative stress. *Plant Physiol*, 127, 159–172.

Ding, Z. (2018). Hydrogen peroxide positively regulates brassinosteroid signaling through oxidation of the BRASSINAZOLE-RESISTANT1 transcription factor. *Nat Commun*, 9(1), 1063.

Duan, Q., Kita, D., Johnson, E. A., Aggarwal, M., Gates, L., Wu, H.-M., & Cheung, A. Y. (2014). Reactive oxygen species mediate pollen tube rupture to release sperm for fertilization in Arabidopsis. *Nat Commun*, 5, 3129. https://doi.org/10.1038/ncomms4129.

Dunand, C., Crèvecoeur, M., & Penel, C. (2007). Distribution of superoxide and hydrogen peroxide in Arabidopsis root and their influence on root development: Possible interaction with peroxidases. *New Phytol*, 174(2), 332–341. https://doi.org/10.1111/j.1469-8137.2007.01995.x.

Eljebbawi, A., Guerrero, Y. D. C. R., Dunand, C., & Estevez, J. M. (2021). Highlighting reactive oxygen species as multitaskers in root development. *iScience*, 24(1), 101978. https://doi.org/10.1016/j.isci.2020.101978.

Exposito-Rodriguez, M., Laissue, P. P., Yvon-Durocher, G., Smirnoff, N., & Mullineaux, P. M. (2017). Photosynthesis-dependent H_2O_2 transfer from chloroplasts to nuclei provides a high-light signalling mechanism. *Nat Commun*, 8(1), Article 1. https://doi.org/10.1038/s41467-017-00074-w.

Fahnenstich, H., Scarpeci, T. E., Valle, E. M., Flügge, U.-I., & Maurino, V. G. (2008). Generation of hydrogen peroxide in chloroplasts of Arabidopsis overexpressing glycolate oxidase as an inducible system to study oxidative stress. *Plant Physiol*, 148(2), 719–729. https://doi.org/10.1104/pp.108.126789.

Fan, Y., & Huang, Y. (2012). The effective peroxidase-like activity of chitosan-functionalized $CoFe_2O_4$ nanoparticles for chemiluminescence sensing of hydrogen peroxide and glucose. *Analyst*, 137(5), 1225–1231. https://doi.org/10.1039/C2AN16105B.

Fariduddin, Q., Khan, T. A., & Yusuf, M. (2014). Hydrogen peroxide mediated tolerance to copper stress in the presence of 28-homobrassinolide in Vigna radiata. *Acta Physiol Plant*, 36(10), 2767–2778. https://doi.org/10.1007/s11738-014-1647-0.

Farooq, M., Nawaz, A., Chaudhary, M. A., & Rehman, A. (2017). Foliage-applied sodium nitroprusside and hydrogen peroxide improves resistance against terminal drought in bread wheat. *J Agron Crop Sci*, 203(6), 473–482. https://doi.org/10.1111/jac.12215.

Farooq, M., Wahid, A., Kobayashi, N., Fujita, D., & Basra, S. (2009). Plant drought stress: Effects, mechanisms and management. *Agron Sustain Dev*, 29. https://doi.org/10.1051/agro:2008021.

Foreman, J., Demidchik, V., Bothwell, J. H. F., Mylona, P., Miedema, H., Torres, M. A., Linstead, P., Costa, S., Brownlee, C., Jones, J. D. G., Davies, J. M., & Dolan, L. (2003). Reactive oxygen species produced by NADPH oxidase regulate plant cell growth. *Nature*, 422(6930), 442–446. https://doi.org/10.1038/nature01485.

Foyer, C. H. (2021). Redox control of flowering. *Nat Chem Biol*, 17(5), 504–505. https://doi.org/10.1038/s41589-021-00758-x.

Gao, Y., Guo, Y.-K., Lin, S.-H., Fang, Y.-Y., & Bai, J.-G. (2010). Hydrogen peroxide pretreatment alters the activity of antioxidant enzymes and protects chloroplast ultrastructure in heat-stressed cucumber leaves. *Sci Hortic*, 126(1), 20–26. https://doi.org/10.1016/j.scienta.2010.06.006.

Ge, X.-M., Cai, H.-L., Lei, X., Zhou, X., Yue, M., & He, J.-M. (2015). Heterotrimeric G protein mediates ethylene-induced stomatal closure via hydrogen peroxide synthesis in Arabidopsis. *Plant J Cell Mol Biol*, 82(1), 138–150. https://doi.org/10.1111/tpj.12799.

Gechev, T. S., & Hille, J. (2005). Hydrogen peroxide as a signal controlling plant programmed cell death. *J Cell Biol*, 168(1), 17–20. https://doi.org/10.1083/jcb.200409170.

Gechev, T., Minkov, I., & Hille, J. (2005). Hydrogen peroxide-induced cell death in Arabidopsis, transcriptional and mutant analysis reveals a role of an oxoglutarate- dependent dioxygenase gene in the cell death process. *IUBMB Life*, 57(3), 181–188.

Geros, H., Chaves, M., & Delrot, S. (2012). *The Biochemistry of the Grapefruit*. Bentham Science Publishers, USA, ISBN: 9781608053605.

Gondim, F., Miranda, R., Prisco, J. T., & Gomes-Filho, E. (2013). Enhanced salt tolerance in maize plants induced by H_2O_2 leaf spraying is associated with improved gas exchange rather than with non-enzymatic antioxidant system. *Theor Exp Plant Physiol*, 25, 251–260. https://doi.org/10.1590/S2197-00252013000400003.

Grant, J.J., Yun, B.W., & Loake, G.J. (2000). Oxidative burst and cognate redox signaling reported by luciferase imaging: Identification of a signal network that functions independently of ethylene, SA and Me-JA but is dependent on MAPKK activity. *Plant J*, 24, 569–582.

Guzel, S., & Terzi, R. (2013). Exogenous hydrogen peroxide increases dry matter production, mineral content and level of osmotic solutes in young maize leaves and alleviates deleterious effects of copper stress. *Bot Stud*, 54, 26. https://doi.org/10.1186/1999-3110-54-26.

Habib, N., Ali, Q., Ali, S., Javed, M. T., Zulqurnain Haider, M., Perveen, R., Shahid, M. R., Rizwan, M., Abdel-Daim, M. M., Elkelish, A., & Bin-Jumah, M. (2020). Use of nitric oxide and hydrogen peroxide for better yield of wheat (Triticum aestivum L.) under water deficit conditions: Growth, osmoregulation, and antioxidative defense mechanism. *Plants*, 9(2), 285. https://doi.org/10.3390/plants9020285.

Hajivar, B., & Zare-Bavani, M. R. (2019). Alleviation of salinity stress by hydrogen peroxide and nitric oxide in tomato plants. *Adv Hortic Sci*, 33(3), 409–416.

Hasan, S. A., Irfan, M., Masrahi, Y. S., Khalaf, M. A., & Hayat, S. (2016). Growth, photosynthesis, and antioxidant responses of Vigna unguiculata L. treated with hydrogen peroxide. *Cogent Food Agric*, 2(1), 1155331. https://doi.org/10.1080/23311932.2016.1155331.

Hasanuzzaman, M., Nahar, K., Alam, Md. M., Roychowdhury, R., & Fujita, M. (2013). Physiological, biochemical, and molecular mechanisms of heat stress tolerance in plants. *Int J Mol Sci*, 14(5), 9643–9684. https://doi.org/10.3390/ijms14059643.

Hernández-Barrera, A., Velarde-Buendía, A., Zepeda, I., Sanchez, F., Quinto, C., Sánchez-Lopez, R., Cheung, A. Y., Wu, H.-M., & Cardenas, L. (2015). Hyper, a hydrogen peroxide sensor, indicates the sensitivity of the Arabidopsis root elongation zone to aluminum treatment. *Sensors (Basel, Switzerland)*, 15(1), 855–867. https://doi.org/10.3390/s150100855.

Hossain, M. A., Bhattacharjee, S., Armin, S.-M., Qian, P., Xin, W., Li, H.-Y., Burritt, D. J., Fujita, M., & Tran, L.-S. P. (2015). Hydrogen peroxide priming modulates abiotic oxidative stress tolerance: Insights from ROS detoxification and scavenging. *Front Plant Sci*, 6. https://www.frontiersin.org/articles/10.3389/fpls.2015.00420.

Hossain, M. A., Mostofa, M., & Fujita, M. (2013). Heat-shock positively modulates oxidative protection of salt and drought-stressed mustard (Brassica campestris L.) seedlings. *J Plant Sci Mol Breed*, 2, 1–13. https://doi.org/10.7243/2050-2389-2-2.

Hu, Y., Ge, Y., Zhang, C., Ju, T., & Cheng, W. (2009). Cadmium toxicity and translocation in rice seedlings are reduced by hydrogen peroxide pretreatment. *Plant Growth Regul*, 59, 51–61. https://doi.org/10.1007/s10725-009-9387-7.

Huang, X., Chen, S., Li, W., Tang, L., Zhang, Y., Yang, N., Zou, Y., Zhai, X., Xiao, N., Liu, W., Li, P., & Xu, C. (2021). ROS regulated reversible protein phase separation synchronizes plant flowering. *Nat Chem Biol*, 17(5), 549–557. https://doi.org/10.1038/s41589-021-00739-0.

Iqbal, H., Yaning, C., Rehman, H., Shareef, M., & Iqbal, S. (2018). Hydrogen peroxide application improves quinoa performance by affecting physiological and biochemical mechanisms under water-deficit conditions. *J Agron Crop Sci*, 204. https://doi.org/10.1111/jac.12284.

Iqbal, N., Khan, N.A., Ferrante, A., Trivellini, A., Francini, A., & Khan, M.I. (2017). Ethylene role in plant growth, development and senescence, interaction with other phytohormones. *Front Plant Sci*, 8, 475.

Işeri, O., Körpe, D., Sahin, F., & Haberal, M. (2013). Hydrogen peroxide pretreatment of roots enhanced oxidative stress response of tomato under cold stress. *Acta Physiol Plant*, 35. https://doi.org/10.1007/s11738-013-1228-7.

Ishibashi, Y., Aoki, N., Kasa, S., Sakamoto, M., Kai, K., Tomokiyo, R., Watabe, G., Yuasa, T., & Iwaya-Inoue, M. (2017). The interrelationship between abscisic acid and reactive oxygen species plays a key role in barley seed dormancy and germination. *Front Plant Sci*, 8, 275.

Ishibashi, Y., Yamaguchi, H., Yuasa, T., Iwaya-Inoue, M., Arima, S., & Zheng, S.-H. (2011). Hydrogen peroxide spraying alleviates drought stress in soybean plants. *J Plant Physiol*, 168(13), 1562–1567. https://doi.org/10.1016/j.jplph.2011.02.003

Ismail, S., Khandaker, M., Mat, N., & Nasrulhaq Boyce, A. (2015). Effects of hydrogen peroxide on growth, development and quality of fruits: A review. *J Agron*, 14, 331–336. https://doi.org/10.3923/ja.2015.331.336.

Jahan, A. A., & Anis, M. (2017). Changes in antioxidative enzymatic responses during acclimatization of in vitro raised plantlets of cardiospermum halicacabum L. against oxidative stress. *J Plant Physiol Pathol*, 2014. https://doi.org/10.4172/2329-955X.1000137.

Jiang, Y., Cheng, F., Zhou, Y., Xia, X., Mao, W., Shi, K., Chen, Z., & Yu, J. (2012). Hydrogen peroxide functions as a secondary messenger for brassinosteroids-induced CO_2 assimilation and carbohydrate metabolism in Cucumis sativus. *J Zhejiang Univ Sci B*, 13(10), 811–823. https://doi.org/10.1631/jzus.B1200130.

Jira-anunkul, W., & Pattanagul, W. (2021). Effects of hydrogen peroxide application on agronomic traits of rice (Oryza sativa L.) under drought stress. *Plant Soil Environ*, 67(4), 221–229. https://doi.org/10.17221/628/2020-PSE.

Joo, J. H., Bae, Y. S., & Lee, J. S. (2001). Role of auxin-induced reactive oxygen species in root gravitropism. *Plant Physiol*, 126(3), 1055–1060. https://doi.org/10.1104/pp.126.3.1055.

Kang, N. J., Kang, Y. I., Kang, K. H., & Jeong, B. R. (2009). Induction of thermotolerance and activation of antioxidant enzymes in H_2O_2 pre-applied leaves of cucumber and tomato seedlings. *J Jpn Soc Hortic Sci*, 78(3), 320–329. https://doi.org/10.2503/jjshs1.78.320.

Kapoor, D., Sharma, R., Handa, N., Kaur, H., Rattan, A., Yadav, P., Gautam, V., Kaur, R., & Bhardwaj, R. (2015). Redox homeostasis in plants under abiotic stress: Role of electron carriers, energy metabolism mediators and proteinaceous thiols. *Front Environ Sci*, 3. https://www.frontiersin.org/articles/10.3389/fenvs.2015.00013.

Katzman, L., Taylor, A., & Langhans, R. (2001). Seed enhancements to improve spinach germination. *HortScience*, 36. https://doi.org/10.21273/HORTSCI.36.5.979.

Khan, T. A., Yusuf, M., & Fariduddin, Q. (2018). Hydrogen peroxide in regulation of plant metabolism: Signalling and its effect under abiotic stress. *Photosynthetica*, 56(4), 1237–1248. https://doi.org/10.1007/s11099-018-0830-8.

Khan, T., Yusuf, M., & Fariduddin, Q. (2015). Seed treatment with H_2O_2 modifies net photosynthetic rate and antioxidant system in mung bean (Vigna radiata L. Wilczek) plants. *Isr J Plant Sci*, 62. https://doi.org/10.1080/07929978.2015.1060806.

Khandaker, M. M., Boyce, A. N., & Osman, N. (2012). The influence of hydrogen peroxide on the growth, development and quality of wax apple (*Syzygium samarangense*, [Blume] Merrill & L. M. Perry var. *Jambu madu*) fruits. *Plant Physiol Biochem*, 53, 101–110. https://doi.org/10.1016/j.plaphy.2012.01.016.

Kolla, V. A., Vavasseur, A., & Raghavendra, A. S. (2007). Hydrogen peroxide production is an early event during bicarbonate induced stomatal closure in abaxial epidermis of Arabidopsis. *Planta*, 225(6), 1421–1429. https://doi.org/10.1007/s00425-006-0450-6.

Kovtun, Y., Chiu, W.L., Tena, G., & Sheen, J. (2000). Functional analysis of oxidative stress-activated mitogen-activated protein kinase cascade in plants. *Proc Nat Acad Sci USA*, 97, 2940–2945.

Kwak, J.M., Mori, I.C., Pei, Z.M., Leonhardt, N., Torres, M.A., Dangl, J.L., Bloom, R.E., Bodde, S., Jones, J.D., & Schroeder, J.I. (2003). NADPH oxidase AtrbohD and AtrbohF genes function in ROS-dependent ABA signaling in Arabidopsis. *EMBO J*, 22 (11), 2623–2633.

Larkindale, J., & Huang, B. (2004). Thermotolerance and antioxidant systems in Agrostis stolonifera: Involvement of salicylic acid, abscisic acid, calcium, hydrogen peroxide, and ethylene. *J Plant Physiol*, 161(4), 405–413. https://doi.org/10.1078/0176-1617-01239.

Lewis, B.D., Karlin-Neumann, G., Davis, R.W., & Spalding, E.P. (1997). Ca^{2+}-activated anion channels and membrane depolarizations induced by blue light and cold in Arabidopsis seedlings. *Plant Physiol*, 114, 1327–1334.

Li, Z., Xu, J., Gao, Y., Wang, C., Guo, G., Luo, Y., Huang, Y., Hu, W., Sheteiwy, M.S., Guan, Y., & Hu, J. (2017). The synergistic priming effect of exogenous salicylic acid and H_2O_2 on chilling tolerance enhancement during maize (Zea mays L.) seed germination. *Front Plant Sci*, 8, 1153.

Li, S., Xue, L., Xu, S., Feng, H., & An, L. (2007). Hydrogen peroxide involvement in formation and development of adventitious roots in cucumber. *Plant Growth Regul*, 52(2), 173–180. https://doi.org/10.1007/s10725-007-9188-9.

Li, X.P., Xu, Q.Q., Liao, W.B., Ma, Z.J., Xu, X.T., Wang, M., Ren, P.J., Niu, L.J., Jin, X., & Zhu, Y.C. (2016). Hydrogen peroxide is involved in abscisic acid-induced adventitious rooting in cucumber (Cucumis sativus L.) under drought stress. *J Plant Biol*, 59(5), 536–548.

Li, Y., Zhao, H., Duan, B., Korpelainen, H., & Li, C. (2011). Effect of drought and ABA on growth, photosynthesis and antioxidant system of Cotinus coggygria seedlings under two different light conditions. *Environ Exp Bot*, 71(1), 107–113. https://doi.org/10.1016/j.envexpbot.2010.11.005.

Liao, W., Xiao, H., & Zhang, M. (2009). Role and relationship of nitric oxide and hydrogen peroxide in adventitious root development of marigold. *Acta Physiol Plant*, 31, 1279–1289. https://doi.org/10.1007/s11738-009-0367-3.

Liao, W.-B., Xiao, L., & Zhang, M.-L. (2010). Effect of nitric oxide and hydrogen peroxide on adventitious root development from cuttings of ground-cover chrysanthemum and associated biochemical changes. *J Plant Growth Regul*, 29, 338–348. https://doi.org/10.1007/s00344-010-9140-5.

Liao, W.-B., Zhang, M.-L., Huang, G.-B., & Yu, J.-H. (2012). Hydrogen peroxide in the vase solution increases vase life and keeping quality of cut Oriental×Trumpet hybrid lily 'Manissa.' *Sci Hortic*, 139, 32–38. https://doi.org/10.1016/j.scienta.2012.02.040.

Liu, K.-L., Shen, L., Wang, J.-Q., & Sheng, J.-P. (2008). Rapid Inactivation of Chloroplastic Ascorbate Peroxidase is Responsible for Oxidative Modification to Rubisco in Tomato (Lycopersicon esculentum) under Cadmium Stress. *J Integr Plant Biol*, *50*(4), 415–426. https://doi.org/10.1111/j.1744-7909.2007.00621.x

Liu, X.M., Kim, K.E., Kim, K.C., Nguyen, X.C., Han, H.J., Jung, M.S., Kim, H.S., Kim, S.H., Park, C., Yun, D.J., & Chung, W.S. (2010a). Cadmium activates Arabidopsis MPK3 and MPK6 via accumulation of reactive oxygen species. *Phytochemistry*, 71, 614–618.

Liu, C.H., Chao, Y.Y., & Kao, C.H. (2012). Abscisic acid is an inducer of hydrogen peroxide production in leaves of rice seedlings grown under potassium deficiency. *Bot Stud* 53, 229–237.

Liu, J., Macarisin, D., Wisniewski, M., Sui, Y., Droby, S., Norelli, J. L., & Hershkovitz, V. (2013). Production of hydrogen peroxide and expression of ROS-generating genes in peach flower petals in response to host and non-host fungal pathogens. *Plant Pathol*, 62. https://doi.org/10.1111/j.1365-3059.2012.02683.x.

Liu, Y., Ye, N., Liu, R., Chen, M., & Zhang, J. (2010b). H_2O_2 mediates the regulation of ABA catabolism and GA biosynthesis in Arabidopsis seed dormancy and germination. *J Exp Bot*, 61, 2979–2990.

Luna, C. M. (2004). Drought controls on H_2O_2 accumulation, catalase (CAT) activity and CAT gene expression in wheat. *J Exp Bot*, 56(411), 417–423. https://doi.org/10.1093/jxb/eri039.

Ma, F., Wang, L., Li, J., Samma, M. K., Xie, Y., Wang, R., Wang, J., Zhang, J., & Shen, W. (2014). Interaction between HY1 and H_2O_2 in auxin-induced lateral root formation in Arabidopsis. *Plant Mol Biol*, 85(1–2), 49–61. https://doi.org/10.1007/s11103-013-0168-3.

Merewitz, E.B., Gianfagna, T., & Huang, B. (2011). Photosynthesis., water use., and root viability under water stress as affected by expression of SAG12-ipt controlling cytokinin synthesis in Agrostis stolonifera. *J Exp Bot*, 62, 383–395.

Mhamdi, A. (2022). Hydrogen peroxide in plants. In *Advances in Botanical Research*. https://doi.org/10.1016/bs.abr.2022.11.002.

Miao, Y., Lv, D., Wang, P., Wang, X.C., Chen, J., Miao, C., & Song, C.P. (2006). An Arabidopsis glutathione peroxidase functions as both a redox transducer and a scavenger in abscisic acid and drought stress responses. *Plant Cell*, 18(10), 2749–2766.

Migahid, M. M., Elghobashy, R. M., Bidak, L. M., & Amin, A. W. (2019). Priming of Silybum marianum (L.) Gaertn seeds with H_2O_2 and magnetic field ameliorates seawater stress. *Heliyon*, 5(6), e01886. https://doi.org/10.1016/j.heliyon.2019.e01886.

Mignolet-Spruyt, L., Xu, E., Idänheimo, N., Hoeberichts, F. A., Mühlenbock, P., Brosché, M., Van Breusegem, F., & Kangasjärvi, J. (2016). Spreading the news: Subcellular and organellar reactive oxygen species production and signalling. *J Exper Bot*, 67(13), 3831–3844. https://doi.org/10.1093/jxb/erw080.

Mittler, R. (2006). Abiotic stress, the field environment and stress combination. *Trends Plant Sci*, 1, 15–19.

Mittler, R., Vanderauwera, S., Gollery, M., & Van Breusegem, F. (2004). Reactive oxygen gene network of plants. *Trends Plant Sci*, 9(10), 490–498. https://doi.org/10.1016/j.tplants.2004.08.009.

Monroy, A.F., & Dhindsa, R.S. (1995). Low temperature signal transduction: Induction of cold acclimation-specific genes of alfalfa by calcium at 25°C. *Plant Cell*, 7, 321–331.

Moon, S., Derakhshani, B., Gho, Y. S., Kim, E.-J., Lee, S. K., Jiang, X., Lee, C., & Jung, K.-H. (2023). PRX102 participates in root hairs tip growth of rice. *Rice*, 16, 51. https://doi.org/10.1186/s12284-023-00668-7.

Mostofa, M., Yoshida, N., & Fujita, M. (2013). Spermidine pretreatment enhances heat tolerance in rice seedlings through modulating antioxidative and glyoxalase systems. *Plant Growth Regul*, 73. https://doi.org/10.1007/s10725-013-9865-9.

Mukherjee, S., & Corpas, F. J. (2023). H_2O_2, NO, and H_2 S networks during root development and signalling under physiological and challenging environments: Beneficial or toxic? *Plant Cell Environ*, 46(3), 688–717. https://doi.org/10.1111/pce.14531.

Muller, M., & Munne-Bosch, S. (2015). Ethylene response factors, a key regulatory hub in hormone and stress signaling. *Plant Physiol*, 169, 32–41.

Nazir, F., Fariduddin, Q., Hussain, A., & Khan, T. A. (2021). Brassinosteroid and hydrogen peroxide improve photosynthetic machinery, stomatal movement, root morphology and cell viability and reduce Cu-triggered oxidative burst in tomato. *Ecotoxicol Environ Saf*, 207, 111081. https://doi.org/10.1016/j.ecoenv.2020.111081.

Nazir, F., Hussain, A., & Fariduddin, Q. (2019). Hydrogen peroxide modulate photosynthesis and antioxidant systems in tomato (Solanum lycopersicum L.) plants under copper stress. *Chemosphere*, 230, 544–558. https://doi.org/10.1016/j.chemosphere.2019.05.001.

Neill, S.J., Desikan, R., & Hancock, J. (2002). Hydrogen peroxide signaling. *Curr Opin Plant Biol*, 5, 388–395. https://doi.org/10.1016/s1369-5266(02)00282-0.

Nie, W.F., Wang, M.M., Xia, X.J., Zhou, Y.H., Shi, K., Chen, Z., & Yu, J.Q. (2013). Silencing of tomato RBOH1 and MPK2 abolishes brassinosteroid-induced H_2O_2 generation and stress tolerance. *Plant Cell Environ*, 36, 789–803.

Nishiyama, R., Watanabe, Y., Fujita, Y., Le, D.T., Kojima, M., Werner, T., Vankova, R., Yamaguchi-Shinozaki, K., Shinozaki, K., Kakimoto, T., & Sakakibara, H. (2011). Analysis of cytokinin mutants and regulation of cytokinin metabolic genes reveals important regulatory roles of cytokinins in drought., salt and abscisic acid responses., and abscisic acid biosynthesis. *Plant Cell*, 23, 2169–2183.

Niu, L., & Liao, W. (2016). Hydrogen peroxide signaling in plant development and abiotic responses: Crosstalk with nitric oxide and calcium. *Front Plant Sci*, 7, 230. https://doi.org/10.3389/fpls.2016.00230.

Noctor, G., & Foyer, C. H. (2016). Intracellular redox compartmentation and ROS-related communication in regulation and signaling. *Plant Physiol*, 171(3), 1581–1592. https://doi.org/10.1104/pp.16.00346.

Nogués, S., & Baker, N. R. (2000). Effects of drought on photosynthesis in Mediterranean plants grown under enhanced UV-B radiation. *J Exper Bot*, 51(348), 1309–1317. https://doi.org/10.1093/jxb/51.348.1309.

Ogawa, K., & Iwabuchi, M. (2001). A Mechanism for Promoting the Germination of Zinnia elegans Seeds by Hydrogen Peroxide. *Plant Cell Physiol*, 42(3), 286–291. https://doi.org/10.1093/pcp/pce032

Oguz, M. C., Aycan, M., Oguz, E., Poyraz, I., & Yildiz, M. (2022). Drought stress tolerance in plants: Interplay of molecular, biochemical and physiological responses in important development stages. *Physiologia*, 2(4), Article 4. https://doi.org/10.3390/physiologia2040015.

Orabi, S.A., Dawood, M.G., & Salman, S.R. (2015). Comparative study between the physiological role of hydrogen peroxide and salicylic acid in alleviating the harmful effect of low temperature on tomato plants grown under sand-ponic culture. *Sci Agric*, 9(1), 49–59.

Orabi, S., Moursy, M., Zaki, S.-N., & Sharara, F. (2018). Influence of hydrogen peroxide on growth, yield and biochemical constituents of canola plants grown under different irrigation intervals. *Curr Sci*, *07*, 407–418.

Overmyer, K., Brosché, M., & Kangasjärvi, J. (2003). Reactive oxygen species and hormonal control of cell death. *Trends Plant Sci*, *8*(7), 335–342. https://doi.org/10.1016/S1360-1385(03)00135-3

Pandey, V., & Shukla, A. (2015). Acclimation and tolerance strategies of rice (Oryza sativa L.) under drought stress. *Rice Sci*, 22. https://doi.org/10.1016/j.rsci.2015.04.001.

Pei, Z.M., Murata, Y., Benning, G., Thomine, S., Klusener, B., Allen, G.J., Grill, E., & Schroeder, J.I. (2000). Calcium channels activated by hydrogen peroxide mediate abscisic acid signalling in guard cells. *Nature*, 406, 731–734.

Peng, H.P., Lin, T.Y., Wang, N.N., & Shih, M.C. (2005). Differential expression of genes encoding 1-aminocyclopropane-1-carboxylate synthase in Arabidopsis during hypoxia. *Plant Mol Biol*, 58(1), 15–25.

Petrov, V. D., & Van Breusegem, F. (2012). Hydrogen peroxide—A central hub for information flow in plant cells. *AoB Plants*, 2012, pls014. https://doi.org/10.1093/aobpla/pls014.

Polisensky, D.H., & Braam, J. (1996). Cold-shock regulation of the Arabidopsis TCH genes and the effects of modulating intracellular calcium levels. *Plant Physiol*, 111, 1271–1279.

Qu, Y., Wang, Q., Guo, J., Wang, P., Song, P., Jia, Q., Zhang, X., Kudla, J., Zhang, W., & Zhang, Q. (2017). Peroxisomal CuAOζ and its product H_2O_2 regulate the distribution of auxin and IBA-dependent lateral root development in Arabidopsis. *J Exper Bot*, 68(17), 4851–4867. https://doi.org/10.1093/jxb/erx290.

Quan, L.-J., Zhang, B., Shi, W.-W., & Li, H.-Y. (2008). Hydrogen peroxide in plants: A versatile molecule of the reactive oxygen species network. *J Integr Plant Biol*, 50(1), 2–18. https://doi.org/10.1111/j.1744-7909.2007.00599.x.

Radha, B., Sunitha, N. C., Sah, R. P., Md Azharudheen, T. P., Krishna, G. K., Umesh, D. K., Thomas, S., Anilkumar, C., Upadhyay, S., Kumar, A., Manikanta Ch, L. N., Behera, S., Marndi, B.C., & Siddique, K. H. M. (2023). Physiological and molecular implications of multiple abiotic stresses on yield and quality of rice. *Front Plant Sci*, 13. https://www.frontiersin.org/articles/10.3389/fpls.2022.996514.

Reis, A. D. P., Carvalho, R. F., Costa, I. B., Girio, R. J. S., Gualberto, R., Spers, R. C., & Gaion, L. A. (2022). Hydrogen peroxide is involved in drought stress long-distance signaling controlling early stomatal closure in tomato plants. *Braz J Biol Rev Bras Biol*, 82, e267343. https://doi.org/10.1590/1519-6984.267343.

Rentel, M. C., Lecourieux, D., Ouaked, F., Usher, S. L., Petersen, L., Okamoto, H., Knight, H., Peck, S. C., Grierson, C. S., Hirt, H., & Knight, M. R. (2004). OXI1 kinase is necessary for oxidative burst-mediated signalling in Arabidopsis. *Nature*, 427(6977), 858–861. https://doi.org/10.1038/nature02353.

Rizhsky, L. (2004). The zinc finger protein Zat12 is required for cytosolic ascorbate peroxidase 1 expression during oxidative stress in Arabidopsis. *J Biol Chem*, 279, 11736–11743.

Sachdev, S., Ansari, S. A., Ansari, M. I., Fujita, M., & Hasanuzzaman, M. (2021). Abiotic stress and reactive oxygen species: Generation, signaling, and defense mechanisms. *Antioxidants (Basel, Switzerland)*, 10(2), 277. https://doi.org/10.3390/antiox10020277.

Sakamoto, H., Maruyama, K., Sakuma, Y., Meshi, T., Iwabuchi, M., & Shinozaki, K. (2004). Arabidopsis Cys2/His2-type zinc-finger proteins function as transcription repressors under drought, cold, and high-salinity stress conditions. *Plant Physiol*, 136, 2734–2746.

Sanders, D., Brownlee, C., & Harper, J.F. (1999). Communicating with calcium. *Plant Cell*, 11, 691–706.

Sanders, D., Pelloux, J., Brownlee, C., & Harper, J.F. (2002). Calcium at the crossroads of signalling. *Plant Cell*, 14, S401–S417.

Sarath, G., Hou, G., Baird, L. M., & Mitchell, R. B. (2007). Reactive oxygen species, ABA and nitric oxide interactions on the germination of warm-season C4-grasses. *Planta*, *226*(3), 697–708. https://doi.org/10.1007/s00425-007-0517-z

Sarwar, M., Saleem, M. F., Ullah, N., Rizwan, M., Ali, S., Shahid, M. R., Alamri, S. A., Alyemeni, M. N., & Ahmad, P. (2018). Exogenously applied growth regulators protect the cotton crop from heat-induced injury by modulating plant defense mechanism. *Sci Rep*, 8(1), 17086. https://doi.org/10.1038/s41598-018-35420-5.

Saxena, I., Srikanth, S., & Chen, Z. (2016). Cross talk between H_2O_2 and interacting signal molecules under plant stress response. *Front Plant Sci*, 7, 570. https://doi.org/10.3389/fpls.2016.00570.

Schützendübel, A., & Polle, A. (2002). Plant responses to abiotic stresses: Heavy metal-induced oxidative stress and protection by mycorrhization. *J Exper Bot*, 53(372), 1351–1365. https://doi.org/10.1093/jexbot/53.372.1351.

Semida, W. M. (2016). Hydrogen peroxide alleviates salt-stress in two onion (Allium cepa L.) cultivars. *Am Eurasian J Agric Environ Sci*, 16(2), 294–301.

Serrano, I., Romero-Puertas, M. C., Rodríguez-Serrano, M., Sandalio, L. M., & Olmedilla, A. (2012). Peroxynitrite mediates programmed cell death both in papillar cells and in self-incompatible pollen in the olive (Olea europaea L.). *J Exper Bot*, 63(3), 1479–1493. https://doi.org/10.1093/jxb/err392.

Serrano, I., Romero-Puertas, M. C., Sandalio, L. M., & Olmedilla, A. (2015). The role of reactive oxygen species and nitric oxide in programmed cell death associated with self-incompatibility. *J Exper Bot*, 66(10), 2869–2876. https://doi.org/10.1093/jxb/erv083.

Sharma, S. S., & Dietz, K.-J. (2009). The relationship between metal toxicity and cellular redox imbalance. *Trends Plant Sci*, 14(1), 43–50. https://doi.org/10.1016/j.tplants.2008.10.007.

Shi, H., Wang, X., Ye, T., Cheng, F., Deng, J., Yang, P., Zhang, Y., & Chan, Z. (2014). The cysteine2/histidine2-type transcription factor zinc finger of Arabidopsis thaliana 6 modulates biotic and abiotic stress responses by activating salicylic acid-related genes and C-repeat-binding factor genes in Arabidopsis. *Plant Physiol*, 165, 1367–1379.

Shu, H. H., Chun, C. W., Ivanov, S., Alexieva, V., & Chih, W. Y. (2007). Repetition of hydrogen peroxide treatment induces a chilling tolerance comparable to cold acclimation in mung bean. *J Am Soc Hortic Sci*, 132, 770–776. https://doi.org/10.21273/JASHS.132.6.770.

Shu, S., Gao, P., Li, L., Yuan, Y., Sun, J., & Guo, S. (2016). Abscisic acid-Induced H_2O_2 accumulation enhances antioxidant capacity in pumpkin-grafted cucumber leaves under Ca $(NO_3)_2$ stress. *Front Plant Sci*, 7, 1489.

Silva, H. H. B., Azevedo Neto, A. D. D., Menezes, R. V., Silva, P. C. C., & Gheyi, H. R. (2019). Use of hydrogen peroxide in acclimation of basil (Ocimum basilicum L.) to salt stress. *Turk J Bot*, 43(2), 208–217. https://doi.org/10.3906/bot-1807-80.

Slesak, I., Libik, M., Karpinska, B., Karpinski, S., & Miszalski, Z. (2007). The role of hydrogen peroxide in regulation of plant metabolism and cellular signalling in response to environmental stresses. *Acta Biochim Pol*, 54(1), 39–50.

Sohag, A. A. M., Tahjib-Ul-Arif, Md., Brestič, M., Afrin, S., Sakil, Md. A., Hossain, Md. T., Hossain, M. A., & Hossain, Md. A. (2020). Exogenous salicylic acid and hydrogen peroxide attenuate drought stress in rice. *Plant, Soil Environ*, 66(1), 7–13. https://doi.org/10.17221/472/2019-PSE.

Su, C., Liu, L., Liu, H., Ferguson, B. J., Zou, Y., Zhao, Y., Wang, T., Wang, Y., & Li, X. (2016). H_2O_2 regulates root system architecture by modulating the polar transport and redistribution of auxin. *J Plant Biol*, 59(3), 260–270. https://doi.org/10.1007/s12374-016-0052-1.

Suzuki, K., Yano, A., & Shinshi, H. (1999). Slow and prolonged activation of the p47 protein kinase during hypersensitive cell death in a culture of tobacco cells. *Plant Physiol*, 119(4), 1465–1472. https://doi.org/10.1104/pp.119.4.1465.

Takáč, T., Obert, B., Rolčík, J., & Šamaj, J. (2016). Improvement of adventitious root formation in flax using hydrogen peroxide. *New Biotechnol*, 33(5 Pt B), 728–734. https://doi.org/10.1016/j.nbt.2016.02.008.

Terzi, R., Kadioğlu, A., Kalaycioglu, E., & Saglam, A. (2014). Hydrogen peroxide pretreatment induces osmotic stress tolerance by influencing osmolyte and abscisic acid levels in maize leaves. *J Plant Interact*, 9, 559–565. https://doi.org/10.1080/17429145.2013.871077.

Tian, Y., Fan, M., Qin, Z., Lv, H., Wang, M., Zhang, Z., Zhou, W., Zhao, N., Li, X., Han, C., Ding, Z., Wang, W., Wang, Z.-Y., & Bai, M.-Y. (2018). Hydrogen peroxide positively regulates brassinosteroid signaling through oxidation of the BRASSINAZOLE-RESISTANT1 transcription factor. *Nat Commun*, *9*(1), 1063. https://doi.org/10.1038/s41467-018-03463-x

To, J. P., Deruère, J., Maxwell, B. B., Morris, V. F., Hutchison, C. E., Ferreira, F. J., Schaller, G. E., & Kieber, J. J. (2007). Cytokinin Regulates Type-A Arabidopsis Response Regulator Activity and Protein Stability via Two-Component Phosphorelay. *Plant Cell*, *19*(12), 3901. https://doi.org/10.1105/tpc.107.052662

Tsai, Y.C., & Kao, C.H. (2004). The involvement of hydrogen peroxide in abscisic acid induced activities of ascorbate peroxidase and glutathione reductase in rice roots. *Plant Growth Regul*, 43, 207–212.

Uchida, A., Jagendorf, A., Hibino, T., Takabe, T., & Takabe, T. (2002). Effects of hydrogen peroxide and nitric oxide on both salt and heat stress tolerance in rice. *Plant Sci*, 163, 515–523. https://doi.org/10.1016/S0168-9452(02)00159-0.

Van Camp, W. (2005). Yield enhancement genes: Seeds for growth. *Curr Opin Biotechnol*, 16(2), 147–153. https://doi.org/10.1016/j.copbio.2005.03.002.

Wahid, A., Perveen, M., Gelani, S., & Basra, S. M. A. (2007). Pretreatment of seed with H_2O_2 improves salt tolerance of wheat seedlings by alleviation of oxidative damage and expression of stress proteins. *J Plant Physiol*, 164(3), 283–294. https://doi.org/10.1016/j.jplph.2006.01.005.

Wang, H., Liang, X., Huang, J., Zhang, D., Lu, H., Liu, Z., & Bi, Y. (2010). Involvement of ethylene and hydrogen peroxide in induction of alternative respiratory pathway in salt-treated Arabidopsis calluses. *Plant Cell Physiol*, 51(10), 1754–1765.

Wang, P., & Song, C.-P. (2008). Guard-cell signalling for hydrogen peroxide and abscisic acid. *New Phytol*, 178(4), 703–718. https://doi.org/10.1111/j.1469-8137.2008.02431.x.

Wang, S. S., Zhu, X. N., Lin, J. X., Zheng, W. J., Zhang, B. T., Zhou, J. Q., Ni, J., Pan, Z. C., Zhu, S. H., & Ding, W. N. (2018a). $OsNOX_3$, encoding a NADPH oxidase, regulates root hair initiation and elongation in rice. *Biol Plant*, 62(4), 732–740. https://doi.org/10.1007/s10535-018-0714-3.

Wang, W., Wang, X., Huang, M., Cai, J., Zhou, Q., Dai, T., Cao, W., & Jiang, D. (2018b). Hydrogen peroxide and abscisic acid mediate salicylic acid-induced freezing tolerance in wheat. *Front Plant Sci*, 9, 1137.

Wang, Y., Zhang, J., Li, J.-L., & Ma, X.-R. (2014). Exogenous hydrogen peroxide enhanced the thermotolerance of Festuca arundinacea and Lolium perenne by increasing the antioxidative capacity. *Acta Physiol Plant*, 36(11), 2915–2924. https://doi.org/10.1007/s11738-014-1661-2.

White, P. J. (2000). Calcium channels in higher plants. *Biochimica et Biophysica Acta (BBA) - Biomembranes, 1465*(1), 171–189. https://doi.org/10.1016/S0005-2736(00)00137-1

Willekens, H., Chamnongpol, S., Davey, M., Schraudner, M., Langebartels, C., Montagu, M. V., Inzé, D., & Camp, W. V. (1997). Catalase is a sink for H_2O_2 and is indispensable for stress defence in C3 plants. *EMBO J*, 16(16), 4806. https://doi.org/10.1093/emboj/16.16.4806.

Xia, X.J., Gao, C.J., Song, L.X., Zhou, Y.H., Shi, K.A., & Yu, J.Q. (2014). Role of H_2O_2 dynamics in brassinosteroid-induced stomatal closure and opening in Solanum lycopersicum. *Plant Cell Environ*, 37(9), 2036–2050.

Xia, X.-J., Zhou, Y.-H., Shi, K., Zhou, J., Foyer, C. H., & Yu, J.-Q. (2015). Interplay between reactive oxygen species and hormones in the control of plant development and stress tolerance. *J Exper Bot*, 66(10), 2839–2856. https://doi.org/10.1093/jxb/erv089.

Xing, Y., Jia, W., & Zhang, J. (2008). AtMKK1 mediates ABA-induced CAT1 expression and H_2O_2 production via AtMPK6-coupled signalling in Arabidopsis. *Plant J*, 54, 440–451.

Xiong, J., Yang, Y., Fu, G., & Tao, L. (2015). Novel roles of hydrogen peroxide (H_2O_2) in regulating pectin synthesis and demethylesterification in the cell wall of rice (Oryza sativa) root tips. *New Phytol*, 206(1), 118–126. https://doi.org/10.1111/nph.13285.

Xu, F. J., Jin, C. W., Liu, W. J., Zhang, Y. S., & Lin, X. Y. (2011). Pretreatment with H(2) O(2) alleviates aluminum-induced oxidative stress in wheat seedlings. *J Integr Plant Biol*, 53(1), 44–53. https://doi.org/10.1111/j.1744-7909.2010.01008.x.

Xu, Y., Burgess, P., Zhang, X., & Huang, B. (2016). Enhancing cytokinin synthesis by overexpressing ipt alleviated drought inhibition of root growth through activating ROS scavenging systems in Agrostis stolonifera. *J Exp Bot*, 67(6), 1979–1992.

Yang, C.Y. (2014). Ethylene and hydrogen peroxide are involved in hypoxia signalling that modulates AtERF73/HRE1 expression. *Plant Signal Behav*, 9(5), 877.

Yang, T., & Poovaiah, B.W. (2002). Hydrogen peroxide homeostasis: Activation of plant catalase by calcium/calmodulin. *Proc Natl Acad Sci USA*, 99, 4097–4102.

Yıldız, M., Terzi, H., & Bingül, N. (2013). Protective role of hydrogen peroxide pretreatment on defense systems and BnMP1 gene expression in Cr(VI)-stressed canola seedlings. *Ecotoxicology (London, England)*, 22(8), 1303–1312. https://doi.org/10.1007/s10646-013-1117-2.

Zalewski, W., Galuszka, P., Gasparis, S., Orczyk, W., & Nadolska-Orczyk, A. (2010). Silencing of the HvCKX1 gene decreases the cytokinin oxidase/dehydrogenase level in barley and leads to higher plant productivity. *J Exp Bot*, 61, 1839–1851.

Zeinalabedini, M., Majourhat, K., Khayam-Nekoui, M., Hernández, J. A., & Martínez-Gómez, P. (2009). Breaking seed dormancy in long-term stored seeds from Iranian wild almond species. *Seed Sci Technol, 37*(2), 267–275. https://doi.org/10.15258/sst.2009.37.2.01

Zhang, A., Jiang, M., Zhang, J., Tan, M., & Hu, X. (2006). Mitogen-activated protein kinase is involved in abscisic acid-induced antioxidant defense and acts downstream of reactive oxygen species production in leaves of maize plants. *Plant Physiol*, 141, 475–487.

Zhang, M., Smith, J.A.C., Harberd, N.P., & Jiang, C. (2016). The regulatory roles of ethylene and reactive oxygen species (ROS) in plant salt stress responses. *Plant Mol*, 91, 651–659.

Zhang, R., Wang, Y., Hussain, S., Yang, S., Li, R., Liu, S., Chen, Y., Wei, H., Dai, Q., & Hou, H. (2022). Study on the effect of salt stress on yield and grain quality among different rice varieties. *Front Plant Sci*, 13. https://www.frontiersin.org/articles/10.3389/fpls.2022.918460.

Zhang, X., Zhang, L., Dong, F., Gao, J., Galbraith, D.W., & Song, C.P. (2001). Hydrogen peroxide is involved in abscisic acid-induced stomatal closure in Vicia faba. *Plant Physiol*, 126, 1438–1448.

Zhang, Y., Tan, J., Guo, Z., Lu, S., He, S., Shu, W., & Zhou, B. (2009). Increased abscisic acid levels in transgenic tobacco over-expressing 9 cis-epoxycarotenoid dioxygenase influence H_2O_2 and NO production and antioxidant defences. *Plant Cell Environ*, 32(5), 509–519.

Zhongjing, L., Yan-kui, G., Shaohang, L., & Jigang, B. (2009). Effects of exogenous hydrogen peroxide on ultrastructure of chloroplasts and activities of antioxidant enzymes in greenhouse-ecotype cucumber under drought stress. https://www.semanticscholar.org/paper/Effects-of-exogenous-hydrogen-peroxide-on-of-and-of-Zhongjing-Yan-kui/f309a34ddba537d61a4fdc45599b491e98cd713e.

Zhou, B., Li, N., Zhang, Z., Huang, X., Chen, H., Hu, Z., Pang, X., Liu, W., & Lu, Y. (2012). Hydrogen peroxide and nitric oxide promote reproductive growth in Litchi chinensis. *Biol Plant*, 56(2), 321–329. https://doi.org/10.1007/s10535-012-0093-3.

Zhu, T., Deng, X., Zhou, X., Zhu, L., Zou, L., Li, P., Zhang, D., & Lin, H. (2016). Ethylene and hydrogen peroxide are involved in brassinosteroid-induced salt tolerance in tomato. *Sci Rep*, 6, 35392.

7 Role of Hydrogen Peroxide in Relation to Metabolic Networks

Tanveer Ahmad Khan, Bisma Hilal, and Qazi Fariduddin

INTRODUCTION

Among the ROS, hydrogen peroxide (H_2O_2) is a non-radical reactive molecule discovered by Louis Jacques Thenard and has received large attention in the last decades from plant scientists. It is now well-established that H_2O_2 plays a dual role in physiological and metabolic processes as well as in stress tolerance in plants (Petrov and Van Breusegem, 2012; Černý et al., 2018). Plasma membrane-localized NADPH oxidase, which produces the extracellular formation of $O_2^{\bullet-}$ from molecular oxygen (O_2), in which NADPH acts as an electron donor, is the key source of H_2O_2 signaling in plants (Smirnoff and Arnaud, 2019). Plant NADPH oxidases are partially homologous to the mammalian phagocyte gp91phox (NADPH oxidase 2) and have similar $O_2^{\bullet-}$ producing activity which is then converted into H_2O_2 (Keller et al., 1998; Sagi & Fluhr, 2001; Halliwell & Gutteridge, 1999). In *Arabidopsis thaliana* ten NADPH oxidase genes (*Atrboh*) have been identified (Halliwell & Gutteridge, 1999). Superoxide dismutases (SOD) enzyme catalyzes the dismutation of $O_2^{\bullet-}$ to H_2O_2 (Halliwell & Gutteridge, 1999). H_2O_2 can be transported through the cellular membranes by aquaporins (Henzler & Steudle, 2000). Compared to other ROS compounds, H_2O_2 acts as an intercellular as well as intracellular signaling molecule in plants (Allan & Fluhr, 1997; Kovtun et al., 2000), as it is moderately reactive with no unpaired electrons and is relatively stable compared with other ROS (Halliwell & Gutteridge, 1999). In response to stress conditions, the level of H_2O_2 upsurges and induces oxidative stress, but normal metabolism in plant cells also produces H_2O_2. Above a certain limit, ROS imposes oxidative damage and potentially disturbs the redox balance and metabolic state in cells, however, plants have also developed a series of antioxidant enzymes which maintain the steady state of ROS production and scavenging and also upregulating under environmental perturbation. The antioxidant enzymes include superoxidase dismutase, ascorbate peroxidase, catalase, and glutathione peroxidase are the first line or main ROS-scavenging enzymes in plants. These enzymes together with other non-enzymatic antioxidants, forms antioxidant systems and make the cells highly efficient for detoxifying ROS including H_2O_2. The stability between superoxidase dismutase and the diverse H_2O_2-neutralizing enzymes is thought to be important in maintaining the steady state of $O_2^{\bullet-}$ and H_2O_2 in cells (Asada and Takahashi, 1987; Bowler et al., 1991). Furthermore, catalase is the major H_2O_2 scavenger in photosynthetic cells, which catalyzes H_2O_2 into H_2O and O_2 (Scandalios, 1987). Catalase neutralizes H_2O_2 produced during β-oxidation of fatty acids, mitochondrial electron transport, and, particularly in photorespiratory oxidation (Scandalios et al. 1997). Moreover, H_2O_2 is continuously formed by both enzymatic as well as by non-enzymatic pathways and is involved in diverse plant processes such as stomatal movement (Ge et al., 2015), seed germination (Barba-Espín et al., 2010), development of root system (Ma et al., 2014; Hernández-Barrera et al., 2015), flowering (Liu et al., 2013), senescence (Liao et al., 2012) and programmed cell death (Cheng et al., 2016; Vavilala et al., 2015). However, the dual role performed by H_2O_2 in plants depends on the H_2O_2 level and physiological and metabolic processes affected by H_2O_2. Therefore, it is necessary to differentiate the positive (signaling and metabolic regulation)

 DOI: 10.1201/9781032647692-7

and negative (causing damage) roles governed by H_2O_2. It is also important to decipher the involvement of H_2O_2 in plant metabolism and at higher concentrations how it affects plant metabolism. In this chapter, we focus on the dual role of H_2O_2 in influencing plant metabolism under normal and stressed conditions along with the role of exogenous H_2O_2 in modulating plant metabolism.

H_2O_2: BOON OR BANE

Reactive oxygen species (ROS) including hydrogen peroxide (H_2O_2) are unavoidable byproducts of aerobic metabolism and are generated in two ways i.e., either after incomplete reduction of oxygen or energy transfer to its chemically inert triplet ground states, the former process produces superoxide radical— $O_2^{\bullet-}$; H_2O_2; hydroxyl radical—$HO^{\bullet}$, while the latter yields singlet oxygen—$^1O_2$ (Gechev et al., 2006; Kim et al., 2008). H_2O_2 or its precursor superoxide $O_2^{\bullet-}$; is generated as a byproduct of several cellular processes such as photosynthetic linear electron transport, the mitochondrial respiratory chain, or substrate oxidation, for example by the photorespiration enzyme glycolate oxidase (Cheeseman, 2007; Foyer and Noctor, 2016). H_2O_2 is a relatively stable molecule and less reactive compared to other ROS it is thus recognized as a signaling molecule in physiological and metabolic processes (Waszczak et al., 2018). For instance, RBOHD-mediated H_2O_2 production plays a role in local and systemic signaling as in the case of abscisic acid-mediated stomatal movement (Pei et al., 2000; Kwak et al., 2003). However, ROS at higher levels induces oxidative stress and alters or disturbs the metabolic and physiological processes. Because of the strong oxidizing potential of these oxygen derivatives (ROS), they can lead to the impairment of diverse biomolecules such as DNA, RNA, proteins, and lipids, and therefore ROS are undesirable byproducts of normal physiological and metabolic processes in all aerobic organisms (Halliwell, 2006; Khan et al., 2022, 2023a, b). In response to environmental stress conditions, there can be an excessive rise in the levels of ROS which could lead to an oxidative stress (Mittler et al., 2022 Khan et al., 2020, 2023a,b; Hilal et al., 2023). Since plants being sessile cannot escape from the negative environmental stress factors, therefore they have developed an intricate system to regulate ROS levels (Mittler et al., 2011; Hilal et al., 2023). Notably, plants have evolved different routes to utilize lower levels of ROS as signal transducers for a range of plant metabolic, physiological, and developmental processes such as photosynthesis, cell elongation (Foreman et al., 2003), and cellular differentiation (Tsukagoshi et al., 2010), and in responses to abiotic and biotic stresses (Dat et al., 2000; Gapper and Dolan, 2006).

INVOLVEMENT OF H_2O_2 IN PLANT METABOLISM

The cellular metabolism such as carbon fixation, carbon assimilation, synthesis of starch and other sugars, amino acids, lipids, and chlorophylls are under redox control but are also affected by the concentration of H_2O_2 along with its spatial and temporal production (Ozaki et al., 2009; Strand et al., 2015; Wan and Liu, 2008; Asghar et al., 2022). It has been reported in various studies that oxidative stress altered the numbers of metabolites and their levels (Savchenko and Tikhonov, 2021). For instance, the contents of several sugars including fructose, sucrose, raffinose, glucose, and verbascose along with most amino acids enhanced after oxidative stress, while asparagine, glutamic acid, and methionine levels and contents of few tricarboxylic acid cycle intermediates (succinate, 2-oxoglutarate, malate, isocitrate) became lesser. The oxidative state-triggered metabolic modification suggests the rerouting of glycolytic carbon flow into the oxidative pentose phosphate pathway (OPPP) in order to provide NADPH for the antioxidant system (Obata and Fernie, 2012). In a study, oxidative stress induced by methyl viologen in Arabidopsis led to the alteration in carbon metabolism as evident by the suppression of anabolic processes (e.g., Calvin cycle) and the stimulation of catabolic pathways (e.g., OPPP) (Scarpeci and Valle, 2008). The importance of regulation in appropriate metabolic pathway adjustment was demonstrated during the treatment of methyl viologen by studying the difference between the *Rcd*1 (radical-induced cell death1) mutant which was tolerant to methyl viologen and wild-type Arabidopsis (Sipari et al., 2020). In wild-type plants, the contents of glycolytic and tricarboxylic acid cycle compounds were reduced and the processes of photosynthesis and respiration declined, while in the mutants these processes and the energy production pathways were sustained (Sipari et al., 2020). In another study, mutant Arabidopsis plants were developed in which catalase activity was inhibited and resulting in enhanced H_2O_2 concentrations which thereby improved the levels of several amino acids and organic acids compared to wild plants. (Noctor et al., 2015). Ishikawa et al. (2010) also found that menadione-induced and H_2O_2-mediated oxidative stress enhanced the content of various amino acids and reduced the number of many sugars such as glucose 6-phosphate (glucose 6-P) and fructose 6-phosphate (fructose 6-P). The influence of H_2O_2 on metabolic networks including cell defense, redox homeostasis, protein synthesis and degradation, photosynthesis and photorespiration, and carbohydrate/energy metabolism was revealed by protein expression in rice plants. H_2O_2 treatment differentially regulated the above-mentioned metabolic networks at the protein level. Out of 2000 protein spot analyzed, 144 proteins were differentially expressed and 65 protein spots were upregulated, and 79 were down-regulated in H_2O_2-treated rice seedlings (Wan and Liu, 2008). Transcriptome analysis in wheat seedlings demonstrated that H_2O_2 treatment differentially expressed the H_2O_2 responsive genes which were involved in various cellular responses and metabolic processes, with functions toward cell defense, signal transduction, photosynthesis, carbohydrate metabolism, lipid metabolism, redox homeostasis, and transport (Li et al., 2011). Asghar et al. (2022) studied the effect of H_2O_2 on the metabolite profile in wheat seedlings and revealed that H_2O_2 induces the oxidative pentose phosphate pathway (OPPP) which generates reducing power as exhibited by decreased glucose-6-phosphate level. This effect on glucose-6-phosphate can also be related to many amino acids from the intermediate products of glycolysis. Under water stress, H_2O_2 acts as a signaling molecule and led to the up-regulation of ascorbate and glutathione metabolism by increasing the activities of ascorbate peroxidase, glutathione reductase, monodehydroascorbate reductase, dehydroascorbate reductase, L-galactono-1,4-lactone dehydrogenase and gamma-glutamylcysteine synthetase, and the contents of reduced ascorbate, reduced glutathione, total ascorbate and total glutathione (APX, GR, MDHAR, DHAR, GalLDH, γ-ECS and the contents of AsA, GSH) which, in turn, promotes water-stress tolerance in wheat plants (Shan and Ou, 2018).

The beneficial role of H_2O_2 as a signaling hub integrating metabolic and physiological processes during seed germination have been deciphered in earlier studies. Seed germination is the fundamental stage of the plant life cycle, determining plant growth and yield. Germination is a complex process that starts with imbibition and comprises actions connected with the modification of a quiescent dry seed to a metabolically active state (Wojtyla et al., 2016). The emergence of the radicle and plumule completes the process of germination which involve the resumption of metabolism, reactivation of cellular respiration, transcription and translation, and the onset of reserve mobilization among others (Bentsik and Koornneef, 2008; Weitbrecht et al., 2011; Bewley et al., 2013). These metabolic and cellular actions stimulated by water uptake are linked to the production of ROS, particularly H_2O_2 (El-Maarouf-Bouteau and Bailly, 2008). However, H_2O_2 acts as a signaling hub for the regulation of germination. It has been observed that ROS production during the process of seed germination involves reserve mobilization by oxidative changes of available proteins, which may be used by storage organs as signals to mobilize reserves to the fast-growing embryonic axis. The higher amount of seed storage proteins present in seeds, their oxidized forms can neutralize the excess ROS (Job et al., 2005; Barba-Espín et al., 2011). The modification of glycolytic enzymes, aldolase reductase, mitochondrial ATP synthase, and methionine synthase is a positive stimulator of germination, as precise oxidation processes can protect other cellular components against the damaging effects of ROS. Additionally, the disruption of other metabolic activities (e.g., glycolytic enzymes) may lead to the activation of the PPP, providing reducing power for antioxidant enzymes by activating NADPH (Job et al., 2005; Barba-Espín et al., 2011). ROS are also associated with the weakening of endosperm (mechanical barrier of germination) through cell wall loosening in several seed plants. For example, ROS including H_2O_2 facilitates endosperm cap rupture during lettuce seed germination (Zhang et al., 2014). Lariguet et al. (2013) observed that H_2O_2 promotes seed germination in Arabidopsis by regulating gene expression of an enzyme that hydrolyzes the testa and endosperm, which thereby helps in the development of the embryo. It has also been revealed that H_2O_2 regulates differential expression of genes through oxidation of proteins, regulation of kinase transduction cascade, and modifications in the redox state of cysteine residues of transcription factors and thus regulates various metabolic and physiological networks (Job et al., 2005; Barba-Espín et al., 2011; Bykova et al., 2011a, b; Lariguet et al., 2013). In another report, it has been revealed that H_2O_2 regulated the secondary metabolism which manifested in salinity acclimation in perennial ryegrass (Hu et al., 2016). Several studies also evaluated that H_2O_2 regulates metabolites and a number of genes in stress-challenged plants (Shi et al., 2014; Yang et al., 2002; Dittami et al., 2011). For example, accumulation of 22 amino acids, n-3 to n-6 ratio of polyunsaturated fatty acids, and mannitol were modulated by the H_2O_2 treatment in *Ectocarpus siliculosus* (Dittami et al., 2011). Moreover, H_2O_2 acts as a key signaling molecule stimulating gene expression which is involved in metabolite regulation such as proline and sucrose (Verslues et al., 2007). Therefore, H_2O_2 modulates the different metabolic networks and is also involved in regulating secondary metabolism.

METABOLIC ROLE OF H_2O_2 IN PHOTOSYNTHESIS

Linear flow of electrons is the crucial process by which light energy is captured by the photosynthetic pigments in order to drive the electrons and protons extraction from water and transfer them systematically through electron carriers to reduce NADPH in oxygenic photosynthesis. The process of linear electron flow produces an electrochemical proton gradient or proton motive force (pmf) because this process is coupled to proton translocation into the thylakoid lumen. With the help of pmf or electrochemical proton gradient, ATP is formed which is thereby utilized to drive Calvin–Benson–Bassham cycle and other essential metabolic processes in the chloroplast (Müller et al., 2001; Eberhard et al., 2008). However, a few alternative electrons flow pathways have been also proposed in chloroplast to power ATP production, therefore maintaining the ATP/NADPH budget of the chloroplast (Eberhard et al., 2008; Kramer and Evans, 2011) and among these pathways most essential and complex being the cyclic electron flow (CEF), in which electron flow from the acceptor side of PSI is shunted back into the PQ pool, creating extra pmf that can power ATP production without net NADPH production. This CEF pathway is influenced by H_2O_2, which is generated by the light reactions of photosynthesis and it has already been studied to regulate cellular processes in plants (Traverso et al., 2013; O'Brien et al., 2012; Karpinski et al., 2013). *In vitro*, studies demonstrated that H_2O_2 could activate CEF by changing the NADPH dehydrogenase complex (Lascano et al., 2003). Additionally, H_2O_2 can also enhance the expression of NADPH dehydrogenase complex which may further influence the levels of photosynthetic metabolites, therefore indirectly activating CEF (Livingston et al., 2010). Apart from these roles, H_2O_2 acts as a signaling molecule involved in a number of metabolic networks, probably through its capability to oxidize thiols (Buchanan and Balmer, 2005; Hancock et al., 2006). Moreover, H_2O_2 can be generated under several conditions that activate CEF, for instance under ATP deficit conditions, and electrons accumulate around the PSI acceptor pools, resulting in generation of $O_2^{\bullet-}$ that can be catalyzed by superoxide dismutase to H_2O_2 (Asada, 2006). Furthermore, Strand et al. (2015) provided strong evidence that CEF was vigorously enhanced by H_2O_2 through infiltration or *in situ* generation by glycolate oxidase localized in chloroplast, indicating that H_2O_2 could power CEF either directly or indirectly through redox regulation of key enzymes, or by modulating other photosynthetic processes. Recently, it has been also studied that H_2O_2 is directly transferred from chloroplasts to nuclei, skips the cytosol, and regulates photosynthetic gene expression in epidermal cells of *Nicotiana benthamiana* (Exposito-Rodriguez et al., 2017). It has been found that H_2O_2 interacts with reduced sulfhydryl groups which are involved in the light activation of enzymes of the Calvin cycle at the site of fructose- and sedoheptulose bisphophatase, of phosphoribulokinase, as well as in light-inactivation of oxidative pentose phosphate cycle at the site of glucose-6-phosphate dehydrogenase (Kaiser, 1979).

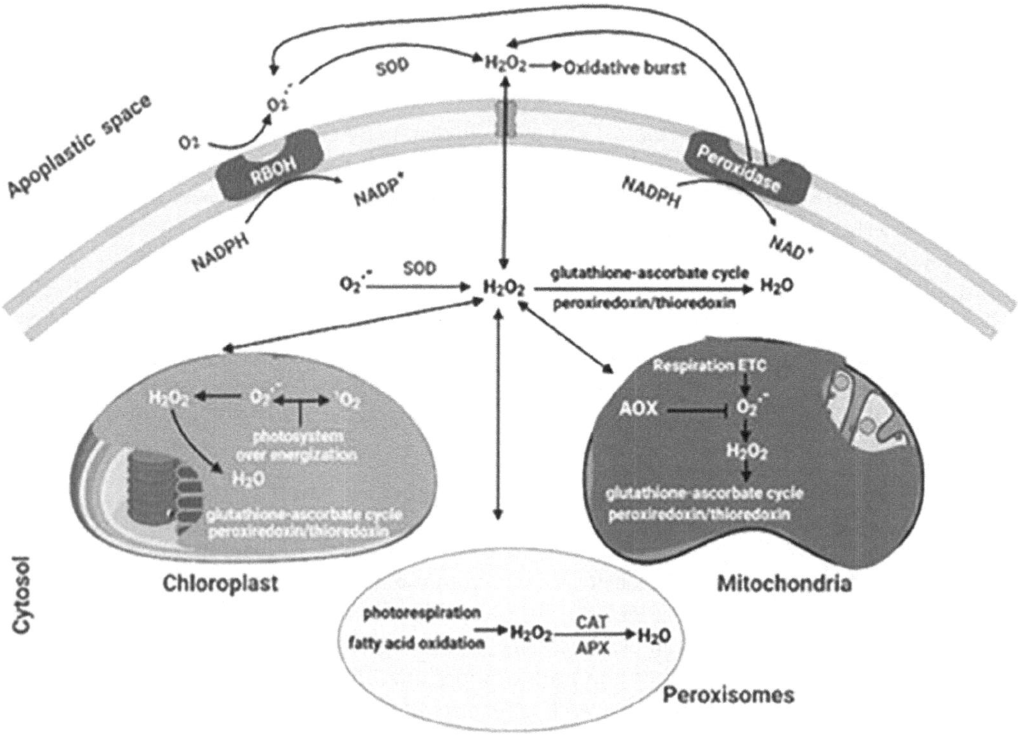

FIGURE 7.1 Depicts production of H_2O_2 and its catabolism in different cellular compartments. The apoplastic ROS generation by NADPH-dependent oxidases (Rboh), and cell wall peroxidases. Superoxide anions are formed by the above-mentioned enzymes and are catalyzed by H_2O_2. ROS are also produced in the chloroplasts where it is turned into water molecules by the antioxidant system such as the glutathione-ascorbate cycle and peroxiredoxins/thioredoxins. Another site of active ROS (H_2O_2) production ismitochondria where a higher accumulation of ROS can be alleviated by an alternative mechanism for electron transport known as alternative oxidase (AOX). Photorespiration and fatty-acid oxidation produce H_2O_2 in peroxisomes at higher levels which are regulated by the catalase enzyme.

It has been reported that H_2O_2 is produced in different cellular locations viz at plasma membranes, chloroplasts, peroxisomes, mitochondria, and cell walls (Figure 7.1) and it has been implicated that H_2O_2 integrates into many signaling networks, which implies H_2O_2 could influence different metabolic processes depending upon the spatial distribution and concentration of H_2O_2.

Transgenic Arabidopsis plants (GO5) that metabolically produce H_2O_2 in chloroplasts and in peroxisomes (which were developed by inhibiting catalase: *cat2-2*) (Fahnenstich et al., 2008; Balazadeh et al., 2012; Queval et al., 2007; Inzé et al., 2012) were grown under non-photorespiratory conditions i.e., under high CO_2 and then shifted to photorespiratory environments (normal air) to stimulate the generation of H_2O_2 in the specific cell organelle. During the process of photorespiration, RuBP oxygenase forms the glycolate. In wild-type plants, peroxisomes having the glycolate which was transported from the chloroplast, where glycolate oxidase catalyzes glycolate into glyoxylate and H_2O_2. H_2O_2 is promptly scavenged by the ample catalase enzyme. Moreover, in catalase-deficient plants, photorespiration impairs H_2O_2 homeostasis in peroxisomes. In the transgenic plants glycolate is already proficiently transformed in glyoxylate and H_2O_2, which overexpresses glycolate oxidase in the chloroplast, therefore disturbing H_2O_2 levels in the chloroplasts. It has been reported that a moderate generation of chloroplastic H_2O_2 activates early signaling responses by regulating gene transcription and also acts as secondary signals, while moderate H_2O_2 generation from peroxisomes stimulates prevailing protein repair responses. Moreover, the results also revealed that H_2O_2

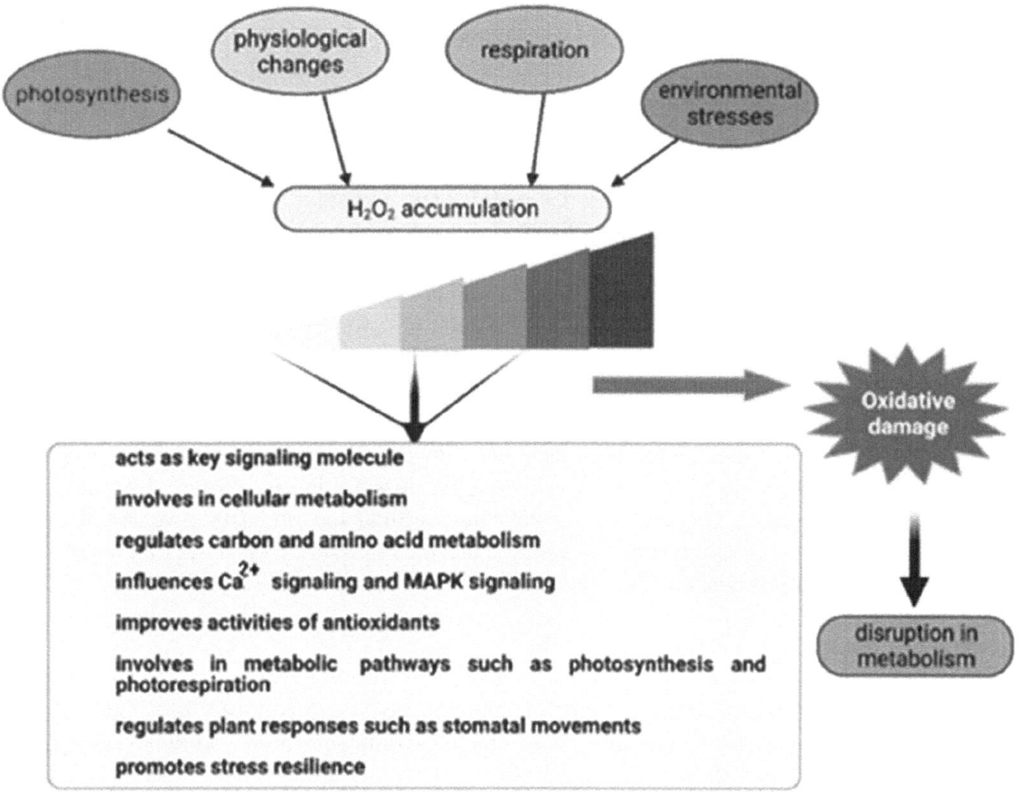

FIGURE 7.2 Endogenous accumulation of hydrogen peroxide (H_2O_2) during normal metabolism (photosynthesis, respiration, growth, stomatal closing) and in response to environmental stress conditions, such as drought stress, salt stress, and heavy metal stress as well as pathogen infection). At lower levels, H_2O_2 acts as a secondary messenger and functions in association with other signaling molecules, and regulates metabolic networks under optimal and suboptimal conditions. However, over-accumulation of H_2O_2 could generate oxidative damage.

production from chloroplast upregulated the genes associated with indolic glucosinolate, camalexin, and stigmasterol biosynthesis. Additionally, H_2O_2 production from chloroplast also accumulates 4-methoxy-indol-3-ylmethyl-glucosinolate and stigmasterol in plants (Sewelam et al., 2014). In summary, H_2O_2 acts as a key signaling molecule and regulates diverse plant metabolic processes as illustrated in Figure 7.2.

METABOLIC ROLE OF H_2O_2 IN ABIOTIC STRESS TOLERANCE

It is largely studied that H_2O_2 plays an important role and is involved in regulating various signaling and metabolic pathways under optimal and suboptimal conditions. For instance, H_2O_2 pre-treatment led to the over-expression of proteins which functions in processes like defense, redox homeostasis, and photosynthesis and therefore alleviated the water stress in wheat cultivars (Bhardwaj et al., 2021). The exogenously applied H_2O_2 has been described to improve stress resilience in other crops (Terzi et al., 2014; Sohag et al., 2020). In *Solanum lycopersicum* L. H_2O_2 treatment improved photosynthesis and metabolic state of the plants by regulating the activity of carbonic anhydrase, and antioxidant systems under stress-free or Cu stress and thereby providing heavy metal tolerance (Nazir et al., 2019). Similarly, Wen et al. (2013) revealed that treatment of H_2O_2 alleviated the Cu stress by regulating the amino acid metabolism through increasing proline content which was associated with the activation of the ornithine and glutamate pathways and a decrease in ProDH. In another study, exogenous treatment of H_2O_2 improved photosynthetic processes and antioxidant metabolism leading to a significant decrease in the drought stress-induced growth inhibition in cucumber seedlings (Sun et al., 2016). Treatment of H_2O_2 has also been shown to promote salt stress tolerance in pistachio seedlings by increasing the activities of ascorbate peroxidase and catalase along with accumulation of non-enzymatic antioxidant (Bagheri et al., 2019). Moreover, supplementation of H_2O_2 to arsenic-treated rice plants increased the activity of superoxide dismutase, glutathione reductase, and ascorbate peroxidase and improved proline metabolism, and reduced glutathione content. H_2O_2 treatment also upregulated the expression of superoxide dismutase, ascorbate peroxidase, and photosynthetic genes *PSBA* and *PSBB* and thereby promoted arsenic-induced heavy metal tolerance (Asgher et al., 2021). The involvement of H_2O_2 in regulating metabolic processes is also coupled with 24-epibrassinolide, leading to increases in CO_2 assimilation by activated carboxylation efficiency of Rubisco and improved total soluble sugar, sucrose, hexose, and starch content as well as promoted the activities of sugar metabolism such as sucrose phosphate synthase, sucrose synthase, and invertase (Jiang et al., 2012). Ozaki et al. (2009) reported that exogenous application of H_2O_2 to melon plants increases the soluble sugar content of melon fruits and photosynthetic responses and the activities of chloroplastic as well as cytosolic fructose-1,6-bisphosphatase, invertases, and sucrose phosphate synthase. The outcomes of the study of Shan and Ou (2018) revealed that water stress-induced H_2O_2 acts as an important signaling molecule that leads to the up-regulation of ascorbate and glutathione metabolism and also enhanced water-stress tolerance in wheat.

CONCLUSION

For normal plant metabolism, a diverse range of physiological processes and the maintenance of redox homeostasis is crucial. The intricate balance between ROS generation and ROS scavenging is very important to regulate cellular redox homeostasis. However, adverse environmental conditions induced the over-accumulation of ROS and disrupt the redox homeostasis of plants by imposing oxidative stress. Upon disruption in cellular redox status, H_2O_2 acts as a cellular messenger and works in coordination with other plant growth regulators to regulate diverse plant responses. For plant metabolism, H_2O_2 seems multifaceted signaling molecule as it is a byproduct of normal plant metabolism particularly in the photosynthetic process. Therefore, H_2O_2 is one of the crucial signals for photosynthetic status, stomatal movements, and in response to abiotic stress to regulate cellular metabolism under optimal and suboptimal conditions. Nevertheless, elevated H_2O_2 concentration also contributes to inducing oxidative stress but at a lower concentration, H_2O_2 regulates the cellular metabolism and shifts the metabolism toward the protecting end. For instance, H_2O_2 influences the activities of the antioxidant system and regulates carbon and amino acid metabolism. Although significant studies have evaluated the involvement of H_2O_2 in plant metabolic processes, there is a pressing need to employ new genetic approaches to delve into the precise metabolic role of H_2O_2. It would be interesting to develop transgenic plants with optimal and deficient H_2O_2-producing plants to decipher the metabolic role of H_2O_2 along with the use of new H_2O_2 probes capable to avail the spatial and temporal production of H_2O_2 and its involvement in metabolic as well as signaling networks.

REFERENCES

Allan, A.C. and Fluhr, R. (1997). Two distinct sources of elicited reactive oxygen species in tobacco epidermal cells. *Plant Cell* 9, 1559–1572.

Asada, K. (2006). Production and scavenging of reactive oxygen species in chloroplasts and their functions. *Plant Physiol* 141(2), 391–396.

Asada, K. and Takahashi, M. (1987). Production and scavenging active oxygen in photosynthesis. In: DJ Kyle, CB Osmond, CJ Amtzen, eds. *Photoinhibition*. Amsterdam: Elsevier. pp. 227–287

Asgher, M., Ahmed, S., Sehar, Z., Gautam, H., Gandhi, S.G. and Khan, N.A., 2021. Hydrogen peroxide modulates activity and expression of antioxidant enzymes and protects photosynthetic activity from arsenic damage in rice (*Oryza sativa* L.). *Journal of Hazardous Materials*, *401*, p.123365.

Asghar, M.A., Balogh, E., Ahres, M., et al. (2022). Ascorbate and hydrogen peroxide modify metabolite profile of wheat differently. *J Plant Growth Regul*. doi: 10.1007/s00344-022-10793-0.

Bagheri, M., Gholami, M., and Baninasab, B. (2019). Hydrogen peroxide-induced salt tolerance in relation to antioxidant systems in pistachio seedlings. *Sci Hortic* 243, 207–213.

Balazadeh, S., Jaspert, N., Arif, M., Mueller-Roeber, B., and Maurino, V.G. (2012). Expression of ROS-responsive genes and transcription factors after metabolic formation of H_2O_2 in chloroplasts. *Front Plant Sci* 3, 234.

Barba-Espín, G., Diaz-Vivancos, P., Clemente-Moreno, M. J., Albacete, A., Faize, L., Faize, M., et al. (2010). Interaction between hydrogen peroxide and plant hormones during germination and the early growth of pea seedlings. *Plant Cell Environ* 33, 981–994. doi: 10.1111/j.1365-3040.2010.02120.x.

Barba-Espín, G., Diaz-Vivancos, P., Job, D., Belghazi, M., Job, C., and Hernández, J.A. (2011). Understanding the role of H_2O_2 during pea seed germination: A combined proteomic and hormone profiling approach. *Plant Cell Environ* 34, 1907–1919. doi: 10.1111/j.1365-3040.2011.02386.x.

Bentsik, L. and Koornneef, M. (2008). Seed dormancy and germination. *Arabidopsis Book* 6, e0119. doi: 10.1199/tab.0119.

Bewley, J.D., Bradford, K.J., Hilhorst, H.M.W., and Nonogaki, H. (2013). *Seeds: Physiology of Development, Germination and Dormancy*. New York: Springer.

Bhardwaj, R.D., Singh, N., Sharma, A., Joshi, R., and Srivastava, P. (2021). Hydrogen peroxide regulates antioxidant responses and redox related proteins in drought stressed wheat seedlings. *Physiol Mol Biol Plants* 27, 151–163. doi: 10.1007/s12298-021-00937-z.

Bowler, C., Slooten, L., Vandenbranden, S., De Rycke, R., Botterman, J., Sybesma, C., Van Montagu, M., and Inzé, D. (1991). Manganese superoxide dismutase can reduce cellular damage mediated by oxygen radicals in transgenic plants. *EMBO J* 10(7), 1723–1732.

Buchanan, B.B. and Balmer, Y. (2005). Redox regulation: A broadening horizon. *Annu Rev Plant Biol* 56, 187–220.

Bykova, N.V., Hoehn, B., Rampitsch, C., Banks, T., Stebbing, J.A., Fan, T. and Knox, R., (2011a). Redox-sensitive proteome and antioxidant strategies in wheat seed dormancy control. *Proteomics* 11, 865–882. doi: 10.1002/pmic.200900810.

Bykova, N.V., Hoehn, B., Rampitsch, C., Hu, J.J., Stebbing, J.A., and Knox, R. (2011b). Thiol redox-sensitive seed proteome in dormant and non-dormant hybrid genotypes of wheat. *Phytochemistry* 72, 1162–1172. doi: 10.1016/j.phytochem.2010.12.021.

Černý, M., Habánová, H., Berka, M., Luklová, M., Brzobohatý, B. (2018) Hydrogen peroxide: Its role in plant biology and crosstalk with signalling networks. *Int J Mol Sci* 19, 2812. doi: 10.3390/ijms19092812.

Cheeseman, J. (2007). Hydrogen peroxide and plant stress: A challenging relationship. *Plant Stress* 1, 4–15.

Cheng, X.X., Yu, M., Zhang, N., Zhou, Z.Q., Xu, Q.T., Mei, F.Z., and Qu, L.H. (2016). Reactive oxygen species regulate programmed cell death progress of endosperm in winter wheat (*Triticum aestivum* L.) under waterlogging. *Protoplasma*, 253, 311–327.

Dat, J., Vandenabeele, S., Vranová, E., Van Montagu, M., Inzé, D., and Van Breusegem, F. (2000). Dual action of the active oxygen species during plant stress responses. *Cell Mol Life Sci* 57, 779–795.

Dittami, S.M., Gravot, A., Renault, D., Goulitquer, S., Eggert, A., Bouchereau, A., Boyen, C., and Tonon, T. (2011). Integrative analysis of metabolite and transcript abundance during the short-term response to saline and oxidative stress in the brown alga *Ectocarpus siliculosus*. *Plant Cell Environ* 34(4), 629–642. doi: 10.1111/j.1365-3040.2010.02268.x.

Eberhard, S., Finazzi, G., and Wollman, F.A. (2008). The dynamics of photosynthesis. *Annu Rev Genet* 42, 463–515.

El-Maarouf-Bouteau, H. and Bailly, C. (2008). Oxidative signaling in seed germination and dormancy. *Plant Signal Bechav* 3, 175–182. doi: 10.4161/psb.3.3.5539.

Exposito-Rodriguez, M., Laissue, P.P., Yvon-Durocher, G., Smirnoff, N., and Mullineaux, P.M. (2017). Photosynthesis-dependent H_2O_2 transfer from chloroplasts to nuclei provides a high-light signalling mechanism. *Nature Commun* 8(1), 49. doi: 10.1038/s41467-017-00074-w.

Fahnenstich, H., Scarpeci, T.E., Valle, E.M., Flügge, U.-I., and Maurino, V.G. (2008). Generation of hydrogen peroxide in chloroplasts of Arabidopsis overexpressing glycolate oxidase as an inducible system to study oxidative stress. *Plant Physiol* 148, 719–729.

Foreman, J., Demidchik, V., Bothwell, J.H., Mylona, P., Miedema, H., Torres, M.A., Linstead, P., Costa, S., Brownlee, C., Jones, J.D., Davies, J.M., and Dolan, L. (2003). Reactive oxygen species produced by NADPH oxidase regulate plant cell growth. *Nature* 422, 442–446.

Foyer, C.H. and Noctor, G. (2016). Stress-triggered redox signalling: What's in prospect? *Plant Cell Environ* 39, 951–964.

Gapper, C. and Dolan, L. (2006). Control of plant development by reactive oxygen species. *Plant Physiol* 141, 341–345.

Ge, X.M., Cai, H.L., Lei, X., Zou, X., Yue, M., and He, J.M. (2015). Heterotrimeric G protein mediates ethylene-induced stomatal closure via hydrogen peroxide synthesis in Arabidopsis. *Plant J* 82, 138–150. doi: 10.1111/tpj.12799.

Gechev, T.S., Van Breusegem, F., Stone, J.M., Denev, I., and Laloi, C. (2006). Reactive oxygen species as signals that modulate plant stress responses and programmed cell death. *BioEssays* 28, 1091–1101.

Halliwell, B. (2006). Reactive species and antioxidants. Redox biology is a fundamental theme of aerobic life. *Plant Physiol* 141, 312–322.

Halliwell, B. and Gutteridge, J.M.C. (1999). *Free Radicals in Biology and Medicine*. Oxford: Oxford University Press.

Hancock, J., et al. (2006). Doing the unexpected: Proteins involved in hydrogen per oxide perception. *J Exp Bot* 57(8), 1711–1718.

Henzler, T. and Steudle, E. (2000). Transport and metabolic degradation of hydrogen peroxide in *Chara corallina*: Model calculations and measurements with the pressure probe suggest transport of H_2O_2 across water channels. *J Exp Bot* 51, 2053–2066.

Hernández-Barrera, A., Velarde-Buendía, A., Zepeda, I., Sanchez, F., Quinto, C., Sánchez-Lopez, R., Cheung, A.Y., Wu, H.M., and Cardenas, L. (2015). Hyper, a hydrogen peroxide sensor, indicates the sensitivity of the Arabidopsis root elongation zone to aluminum treatment. *Sensors* 15(1), 855–867. doi: 10.3390/s150100855.

Hilal, B., Khan, T.A., and Fariduddin, Q. (2023). Recent advances and mechanistic interactions of hydrogen sulfide with plant growth regulators in relation to abiotic stress tolerance in plants. *Plant Physiol Biochem* 196, 1065–1083.

Hu, T., Chen, K., Hu, L., et al. (2016). H_2O_2 and Ca^{2+}-based signaling and associated ion accumulation, antioxidant systems and secondary metabolism orchestrate the response to NaCl stress in perennial ryegrass. *Sci Rep* 6, 36396. doi: 10.1038/srep36396.

Inzé, A., Vanderauwera, S., Hoeberichts, F.A., Vandorpe, M., Van Gaever, T., and Van Breusegem, F. (2012). A subcellular localization compendium of hydrogen peroxide-induced proteins. *Plant Cell Environ* 35, 308–320.

Ishikawa, T., Takahara, K., Hirabayashi, T., et al. (2010). Metabolome analysis of response to oxidative stress in rice suspension cells overexpressing cell death suppressor bax inhibitor-1. *Plant Cell Physiol* 51, 9–20. doi: 10.1093/pcp/pcp162.

Jiang, Y.P., Cheng, F., Zhou, Y.H., Xia, X.J., Mao, W.H., Shi, K., Chen, Z.X., and Yu, J.Q. (2012). Hydrogen peroxide functions as a secondary messenger for brassinosteroids-induced CO_2 assimilation and carbohydrate metabolism in *Cucumis sativus*. *J Zhejiang Univ Sci B* 13, 811–823.

Job, C., Rajjou, L., Lovigny, Y., Belghazi, M., and Job, D. (2005). Patterns of protein oxidation in Arabidopsis seeds and during germination. *Plant Physiol* 138, 790–802. doi: 10.1104/pp.105.062778.

Kaiser, W.M. (1979). Reversible inhibition of the Calvin cycle and activation of oxidative pentose phosphate cycle in isolated intact chloroplasts by hydrogen peroxide. *Planta* 145, 377–382.

Karpinsk, S., Szechynska-Hebda, M., Wituszy nska, W., and Burdiak, P. (2013) Light acclimation, retrograde signalling, cell death and immune defences in plants. *Plant Cell Environ* 36(4), 736–744.

Keller, T., Damude, H.G., Werner, D., Doerner, P., Dixon, R.A., and Lamb, C. (1998). A plant homolog of the neutrophil NADPH oxidase gp91phox subunit gene encodes a plasma membrane protein with Ca^{2+} binding motifs. *Plant Cell* 10, 255–266.

Khan, T.A., Fariduddin, Q., Nazir, F., and Saleem, M. (2020). Melatonin in business with abiotic stresses in plants. *Physiol Mol Biol Plants* 26, 1931–1944.

Khan, T.A., Saleem, M., and Fariduddin, Q. (2022). Recent advances and mechanistic insights on melatonin-mediated salt stress signaling in plants. *Plant Physiol Biochem* 188, 97–107.

Khan, T.A., Saleem, M., and Fariduddin, Q. (2023a). Melatonin influences stomatal behavior, root morphology, cell viability, photosynthetic responses, fruit yield, and fruit quality of tomato plants exposed to salt stress. *J Plant Growth Regul* 42(4), 2408–32.

Khan, T.A., Saleem, M., and Fariduddin, Q. (2023b). Differential drought stress tolerance in five varieties of tomato (*Solanum lycopersicum* L.): An evaluation of photosynthesis and antioxidant system. *J Soil Sci Plant Nutr* 23(2), 2810–31.

Kim, C., Meskauskiene, R., Apel, K., and Laloi, C. (2008). No single way to understand singlet oxygen signalling in plants. *EMBO Rep* 9, 435–439.

Kovtun, Y., Chiu, W.L., Tena, G., and Sheen, J. (2000). Functional analysis of oxidative stress-activated mitogen-activated protein kinase cascade in plants. *Proc Natl Acad Sci USA* 97, 2940–2945.

Kramer, D.M. and Evans, J.R. (2011). The importance of energy balance in improving photosynthetic productivity. *Plant Physiol* 155(1), 70–78.

Kwak, J.M., Mori, I.C., Pei, Z.M., Leonhardt, N., Torres, M.A., Dangl, J.L., Bloom, R.E., Bodde, S., Jones, J.D., and Schroeder, J.I. (2003) NADPH oxidase AtrbohD and AtrbohF genes function in ROS-dependent ABA signaling in Arabidopsis. *EMBO J* 22, 2623–2633.

Lariguet, P., Ranocha, P., De Meyer, M., Barbier, O., Penel, C., and Dunand, C. (2013). Identification of a hydrogen peroxide signalling pathway in the control of light-dependent germination in Arabidopsis. *Planta* 238, 381–395. doi: 10.1007/s00425-013-1901-5.

Lascano, H.R., Casano, L.M., Martín M., and Sabater, B. (2003). The activity of the chloroplastic Ndh complex is regulated by phosphorylation of the NDH-F subunit. *Plant Physiol* 132(1), 256–262.

Li, A., Zhang, R., Pan, L., et al. (2011). Transcriptome analysis of H_2O_2-treated wheat seedlings reveals a H_2O_2-responsive fatty acid desaturase gene participating in powdery mildew resistance. *PLoS One* 6, e28810. doi: 10.1371/journal.pone.0028810.

Liao, W.B., Zhang, M.L., Huang, G.B., and Yu, J.H. (2012). Hydrogen peroxide in the vase solution increases vase life and keeping quality of cut Oriental×Trumpet hybrid lily 'Manissa'. *Sci Horticult* 139, 32–38. doi: 10.1016/j.scienta.2012.02.040.

Liu, J., Macarisin, D., Wisniewski, M., Sui, Y., By, S., Norelli, J., et al. (2013). Production of hydrogen peroxide and expression of ros-generating genes in peach flower petals in response to host and non-host fungal pathogens. *Plant Pathol* 62, 820–828. doi: 10.1111/j.1365-3059.2012.02683.x.

Livingston, A.K., Kanazawa, A., Cruz, J.A., and Kramer, D.M. (2010). Regulation of cyclic electron flow in C_3 plants: Differential effects of limiting photosynthesis at ribulose-1,5- bisphosphate carboxylase/oxygenase and glyceraldehyde-3-phosphate dehydrogenase. *Plant Cell Environ* 33(11), 1779–1788.

Ma, F., Wang, L.J., Li, J.L., Samma, M.K., Xie, Y.J., Wang, R., et al. (2014). Interaction between HY1 and H_2O_2 in auxin-induced lateral root formation in Arabidopsis. *Plant Mol Biol* 85, 49–61. doi: 10.1007/s11103-013-0168-3.

Mittler, R., Vanderauwera, S., Suzuki, N., Miller, G., Tognetti, V.B., Vandepoele, K., Gollery, M., Shulaev, V., and Van Breusegem, F. (2011). ROS signaling: The new wave? *Trends in Plant Sci* 16, 300–309.

Müller, P., Li, X.P., and Niyogi, K.K. (2001). Non-photochemical quenching. A response to excess light energy. *Plant Physiol* 125(4), 1558–1566.

Mittler, R., Zandalinas, S.I., Fichman, Y. and Van Breusegem, F., 2022. Reactive oxygen species signalling in plant stress responses. *Nature Reviews Mol Cell Biology*, *23*(10), pp.663-679.

Nazir, F., Hussain, A., and Fariduddin, Q. (2019). Hydrogen peroxide modulate photosynthesis and antioxidant systems in tomato (*Solanum lycopersicum* L.) plants under copper stress. *Chemosphere* 230, 544–558.

Noctor, G., Lelarge-Trouverie, C., and Mhamdi, A. (2015). The metabolomics of oxidative stress. *Phytochemistry* 112, 33–53. doi: 10.1016/j.phytochem.2014.09.002.

O'Brien, J.A., Daudi, A., Butt, V.S., and Bolwell, G.P. (2012). Reactive oxygen species and their role in plant defence and cell wall metabolism. *Planta* 236(3), 765–779.

Obata, T. and Fernie, A.R. (2012). The use of metabolomics to dissect plant responses to abiotic stresses. *Cell Mol Life Sci* 69, 3225–3243. doi: 10.1007/s00018-012-1091-5.

Ozaki, K., et al. (2009). Enrichment of sugar content in melon fruits by hydrogen peroxide treatment. *J Plant Physiol* 166, 569–578. doi: 10.1016/j.jplph.2008.08.007.

Pei, Z.M., Murata, Y., Benning, G., Thomine, S., Klusener, B., Allen, G.J., Grill, E., and Schroeder, J.I. (2000). Calcium channels activated by hydrogen peroxide mediate abscisic acid signalling in guard cells. *Nature* 406, 731–734.

Petrov, V.D. and Van Breusegem, F. (2012). Hydrogen peroxide—a central hub for information flow in plant cells. *AoB Plants* 2012, pls014, doi: 10.1093/aobpla/pls014.

Queval, G., Issakidis-Bourguet, E., Hoeberichts, F.A., Vandorpe, M., Gakière, B., Vanacker, H., Miginiac-Maslow, M., Van Breusegem, F., and Noctor, G. (2007). Conditional oxidative stress responses in the Arabidopsis photorespiratory mutant cat2 demonstrate that redox state is a key modulator of daylength-dependent gene expression, and define photoperiod as a crucial factor in the regulation of H_2O_2-induced cell death. *Plant J.* 52, 640–657.

Sagi, M. and Fluhr, R. (2001). Superoxide production by plant homologues of the gp91phox NADPH oxidase. Modulation of activity by calcium and by tobacco mosaic virus infection. *Plant Physiol* 126, 1281–1290.

Savchenko, T. and Tikhonov, K. (2021). Oxidative stress-induced alteration of plant central metabolism. *Life* 11, 304. doi: 10.3390/life11040304.

Scandalios, J.G. (1987). The antioxidant enzyme genes CAT and SOD of maize: Regulation, functional significance, and molecular biology. *Isozymes Curr Topics Biol Med Res* 14, 19–44.

Scandalios, J.G., Guan, L.M., and Polidoros, A. (1997). *Oxidative Stress and the Molecular Biology of Antioxidant Defenses*. Cold Spring Harbor Lab. Press Planview, New York, pp. 343–406.

Scarpeci, T.E. and Valle, E.M. (2008). Rearrangement of carbon metabolism in *Arabidopsis thaliana* subjected to oxidative stress condition: an emergency survival strategy. *Plant Growth Regul* 54, 133–142. doi: 10.1007/s10725-007-9236-5.

Sewelam, N., Jaspert, N., Van Der Kelen, K., Tognetti, V.B., Schmitz, J., Frerigmann, H., Stahl, E., Zeier, J., Van Breusegem, F., and Maurino, V.G. (2014). Spatial H_2O_2 signaling specificity: H_2O_2 from chloroplasts and peroxisomes modulates the plant transcriptome differentially. *Mol Plant* 7(7), 1191–1210.

Shan, C. and Ou, X. (2018). Hydrogen peroxide is involved in the regulation of ascorbate and glutathione metabolism in wheat leaves under water stress. *Cereal Res Commun* 46, 21–30. doi: 10.1556/0806.45.2017.053.

Shi, H., Ye, T., Zhong, B., Liu, X., and Chan, Z. (2014). Comparative proteomic and metabolomic analyses reveal mechanisms of improved cold stress tolerance in bermudagrass (Cynodon dactylon (L.) Pers.) by exogenous calcium. *J Integr Plant Biol* 56, 1064–1079. doi: 10.1111/jipb.12167.

Sipari, N., Lihavainen, J., Shapiguzov, A., et al. (2020). Primary metabolite responses to oxidative stress in early-senescing and paraquat resistant *Arabidopsis thaliana* rcd1 (radical-induced cell death1). *Front Plant Sci* 11, 194. doi: 10.3389/fpls.2020.00194.

Smirnoff, N., and Arnaud, D. (2019) Hydrogen peroxide metabolism and functions in plants. *New Phytol* 221(3): 1197–1214

Sohag, A.A., Tahjib-Ul-Arif, M., Brestic, M., Afrin, S., Sakil, M.A., Hossain, M.T., Hossain, M.A., and Hossain, M.A. (2020). Exogenous salicylic acid and hydrogen peroxide attenuate drought stress in rice. *Plant Soil Environ* 66, 7–13.

Strand, D.D., Livingston, A.K., Satoh-Cruz, M., Froehlich, J.E., Maurino, V.G., and Kramer, D.M. (2015). Activation of cyclic electron flow by hydrogen peroxide in vivo. *Proc Natl Acad Sci* 112(17), 5539–5544.

Sun, Y., Wang, H., Liu, S., and Peng, X. (2016). Exogenous application of hydrogen peroxide alleviates drought stress in cucumber seedlings. *S Afr J Bot* 106, 23–28.

Terzi, R., Kadioglua, A., Kalaycioglua, E., and Saglamb, A. (2014). Hydrogen peroxide pretreatment induces osmotic stress tolerance by influencing osmolyte and abscisic acid levels in maize leaves. *J Plant Interact* 9, 559–565.

Traverso, J.A., Pulido, A., Rodríguez-García, M.I., and Alché, J.D. (2013). Thiol-based redox regulation in sexual plant reproduction: New insights and perspectives. *Front Plant Sci* 4, 465.

Tsukagoshi, H., Busch, W., and Benfey, P.N. (2010). Transcriptional regulation of ROS controls transition from proliferation to differentiation in the root. *Cell* 143, 606–616.

Vavilala, S.L., Gawde, K.K., Sinha, M., and D'Souza, J.S. (2015). Programmed cell death is induced by hydrogen peroxide but not by excessive ionic stress of sodium chloride in the unicellular green alga Chlamydomonas reinhardtii. *Eur J Phycol* 50, 422–438. doi: 10.1080/09670262.2015.1070437

Verslues, P.E., Batelli, G., Grillo, S., Agius, F., Kim, Y.S., Zhu, J., Agarwal, M., Katiyar-Agarwal, S., and Zhu, J.K. (2007). Interaction of SOS2 with nucleoside diphosphate kinase 2 and catalases reveals a point of connection between salt stress and H_2O_2 signaling in *Arabidopsis thaliana*. *Mol Cell Biol* 27(22), 7771–7780. doi: 10.1128/mcb.00429-07.

Wan, X.-Y. and Liu, J.-Y. (2008). Comparative proteomics analysis reveals an intimate protein network provoked by hydrogen peroxide stress in rice seedling leaves. *Mol Cell Proteomics* 7, 1469–1488. doi: 10.1074/mcp.M700488-MCP200.

Waszczak, C., Carmody, M. and Kangasjarvi, J. (2018). Reactive oxygen species in plant signaling. *Ann Rev Plant Biol* 69, 209–236.

Weitbrecht, K., Müller, K., and Leubner-Metzger, G. (2011). First off the mark: Early seed germination. *J Exp Bot* 62, 3289–3309. doi: 10.1093/jxb/err030.

Wen, J.F., Gong, M., Liu, Y., Hu, J.L., and Deng, M.H. (2013). Effect of hydrogen peroxide on growth and activity of some enzymes involved in proline metabolism of sweet corn seedlings under copper stress. *Sci Hortic* 164, 366–371.

Wojtyla, Ł., Lechowska, K., Kubala, S., and Garnczarska, M. (2016). Different modes of hydrogen peroxide action during seed germination. *Front Plant Sci* 7, 66.

Yang, T. and Poovaiah, B.W. (2002). Hydrogen peroxide homeostasis: activation of plant catalase by calcium/calmodulin. *PNAS* 99, 4097–4102. doi: 10.1073/pnas.052564899.

Zhang, Y., Chen, B., Xu, Z., Shi, Z., Chen, S., Huang, X., Chen, J., and Wang, X. (2014). Involvement of reactive oxygen species in endosperm cap weakening and embryo elongation growth during lettuce seed germination. *J Exper Bot* 65(12), 3189–3200. doi: 10.1093/jxb/eru167.

8 Role of Hydrogen Peroxide in Regulation of Photosynthesis and Its Crosstalk with Other Phytohormones under Abiotic Stress

Mohammad Yusuf, Mohd Tanveer Alam Khan, Taiba Saeed, Mohammad Faizan, and Mayank Anand Gururani

INTRODUCTION

A century ago, Louis Jacques Thenard made the discovery of hydrogen peroxide (H_2O_2), a chemical compound with characteristics that could warrant its classification as a phytohormone (Cerny et al. 2018). Today, H_2O_2 is recognized as a type of reactive oxygen species and is regarded as a ubiquitous cellular metabolite. Plants produce this molecule through multiple sources, encompassing both enzymatic and non-enzymatic routes, involving NADPH oxidases and peroxidases, as well as metabolic processes like photorespiration (Niu and Liao 2016). Moreover, its production is elevated during stress conditions and is also a crucial component in various growth and developmental processes. However, H_2O_2 has the potential to cause oxidative damage in excess, its controlled production and regulated levels serve important functions in plant physiology, helping plants adapt to environmental challenges and maintain their overall health. On the other hand, when present in minute nanomolar concentrations, H_2O_2 acts as a signaling molecule, garnering increased attention for its influence on plant growth and development (Htet et al. 2019). Studies have unveiled that H_2O_2 plays a pivotal role in regulating stress defenses and secondary metabolic functions in plants (Ellouzi et al. 2017). By modulating the proteome model, H_2O_2 also holds the potential to significantly modulate the morpho-physiological and biochemical responses in plants (Khan et al. 2019). H_2O_2 exhibits the potential to alter the activation state of RuBisCO (Jamaludin et al. 2020). In a recent study by Jamaludin et al. (2020), foliar application of H_2O_2 was found to enhance the expression of the plastid DNA-origin rbcL gene, associated with the RuBisCO component, in *F. deltoideae*, thereby boosting photosynthesis. H_2O_2 has firmly established its significant contributions to plant responses to various environmental factors, including heavy metals (Hasanuzzaman et al. 2017; Nazir et al. 2019), waterlogging (Andrade et al. 2018), salinity (Bagheri et al. 2019), and low-temperature stress (Khan et al. 2019). The researchers propose that H_2O_2 influences gene expression through protein oxidation, kinase activation, and modulation of transduction cascades. Moreover, H_2O_2 foliar application increased both electron transport (J) and triose phosphate (TPU) in water-deficient plants, which positively impacted the photosynthetic dynamics and resulted in improved growth compared to well-watered plants. Stomatal closure, a primary limiting factor for photosynthesis in moderately water-deficient plants, reduces CO_2 concentration available for the process (Barzotto et al. 2023). As assimilation decreases with declining CO_2 concentration, there is a subsequent reduction in sucrose-phosphate synthase enzyme activity (Barzotto et al. 2023), potentially leading to decreased TPU. Sharkey (2007 suggested that restricting TPU capacity by

DOI: 10.1201/9781032647692-8

eliminating the reproductive component causes sucrose accumulation and negative feedback for photosynthesis. The condition of water deficiency and the need for osmotic adjustment likely constrain drainage strength, consequently diminishing growth capacity. Higher TPU is correlated with increased sucrose-phosphate synthase enzyme activity, as demonstrated by Ozaki et al. (2009) following H_2O_2 administration to *Cucumis melo* plants. It is widely acknowledged that H_2O_2 exerts a substantial influence on plant stress responses, primarily by serving as a mediator that orchestrates the metabolically or ecologically monitored activation of specific tolerance genes (Phukan et al. 2016; Pornsiriwong et al. 2017; Qi et al. 2018). Research on protein kinases indicates that mitogen-activated protein kinases (MAPKs) undergo activation in response to H_2O_2 in both animal and plant systems, potentially influencing gene expression (Torres and Forman, 2003). The role of H_2O_2 in plant signaling during abiotic stress suggests its potential exogenous use to induce metabolic shifts and facilitate acclimation (Barzotto et al. 2023). This process, akin to defense priming, enhances the capacity for improved stress response following initial signaling (Torres and Forman 2023). H_2O_2 induces the oxidation of cysteine residues, serving as a recognized molecular switch implicated in numerous signaling pathways. This leads to the formation of disulfides in conjunction with glutathione, believed to protect proteins from further oxidation and modulate interactions with DNA. Through control of gene expression, this signaling mechanism enables swift responses without the need for de novo mRNA synthesis and nuclear export (Lennicke et al. 2015). Crucial to these responses within the plant innate immune system is a signaling pathway wherein NADPH oxidase serves as a key mediator, regulating H_2O_2 production and influencing the intensity of defensive responses (Rohman et al. 2024). In the upcoming sections, this chapter will examine the origin, prevalence, and dispersion of H_2O_2 in plants. It will delve into the physiological responses of plants, with particular attention to how the photosynthetic machinery responds to H_2O_2 in stressful environmental conditions. Additionally, it will analyze the signaling pathways triggered by H_2O_2 in plant cells under abiotic stress. Lastly, it will investigate the interactions between H_2O_2 and other phytohormones during periods of abiotic stress.

ORIGIN, OCCURRENCE, AND DISTRIBUTION OF H_2O_2

Origin: H_2O_2 is a common molecule found in plant cells. It plays a crucial role in various physiological processes such as growth, development, and stress responses. The origin of hydrogen peroxide in plants can be traced back to both enzymatic and non-enzymatic processes. Enzymatic production of hydrogen peroxide occurs via the activity of various enzymes such as NADPH oxidases, peroxidases, and amine oxidases. NADPH oxidases are responsible for generating H_2O_2 during growth and development in different areas of the plant such as the root, shoot, leaf, and flower (Cerny et al. 2018). Peroxidases, on the other hand, play a vital role in stress responses such as pathogen attack, wounding, and exposure to abiotic stress. They catalyze the oxidation of different macromolecules such as lignin and phenolics, which leads to the formation of H_2O_2 (Quan et al. 2008). Additionally, amine oxidases contribute to H_2O_2 generation during cell wall maturation and ripening of fruits. Non-enzymatic processes are also involved in the production of H_2O_2 in plants. This includes the auto-oxidation of various molecules such as flavonoids and polyphenols. These molecules can undergo oxidation in the presence of atmospheric oxygen, which results in the formation of H_2O_2 (Quan et al. 2008). Moreover, the photochemical reactions that take place during photosynthesis can also lead to the formation of H_2O_2. Once produced, H_2O_2 is involved in various physiological processes in plants such as programmed cell death, pathogen defense, and cell signaling. It is also involved in redox signaling, which is the process of regulating cellular processes through the transfer of electrons between different molecules. H_2O_2 functions as a signaling molecule in plants, regulating various processes such as growth, development, and stress responses.

Occurrence: In plants, flavin-containing proteins have the potential to catalyze the reduction of oxygen in a manner dependent on blue light. H_2O_2 production stemming from cryptochrome, a blue light photoreceptor, has been observed in Arabidopsis (Arthaut et al. 2017). Additionally, other flavin-containing enzymes may generate $O_2^{\cdot-}$ and subsequently H_2O_2 in the presence of light. Evidence suggests that a yeast mutant lacking peroxisomal acyl-CoA oxidase exhibits impairment in light-induced H_2O_2 production and downstream signaling (Bodvard et al. 2017). In plants, $O_2^{\cdot-}$ / H_2O_2 forming oxidases are involved in photorespiration (glycolate oxidase), fatty acid oxidation (acyl-CoA oxidase), polyamine and purine catabolism, and the synthesis of some hormones. Large fluxes of H_2O_2 are particularly likely for photorespiration (Queval et al. 2007) and fatty acid utilization during germination of oilseeds in glyoxysomes, a specialized type of peroxisome. Peroxisomes appear to be a source of NO as well, providing the possibility of peroxinitrite formation. Chloroplasts and mitochondria are sites of intense electron transport activity, coupled with the production of large electro-chemical gradients across energy transducing membranes. These conditions provide opportunities for H_2O_2 generation via $O_2^{\cdot-}$ dismutation. In chloroplasts, there are multiple potential sites of $O_2^{\cdot-}$ production, such as photosystems I and II (PSI and PSII) and the electron transport chain (ETC). Oxygen photoreduction producing $O_2^{\cdot-}$ and H_2O_2 has been documented *in situ*, in isolated chloroplasts and in thylakoid membranes (Mubarakshina Borisova et al. 2012). The stomatal and thylakoid membrane located SOD assists conversion to H_2O_2, followed by reduction to water by peroxidases. Although it is tacitly assumed that the Mehler reaction does not involve specific proteins, this may repay consideration. Alternative routes of oxygen photoreduction have been proposed which could act as 'valves' to decrease H_2O_2 production. Flavodi iron proteins (FDPs) stake electrons from PSI and reduce oxygen to water instead of H_2O_2. Relevant to this, H_2O_2 production activates NADPH-dependent CET (Strand al. 2015). Mitochondria generate $O_2^{\cdot-}$ and H_2O_2 through the ETC, particularly through respiratory complexes I, II and II and Flavin-containing dehydro-genases (Schwarzlander et al. 2009; Riemer et al. 2015; S. Huang et al. 2016). NOXs are plasma membrane enzymes which use electrons from cytosolic NADPH to reduce oxygen to $O_2^{\cdot-}$ in the apoplast. $O_2^{\cdot-}$ production is followed by its rapid dismutation to H_2O_2. This proton-requiring reaction is favored by low apoplastic pH, but there is also evidence for apoplastic SOD activity, perhaps using germin-like proteins. The proton consumption by $O_2^{\cdot-}$ dismutation could increase apoplastic pH, and the RBOH-mediated transport of electrons to the extracellular space will depolarize the membrane, so that NOX function might not just be about $O_2^{\cdot-}$ (Segal 2016). Given that $O_2^{\cdot-}$ is both short lived and poorly membrane permeable, the proposed signaling roles for NOXs either require aquaporin-mediated H_2O_2 transport to the cytosol or interaction of its products with plasma membrane sensor kinases. NOX has received high profile attention, but there are other sources of apoplastic H_2O_2. The cell wall contains numerous type III peroxidase isoenzymes with functions in the cross-linking of cell wall polymers. However, peroxidases have been reported to variously generate $O_2^{\cdot-}$, H_2O_2 and hydroxyl radicals, depending on the chemical environment (Daudi et al. 2012; Kimura et al. 2014). The H_2O_2-generating reaction is not fully understood but is favored by high pH and requires a reductant which has not been identified. The ER is an oxidizing compartment for glutathione and thiol groups which facilitates the protein folding processes that depend on the formation of disulfide bonds. The two means of oxidative cross-linking via protein disulfide isomerases (PDIs) involve H_2O_2 (Bulleid 2012). In the first, a FAD-containing ER oxidase (ERO) produces an endo-disulfide by reducing oxygen to H_2O_2. Oxidized ERO then oxidizes PDI, which, in turn, transfers the disulfide bond to target proteins to induce correct folding. This process therefore generates H_2O_2. However, there is strong evidence for the ERO pathway in developing soybean seedlings, where storage protein synthesis involves the extensive formation of disulfide links (Matsusaki et al. 2016). The possibility of a GPX-like (GPXL) pathway in plants is supported by the observation that rice mutants in OsGPXL5, which is ER/chloroplast localized, shown noticeably decreased grain filling, consistent with a role in storage protein synthesis (Matsusaki et al. 2016). Arabidopsis GPXL3 I sattached to the ER membrane (Attacha et al. 2017). GPXLs or Prx scould be involved in the removal of H_2O_2 produced by the ERO system. Finally, but not investigated, the presence of DHA, the oxidized form of ascorbate, in the ER could also facilitate disulfide bond formation. Further investigation of the oxidative processes involved in protein folding in plants is required.

Distribution: H_2O_2 is a reactive oxygen species (ROS) that plays a crucial role in many physiological and developmental processes in plants. However, H_2O_2 must be distributed throughout the plant effectively in order to perform its functions in signaling, stress responses, and other processes. The distribution of H_2O_2 in plants is regulated by several mechanisms, including synthesis, degradation, and transport. H_2O_2 is synthesized through various enzymatic and non-enzymatic reactions, including the respiratory burst oxidase homolog (RBOH) family of proteins, peroxisomal oxidases, and polyamine metabolism (Cerny et al. 2018). Once H_2O_2 is produced, it is rapidly decomposed by ROS-scavenging enzymes such as catalase, ascorbate peroxidase, and glutathione peroxidase. These enzymes play a critical role in modulating the distribution of H_2O_2 and maintaining a balance between its synthesis and degradation. H_2O_2 transport in plants is facilitated by aquaporins, which are channel proteins that allow for the diffusion of small molecules including H_2O_2 and other signaling molecules across cell membranes (Bienert et al. 2014). Aquaporins are present on various membranes including plasma membrane, tonoplast, and potentially chloroplast inner envelope, and evidence suggests that they are required for H_2O_2-mediated signaling and transport. Specific isoforms of aquaporins may have greater selectivity for H_2O_2, and mutations of specific aquaporins have been shown to impair H_2O_2 movement, affecting intracellular signaling and causing increased bacterial growth. The distribution of H_2O_2 in plants is also influenced by its source and target cells. Once produced, H_2O_2 can diffuse across cell membranes to reach target cells where it can act to induce gene expression, aid in stomatal closure, or play a role in other physiological processes. The distribution of H_2O_2 can also be influenced by various stresses, including drought, salinity, and heavy metal toxicity, which can lead to changes in aquaporin expression and other mechanisms of transport (Hossain et al. 2015). In addition to transport through aquaporins, recent studies have suggested that vesicular transport may also play a role in H_2O_2 distribution in plants. Endosomes and lysosomes have been implicated in H_2O_2 distribution and signaling, as they can internalize H_2O_2 and regulate its availability to target compartments. Endosomes have been shown to be involved in the internalization of aquaporins in response to H_2O_2 concentration, suggesting that endosomal transport may play a key role in the regulation of H_2O_2 transport and signaling between organelles and other cellular compartments. The role of chloroplasts in H_2O_2 distribution is particularly important, as chloroplast derived H_2O_2 has been shown to influence nuclear H_2O_2 levels and gene expression, playing a key role in light signaling and other processes. Chloroplast inner envelope aquaporins have been implicated in the transport of chloroplast derived H_2O_2, and their expression or gating could influence the availability of H_2O_2 for signaling and other processes (Nazir et al. 2020). Overall, the distribution of H_2O_2 in plants is a complex and highly regulated process that involves the synthesis, degradation, and transport of the ROS. Aquaporins play a crucial role in H_2O_2 transport, and different isoforms may have greater selectivity for H_2O_2 or other signaling molecules. The different types of cells and organelles in plants also play a role in the distribution of H_2O_2, with specific organelles being involved in its production and signaling. Future studies are needed to fully understand the mechanisms underlying H_2O_2 distribution in plants and its role in various physiological and developmental processes.

EFFECT OF H_2O_2 ON PHYSIOLOGICAL PERFORMANCE OF PLANTS

H_2O_2 holds a central role in the physiological processes of plants, exerting a range of effects on their growth, development, and reactions to environmental pressures. Gaining insight into the intricate relationship between H_2O_2 and plant physiology is essential for comprehending mechanisms of plant adaptation and devising strategies to bolster crop yield and resilience to stressors. One primary function of H_2O_2 in plant physiology is its participation in cellular signaling pathways (Nazir et al. 2020). Plants generate H_2O_2 in response to various biotic and abiotic stresses, such as pathogen intrusion, drought, salinity, and extreme temperatures. Upon detecting stress cues, plants trigger enzymatic pathways that lead to the production of H_2O_2, serving as a secondary messenger in signal transduction cascades. H_2O_2 plays a role in regulating gene expression, ion fluxes, stomatal conductance, and antioxidant defense mechanisms, thereby coordinating adaptive responses to stress. The impact of H_2O_2 on plant growth and development is multifaceted. At low concentrations, H_2O_2 acts as a signaling molecule promoting cell proliferation, root elongation, seed germination, and leaf expansion. H_2O_2 influences plant growth in a concentration-dependent manner (Xiong et al. 2015). Adding 100–500 μM of H_2O_2 to rice plants resulted in the inhibition of root elongation; however, a lower concentration (0–100 μM) led to enhanced cell expansion and increased root diameter (Xiong et al. 2015). In *Solanum lycopersicum*, a root dipping treatment with lower doses of H_2O_2 (0.1 or 0.5 mM) holds potential advantages in influencing plant development (Nazir et al. 2019). Furthermore, H_2O_2 triggers the activation of MAPKs and cell cycle genes during the development of adventitious and lateral roots (Pan et al. 2012). Specifically, H_2O_2 strongly activates MAPK4, MAPK12 (Mittler et al. 2011), MPK1/2 (Zhou et al. 2014), MPK6, and nucleotide diphosphate kinase 2 (NDPK2), which interacts directly with MPK3 and MPK6, thereby regulating redox homeostasis (Moons 2003). Recent studies indicate that H_2O_2 induces the activation of genes responsible for various plant signaling pathways, cell expansion, elasticity, contraction, and cell cycle regulation. These include 14-3-3-like protein, profiling 4, proteasome-alpha-type-5, translationally monitored homolog of tumor protein, and reductase benzoquinone. However, genes encoding albumin-2, lectin, abscisic acid (ABA)-responsive protein ABR18, and cell-autonomous heat shock cognate protein 70 show decreased activation in response to H_2O_2 (Barba-Espin et al. 2010). Application of H_2O_2 to *Brassica juncea* plants via foliar spraying has been shown to enhance photosynthesis (Khan et al. 2016). Similarly, treating wheat seedlings with H_2O_2 leads to improvements in the net photosynthetic rate, chlorophyll content, and stomatal conductance (Ashfaque et al. 2014). Moreover, H_2O_2 boosts the activation of genes encoding pyrroline-5-carboxylate synthase, sucrose-phosphate synthase, and small heat shock proteins (Uchida et al. 2002). In subsidiary cells, H_2O_2 acquisition encompasses regulatory indicators of stomatal closure in maize leaves (Yao et al. 2013). Pre-treatment of seeds with 80 mM H_2O_2 enhanced the seed germination index and photosynthetic efficiency of *Bombax ceiba* seedlings, likely due to the activation of superoxide dismutase activity by H_2O_2 (Zheng et al. 2018). However, seeds pre-treated with a higher concentration of 120 mM H_2O_2 experienced accelerated chlorophyll degradation and reduced photosynthetic efficiency in the seedlings (Zheng et al. 2018). The application of H_2O_2 also increased chlorophyll content, photosynthetic rate, and CO_2 accumulation in *Vigna radiata* (Khan et al. 2015). These findings suggest that lower levels of H_2O_2 enhance photosynthesis, whereas excessive levels impede it. In wheat, foliar spray treatment with H_2O_2 affected stomatal opening, correlated with reduced stomatal size and development (Kumari and Verma 2019). It has been shown that H_2O_2-induced stomatal closure is due to a decrease in K^+ concentration and increased K^+ efflux at the guard cell plasma membrane via K^+ channels (An et al. 2016). Zhang et al. (2001) found that H_2O_2 had reversible effects on stomatal morphology at concentrations below $10–^5$ M, while effects were irreversible at concentrations above 10^{-5} M. This suggests that the impacts of low-dose H_2O_2 may be due to signaling cascade activation, while high-dose impacts may result from alterations in membrane stability. H_2O_2-mediated signaling pathways regulate the activity of transcription factors, hormone biosynthesis, and nutrient uptake, facilitating growth under favorable conditions. Conversely, excessive accumulation of H_2O_2 can induce oxidative damage to cellular components, leading to growth inhibition, leaf senescence, and programmed cell death. In conclusion, H_2O_2 significantly impacts plant physiological performance, acting as a signaling molecule that regulates growth, development, and stress responses. Understanding the complex mechanisms of H_2O_2-mediated signaling pathways is crucial for leveraging its potential to enhance crop productivity and improve plant resilience to environmental challenges. Future research aimed at uncovering the molecular basis of H_2O_2 signaling in plants promises to advance sustainable agriculture and mitigate the adverse effects of climate change on global food security.

EFFECT OF H_2O_2 ON PHOTOSYNTHETIC MACHINERY UNDER ABIOTIC STRESS

Photosynthesis is the cornerstone of life on Earth, providing materials and energy to nearly all organisms either directly or indirectly. However, higher plants only utilize 1%–2% of sunlight for photosynthesis (Ooms et al. 2017). The photosystem, which is the functional unit of light absorption in photosynthesis, comprises chlorophyll, carotenoids, lipids, and proteins. There are two photosystems in the thylakoid membranes of chloroplasts: photosystem II (PSII) and photosystem I (PSI), with maximum absorption wavelengths of 680 and 700 nm, respectively. Light energy absorbed by antenna pigments is transferred to the PSII reaction center pigment P680 and the PSI reaction center pigment P700 through fluorescence resonance energy transfer between pigment molecules (Chitnis 2001; Zhang and Reisner 2020). Photosynthesis is significantly affected by various abiotic stresses (Sharma et al. 2019; Dey et al. 2022), which alter the chemical reactions mediated by PSI and PSII and affect chlorophyll biosynthesis (Sharma et al. 2019). These stressors also hinder photosynthesis by preventing the assimilation of CO_2 through the Ribulose-1,5-bisphosphate carboxylase/oxygenase (RuBisCO) pathway (Kohli et al. 2018). Abiotic stress induces the production of high levels of ROS in plant organelles like plastids, peroxisomes, and mitochondria. These ROS act as crucial signaling molecules in major physico-chemical processes, severely affecting overall plant growth and development (Waszczak et al. 2018). Therefore, maintaining ROS homeostasis in these organelles is essential to mitigate the adverse effects of abiotic stress. Specific ROS signatures, such as cysteine-rich peptides and cysteine-rich receptor-like kinases, have been identified under stress conditions (Hasanuzzaman et al. 2020). Plants detect stress signals through receptors that initiate molecular cascades, transmitting these signals to regulatory systems via ion channels, signaling proteins, and secondary messengers (Schmutz et al. 2010; Choudhary et al. 2012; Le et al. 2012). These regulatory systems consist of various components, including phytohormones, transcription factors (TFs), MAPKs, and photosynthetic protein kinases and phosphatases, all of which regulate the expression of stress-responsive genes (Foyer and Shigeoka 2011; Osakabe et al. 2014). While the generation of ROS due to abiotic stress and their detoxification have been extensively studied, the specific mechanisms through which H_2O_2 mediates ROS distribution in particular cellular compartments and the entire photosynthetic apparatus remain unclear. H_2O_2, the most stable ROS, has traditionally been associated with chloroplasts, but recent studies suggest that peroxisomes may produce more H_2O_2 than other organelles (Noctor et al. 2014). In leaf tissues, the production of ROS is regulated by the capture and distribution of light energy to photosynthetic machinery (Tikkanen et al. 2014). ROS are generated in chloroplasts within the electron transport chains (ETCs) of PSII and PSI during light reactions, with production increasing under stress conditions when CO_2 is limited and ATP synthesis is impaired (Takahashi and Murata 2008; Yamamoto et al. 2008; Nishiyama and Murata 2014; Noctor et al. 2014). Elevated levels of intracellular ROS cause oxidative stress, leading to damage to DNA, lipids, and proteins (Schmutz et al. 2010; Foyer and Shigeoka 2011). Accumulated ROS in chloroplasts can inhibit the de novo synthesis of the D1 protein (also known as photosystem II reaction center protein A or PsbA), which is necessary for PSII repair (Miyao 1994; Tikkanen et al. 2011; Nishiyama et al. 2011), suppress ROS-responsive chloroplastic enzymes (Yoshioka et al. 2006; Kato and Sakamoto 2014), and disrupt thylakoid architecture (Gratao et al. 2009).

Recent studies in various organisms indicate that ROS also play crucial roles as signaling molecules in normal biochemical and physiological responses (Schmutz et al. 2010; Foyer and Shigeoka 2011; Tikkanen et al. 2014). Among ROS compounds, H_2O_2 has garnered the most attention from the scientific community over the past decade. H_2O_2 is produced through a two-step reduction of molecular oxygen, with the first step generating a superoxide radical. It has a relatively long lifespan compared to other ROS, with a half-life of approximately 1 millisecond. This long half-life and its small molecular size enable H_2O_2 to traverse cellular membranes and migrate between different compartments, facilitating it signaling functions (Bienert et al. 2006). It is now well established that H_2O_2 regulates a multitude of physiological processes, including resistance acquisition, cell wall strengthening, senescence, phytoalexin production, photosynthesis, stomatal movement, and the cell cycle. The multifunctionality of H_2O_2, along with its potential toxicity at elevated concentrations, necessitates strict regulation of its levels within plant cells. H_2O_2 is involved in ABA-induced stomatal opening and closing. Kim and Portis (2004) reported that H_2O_2 is produced as a side reaction of RuBisCO oxygenation, occurring more frequently at higher temperatures. Rubisco, the most abundant protein on Earth, catalyzes both photosynthetic carbon fixation and photorespiratory oxygen incorporation in plants. During photorespiration, RuBisCO reacts with oxygen to produce fixed CO_2, NH_3, and energy. H_2O_2 regulates gene expression through protein oxidation and the activation of kinase transduction cascades (Lariguet et al. 2013). A study found that foliar application of H_2O_2 to tomato plants enhances metabolic acclimation in water-stressed plants and boosts the RuBisCO carboxylation rate and growth in well-watered plants. Other studies have shown that foliar H_2O_2 promotes tolerance to abiotic stresses and stimulates development in plants grown under optimal water conditions (Barzotto et al. 2023). H_2O_2 foliar treatment induced changes in tomato photosynthetic metabolism regardless of water status, but growth enhancement was observed only in well-watered plants. The application of H_2O_2 increased Vcmax in tomato plants (Barzotto et al. 2023), a finding also confirmed in other species, suggesting that H_2O_2 can alter the RuBisCO activation state (Khan et al. 2018). A more recent study by Jamaludin et al. (2020) showed that H_2O_2 foliar application increased the expression of the plastid DNA-encoded rbcL gene, which is related to the RuBisCO component, in *F. deltoideae*, thus enhancing photosynthesis. The researchers proposed that H_2O_2 influences gene expression through protein oxidation, kinase activation, and modulation of transduction cascades. Charles and Halliwell (1981) demonstrated that high H_2O_2 concentrations can alter the redox state of fructose-1,6-bisphosphatase (FBPase), a key enzyme in the Calvin-Benson cycle, oxidizing it and rendering it inactive, which reduces the rate of triose phosphate formation and RuBP regeneration. Barzotto et al. (2023) found that foliar application of 1 mM H_2O_2 aided tomato photosynthetic adjustment by increasing the maximum rate of RuBisCO carboxylation by 69% in continuously well-watered plants. Additionally, applying H_2O_2 at a concentration of 20 μM promoted greater stomatal conductance compared to the control treatment (0 μM H_2O_2) across all salinity levels at 110 days post-transplantation. Variable fluorescence and the quantum efficiency of photosystem II benefited from H_2O_2 applications at estimated concentrations of 16 and 13 μM, respectively. This suggests that low concentrations of H_2O_2 contribute to greater efficiency in the photoreduction of quinone A and electron flow between photosystems, promoting proper PSII activity in the thylakoid membrane and indicating that PSII was not damaged. When the photosynthetic apparatus is intact, Fv/Fm values typically range between 0.75 and 0.85 (Reis et al. 2011). These results are similar to those reported by de S. A. Veloso et al. (2020), who evaluated the photochemical efficiency and growth of soursop rootstocks subjected to salt stress (ECw ranging from 0.6 to 3.0 dS/m) and hydrogen peroxide (0 and 20 μM) and found that applications of H_2O_2 at 20 μM minimized the negative effects of salinity on initial fluorescence and favored variable fluorescence and quantum efficiency of photosystem II at 120 days after sowing.

H_2O_2-INDUCED SIGNALING TOWARD ABIOTIC STRESS TOLERANCE

Abiotic stresses such as drought, salinity, extreme temperatures, and heavy metal toxicity pose significant threats to agricultural productivity worldwide. Plants, being sessile organisms, have evolved intricate signaling mechanisms to perceive and respond to these environmental challenges. Among these signaling molecules, H_2O_2 emerges as a pivotal player in orchestrating plant responses to abiotic stress. H_2O_2, an ROS, was long perceived solely as a toxic byproduct of cellular metabolism. However, recent research has unveiled its dual role as a signaling molecule in plants. H_2O_2 acts as a secondary messenger in transducing various environmental cues into adaptive responses, including those triggered by abiotic stresses (Saxena et al. 2016). It plays a crucial role in cellular signaling due to its ability to diffuse across cellular membranes, enabling rapid and spatially controlled signaling necessary for coordinating stress responses. Abiotic stresses trigger a burst of H_2O_2

production in plant cells, primarily through NADPH oxidases and peroxisomal processes. This transient increase in H_2O_2 levels serves as an early signaling event, initiating a cascade of molecular responses. H_2O_2 acts upstream of numerous stress-responsive genes, TFs, and protein kinases, modulating gene expression and enzymatic activities essential for stress tolerance. Notably, H_2O_2-mediated post-translational modifications, such as the oxidation of cysteine residues in proteins, are vital in fine-tuning stress-responsive gene expression and cellular processes. In plants, MAPKs are involved in various functions such as growth, development, and acclimation to environmental changes. Exogenously applied H_2O_2, its accumulation, or H_2O_2-induced oxidative bursts can activate the MAPK pathway, relaying the intracellular H_2O_2 signal. H_2O_2 induces signaling cascades by modulating the expression of downstream genes, essentially switching on or off the transcription machinery. Transcription factor families such as NAC, Zinc Finger of *Arabidopsis Thaliana* (ZAT), WRKY, ethylene response factor (ERF), MYB, Dehydration Responsive Element Binding (DREB), and Basic Leucine Zipper (bZIP) are linked with H_2O_2-induced signaling. The complexity and multidirectionality of the MAPK cascades are highlighted by the upstream or downstream application of H_2O_2. H_2O_2 activates the ANP1 (MAPKKK) protein, which subsequently induces phosphorylation of AtMPK3 and AtMPK6 (Kovtun et al. 2000). Some MAPK pathways, such as MEK2, act upstream of Rboh genes, ultimately producing H_2O_2 (Yoshioka et al. 2003). In Arabidopsis, the ROS-mediated MAPK pathway includes the MEKK1–MKK1/2–MPK4 cascade, where the MAPK kinase kinase MEKK1 is controlled by H_2O_2 and various stresses in a proteasome-dependent manner (Pitzschke et al. 2009), activating downstream kinases MPK3, MPK4, and MPK6. A unique MAPK pathway involves MEKK1 directly interacting with and phosphorylating the TF WRKY53, which then induces stress and defense-related target genes through its DNA-binding activity (Miao et al. 2007). In Arabidopsis, MPK3 and MPK6 are associated with H_2O_2 signaling (Nakagami et al. 2005), while similar regulatory mechanisms are observed in their homologs SIPK and WIPK in tobacco (Asai and Yoshioka 2008). H_2O_2 treatment has been shown to enhance the expression of tomato SlMAPK1/2/3 and SlCBF1 genes, improving chilling tolerance by modulating phytohormone concentrations (Wang et al. 2017). H_2O_2 also confers tolerance to drought and salinity through MAP kinase pathways (Wang et al. 2010a). The cotton MKK1 (GhMKK1) is upregulated by H_2O_2 and enhances stress tolerance in *Nicotiana benthamiana* transgenics (Lu et al. 2013). H_2O_2 activates the rice salt-responsive ERF1 (SERF1) TF, which binds to MAP3K6 and MAPK5, playing a role in the initial phase of salt stress and subsequent tolerance (Schmidt et al. 2013). MAPK cascade genes also reduce H_2O_2 content in transgenic plants during abiotic stress, providing tolerance. A poplar PtMKK4 gene enhances drought tolerance by reducing H_2O_2 levels and increasing antioxidant enzyme activity (Wang et al. 2010b). Similarly, the cotton MAPK (GhMPK17) transgenics show reduced H_2O_2 levels and tolerance to salt and drought (Zhang et al. 2014). H_2O_2 upregulates several senescence-associated NAC TFs, such as ANAC032 and ANAC042, which are involved in H_2O_2-induced relay cascades via OXI1/MPK3/6 activation (Gechev and Hille 2005). The JUNGBRUNNEN1 (JUB1) gene, a member of the NAC TF family, delays senescence and confers abiotic stress tolerance by reducing intracellular H_2O_2 in transgenic Arabidopsis (Wu et al. 2012). The Litchi chinensis LcNAC1 gene is positively activated by ABA and H_2O_2 treatments, reducing senescence time by regulating ROS-related genes and energy metabolism, while LcWRKY1 antagonistically regulates LcAOX1a (Jiang et al. 2017). Overexpression of MusaSNAC1 increases H_2O_2 content in guard cells, augmenting stomatal closure to facilitate drought tolerance (Negi et al. 2018). Enhanced chilling tolerance is observed in tomato SlNAM1 TF overexpression in tobacco, where increased H_2O_2 content is reduced by antioxidative enzymes (Li et al. 2016). The Aeluropus AlNAC4 TF shows significant expression with H_2O_2 and dehydration treatments. Overexpression of AlNAC4 in tobacco confers oxidative stress tolerance by reducing H_2O_2 content through ROS homeostasis and regulating stress-related downstream genes (Khedia et al. 2018). The tomato SlNAC2 TF enhances abiotic stress tolerance by increasing glutathione biosynthetic gene expression, improving antioxidative response, and reducing ROS accumulation (Borgohain et al. 2019). H_2O_2 strongly induces the zinc finger TF family, especially ZAT family members, at the transcript level (Gadjev et al. 2006).

ZAT12 transduces signals to ZAT7 and WRKY25 TFs, expressing cytosolic APX1 during elevated H_2O_2 levels in Arabidopsis (Rizhsky et al. 2004). The ZAT12 mutant plants show reduced ZAT7 expression upon H_2O_2 treatment, confirming ZAT12's regulatory role over ZAT7. The AtZAT6 TF plays a pivotal role in H_2O_2-mediated secondary metabolite synthesis, such as anthocyanin, by directly binding to promoters of anthocyanin synthesis genes (Shi et al. 2018). Studies support the antioxidant role of anthocyanin in reducing photooxidative damage under high irradiance (Zeng et al. 2010; Zhang et al. 2012; Maruta et al. 2014) and wounding (Gould et al. 2002). In *Populus euphratica*, the PeSTZ1 TF confers freezing tolerance through direct regulation of PeAPX2 and modulation of ROS scavenging (He et al. 2019). WRKY TFs also play significant roles in biotic, abiotic, and oxidative stresses (Eulgem and Somssich 2007), with improved expression of several WRKYs in Arabidopsis upon H_2O_2 treatment (Chen et al. 2010), regulating senescence and defense-related gene expression (Besseau et al. 2012). WRKY52 and WRKY70 are widely reported H_2O_2-inducible TFs, from which WRKY52 is a senescence-associated and regulates functions of various stress and defense-responsive genes (Miao et al. 2004). Oryza sativa SALT-RESPONSIVE ERF1 (SERF1) a root-specifc TF induces during salinity and H_2O_2 treatments. It promotes the expression of the H_2O_2-responsive genes, such as MAPK KINASE KINASE6 (MAP3K6), MAPK KINASE5 (MAPK5), DREB2A, and ZFP179 (Schmidt et al. 2013). Ectopic expression of tomato TERF1 in *Nicotiana tabacum* showed enhanced H_2O_2 stress tolerance in seedlings as well as activation of downstream genes involved in oxidative responses, including carbonic anhydrase (in hypersensitive defense), catalase, glutathione peroxidase (in antioxidant response) and GDP-d-mannose pyrophosphorylase (in ascorbic acid biosynthesis) (Zhang et al. 2009. SbMYB15 TF from *Salicornia brachiata* conferred heavy metal (Cd^{2+} and Ni^{2+}) tolerance in transgenic tobacco with reduced level of H_2O_2 content and increased ROS-scavenging antioxidative enzyme activities (Sapara et al. 2019). In conclusion, hydrogen peroxide-mediated signaling represents a central hub in the intricate network of plant responses to abiotic stresses. Its multifaceted role in stress perception, signal transduction, and regulation of antioxidant defense mechanisms underscores its significance in plant adaptation to adverse environmental conditions. Understanding the intricacies of H_2O_2-induced signaling pathways not only sheds light on fundamental aspects of plant stress physiology but also provides opportunities for developing innovative strategies to enhance crop resilience and ensure global food security in the face of escalating climate challenges.

INTRICATE INTERPLAY BETWEEN HYDROGEN PEROXIDE AND PHYTOHORMONES UNDER ABIOTIC STRESS

In the realm of plant biology, abiotic stresses such as drought, salinity, extreme temperatures, and heavy metals pose significant challenges to plant growth, development, and productivity. These stresses disrupt cellular homeostasis, leading to physiological and biochemical alterations that can compromise plant survival. However, plants have evolved sophisticated mechanisms to cope with such adversities, among which phytohormones play a pivotal role. Phytohormones, also known as plant hormones, are signaling molecules that regulate various aspects of plant growth and development. In the context of abiotic stress, phytohormones orchestrate adaptive responses, enabling plants to withstand adverse environmental conditions. Phytohormones act as chemical messengers, mediating plant responses to environmental cues, including abiotic stresses. They regulate a myriad of physiological processes involved in stress perception, signal transduction, and the activation of stress-responsive genes. Key phytohormones involved in abiotic stress responses include auxin, gibberellins, cytokinin, ABA, ethylene, brassinosteroids, jasmonic acid, nitric oxide, and salicylic acid. This section elucidates the dynamic interplay between H_2O_2 and phytohormones in mediating abiotic stress responses, underscoring their synergistic and antagonistic interactions.

H_2O_2, a ROS, has become recognized as a vital signaling molecule in plants, orchestrating a range of physiological processes, particularly stress responses. In response to abiotic stresses, plants generate H_2O_2 through both enzymatic and non-enzymatic pathways, triggering signaling cascades that lead to adaptive responses. As a secondary messenger, H_2O_2 modulates gene expression, enzyme activities, and protein functions to enhance stress tolerance. Concurrently, phytohormones act as master regulators of plant growth and development, integrating environmental signals to coordinate adaptive responses. Key phytohormones involved in abiotic stress tolerance include ABA, salicylic acid (SA), jasmonic acid (JA), ethylene, and cytokinins. These hormones regulate various processes such as stomatal closure, osmotic adjustment, antioxidant defense, and metabolic reprogramming to mitigate the negative effects of abiotic stresses. The interaction between H_2O_2 and phytohormones forms a complex regulatory network that fine-tunes plant responses to abiotic stresses. This crosstalk occurs at multiple levels, including hormone biosynthesis, perception, and downstream signaling events, creating a sophisticated system for managing stress responses.

Auxin is renowned for its ability to regulate numerous growth and development processes, including cell division and differentiation, cell enlargement, and apical dominance (Weijers et al. 2018). Exogenously applied H_2O_2 has been shown to induce adventitious root formation in *Linum usitatissimum* by modulating endogenous auxin levels (Takac et al. 2016). Studies also indicate that dose-dependent application of indole-acetic acid (IAA) increases endogenous H_2O_2 levels in the root tips of wild-type *Solanum lycopersicum*, demonstrating that H_2O_2 is controlled by auxin in the growing parts of the root (Ivanchenko et al. 2013). Additionally, naphthalene acetic acid (NAA) and indole-3-butyric acid (IBA) supplementation enhance H_2O_2 production in Azalea microshoots in vitro (Elmongy et al. 2018). In tomato, red light exposure of the shoot apex activates phyB-induced auxin synthesis in the apex, leading to H_2O_2 production in systemic leaves. This increase in H_2O_2 stimulates cyclic electron flow and ATP production, enhancing the assimilation of photosynthetic CO_2 in the systemic leaves, thereby enabling the plant to adapt more effectively to dynamic light environments (Guo et al. 2016). H_2O_2-auxin interaction has been observed in Arabidopsis, where it mediates lateral root development (Qu et al. 2017) and significantly reduces primary root gravitropism (Zhou et al. 2018). H_2O_2 maintains root structure by regulating the polar transport and redistribution of auxin (Su et al. 2016). Further exploration of the molecular-level H_2O_2-auxin interaction is needed to fully understand their signaling mechanisms. Transcriptomic analyses have shown that apoplastic ROS affect various aspects of auxin stability and signaling (Blomster et al. 2011). A notable example is a H_2O_2-dependent MAPK cascade that reduces auxin sensitivity by decreasing auxin-induced gene activation (Nakagami et al. 2006). Changes in auxin metabolism, elevated H_2O_2 levels, and structural modifications triggered by ROS also suggest a complex interplay between ROS and auxin signaling (Potters et al. 2007, 2009). H_2O_2-induced changes in PINOID gene expression can influence the polar transport of auxin, altering auxin homeostasis (Pasternak et al. 2005). Moreover, oxidative degradation of auxin by H_2O_2-related peroxidases affects auxin homeostasis by activating genes involved in Trp-dependent auxin biosynthesis and conjugation (Woodward and Bartel 2005). The production of apoplastic H_2O_2 by the RBOH1 (RESPIRATORY BURST OXIDASE HOMOLOG 1) and WFI1 (WHITEFLY INDUCED 1) genes stimulates auxin biosynthesis in the apex of tomato plants, reducing cytokinin biosynthesis in the stem and consequently diminishing axillary bud outgrowth. These findings provide new insights into the role of H_2O_2 in lateral bud outgrowth through its crosstalk with auxin signaling in tomato (Chen et al. 2016).

Gibberellins (GAs) play a crucial role in promoting seed germination, stem elongation, floral initiation, and fruit development (Gantait et al. 2015; Huang et al. 2017). H_2O_2 has been reported to interact with GAs to modulate various morphophysiological attributes, such as breaking dormancy, sprouting, and inhibiting hypocotyl and primary root development during de-etiolation (Zhong et al. 2015; Hejazi et al. 2018). Additionally, ROS, particularly H_2O_2, are involved in GA signaling in barley aleurone cells, promoting GA-stimulated amylase production. Thus, H_2O_2 appears to be involved in GA signaling, controlling the synthesis of α-amylase in barley aleurone cells (Ishibashi et al. 2012b). Interestingly, negative interactions between H_2O_2 and GA have also been reported. GAs regulate DELLA proteins, which are negative regulators of GA signaling. In GA-deficient ga1–3 mutants of Arabidopsis, higher concentrations of DELLA proteins make the plants less susceptible to ROS-induced cell death (Achard et al. 2008). Under stress conditions (drought or excess salt), increased GA levels promote the production of DELLA proteins and elevate ROS concentrations (Considine and Foyer 2014). Furthermore, the GA-activated gene GASA14 in Arabidopsis regulates leaf development and abiotic stress resilience by modulating ROS accumulation (Sun and Foyer 2014). Li et al. (2018) found that the combined action of H_2O_2 and GA is crucial for tobacco seed germination, primarily due to the downregulation of GA and the mobilization of reserves to facilitate germination. H_2O_2 influences GA metabolism and signal transduction by regulating gene expression. In Arabidopsis, maize, and sunflower seeds, exogenous H_2O_2 enhances seed germination by stimulating GA biosynthesis and ABA catabolism (Huang et al. 2017; Li et al. 2017). GA also regulates the germination of dormant caryopses by influencing ABA levels and ROS-antioxidant activity (Cembrowska-Lech et al. 2015). In aleurone layers, GA induces programmed cell death by regulating H_2O_2 production (Aoki et al. 2014). Surprisingly, the exogenous application of GA upregulates the expression of NtRBOH (*Nicotiana tabacum* respiratory burst oxidase homolog) genes involved in NADPH oxidase biosynthesis, while H_2O_2 suppresses their expression (Li et al. 2018). In Arabidopsis, H_2O_2 enhances GA biosynthesis by upregulating GA biosynthesis genes such as GA_3ox and $GA_{20}ox$, which are crucial for seed germination (Liu et al. 2010).

Cytokinins (CKs) are integral to various physiological, cognitive, and metabolic functions in plants, including cell division, leaf senescence, and light-regulated development (Chang et al. 2016; Hönig et al. 2015). H_2O_2 helps maintain the appropriate concentration of CKs needed for plant development and enhances cytokinin oxidase activity and gene expression (Causin et al. 2009). In wheat, CKs interact synergistically with H_2O_2 to regulate shade-induced leaf senescence (Causin et al. 2009). However, H_2O_2 can also suppress CK synthesis, increase degradation, and reduce CK concentrations in roots and shoots (Nishiyama et al. 2011). Transgenic creeping bentgrass (S41) roots exhibited significantly higher CK content and lower H_2O_2 levels compared to wild-type (WT) roots (Xu et al. 2016). Increasing CK content has been shown to enhance growth and fibrous root systems in creeping bent grass and monocot species like barley (*Hordeum vulgare*) through the overexpression of the isopentyl transferase (IPT) gene and RNAi silencing of HvCKX1, which encodes cytokinin oxidase responsible for CK degradation (Zalewski et al. 2010; Merewitz et al. 2011). In *Arabidopsis thaliana*, exogenously applied zeatin causes stomatal closure, which is associated with increased H_2O_2 accumulation in guard cells and the activation of the HISTIDINE KINASE3 (AHK3) and RESPONSE REGULATOR2 (ARR2) receptors (Arnaud et al. 2017). CKs play a crucial role in cellular signal transduction processes, initiating transcription, governing protein standardization and stabilization, and regulating phosphorelay activity (To et al. 2007). Additionally, H_2O_2 has been shown to upregulate the C2H2 zinc finger gene ZAT11, which plays a significant role in stress responses and programmed cell death (Gechev et al. 2005).

ABA is a pivotal phytohormone with significant roles across various environmental conditions. It accumulates in response to stressors such as dehydration, cold temperatures, or shortened day lengths. ABA application is fundamental in diverse physiological processes and biological pathways, including seed dormancy, germination delay, seed development, stomatal closure promotion, embryo morphogenesis, storage protein and lipid synthesis, leaf senescence, and pathogen defense (Swamy and Smith 1999). The ABA signaling pathway serves as a central regulator of plant responses to abiotic stress, instigating substantial changes in gene expression and adaptive physiological responses. Recent studies have highlighted the involvement of MAPK cascades in ABA signaling. External ABA treatment induces transcriptional regulations, protein accumulation, stability, and kinase activity of various components of distinct MAPK signaling cascades across multiple plant species, suggesting the active participation of MAPK cascades in various ABA responses, including abiotic stress defense mechanisms and guard cell signaling (Zhang et al. 2007; Jammes et al. 2009; Zong et al. 2009). Rodriguez et al. (2010) reported that the MAPK cascade is activated by exogenous H_2O_2, mediated by hormones such as ABA, JAs, and SA. Additionally, ABA plays a crucial role under drought and salinity stress conditions and interacts with H_2O_2 for signaling through MAPK cascades. In ABA transduction, H_2O_2 serves as a significant secondary transmitter (Miao et al. 2006). The accumulation of H_2O_2 in *Vicia faba* triggers plasma membrane calcium channels, mediating ABA-mediated stomatal behavior (Zhang et al. 2001). NADPH oxidase catalytic subunit genes, Atrboh D and Atrboh F, embedded in the plasma membrane, are essential to produce ABA-triggered stomatal ROS (Kwak et al. 2003). NADPH oxidase can initiate calcium and MAPK signaling routes and activate hormone signal attenuation pathways involving SA, JA, and ethylene (Overmyer et al. 2003). In barley aleurone cells, H_2O_2 enhances GA/ABA signaling (Ishibashi et al. 2012a). Calcium and calmodulin (CaM)-dependent protein kinase (CCaMK), triggered by free Ca^{2+}, and Ca^{2+}-bound CaM, are involved in ABA-induced antioxidant system regulation by acting upstream of ABA-activated MAPK and ABA signaling during abiotic stress (Shi et al. 2014). The interplay between ABA and H_2O_2 provides acclimation to water stress in plants (Li et al. 2016). Increased ABA and H_2O_2 concentrations were observed in *Zea mays* leaves under drought stress, and exogenous ABA treatment further enhanced H_2O_2 generation in rice seedling leaves grown in potassium-rich soil (Liu et al. 2012). NADPH oxidase has been identified as the enzyme involved in H_2O_2 production in rice seedling roots and leaves (Hung and Kao 2004; Tsai and Kao 2004). In pumpkin-grafted cucumber leaves, induced ABA accumulation triggered H_2O_2 production, upregulating the antioxidant defense system and activating antioxidant enzyme encoding genes (SOD, POD, and cAPX) (Shu et al. 2016). In tobacco, ABA enrichment activates the 9-cis epoxycarotenoid dioxygenase gene (SgNCED1), increasing H_2O_2 and NO levels in guard and mesophyll cells, leading to the activation of various antioxidant enzymes (Zhang et al. 2009). Elevated H_2O_2 influences ABA output, resulting in subsequent H_2O_2 production through NADPH oxidase stimulation, contributing to prolonged stress resilience. H_2O_2 and ABA thus establish a feedback mechanism regulating freezing tolerance triggered by SA in wheat plants. Elevated H_2O_2 and ABA independently activate protective mechanisms, enabling the activation of cold-responsive gene expression and antioxidant enzyme activity to mitigate ion leakage, lipid peroxidation, and enhance photosynthesis efficiency (Fv/Fm and ETR) in wheat plants subjected to freezing stress (Wang et al. 2018). Ishibashi et al. (2017) focused on the relationship between H_2O_2 and ABA in quiescent (dormant) and active (non-dormant) barley seeds. After imbibition, non-dormant seeds exhibited higher H_2O_2 concentrations in embryos compared to dormant seeds. In embryos, ABA content was regulated by H_2O_2 through an ABA-80-hydroxylase catabolic enzyme. NADPH oxidase activity producing ROS was lower in dormant seeds compared to non-dormant seeds, while catalase activity, involved in H_2O_2 detoxification, was higher due to the activation of HvCAT2 genes by the ABA-responsive transcription factor HvABI5. These findings suggest that the balance between ABA and ROS in barley seed embryos after imbibition regulates dormancy and germination. Additionally, exogenous inhibitors of H_2O_2 and NADPH oxidase improved ABA catabolism by upregulating CYP707A gene expression encoding ABA 80-hydroxylases and promoted GA synthesis gene activation in dormant Arabidopsis seeds (Liu et al. 2010). However, in sunflower seeds, H_2O_2 interacts with the ABA signaling pathway at the transcriptional level (El-Maarouf-Bouteau et al. 2015). In *Cucumis sativus*, H_2O_2 is implicated in ABA-mediated adventitious root development, enhancing chlorophyll and relative water content, quantum yield, carbohydrate and nitrogen content, and stimulating enzyme activity (Li et al. 2016).

Ethylene has been recognized as a stress hormone for a long time (Morgan and Drew 1997). It plays a crucial role not only in plant growth and development but also in plant responses to biotic stress (such as pathogen attack) and abiotic stress (such as wounding, ozone, and salinity) (Abeles et al. 1992; Wang et al. 2009). Ethylene regulates numerous processes in plants and has shown involvement in defense mechanisms as well (Ecker 1995). Studies using mutants have been conducted to understand the defensive role of ethylene under various environmental conditions, focusing on its signal transduction pathways. Both H_2O_2 and ethylene can affect stress responses or productivity either synergistically or antagonistically (Zhu et al. 2016). The interaction between H_2O_2 and ethylene primarily involves ETHYLENE INSENSITIVE3 (EIN3), an ethylene-activated transcription factor, and its homolog EIN3-Like1 (EIL1) (Zhang et al. 2016b). ERFs play a significant role in promoting ethylene signaling, which helps reduce abiotic stress tolerance in plants. For example, in

Nicotiana tabacum, TERF1 is activated by exogenous ROS-generating substrates and is regulated by upstream EIN3-like regulators LeEIL3 and LeEIL4 (Zhang et al. 2016). Activation of TERF1 in tobacco seedlings influences a variety of stress-responsive genes and increases resilience to H_2O_2 by linking their promoters with specific cis-elements (Muller and Munne-Bosch 2015). It is suggested that ethylene and H_2O_2 act synergistically in tomatoes, where H_2O_2 is a major contributor to ethylene production, mediating cell death caused by Cd (Liu et al. 2008). Under salt stress, H_2O_2 and ethylene are implicated in regulating the alternative respiratory pathway (AP) (Wang et al. 2010b). Zhu et al. (2016) demonstrated that ethylene and H_2O_2 play significant roles in triggering BR-dependent endurance in tomato plants under salt stress. Oxygen deficiencies in plants shift energy metabolism from aerobic to anaerobic, influencing water and nutrient absorption and inducing the production of both H_2O_2 and ethylene through hypoxia signaling. In Arabidopsis, the gene expression levels of AtERF73/HRE1 and ADH1 are affected by the interaction of H_2O_2 and ethylene during early hypoxia signaling (Yang 2014). Ethylene and H_2O_2 also affect sterol-enriched domain production in developing Vicia faba cotyledons, crucial for nutrient transport (Zhang et al. 2019). Additionally, ROS and ethylene are part of the signal cascade induced by Nod factor, necessary for nodule formation in legumes (DHaeze et al. 2003). Ethylene perception is mediated by receptors that act as negative regulators of ethylene signaling (Wang et al. 2013). The negative regulator CTR1 controls a cascade of MAPKs that transmit signals to the transcription factor EIN3, linking GCC-boxes in the promoters of ethylene-responsive genes (Muller and Munne-Bosch 2015). In stomatal guard cells, the ethylene receptor ETR1 plays a significant role in the crosstalk between ethylene and H_2O_2 signaling (Desikan et al. 2006). Both ethylene and H_2O_2 share the MAPK cascade for signaling (Opdenakker et al. 2012). In Arabidopsis, mutations in the ethylene receptor ETR1 and transcription factor EIN3 may increase ROS tolerance (Desikan et al. 2006). Song et al. (2007) identified an antagonistic relationship between exogenous H_2O_2 and ethylene in mungbean hypocotyls, where H_2O_2 increased ethylene accumulation but exogenously applied ACC did not alter ROS generation. In soybean seeds, H_2O_2 in the embryonic axis signals endogenous ethylene production, stimulating seed germination and root tip cell elongation (Ishibashi et al. 2012a). H_2O_2, alongside ethylene biosynthesis enzymes, acts as a significant plant growth regulator in the biosynthesis of 1-aminocyclopropane-1-carboxylic acid (ACC), a major ethylene precursor. These findings underscore the complex interactions between ethylene and H_2O_2 in various stress responses and developmental processes in plants.

BRs signaling is essential for plant development, while ABA signaling is critical for plant survival under stress conditions. Both H_2O_2 and BRs function as signaling messengers in plant developmental processes, making it important to understand their crosstalk in plants. Previous research has indicated that H_2O_2 regulates photosynthesis through an EBR-mediated mechanism, with the interplay between EBR and H_2O_2 affecting sugar metabolism and the Calvin cycle in *Cucumis sativus* via a redox signaling pathway. Tian et al. demonstrated that BR treatment significantly increased H_2O_2 content in *Arabidopsis thaliana* seedlings, with the rise in H_2O_2 levels being induced through an NADPH-dependent pathway. They further investigated the potential role of H_2O_2 in BR-mediated seedling development, finding that treatment with diphenylene iodonium (DPI), an NADPH oxidase inhibitor, reduced H_2O_2 levels and markedly inhibited BR-induced hypocotyl elongation. High concentrations of DPI also caused *Arabidopsis thaliana* seedlings to become insensitive to BRs. Generally, BRs and H_2O_2 might boost crop yield by regulating photosynthesis and sugar metabolism. BRs may also enhance endogenous H_2O_2 levels in plants, promoting BR-mediated cell elongation. Co-treatment with BRs and H_2O_2 can improve plant resistance to abiotic stresses. In *Lycopersicon esculentum*, EBR and H_2O_2 application significantly increased SPAD chlorophyll, net photosynthetic rate, and the activity of carbonic anhydrase and various antioxidant enzymes (CAT and SOD) under cold stress. High concentrations of heavy metals are detrimental to plant tissues and organs. Nazir et al. (2019) explored whether the combination of BRs and H_2O_2 could mitigate Cu toxicity in *Solanum lycopersicum*. They discovered that co-treatment with EBR and H_2O_2 significantly elevated chlorophyll content and Fv/Fm compared to EBR or H_2O_2 alone.

This combination also enhanced the net photosynthetic rate and related traits (internal carbon dioxide concentration, stomatal conductance, and transpiration rate) while reducing electrolyte leakage. Cu treatment reduced the leaf area and dry mass of shoots and roots in tomato seedlings, but the combined application of EBR and H_2O_2 significantly increased these parameters. Furthermore, EBR and H_2O_2 co-treatment modified chloroplast ultrastructure and stomatal behavior, boosted total protein content, and increased the activities of antioxidant enzymes and carbonic anhydrase in Cu-treated tomato seedlings under Cu stress. Thus, the interaction between BRs and H_2O_2 may enhance photosynthetic capacity and total protein content, maintain the antioxidant system and plasma membrane, and increase plant resistance to abiotic stress. H_2O_2 plays a crucial role in BR-induced growth, development, and stress responses and might act as a downstream signaling messenger in BR signaling within plant growth and stress response networks. However, many theoretical mechanisms underlying the interaction between BRs and H_2O_2 remain unclear, necessitating further research and discoveries.

Methyl jasmonate (MJ) is a plant growth regulator that supports various developmental processes, including growth, photosynthesis, fruit maturation, and senescence, under different ecological conditions (Huang et al. 2017). Evidence indicates a synergistic interaction between H_2O_2 and JAs in plants (Hung et al. 2006). Hu et al. (2003) found that both H_2O_2 and JA are key signaling molecules in the cellular processes involved in oligogalacturonic acid (OGA)-induced saponin biosynthesis, with H_2O_2 upregulating JA. In detached leaves of *Oryza sativa*, MJ application induced the accumulation of H_2O_2, contributing to jasmonate-induced leaf senescence (Hung et al. 2006). MJ has also been studied for its role in H_2O_2 acquisition in parsley suspension-cultured cells (Kauss et al. 1994), and its ability to elicit defensive genes in tomatoes (Orozco-Cardenas et al. 2001). Supplementation with methyl jasmonate (MeJA) and H_2O_2 led to the differential activation of antioxidant stilbenoids (polyphenolic phytoalexins) in hairy roots of *Vitis rotundifolia* (Nopo-Olazabal et al. 2014). In wheat callus, H_2O_2 acts as a signaling molecule that activates the antioxidant defense system triggered by MeJA (Ma et al. 2015). JA also stimulates glutathione, which is crucial for maintaining redox homeostasis, and it mediates the activation of nuclear factor erythroid 2-related factor 2 (NrF2), reducing ROS levels caused by H_2O_2 (Taki-Nakano et al. 2014). Dai et al. (2015) reported that JA-induced H_2O_2 leads to MEK1/2 phosphorylation and activation, which in turn activates metabolic enzymes (AsA and GSH), enhancing the AsA and GSH redox state. The use of H_2O_2 synthesis inhibitor (DPI) and scavenger (DMTU) significantly inhibited H_2O_2 production caused by MeJA (Dai et al. 2015). In pea (*Pisum sativum* L.) leaves, both wounding and JA application significantly increased the activities of phenylalanine ammonia-lyase and plasma membrane NADPH oxidase. This increase was disrupted by the addition of an H_2O_2 scavenger, indicating that H_2O_2 acts downstream of JA (Liu et al. 2008).

SA is a primary phenolic signaling molecule and intracellular growth regulator essential for various physiological processes (Garg and Bharti 2018). The relationship between H_2O_2 and SA can be either positive or negative, depending on the conditions (Cocetta and Ferrante 2018). In *Salvia miltiorrhiza* cell cultures, SA is an effective elicitor that induces the accumulation of rosmarinic acid (RA), which can interact individually or synergistically with H_2O_2. Intracellular H_2O_2 acts as a signaling molecule in response to SA stimulation and contributes to SA accumulation (Hao et al. 2014). Seed soaking with SA and H_2O_2 has been shown to synergistically enhance signal transduction, hormone metabolism, energy supply, and the activation of antioxidative enzymes in maize seeds (Li et al. 2017). The interaction between SA and H_2O_2 involves the benzoic acid pathway (the complex precursor of SA), where H_2O_2-dependent benzoic acid 2-hydroxylase is produced (Leon et al. 1995). H_2O_2 stimulation in cell organelles (peroxisome and chloroplast) induces SA synthesis, contributing to defense mechanisms such as stomatal closure and programmed cell death, while SA also promotes H_2O_2 generation (Yang et al. 2013). SA-mediated H_2O_2 production upregulates the transcription of heat shock protein (Hsp) genes via AtHsfA2, a crucial element in providing thermotolerance in Arabidopsis (Nie et al. 2015). Furthermore, Hou et al. (2018) reported that SA and the alternative respiratory pathway could enhance PSII modification induced by H_2O_2 under excessive light conditions. Several studies have shown an antagonistic interaction between the signaling pathways of H_2O_2 and SA (Peleg-Grossman et al. 2010; Yang et al. 2013). Yang et al. (2013) demonstrated that SA plays a specific role in inducing adventitious rooting through H_2O_2 acquisition. It was found that the acquisition of free H_2O_2 is a downstream action related to SA-induced adventitious root development in mung bean seedlings, affecting the antioxidant system status. The nuclear translocation of the SAR-linked transcription activator NPR1 (non-expressor of pathogenesis-related gene 1) is a key element modulating the signaling of SA and H_2O_2. It was shown that cytoplasmic H_2O_2 inhibits the SA-triggered transfer of NPR1 to the nucleus, thereby impeding the function of pathogenesis-related (PR) genes (Peleg-Grossman et al. 2010).

Numerous studies have revealed the interplay between NO and H_2O_2 signaling pathways, indicating they function in a similar or parallel manner. Research has shown that H_2O_2 treatment triggers NO production, which plays a crucial role in enhancing the efficiency of alternative oxidase (AOX) and mitigating photosystem damage under stress conditions (Arfan et al. 2019). Additionally, H_2O_2-dependent NO production regulates anthocyanin biosynthesis in the hypocotyls of radish sprouts (Wu et al. 2016). Exogenous H_2O_2 also induces NO production in the guard cells and other leaf cells of mung beans (Lum et al. 2002). In wild-type Arabidopsis roots, NO levels increased eightfold following H_2O_2 treatment (Wang et al. 2010b). The regulation of NO production by H_2O_2 is evident from studies showing that inhibitors of NO production in the guard cells of *Vicia faba* reduce the H_2O_2-triggered synthesis of NO (She et al. 2004; He et al. 2005). Interestingly, calcium channel blockers also impede H_2O_2-induced NO production, suggesting that Ca^{2+} ions likely modulate the impact of H_2O_2 on NO generation (Gonzalez et al. 2012). Furthermore, NO inhibits the accumulation of H_2O_2 during leaf senescence by suppressing the activation of several senescence-associated genes (SAGs) and downregulating photosynthetic genes, which are central regulators of H_2O_2-mediated photosynthetic efficiency (De Michele et al. 2009). Moreover, H_2O_2 stimulates MAP Kinase 6, mediating NO production and signal transduction during Arabidopsis root development (Wang et al. 2010b). These findings provide new insights into the molecular mechanisms involved in signaling cascades associated with stress and growth, which can be utilized to control gene expression in plants (Wang et al. 2010c) (Figure 8.1).

The H_2S donor NaHS can enhance the activity of the plasma membrane H^+-ATPase in cucumber under salt or low-temperature stress, increasing the transcript levels of several proton pump-related genes, including CsHA2, CsH4, CsH8, CsH9, and CsHA10. However, NO and H_2O_2 only increased the expression of CsHA1. Thus, H_2S, NO, and H_2O_2 can mitigate salt stress by regulating the plasma membrane proton pump to different extents. Typically, salt stress induces stomatal closure, but H_2S scavengers such as HT, AOA, hydroxylamine (NH_2OH), potassium pyruvate ($C_3H_3KO_3$), ammonia (NH_3), H_2O_2, ascorbic acid (AsA), CAT, and diphenyl iodide (DPI) suppressed stomatal closure in *V. faba* L., indicating that both H_2S and H_2O_2 regulate stomatal movement under salt stress. Furthermore, salt treatment in guard cells increased endogenous H_2S and H_2O_2 accumulation and the activities of LCD and DCD, but these effects were inhibited by H_2O_2 and H_2S scavengers. Exogenous H_2O_2 scavengers prevented the increase in endogenous H_2S levels and stomatal closure, whereas H_2S scavengers had little effect on H_2O_2 generation in guard cells under salt stress. This suggests that H_2S may act downstream of H_2O_2 in alleviating salt stress. In response to dehydration, H_2S may serve as a downstream signaling molecule interacting with NO and H_2O_2. UV-B radiation is a common stressor in agriculture, reducing the levels of electrolyte leakage, MDA, and ultraviolet-absorbing compounds, and decreasing the activities of antioxidant enzymes, GSH, and AsA. However, exogenous H_2S, H_2O_2, and putrescine (Put) can alleviate the negative effects of UV-B stress. The protective role of Put against UV-B radiation damage was diminished by inhibitors of H_2S, H_2O_2, and Put. Moreover, exogenous H_2S increased H_2O_2 levels, promoting the accumulation of UV-absorbing compounds in hulless barley seedlings, thereby maintaining redox balance and enhancing UV-B tolerance. Extreme temperatures also significantly impact plant growth and development. H_2S, NO, and H_2O_2 significantly modulate the plasma membrane proton pump in cucumber roots in response to low temperatures (10°C). Additionally, H_2O_2 treatment improves heat resistance in maize (*Z. mays*) seedlings, an effect

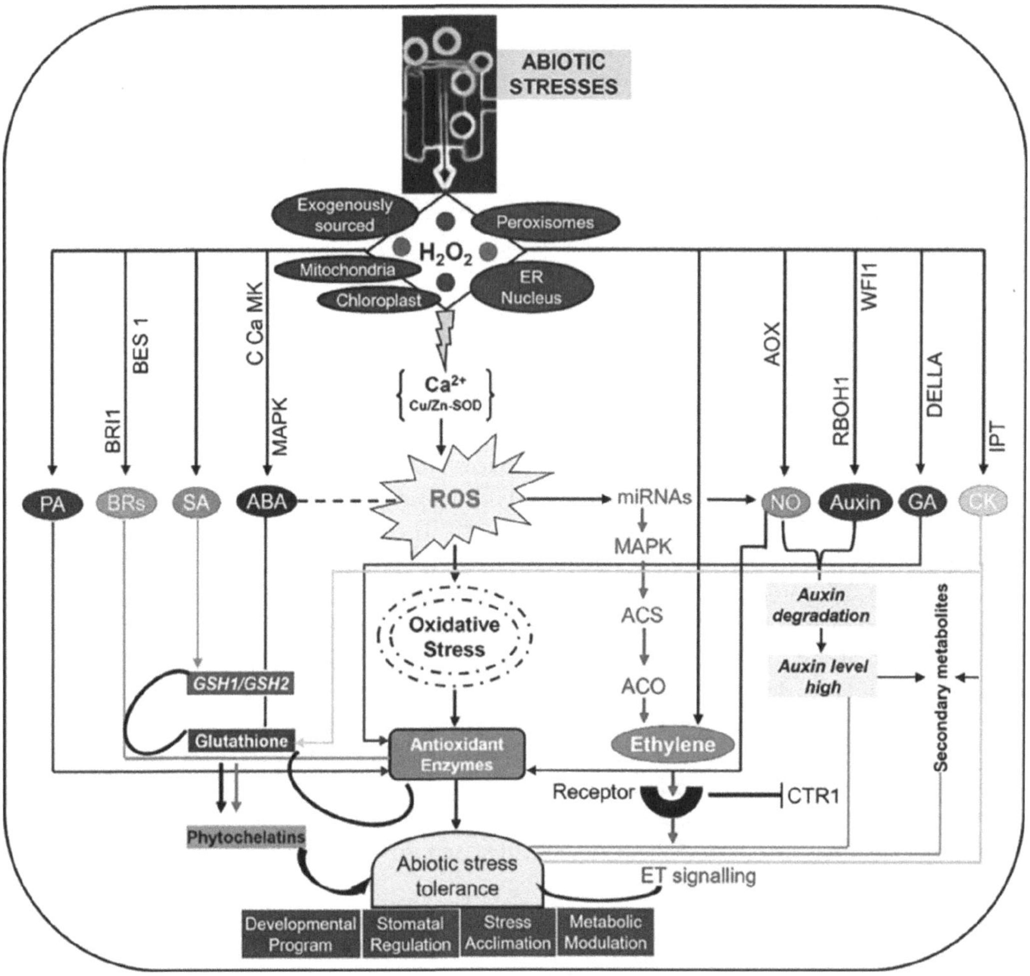

FIGURE 8.1 A visual representation illustrating the interplay and communication between hydrogen peroxide and phytohormones such as abscisic acid (ABA), salicylic acid (SA), brassinosteroids (BRs), polyamines (PA), auxin, nitric oxide (NO), gibberellic acid (GA), and cytokinin (CK) under exposure to abiotic stresses.

enhanced by NO and H_2S donors but negated by NO and H_2S scavengers or synthesis inhibitors. This suggests that NO and H_2S act downstream of H_2O_2 in the acquisition of heat resistance in plants.

Understanding the complex interplay between H_2O_2 and phytohormones provides insights into the molecular mechanisms underlying plant adaptation to abiotic stresses. Manipulating these signaling pathways holds promise for engineering stress-tolerant crops with enhanced yield stability and resilience to changing environmental conditions. Strategies such as genetic engineering, hormone priming, and exogenous hormone application can be employed to optimize plant responses to abiotic stresses while minimizing yield losses. Moreover, the intricate interactions between hydrogen peroxide and phytohormones form a sophisticated signaling network that governs plant responses to abiotic stresses. The dynamic crosstalk between H_2O_2 and phytohormone pathways orchestrates diverse physiological and biochemical changes, enabling plants to withstand adverse environmental conditions. Unraveling the complexities of these interactions not only deepens our understanding of stress physiology but also offers avenues for developing innovative strategies to enhance crop resilience and ensure global food security in the face of escalating climate challenges.

CONCLUSIONS AND FUTURE PROSPECTIVES

H_2O_2, recognized as a powerful molecule, plays a crucial role in enhancing various plant functions, yet much about it remains elusive. Its significant role as a signaling molecule in plants is increasingly evident. Research efforts have focused on understanding how H_2O_2 effectively mitigates abiotic stress-induced changes in developmental processes, including plant growth, photosynthesis, stomatal behavior, pollen and fruit development, flowering, and leaf senescence. H_2O_2 intricate orchestration of reactions responds to a wide array of signals, interacting with other plant growth regulators to modulate responses under both normal and stressful conditions. While H_2O_2 provides protection against environmental stresses, its complex interactions necessitate a thorough evaluation to understand its regulation of various physiological processes alongside nitric oxide, hydrogen sulfide, and brassinosteroids. Recent molecular insights reveal H_2O_2 involvement in modulating signal transduction via MAPKs and TFs, offering a vital strategy for adapting to environmental challenges and maintaining cellular homeostasis. It is hypothesized that H_2O_2 significantly contributes to plant growth and metabolism, with data on the physiological and molecular aspects of plant resilience to heavy metal stress being crucial for breeding resilient plant varieties. Future research should explore the specific roles of H_2O_2 signaling in crop-ecological interactions. Despite advancements in understanding H_2O_2-influenced physiology, many aspects of it signaling system remain enigmatic, holding both fundamental and practical significance for future exploration. Given the diverse outcomes suggesting multifaceted roles of H_2O_2, it is imperative to prioritize future studies aimed at elucidating clear signaling pathways during stress responses across diverse conditions. Cataloging interactions among various signaling molecules and their biological functions within different pathways remains paramount. Additionally, attention should be directed toward elucidating the roles and localizations of enzymes implicated in these signaling pathways. Key areas for future research include identifying the cellular sites of H_2O_2 production and elucidating the primary factors influencing its interactions with other signaling molecules.

In conclusion, the intricate role of hydrogen peroxide in mediating photosynthetic responses and abiotic stress tolerance, as well as its crosstalk with other phytohormones, underscores its significance in plant biology. Its dual nature as a signaling molecule and an ROS highlights its multifaceted involvement in plant adaptation and resilience. Further exploration of these interactions promises valuable insights into enhancing crop productivity and stress resilience in the face of changing environmental conditions.

REFERENCES

Abeles FB, Morgan PW, Saltveit Jr ME (1992) *Ethylene in Plant Biology*, Ed 2. Academic Press, San Diego, CA.

Achard P, Renou JP, Berthome R, Harberd NP, Genschik P (2008) Plant DELLAs restrain growth and promote survival of adversity by reducing the levels of reactive oxygen species. *Curr Biol* 18: 656–660.

An Y, Feng X, Liu L, Xiong L, Wang L (2016) ALA-induced flavonols accumulation in guard cells is involved in scavenging H_2O_2 and inhibiting stomatal closure in *Arabidopsis* cotyledons. *Front Plant Sci* 7: 1713.

Andrade CA, Kamila Dázio de Souza R, de Oliveira Santos M, da Silva DM, Donizeti Alves J (2018) Hydrogen peroxide promotes the tolerance of soybeans to waterlogging. *Sci Horti* 232: 40–45.

Aoki N, Ishibashi Y, Kai K, Tomokiyo R, Yuasa T, Iwaya-Inoue M (2014) Programmed cell death in barley aleurone cells is not directly stimulated by reactive oxygen species produced in response to gibberellin. *J Plant Physiol* 171: 615–618.

Arfan M, Zhang DW, Zou LJ, Luo SS, Taw WR, Zhu T, Lin HH (2019) Hydrogen peroxide and nitric oxide crosstalk mediates brassinosteroids induced cold stress tolerance in *Medicago truncatula*. *Int J Mol Sci* 20: 144.

Arnaud D, Lee S, Takebayashi Y, Choi D, Choi J, Sakakibara H, Hwang I (2017) Cytokinin- mediated regulation of reactive oxygen species homeostasis modulates stomatal immunity in Arabidopsis. *Plant Cell* 29: 543–559.

Arthaut LD, Jourdan N, Mteyrek A, Procopio M, El-Esawi M, D'Harlingue A, Bouchet PE, Witczak J, Ritz T, Klarsfeld A et al. (2017) Blue-light induced accumulation of reactive oxygen species is a consequence of the *Drosophila* cryptochrome photocycle. *PLoS One* 12: e0171836.

Asai S, Yoshioka H (2008). The role of radical burst via MAPK signalling in plant immunity. *Plant Signal Behav* 3: 920–922.

Ashfaque F, Khan MI, Khan NA (2014) Exogenously applied H_2O_2 promotes proline accumulation., water relations., photosynthetic efficiency and growth of wheat (*Triticum aestivum* L.) under salt stress. *Annu Res Rev Biol* 4: 105–120.

Attacha S, Solbach D, Bela K, Moseler A, Wagner S, Schwarzlander M, Aller I, Muller SJ, Meyer AJ (2017) Glutathione peroxidase-like enzymes cover five distinct cell compartments and membrane surfaces in *Arabidopsis thaliana. Plant Cell Environ* 40: 1281–1295.

Bagheri, M., Gholami, M., Baninasab, B (2019) Hydrogen peroxide-induced salt tolerance in relation to antioxidant systems in pistachio seedlings. *Sci Hortic* 243: 207–213.

Barba-Espin G, Diaz-Vivancos P, Clemente-Moreno MJ, Albacete A, Faize L, Faize M (2010) Interaction between hydrogen peroxide and plant hormones during germination and the early growth of pea seedlings. *Plant Cell Environ* 33: 981–994.

Barzotto GR, Cardoso CP, Jorge LG, et al. (2023) Hydrogen peroxide signal photosynthetic acclimation of *Solanum lycopersicum* L. cv Micro-Tom under water deficit. *Sci Rep* 13: 13059.

Besseau S, Li J, Palva ET (2012) WRKY54 and WRKY70 co-operate as negative regulators of leaf senescence in *Arabidopsis thaliana. J Exp Bot* 63: 2667–2679.

Bienert GP, Heinen RB, Berny MC, Chaumont F (2014) Maize plasma membrane aquaporin ZmPIP2, but not ZmPIP1 facilitates transmembrane diffusion of hydrogen peroxide. *Biochim Biophys Acta Biomembr* 1838: 216–222.

Bienert GP, Schjoerring JK, Jahn TP (2006) Membrane transport of hydrogen peroxide. *Biochim Biophys Acta* 1758: 994–1003.

Blomster T, Salojearvi J, Sipari N, Brosche M, Ahlfors R, Keineanen M, Overmyer K, Kangasjearvi J (2011) Apoplastic reactive oxygen species transiently decrease auxin signaling and cause stress-induced morphogenic response in Arabidopsis. *Plant Physiol* 157: 1866–1883.

Bodvard K, Peeters K, Roger F, Romanov N, Igbaria A, Welkenhuysen N, Palais G, Reiter W, Toledano MB, Kall M, et al. (2017). Light-sensing via hydrogen peroxide and a peroxiredoxin. *Nat Commun* 8: 14791.

Borgohain MP, Haridhasapavalan KK, Dey C, Adhikari P, Thummer RP (2019) An insight into DNA-free reprogramming approaches to generate integration-free induced pluripotent stem cells for prospective biomedical applications. *Stem Cell Rev Rep* 15: 286–313.

Bulleid NJ (2012) Disulfide bond formation in the mammalian endoplasmic reticulum. *Cold Spring Harb Perspect Biol* 4: 013219.

Causin HF, Roberts IN, Criado MV, Gallego SM, Pena LB, del Carmen Ríos M, Barneix AJ (2009) Changes in hydrogen peroxide homeostasis and cytokinin levels contribute to the regulation of shade-induced senescence in wheat leaves. *Plant Sci* 177: 698–704.

Cembrowska-Lech D, Koprowski M, Kepczynski J (2015) Germination induction of dormant *Avena fatua* caryopses by KAR1 and GA_3 involving the control of reactive oxygen species (H_2O_2 and O_2) and enzymatic antioxidants (superoxide dismutase and catalase) both in the embryo and the aleurone layers. *J Plant Physiol* 176: 169–179.

Cerny M, Habanova H, Berka M, Luklova M, Brzobohaty B (2018) Hydrogen peroxide: Its role in plant biology and crosstalk with signalling networks. *Int J Mol Sci* 19: 2812.

Chang Z, Liu Y, Dong H, Teng K, Han L, Zhang X (2016) Effects of cytokinin and nitrogen on drought tolerance of creeping bentgrass. *PLoS One* 11: e0154005.

Charles SA, Halliwell B (1981) Light activation of fructose bisphosphatase in isolated spinach chloroplasts and deactivation by hydrogen peroxide: a physiological role for the thioredoxin system. *Planta* 151: 242–246.

Chen L, Zhang L, Yu D (2010) Wounding-induced WRKY8 is involved in basal defense in *Arabidopsis. Mol Plant Microbe Interact* 23: 558–565.

Chen XJ, Xia XJ, Guo X, Zhou YH, Shi K, Zhou J, Yu JQ (2016) Apoplastic H_2O_2 plays a critical role in axillary bud outgrowth by altering auxin and cytokinin homeostasis in tomato plants. *New Phytol* 211: 1266–1278.

Chitnis PR (2001) Photosystem I: Function and physiology. *Annu Rev Plant Physiol Plant Mol Biol* 52: 593–626.

Choudhary SP, Yu JQ, Yamaguchi-Shinozaki K, Shinozaki K, Tran LS (2012) Benefits of brassinosteroid crosstalk. *Trends Plant Sci* 17:594–605.

Cocetta G, Ferrante A (2018) Postharvest application of hydrogen peroxide and salicylic acid differently affects the quality and vase life of cut rose (*Rosa hybrida* L.) petals and leaves. *Adv Hortic Sci* 32: 371–378.

Considine MJ, Foyer CH (2014) Redox regulation of plant development. *Antioxidants Redox Signal* 21: 1305–1326.

Dai H, Jia G, Sh C (2015) Jasmonic acid-induced hydrogen peroxide activates MEK1/2 in upregulating the redox states of ascorbate and glutathione in wheat leaves. *Acta Physiol Plant* 37: 200.

Daudi A, Cheng Z, O'Brien JA, Mammarella N, Khan S, Ausubel FM, Bolwell GP (2012) The apoplastic oxidative burst peroxidase in Arabidopsis is a major component of pattern-triggered immunity. *Plant Cell* 24: 275–287.

de S. A. Veloso LL., da Silva AAR, Capitulino JD, de Lima GS, de Azevedo CAV, Gheyi HR, Nobre RG, Fernandes PD (2020) Photochemical efficiency and growth of soursop rootstocks subjected to salt stress and hydrogen peroxide. *AIMS Agric Food* 5: 1–13.

Desikan R, Last K, Harrett-Williams R, Tagliavia C, Harter K, Hooley R, Hancock JT, Neill SJ (2006) Ethylene-induced stomatal closure in Arabidopsis occurs via AtrbohF-mediated hydrogen peroxide synthesis. *Plant J* 47: 907–916.

Dey P, Datta D, Pattnaik D, Dash D, Saha D, Panda D, Bhatta BB, Parida S, Mishra UN, Chauhan J, Pandey H (2022) Physiological, biochemical, and molecular adaptation mechanisms of photosynthesis and respiration under challenging environments. In: T. Aftab and A. Roychoudhury (eds.), *Plant Perspectives to Global Climate Changes*, Academic Press, Cambridge, MA, pp. 79–100.

DHaeze W, De Rycke R, Mathis R, Goormachtig S, Pagnotta S, Verplancke C, Capoen W, Holsters M (2003) Reactive oxygen species and ethylene play a positive role in lateral root base nodulation of a semiaquatic legume. *Proc Natl Acad Sci USA* 20: 11789–11794.

Ecker JR (1995) The ethylene signal transduction pathway in plants. *Science* 268: 667–675.

Ellouzi H, Sghayar S, Abdelly C (2017) H_2O_2 seed priming improves tolerance to salinity; drought and their combined effect more than mannitol in *Cakile maritima* when compared to *Eutrema salsugineum*. *J Plant Physiol* 210: 38–50.

El-Maarouf-Bouteau H, Sajjad Y, Bazin J, Langlade N, Cristescu SM, Balzergue Baudouin E, Bailly C (2015) Reactive oxygen species, abscisic acid and ethylene interact to regulate sunflower seed germination. *Plant Cell Environ* 38: 364–374.

Elmongy MS, Cao Y, Zhou H, Xia Y (2018) Root development enhanced by using indole-3-butyric acid and naphthalene acetic acid and associated biochemical changes of in vitro Azalea microshoots. *J Plant Growth Regul* 37: 813–825.

Eulgem T, Somssich IE (2007) Networks of WRKY transcription factors in defense signaling. *Curr Opin Plant Biol* 10: 366–371.

Foyer CH, Shigeoka S (2011) Understanding oxidative stress and antioxidant functions to enhance photosynthesis. *Plant Physiol* 155: 93–100.

Gadjev I, Vanderauwera S, Gechev TS, Laloi C, Minkov IN, Shulaev V, Apel K, Inze D, Mittler R, Van Breusegem F (2006) Transcriptomic footprints disclose specificity of reactive oxygen species signaling in Arabidopsis. *Plant Physiol* 141: 436–445.

Gantait S, Rani Sinniah U, Ali N, Chandra Sahu N (2015) Gibberellins-a multifaceted hormone in plant growth regulatory network. *Curr Protein Pept Sci* 16: 406–412.

Garg N, Bharti A (2018) Salicylic acid improves arbuscular mycorrhizal symbiosis., and chickpea growth and yield by modulating carbohydrate metabolism under salt stress. *Mycorrhiza* 28: 727–746.

Gechev T, Minkov I, Hille J (2005) Hydrogen peroxide-induced cell death in Arabidopsis, transcriptional and mutant analysis reveals a role of an oxoglutarate- dependent dioxygenase gene in the cell death process. *IUBMB Life* 57: 181–188.

Gechev TS, Hille J (2005) Hydrogen peroxide as a signal controlling plant programmed cell death. *J Cell Biol* 168: 17–20.

Gould KS, McKelvie J, Markham KR (2002) Do anthocyanins function as antioxidants in leaves? Imaging of H_2O_2 in red and green leaves after mechanical injury. *Plant Cell Environ* 25: 1261–1269.

Gratao PL, Monteiro CC, Rossi ML, Martinelli AP, Peres LEP, Medici LO, Lea PJ, Azevedo RA (2009) Differential ultrastructural changes in tomato hormonal mutants exposed to cadmium. *Environ Exp Bot* 67: 387–394.

Guo Z, Wang F, Xiang X, Ahammed GJ, Wang M, Onac E, Zhou J, Xia X, Shi K, Yin X, Chen K, (2016) Systemic induction of photosynthesis via illumination of the shoot apex is mediated sequentially by phytochrome B., auxin and hydrogen peroxide in tomato. *Plant Physiol* 172: 1259–1272.

Hao W, Guo H, Zhang J, Hu G, Yao Y, Dong J (2014) Hydrogen peroxide is involved in salicylic acid-elicited rosmarinic acid production in *Salvia miltiorrhiza* cell cultures. *Sci World J* 2014: 843764.

Hasanuzzaman M, Bhuyan MHMB, Zulfiqar F, Raza A, Mohsin SM, Mahmud JA, Fujita M, Fotopoulos V. (2020) Reactive oxygen species and antioxidant defense in plants under abiotic stress: Revisiting the crucial role of a universal defense regulator. *Antioxidants (Basel)*. 9: 681.

Hasanuzzaman M, Nahar K, Gill SS, Alharby HF, Razafindrabe B, Fujita M (2017) Hydrogen peroxide pretreatment mitigates cadmium-induced oxidative stress in *Brassica napus* L., an intrinsic study on antioxidant defense and glyoxalase systems. *Front Plant Sci* 10: 108–115.

He F, Li HG, Wang JJ, Su Y, Wang HL, Feng CH, Yang Y, Niu MX, Liu C, Yin W, Xia X (2019) PeSTZ1, a C2H2-type zinc finger transcription factor from *Populus euphratica*, enhances freezing tolerance through modulation of ROS scavenging by directly regulating PeAPX2. *Plant Biotechnol J* 17: 2169–2183.

Hejazi Z, Sadat MI, Ghani M, Raghib AU, Fitrat K (2018) Effect of different concentrations of GA_3, H_2O_2 and bleach solutions on seed germination of guava (*Psidium guajava* L.). *Int J Crop Sci* 6: 2679–2681.

Hönig M, Plíhalová L, Husičková A, Nisler J, Doležal K (2018) Role of cytokinins in senescence, antioxidant Defence and photosynthesis. *Int J Mol Sci* 19: 4045.

Hossain MA, Bhattacharjee S, Armin SM, Qian P, Xin W, Li HY, et al. (2015) Hydrogen peroxide priming modulates abiotic oxidative stress tolerance: Insights from ROS detoxification and scavenging. *Front Plant Sci* 6: 420.

Hou QZ, Wang YP, Liang JY, Jia LY, Feng HQ, Wen J, Ehmet N, Bai JY (2018) H_2O_2 induced acclimation of photosystem II to excess light is mediated by alternative respiratory pathway and salicylic acid. *Photosynthetica* 56: 1154–1160.

Htet Y, Lu Z, Trauger SA, Tennyson AG (2019) Hydrogen peroxide as a hydride donor and reductant under biologically relevant conditions. *Chem Sci* 10: 2025–2033.

Hu X, Bidney DL, Yalpani N, Duvick JP, Crasta O, Folkerts O, Lu G (2003). Overexpression of a gene encoding hydrogen peroxide-generating oxalate oxidase evokes defense responses in sunflower. *Plant Physiol* 133: 170–181.

Hung KT, Kao CH (2004) Hydrogen peroxide is necessary for abscisic acid induced senescence of rice leaves. *J Plant Physiol* 161: 1347–1357.

Hung KT, Hsu YT, Kao CH (2006) Hydrogen peroxide is involved in methyl jasmonate-induced senescence of rice leaves. *Physiol Plantarum* 2: 293–303.

Huang H, Liu B, Liu L, Song S (2017a) Jasmonate action in plant growth and development. *J Exp Bot* 68: 1349–1359.

Huang Y, Lin C, He F, Li Z, Guan Y, Hu Q, Hu J (2017b) Exogenous spermidine improves seed germination of sweet corn via involvement in phytohormone interactions, H_2O_2 and relevant gene expression. *BMC Plant Biol* 17: 1.

Ishibashi Y, Aoki N, Kasa S, Sakamoto M, Kai K, Tomokiyo R, Watabe G, Yuasa T, Iwaya-Inoue M (2017) The interrelationship between abscisic acid and reactive oxygen species plays a key role in barley seed dormancy and germination. *Front Plant Sci* 8: 275.

Ishibashi Y, Koda Y, Zheng SH, Yuasa T, Iwaya-Inoue M (2012a) Regulation of soybean seed germination through ethylene production in response to reactive oxygen species. *Ann Bot* 111: 95–102.

Ishibashi Y, Tawaratsumida T, Kondo K, Kasa S, Sakamoto M, Aoki N, et al. (2012b). Reactive oxygen species are involved in gibberellin/abscisic acid signaling in barley aleurone cells. *Plant Physiol* 158: 1705–1714.

Ivanchenko MG, DenOs D, Monshausen GB, Dubrovsky JG, Bednarova A, Krishnan N (2013) Auxin increases the hydrogen peroxide (H_2O_2) concentration in tomato (*Solanum lycopersicum*) root tips while inhibiting root growth. *Ann Bot* 112: 1107–1116.

Jamaludin R, Mat N, Suryati Mohd K, Afiza Badaluddin N, Mahmud K, Hailmi Sajili M, Khandaker MM (2020) Influence of exogenous hydrogen peroxide on plant physiology, leaf anatomy and rubisco gene expression of the *Ficus deltoidea* Jack var. *Deltoidea*. *Agronomy* 10: 497.

Jammes F, Song C, Shin D, Munemasa S, Takeda K, Gu D et al. (2009) MAP kinases MPK9 and MPK12 are preferentially expressed in guard cells and positively regulate ROS-mediated ABA signaling. *Proc Natl Acad Sci USA* 106: 20520–20525.

Jiang G, Yan H, Wu F, Zhang D, Zeng W, Qu H, Chen F, Tan L, Duan X, Jiang Y (2017) Litchi fruit LcNAC1 is a target of LcMYC2 and regulator of fruit senescence through its interaction with LcWRKY1. *Plant Cell Physiol* 58: 1075–1089.

Kato Y, Sakamoto W (2014) Phosphorylation of photosystem II core proteins prevents undesirable cleavage of D1 and contributes to the fine-tuned repair of photosystem II. *Plant J* 79: 312–321.

Kauss H, Jeblick W, Ziegler J, Krabler W (1994) Pretreatment of parsley (*Petroselinum crispum* L.) suspension cultures with methyl jasmonate enhances elicitation of activated oxygen species. *Plant Physiol* 105: 89–94.

Khan TA, Fariduddin Q, Yusuf M (2018) Effect of exogenously sourced hydrogen peroxide treatments on growth, photosynthesis and antioxidant traits in two contrasting cultivars of tomato: A mode and concentration dependent study. *Acad J Agri Res* 6: 019–029.

Khan MI, Khan NA, Masood A, Per TS, Asgher M (2016) Hydrogen peroxide alleviates nickel- inhibited photosynthetic responses through increase in use efficiency of nitrogen and sulfur, and glutathione production in mustard. *Front Plant Sci* 7: 44.

Khan TA, Yusuf M, Ahmad A, Bashir Z, Saeed T, Fariduddin Q, Hayat S, Mock HP, Wu T (2019). Proteomic and physiological assessment of stress sensitive and tolerant variety of tomato treated with brassinosteroids and hydrogen peroxide under low-temperature stress. *Food Chem* 289: 500–511.

Khan TA, Yusuf M, Fariduddin Q (2015) Seed treatment with H_2O_2 modifies net photosynthetic rate and antioxidant system in mung bean (*Vigna radiata* L. Wilczek) plants. *Isr J Plant Sci* 62: 167–175.

Khedia J, Agarwal P, Agarwal PK (2018) AlNAC4 transcription factor mitigates oxidative stress by maintaining ROS homeostasis in transgenic tobacco. *Front Plant Sci* 9: 1522.

Kim K, Portis AR (2004) Oxygen-dependent H_2O_2 production by Rubisco. *FEBS Lett* 571: 124–128.

Kimura M, Umemoto Y, Kawano T (2014) Hydrogen peroxide-independent generation of superoxide by plant peroxidase: hypotheses and supportive data employing ferrous ion as a model stimulus. *Front Plant Sci* 5: 285.

Kohli SK, Handa N, Sharma A, Gautam V, Arora S, Bhardwaj R, Alyemeni MN, Wijaya L, Ahmad P (2018) Combined effect of 24-epibrassinolide and salicylic acid mitigates lead (Pb) toxicity by modulating various metabolites in *Brassica juncea* L. seedlings. *Protoplasma* 255: 11–24.

Kovtun Y, Chiu WL, Tena G, Sheen J (2000) Functional analysis of oxidative stress-activated mitogen-activated protein kinase cascade in plants. *Proc Natl Acad Sci USA* 197: 2940–5.

Kumari S, Verma VK (2019) Hydrogen peroxide mediates suberization, root thickness and stomatal movement in wheats. *Int J Curr Microbiol App Sci* 8: 2448–2454.

Kwak JM, Mori IC, Pei ZM, Leonhardt N, Torres MA, Dangl JL, Bloom RE, Bodde S, Jones JD, Schroeder, JI (2003) NADPH oxidase AtrbohD and AtrbohF genes function in ROS-dependent ABA signaling in Arabidopsis. *EMBO J* 22: 2623–2633.

Lariguet P, Ranocha P, De Meyer M, Barbier O, Penel C, Dunand C (2013) Identification of a hydrogen peroxide signalling pathway in the control of light-dependent germination in Arabidopsis. *Planta* 238: 381–395.

Le DT, Nishiyama R, Watanabe Y, Vankova R, Tanaka M, Seki M, Ham le H, Yamaguchi-Shinozaki K, Shinozaki K, Tran LS (2012) Identification and expression analysis of cytokinin metabolic genes in soybean under normal and drought conditions in relation to cytokinin levels. *PLoS One* 7: e42411.

Lennicke C, Rahn J, Lichtenfels R, Wessjohann LA, Seliger B (2015) Hydrogen peroxide - production, fate and role in redox signaling of tumor cells. *Cell Commun Signal* 13:39.

Leon J, Lawton MA, Raskin I (1995) Hydrogen peroxide stimulates salicylic acid biosynthesis in tobacco. *Plant Physiol* 108: 1673–1678.

Li Z, Gao Y, Zhang Y, Lin C, Gong D, Guan Y, Hu J (2018) Reactive oxygen species and gibberellin acid mutual induction to regulate tobacco seed germination. *Front Plant Sci* 9: 1279.

Li XP, Xu QQ, Liao WB, Ma ZJ, Xu XT, Wang M, Ren PJ, Niu LJ, Jin X, Zhu YC (2016) Hydrogen peroxide is involved in abscisic acid-induced adventitious rooting in cucumber (*Cucumis sativus* L.) under drought stress. *J Plant Biol* 59: 536–548.

Li Z, Xu J, Gao Y, Wang C, Guo G, Luo Y, Huang Y, Hu W, Sheteiwy MS, Guan Y, Hu J (2017) The synergistic priming effect of exogenous salicylic acid and H_2O_2 on chilling tolerance enhancement during maize (*Zea mays* L.) seed germination. *Front Plant Sci* 8: 1153.

Liu KL, Shen L, Wang JQ, Sheng JP (2008) Rapid inactivation of chloroplastic ascorbate peroxidase is responsible for oxidative modification to Rubisco in tomato (*Lycopersicon esculentum*) under cadmium stress. *J Integr Plant Biol* 50: 415–426.

Liu N, You J, Shi W, Liu W, Yang Z (2012) Salicylic acid involved in the process of aluminum induced citrate exudation in *Glycine max* L. *Plant Soil* 352: 85–97.

Liu Y, Ye N, Liu R, Chen M, Zhang J (2010) H_2O_2 mediates the regulation of ABA catabolism and GA biosynthesis in Arabidopsis seed dormancy and germination. *J Exp Bot* 61: 2979–2990.

Lu W, Chu X, Li Y, Wang C, Guo X (2013) Cotton GhMKK1 induces the tolerance of salt and drought stress, and mediates defence responses to pathogen infection in transgenic nicotiana benthamiana. *PLoS One* 8: e68503.

Lum HK, Butt YK, Lo SC (2002). Hydrogen peroxide induces a rapid production of nitric oxide in mung bean (*Phaseolus aureus*). *Nitric Oxide* 6: 205–213.

Ma C, Wang Z, Sun M, Zhang L, Kong B, Lin T (2015) Hydrogen peroxide acts as a signalling molecule for the methyl jasmonate-induced antioxidant defense in wheat callus to promote enhanced drought tolerance. *J Agric Sci* 7: 99.

Maruta T, Noshi M, Nakamura M, Matsuda S, Tamoi M, Ishikawa T, Shigeoka S (2014) Ferulic acid 5-hydroxylase 1 is essential for expression of anthocyanin biosynthesis-associated genes and anthocyanin accumulation under photooxidative stress in Arabidopsis. *Plant Sci* 219: 61–68.

Matsusaki M, Okuda A, Masuda T, Koishihara K, Mita R, Iwasaki K, Hara K, Naruo Y, Hirose A, Tsuchi Y, et al. (2016) Cooperative protein folding by two protein thiol disulfide oxidoreductases and ERO1 in soybean. *Plant Physiol* 170: 774–789.

Merewitz EB, Gianfagna T, Huang B (2011) Photosynthesis., water use., and root viability under water stress as affected by expression of SAG12-ipt controlling cytokinin synthesis in Agrostis stolonifera. *J Exp Bot* 62: 383–395.

Miao Y, Laun T, Zimmermann P, Zentgraf U (2004) Targets of the WRKY53 transcription factor and its role during leaf senescence in *Arabidopsis*. *Plant Mol Biol* 55: 853–867.

Miao Y, Laun TM, Smykowski A, Zentgraf U (2007) Arabidopsis MEKK1 can take a short cut: It can directly interact with senescence-related WRKY53 transcription factor on the protein level and can bind to its promoter. *Plant Mol Biol* 65: 63–76.

Miao Y, Lv D, Wang P, Wang XC, Chen J, Miao C, et al. (2006). An Arabidopsis glutathione peroxidase functions as both a redox transducer and a scavenger in abscisic acid and drought stress responses. *Plant Cell* 18: 2749–2766.

Mittler R, Vanderauwera S, Suzuki N, Miller GA, Tognetti VB, Vandepoele K, Gollery M, Shulaev V, Van Breusegem F (2011) ROS signaling, the new wave? *Trends Plant Sci* 16: 300–309.

Miyao M (1994) Involvement of active oxygen species in degradation of the D1 protein under strong illumination in isolated subcomplexes of photosystem II. *Biochemistry* 33: 9722–9730.

Moons A (2003) Osgstu3 and osgtu4, encoding tau class glutathione S-transferases, are heavy metal- and hypoxic stress-induced and differentially salt stress responsive in rice roots. *FEBS Lett* 553: 427–432.

Morgan PW, Drew MC (1997) Ethylene and plant responses to stress. *Physiol Plant* 100: 620–630.

Mubarakshina Borisova MM, Kozuleva MA, Rudenko NN, Naydov IA, Klenina IB, Ivanov BN (2012) Photosynthetic electron flow to oxygen and diffusion of hydrogen peroxide through the chloroplast envelope via aquaporins. *Biochim Biophys Acta* 1817: 1314–1321.

Muller M, Munne-Bosch S (2015) Ethylene response factors, a key regulatory hub in hormone and stress signaling. *Plant Physiol* 169: 32–41.

Nakagami H, Pitzschke A, Hirt H (2005) Emerging MAP kinase pathways in plant stress signalling. *Trends Plant Sci* 10: 339–346.

Nakagami H, Soukupova H, Schikora A, Zarsky V, Hirt H (2006) A mitogen activated protein kinase mediates reactive oxygen species homeostasis in Arabidopsis. *J Biol Chem* 281: 38697–38704.

Nazir F, Fariduddin Q, Khan TA (2020) Hydrogen peroxide as a signalling molecule in plants and its crosstalk with other plant growth regulators under heavy metal stress. *Chemosphere* 252: 126486.

Nazir F, Hussain A, Fariduddin, Q (2019) Hydrogen peroxide modulate photosynthesis and antioxidant systems in tomato (*Solanum lycopersicum* L.) plants under copper stress. *Chemosphere* 230: 544–558.

Negi S, Tak H, Ganapathi TR (2018) A banana NAC transcription factor (MusaSNAC1) impart drought tolerance by modulating stomatal closure and H_2O_2 content. *Plant Mol Biol* 96: 457–471.

Nie S, Yue H, Xing D (2015) A potential role for mitochondrial produced reactive oxygen species in salicylic acid-mediated plant acquired thermotolerance. *Plant Physiol* 1: 00719.

Nishiyama Y, Allakhverdiev SI, Murata N (2011) Protein synthesis is the primary target of reactive oxygen species in the photoinhibition of photosystem II. *Physiol Plant* 142: 35–46.

Nishiyama Y, Murata N (2014) Revised scheme for the mechanism of photoinhibition and its application to enhance the abiotic stress tolerance of the photosynthetic machinery. *Appl Microbiol Biotechnol* 98: 8777–8796.

Niu L, Liao W (2016) Hydrogen peroxide signaling in plant development and abiotic responses: Crosstalk with nitric oxide and calcium. *Front Plant Sci* 7: 230.

Noctor G, Mhamdi A, Foyer CH (2014) The roles of reactive oxygen metabolism in drought: not so cut and dried. *Plant Physiol* 164:1636–1648.

Nopo-Olazabal C, Condori J, Nopo-Olazabal L, Medina-Bolivar F (2014) Differential induction of antioxidant stilbenoids in hairy roots of *Vitis rotundifolia* treated with methyl jasmonate and hydrogen peroxide. *Plant Physiol Biochem* 74: 50–69.

Ooms MD, Graham PJ, Nguyen B, Sargent EH, Sinton D (2017) Light dilution via wavelength management for efficient high-density photobioreactors. *Biotechnol Bioeng* 114: 1160–1169.

Opdenakker K, Remans T, Vangronsveld J, Cuypers A (2012) Mitogen-Activated Protein (MAP) kinases in plant metal stress, regulation and responses in comparison to other biotic and abiotic stresses. *Int J Mol Sci* 13: 7828–7853.

Orozco-Cardenas ML, Narvaez-Vasquez J, Ryan CA (2001) Hydrogen peroxide acts as a second messenger for the induction of defense genes in tomato plants in response to wounding, systemin, and methyl jasmonate. *Plant Cell* 13: 179–191.

Osakabe Y, Yamaguchi-Shinozaki K, Shinozaki K, Tran LS (2014) ABA control of plant macroelement membrane transport systems in response to water deficit and high salinity. *New Phytol* 202: 35–49.

Overmyer K, Brosche M, Kangaskärvi J (2003) Reactive oxygen species and hormonal control of cell death. *Trends Plant Sci* 8, 335–342.

Ozaki K, et al. (2009) Enrichment of sugar content in melon fruits by hydrogen peroxide treatment. *J Plant Physiol* 166: 569–578.

Pan J, Zhang M, Kong X, Xing X, Liu Y, Zhou Y, Liu Y, Sun L, Li D (2012) ZmMPK17, a novel maize group D MAP kinase gene, is involved in multiple stress responses. *Planta* 235: 661–676.

Pasternak T, Potters G, Caubergs R, Jansen MA (2005) Complementary interactions between oxidative stress and auxins control plant growth responses at plant, organ, and cellular level. *J Exp Bot* 56: 1991–2001.

Peleg-Grossman S, Melamed-Book N, Cohen G, Levine A (2010) Cytoplasmic H_2O_2 prevents translocation of NPR1 to the nucleus and inhibits the induction of PR genes in Arabidopsis. *Plant Signal Behav* 5: 1401–1406.

Phukan UJ, Jeena GS, Shukla RK (2016) WRKY transcription factors, molecular regulation and stress responses in plants. *Front Plant Sci* 7: 760.

Pitzschke A, Djamei A, Bitton F, Hirt H (2009) A major role of the MEKK1-MKK1/2-MPK4 pathway in ROS signalling. *Mol Plant* 2: 120–137.

Pornsiriwong W, Estavillo GM, Chan KX, Tee EE, Ganguly D, Crisp PA, Phua SY, Zhao C, Qiu J, Park J, Yong MT, Nisar N, Yadav AK, Schwessinger B, Rathjen J, Cazzonelli CI, Wilson PB, Gilliham M, Chen ZH, Pogson BJ (2017) A chloroplast retrograde signal., 30- phosphoadenosine 50-phosphate., acts as a secondary messenger in abscisic acid signaling in stomatal closure and germination. *eLife* 6: e23361.

Qi J, Song CP, Wang B, Zhou J, Kangasjearvi J, Zhu JK, Gong Z (2018) Reactive oxygen species signaling and stomatal movement in plant responses to drought stress and pathogen attack. *J Integr Plant Biol* 60: 805–826.

Qu Y, Wang Q, Guo J, Wang P, Song P, Jia Q, Zhang X, Kudla J, Zhang W, Zhang Q (2017) Peroxisomal CuAO and its product H_2O_2 regulate the distribution of auxin and IBA- dependent lateral root development in Arabidopsis. *J Exp Bot* 68: 4851–4867.

Quan LJ, Zhang B, Shi WW, Li HY (2008) Hydrogen peroxide in plants: A versatile molecule of the reactive oxygen species network. *J Integr Plant Biol* 50: 2–18.

Queval G, Issakidis-Bourguet E, Hoeberichts FA, Vandorpe M, Gakiere B, Vanacker H, Miginiac-Maslow M, Van Breusegem F. Noctor G (2007) Conditional oxidative stress responses in the Arabidopsis photorespiratory mutant cat2 demonstrate that redox state is a key modulator of daylength-dependent gene expression and define photoperiod as a crucial factor in the regulation of H_2O_2-induced cell death. *Plant J* 52: 640–657.

Reis MO, Necchi O, Colepicolo P, Barros MP (2011) Co-stressors chilling and high light increase photooxidative stress in diuron-treated red alga *Kappaphycus alvarezii* but with lower involvement of H_2O_2. *Pest Biochem Physiol* 99: 7–15.

Riemer J, Schwarzlander M, Conrad M, Herrmann JM (2015) Thiol switches in mitochondria: operation and physiological relevance. *Biol Chem* 396: 465–482.

Rizhsky L, Liang H, Shuman J, Shulaev V, Davletova S, Mittler R (2004) When defense pathways collide: The response of Arabidopsis to a combination of drought and heat stress. *Plant Physiol* 134: 1683–1696.

Rodriguez MC, Petersen M, Mundy J (2010) Mitogen-activated protein kinase signalling in plants. *Annu Rev Plant Biol* 61: 621–649.

Rohman MM, Islam MR, Habib SH, Choudhury DA, Mohi-Ud-Din M (2024) NADPH oxidase-mediated reactive oxygen species, antioxidant isozymes, and redox homeostasis regulate salt sensitivity in maize genotypes. *Heliyon* 10: e26920.

Sapara K, Khedia J, Agarwal P, Gangapur D, Agarwal PK (2019) SbMYB15 transcription factor mitigates cadmium and nickel stress in transgenic tobacco by limiting uptake and modulating antioxidative defence system. *Func Plant Biol* 46: 702–714.

Saxena I, Srikanth S, Chen Z (2016) Cross talk between H_2O_2 and interacting signal molecules under plant stress response. *Front Plant Sci* 7: 570.

Schmidt R, Mieulet D, Hubberten HM, Obata T, Hoefgen R, Fernie AR, Fisahn J, San Segundo B, Guiderdoni E, Schippers JH, Mueller-Roeber B (2013) Salt-responsive ERF1 regulates reactive oxygen species-dependent signaling during the initial response to salt stress in rice. *Plant Cell* 25: 2115–31.

Schmutz J, Cannon SB, Schlueter J, Ma J, Mitros T, Nelson W, Hyten DL, Song Q, Thelen JJ, Cheng J et al. (2010) Genome sequence of the palaeopolyploid soybean. *Nature* 463: 178–183.

Schwarzlander M, Fricker MD, Sweetlove LJ (2009) Monitoring the in vivo redox state of plant mitochondria: Effect of respiratory inhibitors, abiotic stress and assessment of recovery from oxidative challenge. *Biochim Biophys Acta Bioenerg* 1787: 468–475.

Segal AW (2016) NADPH oxidases as electrochemical generators to produce ion fluxes and turgor in fungi, plants and humans. *Open Biol* 6: 160028.

Sharkey TD, Bernacchi CJ, Farquhar GD, Singsaas EL (2007) Fitting photosynthetic carbon dioxide response curves for C3 leaves. *Plant Cell Environ* 30: 1035–1040.

Sharma A, Shahzad B, Kumar V, Kohli SK, Sidhu GPS, Bali AS, Handa N, Kapoor D, Bhardwaj R, Zheng B (2019) Phytohormones regulate accumulation of osmolytes under abiotic stress. *Biomolecules* 9, 285.

Shi HT, Liu GY, Wei YX, Chan ZL (2018) The zinc-finger transcription factorZAT6 is essential for hydrogen peroxide induction of anthocyanin synthesis inArabidopsis. *Plant Mol Biol* 97: 165–176.

Shi H, Ye T, Zhu J-K, Chan Z (2014) Constitutive production of nitric oxide leads to enhanced drought stress resistance and extensive transcriptional reprogramming in Arabidopsis. *J Exp Bot* 65: 4119–4131.

Shu S, Gao P, Li L, Yuan Y, Sun J, Guo S (2016) Abscisic acid-Induced H_2O_2 accumulation enhances antioxidant capacity in pumpkin-grafted cucumber leaves under Ca $(NO_3)_2$ stress. *Front Plant Sci* 7: 1489.

Song YJ, Joo JH, Ryu HY, Lee JS, Bae YS, Nam KH (2007) Reactive oxygen species mediate IAA-Induced ethylene production in mungbean (*Vigna radiata* L.) hypocotyls. *J Plant Biol* 50: 18–23.

Strand DD, Livingston AK, Satoh-Cruz M, Froehlich JE, Maurino VG, Kramer DM (2015) Activation of cyclic electron flow by hydrogen peroxide in vivo. *Proc Natl Acad Sci USA* 112: 5539–5544.

Su C, Liu L, Liu H, Ferguson BJ, Zou Y, Zhao Y, Wang T, Wang Y, Li X (2016). H_2O_2 regulates root system architecture by modulating the polar transport and redistribution of auxin. *J Plant Biol* 59: 260–270.

Sun MJ, Foyer CH (2014) Redox regulation of plant development. *Antioxid Redox Signal* 21: 1305–1326.

Swamy PM, Smith B (1999) Role of abscisic acid in plant stress tolerance. *Curr Sci* 76: 1220–1227.

Takac T, Obert B, Rolcik J, Samaj J (2016) Improvement of adventitious root formation in flax using hydrogen peroxide. *N Biotech* 33: 728–734.

Takahashi S, Murata N (2008) How do environmental stresses accelerate photoinhibition? *Trends Plant Sci* 13: 178–182.

Taki-Nakano N, Ohzeki H, Kotera J, Ohta H (2014) Cytoprotective effects of 12- oxo phytodienoic acid, a plant-derived oxylipin jasmonate, on oxidative stress-induced toxicity in human neuroblastoma SH-SY5Y cells. *Biochm Biophys Acta Gen Subj* 1840: 3413–3422.

Tikkanen M, Grieco M, Aro EM (2011) Novel insights into plant light-harvesting complex II phosphorylation and 'state transitions'. *Trends Plant Sci* 16: 126–131.

Tikkanen M, Mekala NR, Aro EM (2014) Photosystem II photoinhibition-repair cycle protects Photosystem I from irreversible damage. *Biochim Biophys Acta* 1837: 210–215.

To JP, Deruere J, Maxwell BB, Morri VF, Hutchison CE, Ferreira FJ, Schaller GE, Kieber JJ (2007) Cytokinin regulates type-A Arabidopsis response regulator activity and protein stability via two-component phosphorelay. *Plant Cell* 19: 3901–3914.

Torres M, Forman HJ (2003) Redox signaling and the MAP kinase pathways. *Biofactors* 17: 287–96.

Tsai YC, Kao CH (2004) The involvement of hydrogen peroxide in abscisic acid induced activities of ascorbate peroxidase and glutathione reductase in rice roots. *Plant Growth Regul* 43: 207–212.

Uchida A, Jagendorf AT, Hibino T, Takabe T, Takabe T (2002) Effects of hydrogen peroxide and nitric oxide on both salt and heat stress tolerance in rice. *Plant Sci* 163: 515–523.

Wang J, Ding H, Zhang A, Ma F, Cao J, Jiang M (2010a) A novel mitogen-activated protein kinase gene in maize (*Zea mays*), ZmMPK3, is involved in response to diverse environmental cues. *J Integr Plant Biol* 52: 442–452.

Wang P, Du Y, Li Y, Ren D, Song CP (2010b) Hydrogen peroxide emediated activation of MAP kinase 6 modulates nitric oxide biosynthesis and signal transduction in Arabidopsis. *Plant Cell* 22: 2981–2998.

Wang H, Liang X, Huang J, Zhang D, Lu H, Liu Z, Bi Y (2010c) Involvement of ethylene and hydrogen peroxide in induction of alternative respiratory pathway in salt-treated Arabidopsis calluses. *Plant Cell Physiol* 51: 1754–1765.

Wang H, Liang X, Wan Q, Wang X, Bi Y (2009) Ethylene and nitric oxide are involved in maintaining ion homeostasis in Arabidopsis callus under salt stress. *Planta* 230: 293–307.

Wang L, Zhao R, Zheng Y, Chen L, Li R, Ma J, Hong X, Ma P, Sheng J, Shen L (2017) SlMAPK1/2/3 and antioxidant enzymes are associated with H_2O_2-induced chilling tolerance in tomato plants. *J Agric Food Chem.* 65: 6812–20.

Wang W, Wang X, Huang M, Cai J, Zhou Q, Dai T, Cao W, Jiang D (2018) Hydrogen peroxide and abscisic acid mediate salicylic acid-induced freezing tolerance in wheat. *Front Plant Sci* 9: 1137.

Waszczak C, Carmody M, Kangasjarvi J (2018) Reactive oxygen species in plant signaling. *Annu Rev Plant Biol* 69: 209–236.

Weijers D, Nemhauser J, Yang Z (2018) Auxin, small molecule, big impact. *J Exp Bot* 5: 133–136.

Woodward AW, Bartel B (2005) Auxin, regulation, action, and interaction. *Ann Bot* 95: 707–735.

Wu A, Allu AD, Garapati P, Siddiqui H, Dortay H, Zanor MI, et al. (2012) JUNGBRUNNEN1, a reactive oxygen species-responsive NAC transcription factor, regulates longevity in *Arabidopsis*. *Plant Cell* 24: 482–506.

Wu Q, Su N, Zhang X, Liu Y, Cui J, Liang Y (2016) Hydrogen peroxide, nitric oxide a and UV RESISTANCE LOCUS8 interact to mediate UV-B-induced anthocyanin biosynthesis in radish sprouts. *Sci Rep* 6: 29164.

Xiong J, Yang Y, Fu G, Tao L (2015) Novel roles of hydrogen peroxide (H_2O_2) in regulating pectin synthesis and dimethyl esterification in the cell wall of rice (*Oryza sativa*) root tips. *New Phytol* 206: 118–126.

Xu Y, Burgess P, Zhang X, Huang B (2016) Enhancing cytokinin synthesis by over-expressing ipt alleviated drought inhibition of root growth through activating ROS-scavenging systems in *Agrostis stolonifera*. *J Exp Bot* 67: 1979–1992.

Yamamoto Y, Aminaka R, Yoshioka M, Khatoon M, Komayama K, Takenaka D, Yamashita A, Nijo N, Inagawa K, Morita N, et al. (2008) Quality control of photosystem II: Impact of light and heat stresses. *Photosyn Res* 98: 589–608.

Yang CY (2014) Ethylene and hydrogen peroxide are involved in hypoxia signalling that modulates AtERF73/HRE1 expression. *Plant Signal Behav* 9: 877–5.

Yang W, Zhu C, Ma X, Li G, Gan L, Ng D, Xia K (2013) Hydrogen peroxide is a second messenger in the salicylic acid-triggered adventitious rooting process in mung bean seedlings. *PLoS One* 8: e84580.

Yao Y, Liu X, Li Z, Ma X, Rennenberg H, Wang X, Li H (2013) Drought-induced H_2O_2 accumulation in subsidiary cells is involved in regulatory signalling of stomatal closure in maize leaves. *Planta* 238: 217–227.

Yoshioka H, Numata N, Nakajima K, Katou S, Kawakita K, Rowland O, Jones JD, Doke N (2003) *Nicotiana benthamiana* gp91phox homologs NbrbohA and NbrbohB participate in H_2O_2 accumulation and resistance to Phytophthora infestans. *Plant Cell* 15: 706–718.

Yoshioka M, Uchida S, Mori H, Komayama K, Ohira S, Morita N, Nakanishi T, Yamamoto Y (2006) Quality control of photosystem II. Cleavage of reaction center D1 protein in spinach thylakoids by FtsH protease under moderate heat stress. *J Biol Chem* 281: 21660–21669.

Zalewski W, Galuszka P, Gasparis S, Orczyk W, Nadolska-Orczyk A (2010) Silencing of the HvCKX1 gene decreases the cytokinin oxidase/dehydrogenase level in barley and leads to higher plant productivity. *J Exp Bot* 61: 1839–1851.

Zeng XQ, Chow WS, Su LJ, Peng XX, Peng CL (2010) Protective effect of supplemental anthocyanins on Arabidopsis leaves under high light. *Physiol Plant* 138: 215–225.

Zhang A, Jiang, M, Zhang J, Ding H, Xu S, Hu X, et al. (2007) Nitric oxide induced by hydrogen peroxide mediates abscisic acid-induced activation of the mitogen-activated protein kinase cascade involved in antioxidant defense inmaize leaves. *New Phytol* 175: 36–50.

Zhang FH, Li A, Zhang Z, Huang Z, Lu P, Zhang D, Liu X, Zhang ZF, Huang R (2016a) Ethylene response factor TERF1., regulated by ETHYLENE-INSENSITIVE3- like factors., functions in reactive oxygen species (ROS) scavenging in tobacco (*Nicotiana tabacum* L.). *Sci Rep* 6: 29948.

Zhang HM, Devine LB, Xia X, Offler CE, Patrick JW (2019) Ethylene and hydrogen peroxide regulate formation of a sterol-enriched domain essential for wall labyrinth assembly in transfer cells. *J Exp Bot* 70: 1469–1482.

Zhang JZ, Reisner E (2020) Advancing photosystem II photoelectrochemistry for semi-artificial photosynthesis. *Nat Rev Chem* 4: 6–21.

Zhang M, Smith JAC, Harberd NP, Jiang C (2016b) The regulatory roles of ethylene and reactive oxygen species (ROS) in plant salt stress responses. *Plant Mol* 91: 651–659.

Zhang Q, Su LJ, Chen JW, Zeng XQ, Sun BY, Peng CL (2012) The antioxidative role of anthocyanins in Arabidopsis under high irradiance. *Biol Plantarum* 56: 97–104.

Zhang Y, Tan J, Guo Z, Lu S, He S, Shu W, Zhou B (2009) Increased abscisic acid levels in transgenic tobacco over-expressing 9 cis-epoxycarotenoid dioxygenase influence H_2O_2 and NO production and antioxidant defences. *Plant Cell Environ* 32: 509–519.

Zhang X, Zhang L, Dong F, Gao J, Galbraith DW, Song CP (2001) Hydrogen peroxide is involved in abscisic acid-induced stomatal closure in *Vicia faba*. *Plant Physiol* 126: 1438–1448.

Zheng YL, Yin XH, Ma HC (2018) Effects of hydrogen peroxide on seed germination, seedling growth and physiological characteristcs of *Bombax ceiba* after heat shock. *Pakistan J Bot* 50: 1327–1333.

Zhong C, Xu H, Ye S, Wang S, Li L, Zhang S, Wang X (2015) Gibberellic acid stimulated Arabidopsis serves as an integrator of gibberellins, abscisic acid, and glucose signaling during seed germination in Arabidopsis. *Plant Physiol* 169: 2288–2303.

Zhou J, Xia XJ, Zhou YH, Shi K, Chen Z, Yu JQ (2014) OH1- dependent H_2O_2 production and subsequent activation of MPK1/2 play an important role in acclimation-induced cross-tolerance in tomato. *J Exp Bot* 65: 595–607.

Zhou L, Hou H, Yang T, Lian Y, Sun Y, Bian Y, Wang ZC (2018) Exogenous hydrogen peroxide inhibits primary root gravitropism by regulating auxin distribution during Arabidopsis seed germination. *Plant Physiol Biochem* 128: 126–133.

Zhu T, Deng X, Zhou X, Zhu L, Zou L, Li P, Zhang D, Lin H (2016) Ethylene and hydrogen peroxide are involved in brassinosteroid-induced salt tolerance in tomato. *Sci Rep* 6: 35392.

Zong XJ, Li DP, Gu LK, Li DQ, Liu LX, Hu XL (2009) Abscisic acid and hydrogen peroxide induce a novel-maize group CMAP kinase gene, ZmMPK7, which is responsible for the removal of reactive oxygen species. *Planta* 229: 485–495.

Zhang J, Zou D, Li Y, Sun X, Wang NN, Gong SY, Zheng Y, Li XB. (2014) GhMPK17, a cotton mitogen-activated protein kinase, is involved in plant response to high salinity and osmotic stresses and ABA signaling. *PloS One* 9 - 95642.

Potters G, Pasternak TP, Guise Y, Palme KJ, Jansen MAK. (2007) Stress-induced morphogenic responses: growing out of trouble? *Trends in Plant Science* 12: 98–105.

Potters G, Pasternak TP, Guisez Y, Jansen MAK (2009) Different stresses, similar morphogenic responses: integrating a plethora of pathways. *Plant Cell Environ* 32: 158–169.

Huang XZ, Zhang XY, Gong ZZ, Yang SH, Shi YT (2017) ABI4 represses the expression of type-A ARRs to inhibit seed germination in *Arabidopsis*. *Plant J.* 89: 354–365.

Wang Q, Zhang W, Yin Z, Wen CK (2013) Rice CONSTITUTIVE TRIPLERESPONSE2 is involved in the ethylene-receptor signalling and regulation of various aspects of rice growth and development. *J Exp Bot* 64: 4863–4875.

SHE Xiao-Ping, SONG Xi-Gui, HE Jun-Min (2004) Role and Relationship of Nitric Oxide and Hydrogen Peroxide in Light/Dark-regulated Stomatal Movement in Vicia faba[J]. *J Integr Plant Biol.,* 46: 1292-1300.

He JM, Xu H, She XP, Song XG, Zhao WM (2005) The role and the interrelationship of hydrogen peroxide and nitric oxide in the UV-B-induced stomatal closure in broad bean. *Funct. Plant Biol.* 32: 237247.

González A, de los Ángeles Cabrera M, Henríquez MJ, Contreras R A, Morales B, Moenne A (2012). Cross talk among calcium, hydrogen peroxide, and nitric oxide and activation of gene expression involving calmodulins and calcium-dependent protein kinases in Ulva compressa exposed to copper excess. *Plant Physiol.* 158: 1451–1462.

De Michele R, Vurro E, Rigo C, Costa A, Elviri L, Di Valentin M, et al. (2009) Nitric oxide is involved in cadmium-induced programmed cell death in Arabidopsis suspension cultures. *Plant Physiol.* 150: 217–228.

9 Elements that Control Plant Growth and Development include Light and Hydrogen Peroxide

Areeba Rehman, Waheed Akram, Tehmina Anjum, Nasim Ahmad Yasin, and Zill e Huma Aftab

INTRODUCTION

Studies have shown that plants usually possess high synergy potential when providing environmental conditions. This flexibility mainly allows it to function and respond to various environmental factors (Heilmeier, 2019). Many internal and external factors determine plants' germination and growth. Among the multiple factors, light and hydrogen peroxide (H_2O_2) are possibly the most significant components involved in managing the plant's physiological functions (Taiz et al., 2018). Light is energy for photosynthesis and a stimulus or starts for other activities like seed germination, stem development, leaf development, elongation, flowering, and fruiting (Lv et al., 2021). Similarly, the H_2O_2, previously ascribed strictly as a metabolite that arises during aerobic respiration, was identified as a signaling molecule involved in a plant's reaction to abiotic and biotic stress and controls hormone signaling and development (Khorobrykh et al., 2015).

This chapter aims to cover many aspects of light and H_2O_2 in plant biology. It examines how light conditions, duration, intensity, and type affect plant growth, including photosynthesis, germination, shoot and leaf formation, and reproduction. Moreover, plant light perception and signaling that regulate reactions involved in phytochromes regulation and special photoreceptors in plants are also touched on.

The impact of light alongside the H_2O_2 auxin is that it assists the plant in dealing with sudden environmental changes (Bielach et al., 2017). It helps moderate oxidant injury, improves the capacity to defend against infection, and moderates various growth processes. When combined with light and H_2O_2, additional progress in plants' physiological and developmental phases can be achieved as fundamental aspects of agriculture (Niu and Liao, 2016), the primary concerns being growth and protection from extreme climatic conditions can be developed. Such discussion provides general details on the function of light and H_2O_2 in plants, their molecular control, and the prospect of their use in arboriculture. As knowledge continues to emerge on these interactions, we can tap these processes for sustainable crop management and improved plant health. This chapter focuses on the role played by light and H_2O_2 individually and in combination in plant metabolism. These pathways help explain the factors that enhance plant growth and development. Thus, this knowledge can improve farming practices and plant resistance in the mentioned contexts.

FUNCTIONS OF LIGHT IN THE PLANT'S GROWTH AND DEVELOPMENT

Light is one of the best-known environmental factors controlling various processes within the plant and its development (Shi et al., 2022). Light performs the work of energy and signaling and influences photosynthesis, seed germination, stem growth, leaf area, flowering, and fruiting

DOI: 10.1201/9781032647692-9

(Leduc et al., 2014). Thus, photomorphogenesis is vital to plants' existence, development, and reproduction by providing the means through which their cells can detect light (Arsovski et al., 2012). In this consultation, the different uses of light in plant development will be focused on critical physiological functions and the mechanisms involved.

Photosynthesis

Energy Source for Photosynthesis: For instance, light is the primary input energy in photosynthesis, the process through which green plants use light to synthesize foods. Chlorophyll pigments capture light energy, which occurs in plant cells' chloroplasts (Cardona et al., 2018).

The light-dependent reaction occurs in the chloroplast's thylakoid membranes, where energy from light is used to make ATP and NADPH. These energy carriers are then used in the Calvin cycle, which occurs in the stroma and produces glucose from CO_2 and water (Foyer and Harbinson, 2019). The amount of light and its intensity determine the rate of photosynthesis and, therefore, the rate of growth and crop production. It is well understood that an optimal light condition has a favorable effect on the biomass and potency of the plant (Ferrante and Mariani, 2018).

Seed Germination

Light as a Germination Trigger: Some seeds are called photoblastic seeds and must be exposed to light before germinating. Light colors, such as blue or red, can help germinate the seeds once exposed (Foschi et al., 2022). Some seeds of certain species have their dormancy in seeds broken by light to be triggered to grow. This mechanism helps ensure the seed germinates only under appropriate light, increasing the seedling survival and formation rate (Penfield, 2017).

Stem Elongation

Phototropism: Phototropism is the plant's growth movement toward the light source. This movement is facilitated by the hormone auxin, which transports toward the shaded side of the stem, elongating the stem in that region and bending it toward the side of the light (Bhatla and Lal, 2023c). Plant growth toward light helps in the perfect arrangement of the plants to capture the light for photosynthesis because energy cannot be created but must be utilized to the maximum (Tian, 2016).

Leaf Expansion

Role of Light in Leaf Development: Light is vital for forming chloroplast organelles and chlorophyll pigments in leaves. Well-lit environments result in many green leaves with high photosynthetic potential (Hart, 2012). Light intensity influences spinner plant leaves' size, shape, and venation pattern. Light intensity and duration are reported to differentiate large petioles and significant areas of the lamina in low-light plants compared with high-light plants, which have formidable, thick lamina with high chlorophyll content (Freschet et al., 2018).

Flowering

Photoperiodism: Photoperiodism is a process of plant response to the proportion of day and night, a method of flowering control. Plants' flowering response to photoperiods can be short, long, or day-neutral (Osnato et al., 2022). This includes phytochromes and cryptochromes, which act as photoreceptors and control the genes related to flowering. The proper photoperiod induces the formation of florigen, a hormone that starts the flowering process (Sharma, 2023).

Fruit Development

Influence on Fruit Growth and Ripening: Appropriate lighting is essential for sugar formation and fruits' color and sweetness. Light affects the carotenoids and anthocyanin metabolism pathways that determine the quality of fruits for the market (Samkumar, 2021). Gas and light influence the synthesis and sensation of ethylene, which fruits use for maturation. Favorable lighting conditions effectively promote ethylene production, leading to synchronous ripening and general desirable qualities of the fruits (Bartholomew and Sanewski, 2018).

Chlorophyll Production

Chlorophyll Synthesis and Maintenance: It was assessed that chlorophyll production is a function of light. Light energy powers the light reactions in photosynthesis, and ATP and NADPH are used in the later dark reactions to combine the carbon and produce glucose (Peck, 2022). Low light harms chlorophyll synthesis, causing a disease called chlorosis. This disease affects plants and causes yellowing or pale leaves. Influencing the amount of light within the environment may have potentially harmful effects on the chlorophyll level and, hence, the health of the plants (Therby-Vale et al., 2022).

Practical Applications in Agriculture

Controlled Environment Agriculture (CEA): Artificial lighting comprising fluorescent and LED grow lights is used in greenhouses and vertical farms. Different light spectra and intensities depend on plant type and development stage (Nájera et al., 2023). These systems enable accurate control of the light conditions, improving food production. Recognizing the specific light demands of the particular crops increases the effectiveness of light conditions and, therefore, plant productivity and efficiency of used resources (Simkin et al., 2019).

Field Management: This knowledge helps select appropriate varieties of crops for given growing seasons and areas and manage light responses for crop improvement. Planting date and canopy management significantly influence light harvesting, reducing the light shutout (Montgomery, 2021). Different breeding programs can choose less inefficient genotypes in utilizing light and stress tolerance. Genetic engineering can complement this by focusing on particular light signaling pathways, which could further improve the traits and, hence, productivity and stress resistance (Ahmar et al., 2020).

Photosensors and Signaling through Light

Since plants are immobile organisms, they have developed intricate ways to gain information about their light environment (Fitter and Hay, 2012). Plants in photosynthesis use light and work as a signal that manages their developmental and physiological statuses. This regulation is done through Specialized photoreceptors and a complex signaling pathway (Wu et al., 2019). This chapter reveals aspects such as the classification of photosensors, their molecular function, and the signaling cascades created to regulate plant growth and development.

TYPES OF PHOTORECEPTORS

Plants contain several classes of photoreceptors, and all can put their senses in specific wavelengths of visible light. Thus, the plants can sense and respond to the range of light available in sunlight (Kharshiing et al., 2019). The primary classes of photoreceptors are:

Phytochromes: These reversible red (600–700 nm) and far-red light receptors are involved in the regulation of germination of seed, response to shade, and flowering (Friesen, 2024). Phytochromes exist in two interconvertible forms:

- The protein form active in the dark and which absorbs red light is called Pr (phytochrome red) (Viczián et al., 2017).
- The one active in the light and which absorbs far-red light is known as Pfr (Phytochrome far-red). This reversible photo conversion enables the plants to perceive the changes in the quality and quantity of the lights (Bojka, 2020).
  1. **Cryptochromes:** Promoting circadian rhythm, seedling de-etiolation, and flowering in response to blue light (400–500 nm) and ultraviolet-A light (315–400 nm) (Yavari, 2020). Through this process, cryptochromes change their conformation, thus starting signal transduction cascades important for downstream events.
  2. **Phototropins:** These blue light receptors control phototropism, which is the growth of a plant in the direction of light, the position of the chloroplast, and the opening of the stomata. Phototropins consist of LOV (light-oxygen-voltage) domains, which undergo conformational changes that dictate kinase activity due to blue light reception (Lokhandwala, 2018).
  3. **UVR8:** This receptor specific to Ultraviolet-B (280–315 nm) does not require a chromophore to absorb UV-B radiation (Vanhaelewyn et al., 2020). As for UVR8, the monomers of this protein will dimerize in response to conditions influenced by UV-B light, such as the synthesis of flavonoids and DNA repair systems (Ferreyra et al., 2021).

SIGNALING PATHWAYS

Light of specific wavelengths stimulates photoreceptors, which sets a series of molecular changes that alter gene activity and cellular functions in motion (Fan et al., 2022). The key components and pathways involved in light signaling. The essential components and pathways involved in light signaling include:

Phytochrome Signaling: Pr transforms to Pfr when the plant absorbs red light and goes to the nucleus (Brown, 2018). Pfr interacts with PIFs, which are phosphorylated and degraded. This relieves the repression of light-reactive genes, hence facilitating photomorphogenesis (Wang et al., 2022). Phytochromes regulate other genes related to light signal transduction, including the chlorophyll-Apoprotein genes and chlorophyll biosynthesis enzymes (Proctor et al., 2022).

Cryptochrome Signaling: A subclass of cryptochromes called the light-activated cryptochromes work with CONSTITUTIVE PHOTOMORPHOGENIC 1 (COP1), which is an E3 ubiquitin ligase to limit its action (de Lucas and Prat, 2014). COP1's inhibition in context extends the stability of transcription factors such as HY5 that activate light-responsive genes (Wang et al., 2022).

Phototropin Signaling: After absorption of blue light, perceived by the photoreceptor's phototropins, they auto-phosphorylate and trigger downstream signaling (Hart and Gardner, 2021). Phototropin proteins control the orientation of the chloroplasts within the cell, with the help of light capturing and managing stomata's opening regarding gas exchange (Ishishita et al., 2020).

UVR8 Signaling: UV-B action provokes the destruction of UVR8 homodimers and their subsequent binding to specific partners, for example, COP1 (Podolec et al., 2021). UVR8 regulates the expression of genes that prepare the body for UV protection, such as pigments that absorb UV and DNA repair enzymes (Tossi et al., 2019).

CROSSTALK AND INTEGRATION

Thus, the signaling pathways that are involved in photoreceptors do not work independently of each other (Galvão and Fankhauser, 2015). However, they show a lot of interaction and coupling, enabling plants to adapt to a different light environment. For example, phytochrome and cryptochrome are two of the five known photoreceptor pathways that work together to control circadian rhythm and flowering (Voitsekhovskaja, 2019). However, integrating light signaling with other environmental factors like temperature and water availability takes the plant responses to the next level, where they get fully prepared for growth (Czarnocka and Karpiński, 2018).

DURATION, INTENSITY, AND QUALITY OF LIGHT

Light is considered one of the essential microenvironments affecting plant growth and development to a considerable extent. Thus, plants have developed sensitivities and adaptations against light and its intensity (quantity), quality, and duration (Carvalho et al., 2011). All of them are unique, but they are interrelated in controlling the functions and structures of the plant. Plants perceive and respond to these characteristics of light, which is critical in efficient bio-agriculture and the production of durable and productive crops (Lüttge, 2016). This discussion focuses on the significance of light interventions' duration, intensity, and quality on the plant growth and development processes, their functioning, and their impact.

Light Duration: Photoperiodism in Plants and Its Development

Photoperiodism, therefore, refers to the response of plants to either the shortness of the day or the shortness of the night. In plants, photoperiodism influences flowering, tuber formation, and dormancy (Osnato et al., 2022). Photoperiodic responses are classified into three main types:

Short-Day Plants (SDPs): Some plants' flowers open only during the night following a critical duration at night. Some of the varieties are *chrysanthemums*, *poinsettias*, and so on. More extended darkness triggers the production of florigen, a flowering hormone (Proietti et al., 2022).

Long-Day Plants (LDPs): This class of plants will bloom if the night length is less than a certain critical length. These include spinach and lettuce, among others. The increase of some durations stimulates florigen synthesis, which results in flowering (Bhatla and Lal, 2023a).

Day-Neutral Plants (DNPs): These plants are not photoperiod-sensitive to trigger flowering at certain seasons. Rather than using pH to trigger floral organs or other parts, they use other outer factors or internal stimuli to initiate floral organelles. (Contreras Planesas, 2022)

The molecular basis of photoperiodism has two main components: light receptors and circadian clocks. Phytochromes and cryptochromes perceive light signals coordinated with the circadian clock to manage gene activity and developmental changes (Lin et al., 2021).

Light Intensity: Efficiency of Photosynthesis and Organizational Changes

Irradiance is the amount of energy per unit area; hence, light intensity means light energy per unit area. It directly relates to the rate of photosynthesis and, therefore, the growth or biomass production by the plant (Feng et al., 2019). The relationship between light intensity and plant responses can be summarized as follows:

Photosynthesis Rate: Light intensity affects the occurrence of photosynthesis proportional to the light intensity to a certain point that any addition will not affect the rate. Such a saturation point differs across plant species and their physiognomy regarding light conditions (Xu et al., 2024).

Photoinhibition: High light intensities cause photoinhibition since the formation of reactive oxygen species (ROS) harms the photosynthetic machinery. Plants, including NPQ (Non-Photochemical Quenching), possess ways to protect and avoid any harm from extra energy intake (Navyashree, 2024).

Morphological Adaptations: The light intensity impacts the overall morphology of plants; low light or shade implies that the stems are much longer and the leaves are relatively larger but thinner to accommodate the limited light. High light conditions cause plants to have smaller rosettes, thicker and more numerous leaves, and more chlorophyll content due to better adaptation to the availability and utilization of light (Su et al., 2024).

Light Quality: Perturbation to Spectrum: Impact on Growth/Development

Light quality is the wavelength distribution that directly impacts plant physiology and growth processes (Bhatla and Lal, 2023b). The primary spectral regions affecting plants are:

Blue Light (400–500 nm): Blue light stimulates phototropins, which control the plant's directional growth toward the light's source. Blue light causes stomata to open, enlarging the opening to allow the exchange of gases and water vapor to escape. Blue light is essential in chlorophyll formation as well as the formation of chloroplasts (Haga and Sakai, 2023).

Red and Far-Red Light (600–750 nm): Phytochromes are attached to red and far-red light and play a role in seed germination and the transition from the vegetative to the generative phase. The red-to-far-red light ratio affects the shade avoidance response. A low value (indicating canopy shade) leads to elongation growth to overcome competitors for the needed light (Huai and Lin, 2023).

Green Light (500–600 nm): Although chlorophyll does not absorb green light as efficiently, it can penetrate through more layers into the leaf, stimulating photosynthesis in the lower layers (Lambers et al., 2019). Green light affects the other photoreceptors depending on its intensity, thus regulating light sensation and the signaling pathways.

Practical Applications in Agriculture

Understanding the effects of light duration, intensity, and quality on plant growth has practical applications in agriculture and horticulture:

Controlled Environment Agriculture (CEA): This is done by controlling the light conditions in greenhouses and Vertical farms to promote plant growth and increase production throughout the year. LED technology enables switching between light spectra and light intensity settings, such as photoperiods (Paradiso and Proietti, 2022).

Field Management: It helps to choose varieties of crops that are ideal for specific periods of the year or the climate of certain areas. The canopy structure and spacing can be carefully managed to maximize light interception and minimize shade impacts.

Breeding and Biotechnology: Plant breeders can choose them, produce traits that optimize light utilization, and show high and fluctuating light intensity tolerance to enhance production. The traits can be even more advanced with the help of genetic engineering because of the diversity in the light signaling pathways.

HYDROGEN PEROXIDE: HIGHLY COMPLEX MOLECULES

H_2O_2 is an ROS, involved in multiple processes in a plant's system and is discussed in the following two-fold capacity. Moreover, it is essential to note that H_2O_2 was considered a toxic byproduct of aerobic metabolism throughout history. However, the last few decades have proven it is a multifunctional signaling molecule involved in multiple physiological and developmental processes. This chapter focuses on the various roles of H_2O_2 in the plant growth process and development, which include signaling during stress and signaling during development.

Hydrogen Peroxide in Stress Responses

Abiotic Stress Tolerance: H_2O_2 is also involved actively in the defense of plants in abiotic stresses, including drought, salinity, heat, cold, and toxicity of heavy metals, et cetera. The flavonoid works as a signaling molecule that initiates the antioxidant monitoring systems within plants, such as the SOD, CAT, and peroxidases. These enzymes form part of the antioxidant defense system to neutralize ROS to a non-threatening level, thereby regulating the redox status of cells and preventing oxidative stress. H_2O_2 plays the role of a secondary messenger in signal transduction during stress. This can regulate the genes and proteins associated with stress response, allowing the plant to adapt. For instance, H_2O_2 can cause the up-regulation of HSPs (heat shock proteins), which protects cellular proteins under heat stress conditions.

Biotic Stress Resistance: H_2O_2 involves the plant's response to pathogens and is crucial to pathogen infection processes. Specific to pathogen attack, an instant and only at-the-site exposure of pathogen to phagocytes leads to the generation of H_2O_2 known as the oxidative burst. This store of H_2O_2 is helpful in direct antimicrobial effects, cross-linking of the cell wall glycoproteins to increase rigidity, and signal downstream responses. H_2O_2 engages plants and causes systemic acquired resistance (SER) in them. These plants defend against almost all forms of attacks by pathogens. It is a mobile signal moving from the center of infection to the periphery and calling for activating the defense reaction from any part of the plant.

Hydrogen Peroxide in Cell Signaling

Hormonal Interactions: The H_2O_2 affects ABA, a crucial stress and stomatal regulation hormone. By increasing the generation of nitric oxide (NO) and ion channel activation in guard cells, H_2O_2 is implicated in stomatal closure induced by ABA to minimize water loss during drought periods. H_2O_2 affects the signaling of two growth-related hormones, auxins and ethylene. For example, H_2O_2 is involved in root development through modulating auxin signaling and impacts cell prolongation and destiny in the root area.

Programmed Cell Death (PCD): Another in the chain of regulation is H_2O_2; the PCD is an orderly cell death, which plays an essential role in ontogenesis and protection against pathological processes. It initiates PCD during pathogen invasion to minimize the pathogen's spread at the cost of the infected cell. However, the H_2O_2 triggered PCD process also participates in development processes like leaf senescence and aerenchyma formation in root under hypoxia conditions.

Hydrogen Peroxide in Developmental Processes

H_2O_2 is involved in root hair formation, vital in absorbing nutrients and water. It also functions as a signaling molecule in gene expression in root hair initiation and elongation. H_2O_2 affects the emergence of lateral roots, pointing to a change in the concentration of auxin in the root tissues and signaling. This regulation is crucial to help the plant develop the right root system, from which it will have a solid base to support itself.

L-Ascorbate expands and differentiates the cells during the formation of shoots and leaves when it takes the form of H_2O_2. It regulates the activation of enzymes and the synthesis/ release of molecules involved in changes to the cell wall and cell division. Oxidative stress is observed in the form of H_2O_2 accumulation in attracting leaf senescence, a plant's genetically controlled process of degradation of cellular components and recycling of nutrients. H_2O_2 is a pro-senescing signaling molecule that induces the expression of the genes associated with senescence and physiological changes.

H_2O_2 causes changes in the flowering time through its effects on the photoreceptors and hormonal signals. They regulate its expression from the vegetative phase to the reproductive phase of the plant's growth cycle. H_2O_2 is involved in pollen germination and pollen tube emergence, which are significant for fertilization. It controls the cytoskeleton and the discharge of cell boundary materials in progressing pollen tubes.

H_2O_2 and Light Interaction

The reaction of H_2O_2 and light involves a rather complicated process, which plays a crucial role in the development of plants. H_2O_2 and light are well-known signals that modulate numerous biological processes, and their crosstalk facilitates the plant's response to different stimuli. This part examines the interaction between H_2O_2 and the light referring to the signaling networks, stress, and developmental programs.

Light-Induced Production of H_2O_2

Light is an essential energy lighting source for photosynthesis and induces ROS generation such as H_2O_2. Several vital processes illustrate this relationship. In photosynthesis, light energy is used in chloroplast Electron Transport Chain to form ATP and NADPH. However, the transfer of electrons can also lead to an incomplete reduction of oxygen, which leads to the generation of H_2O_2 and other ROS. This is often so, especially when the intensity of light is high. This devastates the photosynthetic machinery and causes oxidative stress.

High light intensities that support high rates of photosynthesis can also increase the rates of photorespiration, which forms H_2O_2. Thus, photorespiration encompasses the reaction of ribulose-1,5-bisphosphate with oxygen by the enzyme ribulose-1,5-bisphosphate carboxylase/oxygenase (Rubisco), resulting in the formation of glycolate and subsequently H_2O_2 in peroxisomes. Various receptors detect light, including phytochromes, cryptochromes, and phototropin. The excitation of these photoreceptors could cause the generation of H_2O_2, which is involved in signaling cascades.

H_2O_2 as a Mediator of Light Signaling

H_2O_2 acts as a secondary messenger in light signaling pathways, influencing various aspects of plant growth and development. They involve the processes of plant growth and development controlled by light signals. Some exposure of cells to light leads to the generation of H_2O_2, which, in turn, triggers signals that affect gene expression profiles and physiological manifestations. For instance, H_2O_2 contains transcription factors that dictate the light-dependent gene that can be involved in chlorophyll synthesis and the synthesis of photosynthetic machinery. Thus, producing H_2O_2 under light is essential in the stomatal opening process. Blue light through phototropins leads to the synthesis of H_2O_2 in guard cells. This H_2O_2 acts as a signal to gate ion channels, which causes alteration in the guard cell turgor and stomatal opening. This process enhances the rate of diffusion of gases and the efficiency of water utilization depending on the requisite processes of photosynthesis and transpiration, that is, water evaporation through plant surfaces such as leaves and stems. H_2O_2 affects the circadian clock of the plant as light controls it. The circadian clock regulates numerous physiological rhythms like photosynthetic, hormonal, stress, etc. Water is regulated through H_2O_2 to adjust the circadian rhythms of these processes to coincide with the external conditions of light.

Stress Responses and Adaptation

The interaction between H_2O_2 and light is vital for plant adaptation to abiotic and biotic stresses. In the case of high illumination, the formation of H_2O_2 may increase and overcome the ability of the antioxidant mechanisms to prevent oxidative damage. As a result of this stress, plants have developed ways of coping with this by increasing the production of antioxidant enzymes like SOD, CAT, and peroxidases. The antioxidative enzymes and H_2O_2 signaling can be effectively regulated to improve the detoxification of ROS in plant cells. This effect relates to the light conditions, which impact the plant's immune system to pathogens. The H_2O_2 made in response to light can increase the oxidative burst in infection locations, which will help improve the rigidity of the cell walls and stimulate the processes associated with the defense genes. This combined action of light and H_2O_2 increases the plant's ability to fight various pathogens. H_2O_2 and light often interact with hormonal signaling, such as ABA, auxin, and ethylene signaling networks. For example, depending on H_2O_2 production by light, it can affect ABA signaling, which is responsible for drought stress and stomatal regulation. This crosstalk is essential for the cells to respond effectively to environmental stimuli regarding growth and stress.

Developmental Processes

H_2O_2 and light interaction play a significant role in various developmental processes:

There is a crosstalk between light and H_2O_2 signaling pathways to coordinately control seed germination. Phytochromes are involved in the perception of light and cause the production of H_2O_2 that sets off a signaling process, which facilitates the breaking of the dormancy of seeds and germination. The relative concentrations of H_2O_2 and the light regime affect the root and shoot system. For example, through regulating H_2O_2 levels in roots, light influences both root hair development and lateral root development at the distal site. Likewise, the generation of H_2O_2 in response to light in the shoot influences the cell enlargement and differentiation along with the structure of the plant.

CONCLUSION

The light signaling and H_2O_2 control flowering time. Flavin protein phytochromes and cyan/blue-light photoreceptors cryptochromes realize the light signal and emergent H_2O_2 produces the modulation of flowering time genes. This influences flowering under good light conditions, thereby improving the succession pole. The reaction between H_2O_2 and light is one of the most significant processes in plant growth and development, affecting almost all aspects of plant development. H_2O_2 is involved in signaling light, allowing the plant to adjust to the quality and quantity of the light and stress conditions. These two factors interrelate and give an insight into plant biology. It can be of significant use in agriculture, such as enhancing plant resistance to stress and improving optimal growth conditions. Future work in this field will help to provide a greater understanding of processes of H_2O_2 and light describing the existing interrelation and opening new ways of their usage in the management of crop production and plant health.

REFERENCES

Ahmar, S., Gill, R.A., Jung, K.-H., Faheem, A., Qasim, M.U., Mubeen, M., and Zhou, W. (2020). Conventional and molecular techniques from simple breeding to speed breeding in crop plants: Recent advances and future outlook. *International Journal of Molecular Sciences* 21, 2590.

Arsovski, A.A., Galstyan, A., Guseman, J.M., and Nemhauser, J.L. (2012). Photomorphogenesis. *The Arabidopsis Book/American Society of Plant Biologists*. http://doi.org/10.1199/tab.0054.

Bartholomew, D.P., and Sanewski, G.M. (2018). "Inflorescence and fruit development and yield," in *The Pineapple: Botany, Production and Uses*. CAB International, Wallingford, UK, 233–268. http://doi.org/10.1079/9781786393302.0233.

Bhatla, S.C., and Lal, M.A. (2023a). "Flowering," in *Plant Physiology, Development and Metabolism*. Springer, 565–582. https://doi.org/10.1007/978-1-4757-2451-6_8.

Bhatla, S.C., and Lal, M.A. (2023b). "Light perception and transduction," in *Plant Physiology, Development and Metabolism*. Springer, 363–390. https://doi.org/10.1007/978-981-13-2023-1_13.

Bhatla, S.C., and Lal, M.A. (2023c). "Plant movements," in *Plant Physiology, Development and Metabolism*. Springer, 641–659. https://doi.org/10.1007/978-981-13-2023-1_29.

Bielach, A., Hrtyan, M., and Tognetti, V.B. (2017). Plants under stress: Involvement of auxin and cytokinin. *International Journal of Molecular Sciences* 18, 1427.
Bojka, K. (2020). The role of far-red light (FR) in photomorphogenesis and its use in greenhouse plant production. *Acta Agriculturae Slovenica* 116, 93–105.
Brown, J.C. (2018). *Photoreceptor Regulation of Plant Responses to Light and Carbon Dioxide*. University of Sheffield, Sheffield.
Cardona, T., Shao, S., and Nixon, P.J. (2018). Enhancing photosynthesis in plants: The light reactions. *Essays in Biochemistry* 62, 85–94.
Carvalho, R.F., Takaki, M., and Azevedo, R.A. (2011). Plant pigments: The many faces of light perception. *Acta Physiologiae Plantarum* 33, 241–248.
Contreras Planesas, B. (2022). *Exploring Flowering and Photoperiod Response Mechanisms in Legumes*. University of Tasmania, Hobart, TAS.
Czarnocka, W., and Karpiński, S. (2018). Friend or foe? Reactive oxygen species production, scavenging and signaling in plant response to environmental stresses. *Free Radical Biology and Medicine* 122, 4–20.
de Lucas, M., and Prat, S. (2014). PIF s get BR right: PHYTOCHROME INTERACTING FACTOR s as integrators of light and hormonal signals. *New Phytologist* 202, 1126–1141.
Fan, B., Zhang, C., Chi, J., Liang, Y., Bao, X., Cong, Y., Yu, B., Li, X., and Li, G.-Y. (2022). The molecular mechanism of retina light injury focusing on damage from short wavelength light. *Oxidative Medicine and Cellular Longevity* 2022, 8482149.
Feng, L., Raza, M.A., Li, Z., Chen, Y., Khalid, M.H.B., Du, J., Liu, W., Wu, X., Song, C., and Yu, L. (2019). The influence of light intensity and leaf movement on photosynthesis characteristics and carbon balance of soybean. *Frontiers in Plant Science* 9, 1952.
Ferrante, A., and Mariani, L. (2018). Agronomic management for enhancing plant tolerance to abiotic stresses: High and low values of temperature, light intensity, and relative humidity. *Horticulturae* 4, 21.
Ferreyra, M.L.F., Serra, P., and Casati, P. (2021). Recent advances on the roles of flavonoids as plant protective molecules after UV and high light exposure. *Physiologia Plantarum* 173, 736–749.
Fitter, A.H., and Hay, R.K. (2012). *Environmental Physiology of Plants*. Academic Press, Cambridge, MA.
Foschi, M.L., Juan, M., Pascual, B., and Pascual-Seva, N. (2022). Influence of lighting and laser irradiation on the germination of caper seeds. *Agriculture* 12, 1612.
Foyer, C.H., and Harbinson, J. (2019). "Oxygen metabolism and the regulation of photosynthetic electron transport," in *Causes of Photooxidative Stress and Amelioration of Defense Systems in Plants*. CRC Press, 1–42. https://doi.org/10.1201/9781351070454-1.
Freschet, G.T., Violle, C., Bourget, M.Y., Scherer-Lorenzen, M., and Fort, F. (2018). Allocation, morphology, physiology, architecture: The multiple facets of plant above-and below-ground responses to resource stress. *New Phytologist* 219, 1338–1352.
Friesen, P. (2020). "How far-red photons affect plant growth and development". A guide to optimize the amount and production of far-red under sole-source electric lights by BioChambers. Reterived from: www.biochambers.com
Galvão, V.C., and Fankhauser, C. (2015). Sensing the light environment in plants: Photoreceptors and early signaling steps. *Current Opinion in Neurobiology* 34, 46–53.
Haga, K., and Sakai, T. (2023). Photosensory adaptation mechanisms in hypocotyl phototropism: How plants recognize the direction of a light source. *Journal of Experimental Botany* 74, 1758–1769.
Hart, J.W. (2012). *Light and Plant Growth*. Springer Science & Business Media, Dordrecht.
Hart, J.E., and Gardner, K.H. (2021). Lighting the way: Recent insights into the structure and regulation of phototropin blue light receptors. *Journal of Biological Chemistry* 296, 100594.
Heilmeier, H. (2019). Functional traits explaining plant responses to past and future climate changes. *Flora* 254, 1–11.
Huai, J., and Lin, R. (2023). Green innovation for green light. *Science China Life Sciences* 66, 2443–2444.
Ishishita, K., Higa, T., Tanaka, H., Inoue, S.-I., Chung, A., Ushijima, T., Matsushita, T., Kinoshita, T., Nakai, M., and Wada, M. (2020). Phototropin2 contributes to the chloroplast avoidance response at the chloroplast-plasma membrane interface. *Plant Physiology* 183, 304–316.
Kharshiing, E., Sreelakshmi, Y., and Sharma, R. (2019). "The light awakens! Sensing light and darkness," in *Sensory Biology of Plants*, 21–57. https://doi.org/10.1007/978-981-13-8922-1_2.
Khorobrykh, S.A., Karonen, M., and Tyystjärvi, E. (2015). Experimental evidence suggesting that H_2O_2 is produced within the thylakoid membrane in a reaction between plastoquinol and singlet oxygen. *FEBS Letters* 589, 779–786.

Lambers, H., Oliveira, R.S., Lambers, H., and Oliveira, R.S. (2019). "Photosynthesis, respiration, and long-distance transport: Photosynthesis," in *Plant Physiological Ecology*, 11–114. https://doi.org/10.1007/978-3-030-29639-1_2.

Leduc, N., Roman, H., Barbier, F., Péron, T., Huché-Thélier, L., Lothier, J., Demotes-Mainard, S., and Sakr, S. (2014). Light signaling in bud outgrowth and branching in plants. *Plants* 3, 223–250.

Lin, X., Liu, B., Weller, J.L., Abe, J., and Kong, F. (2021). Molecular mechanisms for the photoperiodic regulation of flowering in soybean. *Journal of Integrative Plant Biology* 63, 981–994.

Lokhandwala, J. (2018). Structural and Functional Characterization of Plant and Fungal LOV Proteins. (Doctoral dissertation, Southern Methodist University). Retrieved from SMU Scholar. https://scholar.smu.edu/hum_sci_chemistry_etds/5

Lüttge, U. (2016). Plants shape the terrestrial environment on earth: Challenges of management for sustainability. *Progress in Botany* 77, 187–217.

Lv, B., Zhu, J., Kong, X., and Ding, Z. (2021). Light participates in the auxin-dependent regulation of plant growth. *Journal of Integrative Plant Biology* 63, 819–822.

Montgomery, B.L. (2021). *Lessons from Plants.* Harvard University Press, Cambridge, MA.

Nájera, C., Gallegos-Cedillo, V.M., Ros, M., and Pascual, J.A. (2023). Role of spectrum-light on productivity, and plant quality over vertical farming systems: bibliometric analysis. *Horticulturae* 9, 63.

Navyashree, R. (2024). "Antioxidants in plants," in *Advanced Botany*, vol. 5, 1–34. H. R. Aglave & S. Baruah (editors), AkiNik Publications; New Delhi-110085, India. https://doi.org/10.22271/ed.book.2647

Niu, L., and Liao, W. (2016). Hydrogen peroxide signaling in plant development and abiotic responses: Crosstalk with nitric oxide and calcium. *Frontiers in Plant Science* 7, 230.

Osnato, M., Cota, I., Nebhnani, P., Cereijo, U., and Pelaz, S. (2022). Photoperiod control of plant growth: Flowering time genes beyond flowering. *Frontiers in Plant Science* 12, 805635.

Paradiso, R., and Proietti, S. (2022). Light-quality manipulation to control plant growth and photomorphogenesis in greenhouse horticulture: The state of the art and the opportunities of modern LED systems. *Journal of Plant Growth Regulation* 41, 742–780.

Peck, B. (2022). "The Chemistry of Photosynthesis". Bibliotex. Ebook ISBN: 9781984666000

Penfield, S. (2017). Seed dormancy and germination. *Current Biology* 27, R874–R878.

Podolec, R., Demarsy, E., and Ulm, R. (2021). Perception and signaling of ultraviolet-B radiation in plants. *Annual Review of Plant Biology* 72, 793–822.

Proctor, M.S., Sutherland, G.A., Canniffe, D.P., and Hitchcock, A. (2022). The terminal enzymes of (bacterio) chlorophyll biosynthesis. *Royal Society Open Science* 9, 211903.

Proietti, S., Scariot, V., De Pascale, S., and Paradiso, R. (2022). Flowering mechanisms and environmental stimuli for flower transition: Bases for production scheduling in greenhouse floriculture. *Plants* 11, 432.

Samkumar, A. (2021). Effect of light quality on the biosynthesis of flavonoids and sugars in bilberry (Vaccinium myrtillus L.). https://hdl.handle.net/10037/23030.

Sharma, M.K. (2023). How plants adapt to the photoperiod. *Annual Research & Review in Biology* 38, 17–45.

Shi, Y., Ke, X., Yang, X., Liu, Y., and Hou, X. (2022). Plants response to light stress. *Journal of Genetics and Genomics* 49, 735–747.

Simkin, A.J., López-Calcagno, P.E., and Raines, C.A. (2019). Feeding the world: improving photosynthetic efficiency for sustainable crop production. *Journal of Experimental Botany* 70, 1119–1140.

Su, P., Ding, S., Wang, D., Kan, W., Yuan, M., Chen, X., Tang, C., Hou, J., and Wu, L. (2024). Plant morphology, secondary metabolites and chlorophyll fluorescence of Artemisia argyi under different LED environments. *Photosynthesis Research* 159, 153–164.

Taiz, L., Zeiger, E., Møller, I.M., and Murphy, A.S. (2018). *Fundamentals of Plant Physiology*. Oxford University Press, Oxford.

Therby-Vale, R., Lacombe, B., Rhee, S.Y., Nussaume, L., and Rouached, H. (2022). Mineral nutrient signaling controls photosynthesis: Focus on iron deficiency-induced chlorosis. *Trends in Plant Science* 27, 502–509.

Tian, F. (2016). *Study and Optimization of Lighting Systems for Plant Growth in a Controlled Environment.* Université Paul Sabatier-Toulouse III, Toulouse.

Tossi, V.E., Regalado, J.J., Iannicelli, J., Laino, L.E., Burrieza, H.P., Escandón, A.S., and Pitta-Álvarez, S.I. (2019). Beyond Arabidopsis: differential UV-B response mediated by UVR8 in diverse species. *Frontiers in Plant Science* 10, 780.

Vanhaelewyn, L., Van Der Straeten, D., De Coninck, B., and Vandenbussche, F. (2020). Ultraviolet radiation from a plant perspective: The plant-microorganism context. *Frontiers in Plant Science* 11, 597642.

Viczián, A., Klose, C., Ádám, É., and Nagy, F. (2017). New insights of red light-induced development. *Plant, Cell & Environment* 40, 2457–2468.

Voitsekhovskaja, O. (2019). Phytochromes and other (photo) receptors of information in plants. *Russian Journal of Plant Physiology* 66, 351–364.

Wang, P., Abid, M.A., Qanmber, G., Askari, M., Zhou, L., Song, Y., Liang, C., Meng, Z., Malik, W., and Wei, Y. (2022). Photomorphogenesis in plants: The central role of phytochrome interacting factors (PIFs). *Environmental and Experimental Botany* 194, 104704.

Wu, Y., Shi, L., Li, L., Fu, L., Liu, Y., Xiong, Y., and Sheen, J. (2019). Integration of nutrient, energy, light, and hormone signalling via TOR in plants. *Journal of Experimental Botany* 70, 2227–2238.

Xu, Y., Kaste, J., Weise, S., Shachar-Hill, Y., and Sharkey, T. (2024). The effects of photosynthetic rate on respiration in light, starch/sucrose partitioning, and other metabolic fluxes within photosynthesis. https://doi.org/10.21203/rs.3.rs-3995199/v1.

Yavari, N. (2020). *An Integrated Approach to Unravel the Physiological and Molecular Basis of Arabidopsis Thaliana Response to Narrow-Wavelength Light*. McGill University, Montreal.

10 Interaction between Various Plant Hormones and Hydrogen Peroxide in Relation to Plant Progress and Developments

Vinay Kumar Pandey, Shivangi Srivastava, Farina Mujeeb, Bushra Ahmad, Syed Khalida Izhar, and Ahamad Faiz Khan

INTRODUCTION

Hydrogen peroxide (H_2O_2) and plant hormones are two crucial signaling molecules that are crucial in controlling several aspects of plant growth and development. They function as deftly orchestrated systems that combine and react to internal and external stimuli, eventually influencing how the plant reacts to its surroundings (Baryla and Gardeström, 2020). Small organic compounds known as plant hormones or phytohormones serve as chemical messengers in plants. They are created in particular tissues and then transferred to the locations where they work. The different physiological processes that each type of plant hormone affects are influenced by its unique functions. Auxin hormone controls root development, apical dominance, and cell elongation (Benková and Hejátko, 2020). They influence tropisms, directing a plant's growth in response to environmental cues like light and gravity. Cytokinins influence the formation of chloroplasts, shoot and root growth, and cell division and differentiation. Gibberellins influence the growth of fruit and seeds as well as stem lengthening, seed germination, and flowering. As a stress hormone, Abscisic Acid (ABA) controls how the body reacts to environmental challenges like salinity and drought. It controls senescence and seed dormancy (Chen and Song, 2020). The ripening of fruits, the aging of leaves and flowers, and how plants react to infections and mechanical stress are all regulated by ethylene. Reactive oxygen species (ROS), such as hydrogen peroxide (H_2O_2), are byproducts of aerobic metabolism in plants. Despite being harmful, recent study has shown that it functions in plant cells as a signaling molecule. Hormonal signaling, developmental processes, biotic and abiotic stressors, and other stimuli all cause the production of H_2O_2 (Ding et al., 2020). As a secondary messenger, hydrogen peroxide takes part in redox signaling networks that control gene expression and cellular reactions. In the plant's defense against environmental challenges such infections, dehydration, heat, and UV radiation, H_2O_2 plays a crucial role. It stimulates the production of genes that respond to stress and activates antioxidant mechanisms. Hydrogen peroxide has a role in a number of developmental processes, such as fruit ripening, floral development, root and shoot growth, and seed germination (Escariz et al., 2020).

In the fields of plant biology and agriculture, it is crucial to comprehend how hydrogen peroxide (H_2O_2) and plant hormones interact. Knowing how plant hormones and hydrogen peroxide interact enables researchers to fine-tune the regulation of growth and development processes in plants (Li et al., 2020). These signaling molecules serve critical roles in regulating many physiological processes. Scientists may be able to increase crop yields, raise fruit quality, and improve overall plant health by modifying these communication pathways. Hydrogen peroxide and plant hormones play a crucial role in how plants react to environmental challenges such infections, heat, salinity, and drought. Understanding how they interact might help crop breeders create stress-tolerant

DOI: 10.1201/9781032647692-10

crop types that can better tolerate challenging circumstances, promoting sustainable agriculture and maintaining food security (Nambara and Marion-Poll, 2018). Utilizing the understanding of hormone- H_2O_2 interactions can help agriculture become less reliant on chemical inputs. The requirement for pesticides and fertilizers can be reduced by promoting natural defense systems in plants through the modulation of these signaling pathways. Applications in biotechnology are made possible by knowing how plant hormones and hydrogen peroxide interact. It makes it possible to create crops that have been genetically engineered to have specific features, such enhanced yield, better nutrient usage efficiency, and increased stress tolerance (Navrot et al., 2018). For a variety of environmental stressors, hydrogen peroxide has the ability to function as a signaling molecule. Researchers can develop methods to use plants for environmental remediation, such as the detoxification of contaminants in contaminated soils, by studying its involvement in stress reactions. Precision agriculture techniques can be used after the interactions between hormones and H_2O_2 are better understood. This entails customizing agricultural techniques and inputs to meet the unique requirements of plants, maximizing resource efficiency, and reducing environmental impact. Understanding how signaling molecules interact can provide insight into the regulatory processes that affect plant development and growth in natural ecosystems (Peleg-Grossman et al., 2018). Understanding the function of plant hormones and hydrogen peroxide in forming plant communities can help conservation efforts and encourage biodiversity. Understanding hormone- H_2O_2 interactions reveals basic biological processes in plants. These discoveries contribute to our general understanding of how plants perceive and react to their surroundings and have ramifications beyond agriculture. The full analysis and elucidation of the complex interactions between several plant hormones and hydrogen peroxide (H_2O_2) in the context of plant growth, development, and responses to environmental signals are the goals of this chapter study. The goal of this chapter is to give readers a thorough grasp of how these signaling molecules integrate and control numerous physiological functions to affect plant growth.

PLANT HORMONES: TYPES AND FUNCTIONS

Overview of Major Plant Hormones

Phytohormones, or plant hormones, are vital signaling molecules that control a variety of physiological processes in plants. Each hormone has a unique purpose and is essential for coordinating growth, development, and reactions to environmental cues. Auxins, cytokinins, gibberellins, ABA, and ethylene are the principal plant hormones (Riemann et al., 2018).

Auxins

Auxins are a class of hormones that control the growth, division, and differentiation of cells. The most prevalent and extensively researched auxin in plants is called indole-3-acetic acid (IAA). Auxins are created in the shoot apical meristem and are largely distributed polarly downward through the plant. They are essential for the tropic reactions known as phototropism (bending toward light), gravitropism (responding to gravity), and others. When the main shoot is active, auxins play a role in apical dominance by preventing the formation of lateral buds (Ljung et al., 2018). Auxins are also important for root growth and encourage the beginning of roots in stem cuttings. Auxins have polar transport, which means they preferentially migrate from the shoot apex to the plant's base or from its top to its bottom. Specific carrier proteins promote this polar movement, which promotes the coordinated growth and development of various plant components. Auxins are crucial for preserving apical dominance, in which the terminal bud at the shoot apex prevents the development of lateral buds. Auxins make sure that the plant concentrates its resources largely on the main shoot, encouraging vertical growth, while limiting the establishment of lateral branches (Mangano et al., 2018). In plant tissues, auxins control cell division and cell elongation. The stem bends toward the light (phototropism) while the roots bend toward gravity because they encourage cell elongation on the stem's

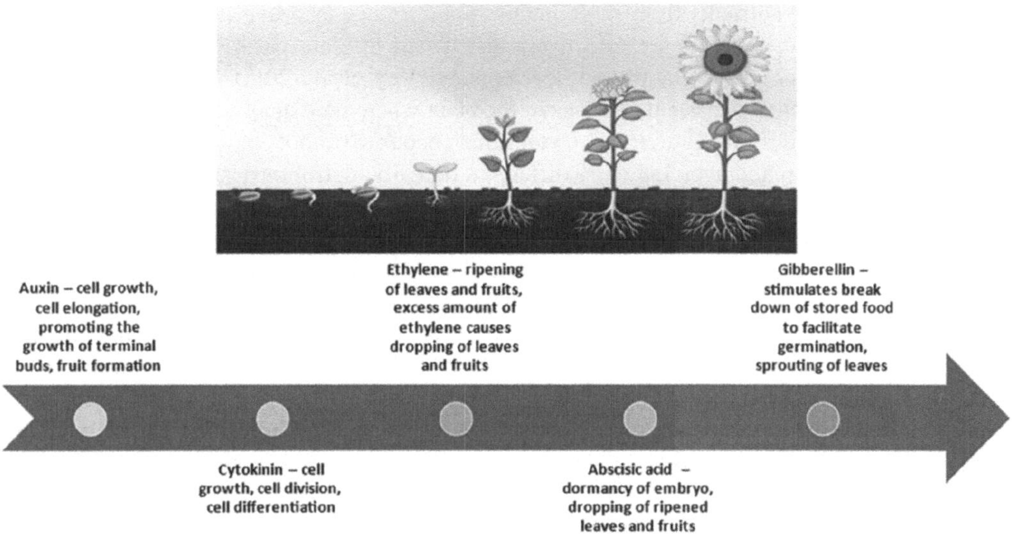

FIGURE 10.1 Schematic representation of plant growth regulators.

shadowed side (gravitropism). This differential growth response enables plants to increase nutrient intake and maximize their light exposure. During vegetative propagation, auxins are essential for starting the production of roots in cuttings. They encourage the emergence of adventitious roots from non-root tissues, facilitating plant proliferation and vegetative reproduction without the use of seeds (Medeiros et al., 2018). Different phases of fruit growth are influenced by auxins. Auxins are delivered to the growing ovary during the early stages of fruit growth, encouraging cell division and expansion. They control the expression of ripening-related genes, which impacts the color, texture, and aroma of the fruit. Auxins play a role in seed development and the establishment of dormancy. They influence the growth of the seed coat and control the buildup of lipids and storage proteins in seeds. Several elements of plant growth and development depend heavily on auxins (Feng et al., 2018). They encourage stem, leaf, and root cell elongation, which affects the height and shape of the entire plant. The growth of lateral branches is likewise regulated by auxins, which helps the plant distribute resources effectively. Auxins also control tropic responses, directing plants toward favorable development environments like light and gravity. Auxins aid in vegetative propagation and the development of new plants from cuttings by promoting root initiation and adventitious rooting. They affect fruit size, shape, and ripening properties, which makes them essential for fruit growth. Auxins help seeds mature and go into dormancy, ensuring that germination takes place in an environment that is favorable for it (Gapper et al., 2018).

Cytokinins

Hormones called cytokinins encourage cell division and differentiation. They are produced in the root tips and moved up the plant by the plant. Together with auxins, cytokinins control a variety of processes in different tissues, such as shoot growth, leaf expansion, and chloroplast development. They enhance the development of axillary buds and postpone leaf senescence. By promoting the growth of lateral branches, cytokinins also aid in the overthrow of apical dominance. When combined with auxins, cytokinins have a synergistic effect that inhibits cell growth. Auxins encourage cell elongation while cytokinins prevent it. The growth and development of various plant tissues and organs is regulated by the delicate balance between auxins and cytokinins (Johansson and Mähler et al., 2018). The normal aging and degeneration of leaves, known as leaf senescence, is known to be postponed by cytokinins. Cytokinins increase a plant's capacity for photosynthesis and nutrient recycling by preserving leaf lifetime, which boosts the plant's general vigor and output.

Cytokinins help move nutrients from more mature plant tissues to younger, actively growing tissues. As a result, vital nutrients are effectively distributed to plant parts with higher metabolic demands, enabling long-term growth and development (Key et al., 2018). The preservation and activity of the shoot apical meristem, the area of the plant where new shoot tissues are continuously generated, depend heavily on cytokinins. Cytokinins encourage shoots to develop indefinitely by promoting cell division and avoiding early differentiation. To control intricate developmental processes, cytokinins interact with other plant hormones like auxins, gibberellins, and ABA. Through these interactions, the plant is better able to respond to environmental cues and optimize its growth and development in response to shifting environmental factors. New tissues and organs are formed as a result of cytokinins' stimulation of cell division and prevention of premature differentiation (Khorassani et al., 2018). The development of the shoot and root systems as well as plant growth in general depend on this process. The development of lateral branches is aided by cytokinins, which encourage lateral bud expansion. Cytokinins support the structure and resource distribution of the plant by controlling branching patterns. By delaying senescence, cytokinins increase the functional lifespan of leaves. This helps to prolong the time that leaves are actively engaged in photosynthetic processes, encouraging plant growth and increasing total production. The flow of nutrients from older, senescent tissues to younger, actively growing tissues is made easier by cytokinins. This nutrient mobilization supports growth and development, especially in times of stress or when resources are scarce. The coordination of numerous growth and developmental processes is achieved by cytokinins in conjunction with other hormones, mainly auxins. Their interaction makes sure that the plant adapts to its surroundings and makes the best use of resources for long-term development and survival (Bellé et al., 2018).

Gibberellins

Gibberellins are hormones that regulate stem elongation, seed germination, and flowering. Over 130 gibberellins have been identified, but gibberellic acid (GA3) is the most extensively studied. Gibberellins promote cell elongation by stimulating cell wall loosening and enhancing the synthesis of enzymes that break down the cell wall components. In seeds, gibberellins promote the synthesis of hydrolytic enzymes that break down stored nutrients, promoting germination (Cui et al., 2018). They are also essential for the transition from vegetative growth to flowering. Regulation of seed germination: Gibberellins play a central role in seed germination. They promote the synthesis of hydrolytic enzymes that break down stored reserves, such as starch and proteins, into usable forms for the growing embryo. By stimulating the mobilization of nutrients, gibberellins trigger the emergence of the embryonic shoot (plumule) and root (radicle), initiating the germination process. Gibberellins are potent regulators of stem elongation. They stimulate cell elongation by promoting the synthesis of enzymes that loosen the cell wall, allowing cells to expand. This is especially significant in the early stages of plant growth when elongation is essential for establishing the plant's stature and reaching sunlight for photosynthesis. Gibberellins influence the transition from vegetative growth to reproductive growth and play a role in flower development. They are involved in the regulation of floral meristem identity genes, influencing the formation and development of flower organs. Gibberellins have the capacity to induce parthenocarpy, a phenomenon where fruit develops without fertilization. Application of gibberellins can result in the formation of seedless fruits, which is of economic importance in agriculture for producing desirable fruit varieties (Kangasjärvi et al., 2012). Gibberellins break the dormancy of buds and seeds, allowing them to resume growth and development. This is particularly crucial in deciduous trees, where gibberellins play a part in the leaf-out process during spring. Gibberellins promote rapid elongation of the stem and root in young seedlings, enabling them to establish themselves quickly after germination. This allows seedlings to emerge from the soil and access light and nutrients for photosynthesis and growth. Gibberellins stimulate cell elongation in the stem, facilitating vertical growth and overall plant height. This is particularly important in the growth of herbaceous plants and young shoots. Gibberellins are key regulators of flower development. They are involved in the formation and growth of floral organs, ensuring the successful transition from vegetative to reproductive growth. Gibberellins influence fruit development by promoting cell division and elongation in the ovary. They are particularly involved in the growth and enlargement of fruits, contributing to fruit size and shape (Alhagdow et al., 2017).

Abscisic Acid (ABA)

An essential part of stress reactions and plant adaptation to environmental challenges is played by the hormone abscisic acid. Water stress triggers the synthesis of ABA, which limits growth and causes stomatal closure to lessen water loss through transpiration. ABA causes dormancy in seeds throughout seed development, delaying germination until suitable conditions are present (Baud et al., 2017). Additionally, ABA controls seed maturity and prevents early germination in unfavorable environments. Response to Stress and Drought: Due to its critical function in regulating plant responses to drought and other environmental stresses, ABA is referred to as the "stress hormone." When there is a water shortage, ABA levels rise and cause the stomata to close, preventing water loss by transpiration. Plants can conserve water and live in harsh situations because to this adaptive reaction. A crucial regulator of seed growth and dormancy is ABA. A buildup of ABA during seed development controls the production of storage compounds and gets the seed ready for dormancy (Anderson et al., 2018). To ensure that seeds germinate under favorable environmental circumstances for optimal growth, ABA-induced dormancy prevents early germination. By preventing the synthesis of gibberellins, which are necessary for germination, ABA prevents seed germination. ABA preserves seed dormancy by suppressing the synthesis of gibberellins until favorable conditions, such as enough moisture and temperature, are present for germination. The regulation of leaf senescence, the aging and degeneration of leaves, is regulated by ABA. Increased ABA levels encourage leaf senescence, which results in the decomposition of chlorophyll and the remobilization of nutrients prior to leaf abscission. ABA aids in plant defense against herbivores and diseases (Arshi and Rastogi, 2018). ABA levels increase in reaction to an attack, which triggers defense mechanisms such the creation of proteins and antibacterial chemicals. The plant's reaction to water stress depends heavily on ABA. ABA aids plants in water conservation and turgor pressure maintenance by causing stomatal closure, which enables them to endure and recover when water becomes accessible once more. For plants to survive in arid and drought-prone areas, this adaptive response is essential (Divi and Krishna, 2010). To guarantee that seeds germinate under ideal circumstances, ABA controls seed dormancy and germination. ABA reduces early germination and seedling establishment when conditions are not favorable for growth by limiting germination at unfavorable times. Senescence of leaves, which is necessary for nutrient remobilization and the effective distribution of resources to other regions of the plant, is regulated by ABA. ABA supports nutrient recycling by encouraging leaf senescence, ensuring that nutrients are available for future growth and development. Due to ABA's role in plant defense mechanisms, plants can efficiently fend off attacks from herbivores and pathogens. ABA helps plants defend themselves against potential threats by triggering defense responses, increasing their chances of survival and growth (Gill and Tuteja, 2010).

Ethylene

A gaseous hormone called ethylene controls different facets of plant growth and development. In reaction to stressors like pathogen infection, mechanical injury, and floods, it is produced. By inducing the expression of ripening-related genes and activating the enzymes that break down cell walls and convert starches to sugars, ethylene encourages fruit ripening. Additionally, ethylene causes the senescence and abscission of leaves and flowers, which results in the loss of leaves and flowers (Hossain et al., 2012). Ethylene has two different effects on plant growth. Low levels of ethylene encourage plant growth by encouraging cell lengthening and expansion, particularly in the roots and stems. High levels of ethylene, on the other hand, prevent growth by preventing cell elongation and causing the senescence of leaves and flowers. Particularly in climacteric fruits, ethylene plays a critical role in controlling fruit ripening. Ethylene production rises as fruit develops, causing a sequence of biochemical adjustments that result in fruit softening, color changes, and the creation of scent and flavor compounds (Balbi et al., 2010). Additionally, ethylene hastens the senescence of leaves and flowers, which results in the loss of leaves and petals. In seedlings, ethylene causes the "hooked" or "apical hook" reaction, which is characterized by a threefold response that comprises a slowing of stem growth, a thickening of the stem, and an increased curving of the stem. This threefold reaction aids seedlings in overcoming obstacles and moving through the soil. Plant responses to various stresses, such as mechanical stress, flooding, and pathogen attacks, are significantly influenced by ethylene. When under stress, ethylene production rises, activating defense mechanisms such the manufacture of protective chemicals and the strengthening of cell walls. Ethylene in low amounts encourages root branching and development (Chehab et al., 2012). The direction in which roots develop in response to gravity is influenced by ethylene's involvement in root gravitropism. Additionally, it affects the growth of root hair, which improves the root's ability to absorb nutrients and water. Fruit ripening is significantly regulated by ethylene. Fruits produce more ethylene as they ripen, which causes the distinctive changes that mark ripening. Additionally, ethylene causes flower senescence, resulting in the wilting and abscission of petals as well as the easier release of pollen. The regulation of seed germination is influenced by ethylene. Depending on the species and environmental factors, exposure to ethylene can either enhance or impede the germination of some seeds. A significant factor in how plants react to stress is ethylene. It aids plants in coping with environmental challenges by triggering defense systems and causing the creation of stress-associated proteins and secondary metabolites, such as those related to pathogen attacks or flooding (Audenaertand Höfte, 2015).

Roles of Plant Hormones in Various Physiological Processes

Phytohormones, or plant hormones, are vital signaling molecules that control a variety of physiological processes in plants. In coordinating growth, development, and reactions to environmental signals, each hormone has a specific function. An overview of the functions of plant hormones in significant physiological processes is given below:

Cell Elongation

The main regulators of cell elongation are auxins. They encourage plant cells to take up more water, which causes the turgor pressure to rise and the cells to expand. This procedure is especially crucial for roots and shoots because it enables the plant to contact sunlight and take up nutrients from the soil. Additionally, gibberellins encourage cell elongation, especially in stems. They stimulate the production of enzymes that disassemble the parts of the cell wall, causing the cell to enlarge and lengthen. Ethylene affects cell elongation in two different ways. In roots and stems, it encourages cell elongation at low concentrations. However, ethylene inhibits cell elongation at high quantities, causing the distinctive triple reaction in seedlings (Bandara and Gardner, 2015).

Cell Division

The main hormones that encourage cell division are cytokinins. In tissues that are actively growing, such as shoot meristems and root tips, they promote cell division. A suitable balance between cell division and elongation is maintained by cytokinins and auxins, supporting healthy tissue growth (Bhattacharya and Sood, 2015).

Flowering

Gibberellins are essential for encouraging flowering. They trigger the production of genes that identify the floral meristem, which starts the development of floral primordia. Gibberellins are necessary for the change from vegetative to reproductive growth in several plant species. Depending on day length and seasonal changes, other hormones, such as phytochrome-mediated responses and flowering locus T (FT) protein, also contribute to flowering regulation (Chen et al., 2015).

Fruit Development

Auxins encourage cell division and growth, which has an impact on early fruit development. They aid in determining the size and shape of the fruit. Gibberellins have a role in the development and expansion of fruit. They promote ovary cell elongation and division, which aids in fruit development. The regulation of fruit ripening involves ABA. It causes the development of enzymes that dissolve the components of the cell wall, softening the fruit and promoting ripening. Fruit ripening is significantly regulated by ethylene. Fruits produce more ethylene as they ripen, which causes the distinctive changes that mark the process of ripening (Müller and Munné-Bosch, 2015).

Stress Responses

Due to its crucial function in regulating plant responses to drought and other environmental stresses, ABA is referred to as the "stress hormone." It causes stomatal closure, which lowers transpirational water loss and aids in water conservation for the plant. Plant responses to various stresses, such as mechanical stress, flooding, and pathogen attacks, are significantly influenced by ethylene. It triggers defense mechanisms such the production of protective chemicals and the fortification of cell membranes. Plant responses to herbivore and pathogen threats involve jasmonates. They trigger defensive mechanisms, like the synthesis of secondary metabolites, to shield the plant from additional harm. Salicylic acid (SA) helps plants defend themselves against infections, especially those that enter through wounds. It enhances the production of pathogenesis-related (PR) proteins and encourages the expression of defense-related genes (Abeles et al., 2012).

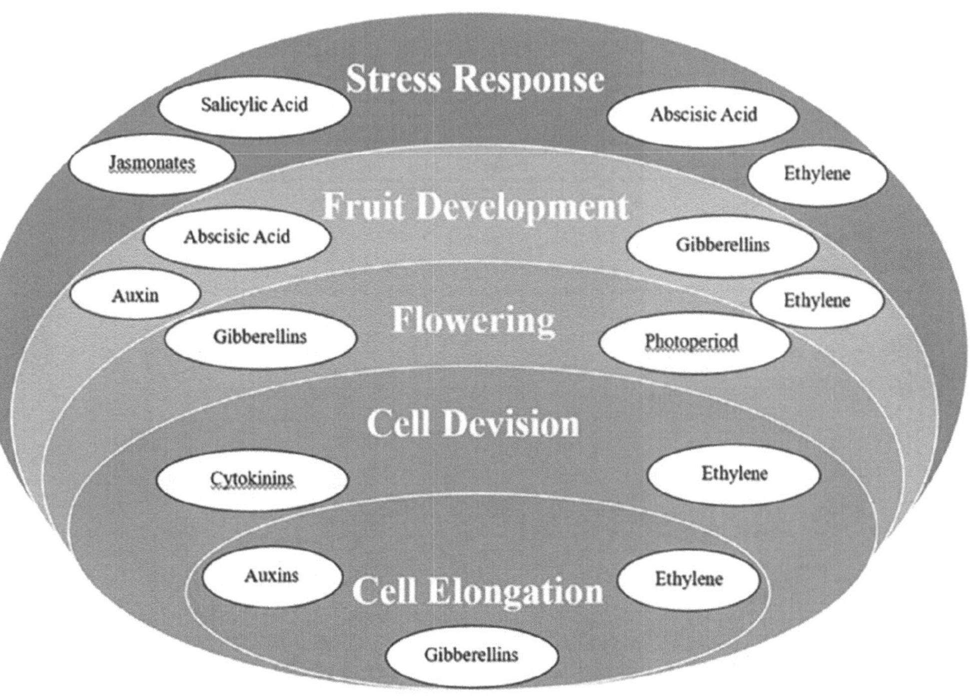

FIGURE 10.2 Diagrammatic representation of plant hormones for growth and development.

HYDROGEN PEROXIDE: A SIGNALING MOLECULE IN PLANTS

Sources and Generation of Hydrogen Peroxide in Plants

ROS including hydrogen peroxide (H_2O_2) are byproducts of several physiological processes in plants. It participates in redox signalling networks that control gene expression and cellular reactions and functions as a signalling molecule. Hydrogen peroxide is a byproduct of the electron transport chain in the chloroplasts during photosynthesis in plants. While antioxidants quickly scavenge the majority of ROS created during photosynthesis, some hydrogen peroxide is still formed. The cellular powerhouses, the mitochondria, are also involved in the production of hydrogen peroxide during cellular respiration (Abeles et al., 2012). In the mitochondrial electron transport chain, some electrons may leak out and interact with oxygen to produce hydrogen peroxide. An oxidative burst takes place as a result of a plant's defensive mechanism when it comes into contact with pathogens or sustains mechanical injury. Several enzymes, including NADPH oxidases, are responsible for this sudden and brief rise in hydrogen peroxide levels in the apoplast. High light levels, dryness, salinity, heavy metals, and other environmental stresses can cause an excess of hydrogen peroxide to be produced in plant cells. Oxidative stress can come from these stressors disrupting the equilibrium between ROS generation and scavenging (Acharya et al., 2013).

Redox Signaling and Cellular Responses Triggered by Hydrogen Peroxide

In plants, hydrogen peroxide (H_2O_2) performs an essential signalling role by taking part in redox signalling pathways that control diverse cellular reactions. Plants create and accumulate hydrogen peroxide in response to stress or particular developmental cues, which affects gene expression and activates particular signalling cascades. A careful balance must be maintained between the production of ROS and their scavenging by antioxidant enzymes in order for hydrogen peroxide to mediate redox signalling. In order to stimulate different cellular responses, hydrogen peroxide can function as a secondary messenger. It can alter the redox state of proteins, changing how they conform and how they function (Bielach et al., 2019). These redox alterations affect the expression of genes that

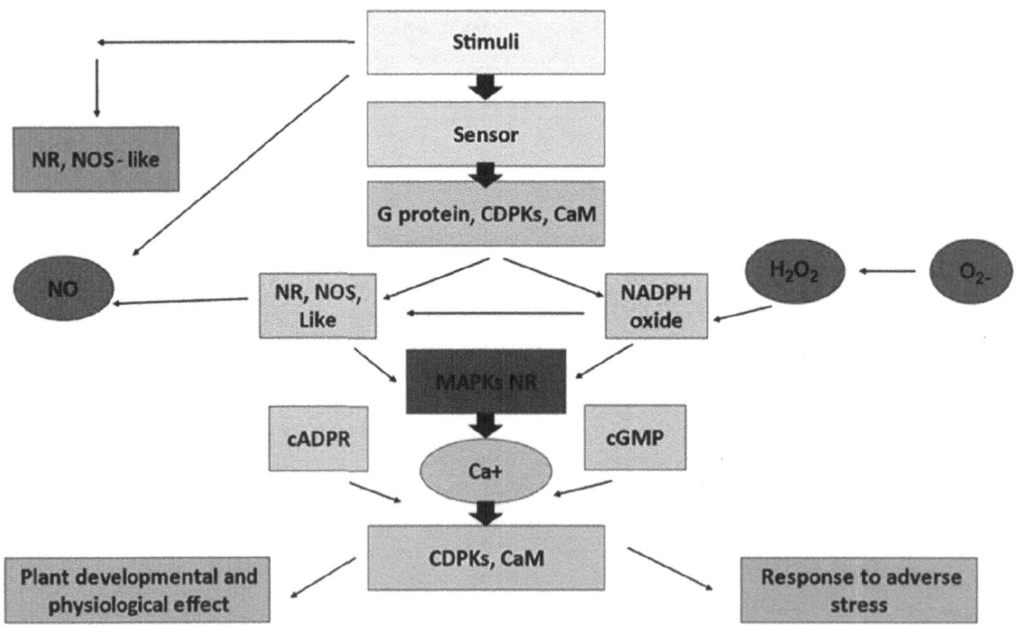

FIGURE 10.3 Schematic model of the interaction among H_2O_2, NO, CA^{2+}.

respond to stress and subsequent signalling cascades through the sensitivity of specific proteins, such as transcription factors and kinases. Hydrogen peroxide-induced redox signalling has a role in a number of physiological processes, including plant development, growth, and defence mechanisms. The buildup of hydrogen peroxide activates defense-related genes, resulting in the creation of antimicrobial chemicals and other defence mechanisms during stressful situations, such as dehydration, bright light, or pathogen attack. Hydrogen peroxide is a key player in redox signalling, which controls how plants grow and develop. It affects activities like fruit ripening, floral growth, root elongation, and seed germination (Gifford et al., 2018). Hydrogen peroxide aids in the coordination of these processes in response to particular environmental signals and developmental phases by modifying the activity of essential regulatory proteins. Hydrogen peroxide buildup that is too great can result in oxidative stress, which harms biological components. Therefore, maintaining cellular homeostasis and guaranteeing adequate redox signalling in plants depend on the strict control of hydrogen peroxide levels and the effective operation of antioxidant defence systems. Overall, hydrogen peroxide-induced redox signalling is essential for coordinating plant responses to environmental cues and enabling adaptability and survival in dynamic environments (Su and Yang, 2019).

Effect of Plant Hormones on Hydrogen Peroxide Production and Scavenging

Hydrogen peroxide (H_2O_2) generation and scavenging in plants are significantly regulated by plant hormones. Plant hormones and H_2O_2 interact in a complex network that affects a variety of physiological functions, including growth, development, and stress reactions. Hydrogen peroxide generation is strongly induced by ethylene. The synthesis of ethylene rises during stress conditions, such as mechanical damage, pathogen attack, or senescence, which causes an oxidative burst and the production of hydrogen peroxide (Niu et al., 2019). As a result of ethylene activating NADPH oxidases, the body produces ROS. In response to abiotic conditions like salinity and drought, ABA can control the generation of hydrogen peroxide. ROS levels, especially hydrogen peroxide, rise as a result of ABA enhancing NADPH oxidases' activity. This acts as an initial stress signalling event for the plant. Jasmonates play a role in controlling the generation of ROS during plant defence reactions. As part of the oxidative burst brought on by herbivore or pathogen attack, they have the ability to induce the production of hydrogen peroxide. The expression and activity of antioxidant enzymes that scavenge hydrogen peroxide and other ROS can be affected by plant hormones. For instance, jasmonates and abscisic acid (ABA) can increase the expression of peroxidases, catalase, and superoxide dismutase (SOD), which improves the plant's ability to neutralise hydrogen peroxide and preserve redox equilibrium. The ascorbate-glutathione cycle, a critical mechanism for hydrogen peroxide scavenging, can be influenced by plant hormones as well. In this cycle, enzymes including glutathione peroxidase (GPX) and ascorbate peroxidase (APX), which aid in the detoxification of hydrogen peroxide, have been demonstrated to be affected by ethylene. Plant hormones have the ability to control the amounts of non-enzymatic antioxidants such glutathione and ascorbate (vitamin C). These compounds effectively neutralise hydrogen peroxide by acting as ROS scavengers (Liu et al., 2018).

Interaction between Specific Plant Hormones and Hydrogen Peroxide in Key Developmental Processes

Hydrogen peroxide (H_2O_2) and some plant hormones interact to influence important plant growth processes. Complex signalling networks that control growth, development, and reactions to environmental signals are created as a result of these interactions. While hydrogen peroxide has the ability to influence seed germination, ABA is well known for its inhibitory effect. By causing seed dormancy, ABA delays germination until optimal circumstances are in place. However, ABA levels drop during seed imbibition, which lowers the ability to scavenge hydrogen peroxide. Hydrogen peroxide buildup then encourages the activation of enzymes responsible for breaking dormancy and starting germination (Rodriguez Shongo et al., 2019). Auxins are the main factors that control root growth, encouraging root extension and branching. The expression of NADPH oxidases is induced by auxins, which causes the root apex to produce hydrogen peroxide. By controlling the function of auxin transporters, hydrogen peroxide in turn modifies auxin signalling, promoting root development and tropic responses. Gibberellins have a crucial role in regulating shoot elongation. NADPH oxidases are expressed as a result of gibberellin induction, which produces hydrogen peroxide. Gibberellin-mediated stem elongation depends on the hydrogen peroxide generation. However, high hydrogen peroxide concentrations can impede growth and trigger stress reactions. The growth of flowers is aided by gibberellins, and a signalling molecule involved in this process is hydrogen peroxide. While hydrogen peroxide functions as a secondary messenger, influencing the activity of transcription factors and other regulatory proteins, gibberellins stimulate the expression of genes essential in flower formation. Fruit ripening is significantly regulated by ethylene, and hydrogen peroxide is involved in this process. Ethylene stimulates an oxidative burst during fruit ripening, which produces hydrogen peroxide. Fruit ripening processes that involve hydrogen peroxide include softening, colour changes, and the creation of scent and flavour components. When under stress, ABA and hydrogen peroxide interact. As ABA levels rise in stressful environments like dryness or salinity, hydrogen peroxide builds up. This hydrogen peroxide functions as a signalling molecule, triggering stress- and defense-related genes and ultimately increasing the plant's resistance to harsh environments. During the senescence of leaves and flowers, ethylene and hydrogen peroxide interact. Hydrogen peroxide is produced by ethylene, and this encourages the breakdown of chlorophyll and starts the senescence process. Genes linked to senescence are regulated by hydrogen peroxide as well (Das et al., 2020).

HORMONE-HYDROGEN PEROXIDE EFFECTS IN ROOT DEVELOPMENT

Hydrogen peroxide (H_2O_2) and plant hormones interact in a complex and precisely controlled manner to influence many elements of root architecture, elongation, and responsiveness to environmental signals. These signaling molecules interact, forming a dynamic regulatory network that ensures ideal root system development and operation (Gizachew et al., 2020).

Synergistic Effects on Root Elongation

Hydrogen peroxide and plant hormones like auxins and gibberellins work together to control root elongation. Auxins promote root growth by stimulating cell elongation. As a secondary messenger, hydrogen peroxide improves the root cells' receptivity to auxins. It alters plasma membrane H^+-ATPase activity, causing a rise in proton outflow into the cell wall. This cell wall acidification encourages the activation of enzymes that loosen the cell wall, promoting cell elongation and root development (Gupta et al., 2020).

Modulation of Root Architecture

Through the growth of lateral roots, root branches, and root hairs, plant hormones and hydrogen peroxide work together to shape root architecture. Auxin gradients have an impact on the beginning of lateral roots, with higher auxin concentrations favoring lateral root growth. By modifying auxin transport, hydrogen peroxide, a signaling molecule, takes part in this process. It affects PIN protein localization by directing auxin flow to lateral root initiation sites, promoting the development of lateral roots. Hydrogen peroxide interacts with cytokinins, which encourage cell division, to control root branching. Cytokinins can stimulate the creation of hydrogen peroxide, which then controls the expression of genes essential for the development of lateral roots. Well-structured root branching is influenced by the harmony between cytokinin-induced cell division and hydrogen peroxide-mediated signaling. Auxins and hydrogen peroxide both play crucial roles in the growth of root hair. Hydrogen peroxide serves as a signaling molecule in the mechanism by which auxins encourage the start of root hairs. Genes involved in the elongation and maturation of root hair are affected by hydrogen peroxide's effects on their expression. Together, they control the entire root hair development process, improving the ability to absorb nutrients and water (Bahuguna et al., 2018).

Environmental Adaptation

Hydrogen peroxide and plant hormones work together to improve a plant's ability to adapt to its environment by assisting root systems in doing so. Hydrogen peroxide interacts with ABA, a hormone implicated in the response to drought, to control root development. Hydrogen peroxide is produced as a result of ABA, which then affects root architecture. It encourages the development of deeper roots, allowing plants to obtain water from deeper soil layers and enhancing their resistance to drought. In conditions of excessive salinity, the hormone ethylene, which is implicated in the response to salt stress, interacts with hydrogen peroxide to control root development. Hydrogen peroxide and other ROS produced by ethylene lead to reactions that encourage root development and saline environment adaption (Chan et al., 2019).

Redox Signaling in Root Development

As a redox signal that interacts with hormonal pathways, hydrogen peroxide is used to fine-tune root growth in response to shifting environmental conditions. Its presence can affect how redox-active proteins are and how genes are expressed, which helps with growth and adaptive responses. A complex interplay between plant hormones and hydrogen peroxide controls a variety of activities in root growth and development, including branching, root hair development, elongation, and environmental adaption. This complex network makes sure that roots can adjust and react to shifting environmental factors, enabling ideal nutrient and water intake and total plant growth. Understanding these relationships offers new perspectives on the biology of plants and presents chances to enhance crop productivity and stress tolerance in agriculture (Yousefi et al. 2020).

HORMONE-HYDROGEN PEROXIDE INTERPLAY DURING FLOWERING

The process of developing a flower is highly coordinated and involves numerous intricate hormonal and molecular interactions. Hydrogen peroxide (H_2O_2) and plant hormones work in concert to control the initiation, differentiation, and maturation of floral organs, among other aspects of flower formation. The sophisticated signaling network that results from this cross-regulation of hormones and H_2O_2 promotes the timely and orderly patterning of floral development (An et al., 2020).

Auxins and Hydrogen Peroxide in Flower Development

Auxins are essential regulators of floral growth, with indole-3-acetic acid (IAA) serving as their main representative. They affect the development, patterning, and initiation of floral organs. Auxin-hydrogen peroxide interaction plays a crucial role in these processes. The development of floral primordia from the floral meristem is aided by auxins. They create gradients of concentration that identify and locate flower organs. As a secondary messenger, hydrogen peroxide increases the auxin response and aids in the formation of these gradients. H_2O_2 modifies the redox states and promotes the activity of important genes involved in floral organ identity, such as homeotic genes, to increase their expression. Auxins are also in charge of the expansion and lengthening of flowers. They promote cell lengthening and division, which causes the growth of floral organs. In this mechanism, hydrogen peroxide affects the activity of the enzymes responsible for loosening cell walls. The flexibility of the cell wall is increased by H_2O_2-mediated redox changes, allowing for cell expansion and promoting the growth of the entire flower (Wang et al., 2020).

Gibberellins and Hydrogen Peroxide in Flower Development

Another group of hormones known as gibberellins (GAs) interacts with hydrogen peroxide to control flower growth. GAs play a role in a number of flowering processes, including floral organ elongation, pollen formation, and control of blooming time. The transition from vegetative growth to reproductive development is influenced by GAs. They work together with hydrogen peroxide to control the pathways that determine when flowers bloom. Hydrogen peroxide, a signaling chemical that activates downstream genes involved in blooming, is produced as a result of GAs. The precise time of flowering is ensured by the careful equilibrium between GAs and H_2O_2 levels. GAs encourage the lengthening of floral organs, which is essential for flower maturity. They encourage the production of enzymes that weaken cell walls and cause cells to enlarge. By serving as a redox signaling molecule that controls the activity of enzymes that change cell walls, hydrogen peroxide increases the lengthening of cells that GA causes (Kaur et al., 2020).

Ethylene and Hydrogen Peroxide in Flower Development

A gaseous hormone called ethylene is important for flower development, especially for senescence and floral organ abscission. These processes are aided by the reaction between ethylene and hydrogen peroxide. Ethylene triggers the creation of hydrogen peroxide, which aids in the promotion of senescence, particularly petal senescence. Hydrogen peroxide functions as a signaling chemical that turns on genes responsible for breaking down chlorophyll, changing the color of the petals. Additionally, the abscission of floral organs and programmed cell death are caused by ethylene-triggered ROS generation. Additionally involved in the pollination and fertilization processes are ethylene and hydrogen peroxide. After pollination, the production of ethylene increases, which also stimulates the creation of hydrogen peroxide. This buildup of ROS contributes to the development of pollen tubes, which direct pollen to the ovule for fertilization. Pollen tube guiding and fertilization success are regulated by interactions between ethylene and H_2O_2 (Sehrawat et al., 2020).

ABA and Hydrogen Peroxide in Flower Development

ABA, which is well known for its role in stress reactions and dormancy, works with hydrogen peroxide to contribute to floral growth. Flowers age as a result of ABA-induced ROS generation, which includes hydrogen peroxide. The expression of genes related to petal senescence is stimulated by ABA, and hydrogen peroxide functions as a secondary messenger, enhancing the effects of ABA. A coordinated process of senescence is ensured by this connection. The interplay between ABA and hydrogen peroxide during blooming is also linked to reactions to stress. Increased ABA levels under unfavorable circumstances cause hydrogen peroxide to build up. Since the resultant ROS generation activates stress-related genes, the plant's resources are switched from reproduction to survival and stress tolerance (Zhang et al., 2020).

Hydrogen peroxide and plant hormone cross-regulation is essential for floral growth. These interactions guarantee the precise timing and patterning of the developmental, differentiational, senescent, and reproductive activities of floral organs. Hormones and H_2O_2 interact to form a complex web of signaling channels that enable plants to successfully reproduce while also adapting to shifting environmental conditions. The ability to alter these processes for horticultural and agricultural objectives is made possible by an understanding of these interactions, which offers insights into the molecular mechanisms that control flower growth (Kaur et al., 2020).

HORMONE-HYDROGEN PEROXIDE NETWORK IN STRESS RESPONSES

Plant Hormone Involvement in Abiotic Stress Tolerance

The coordination of responses to abiotic conditions including drought, salt, and severe temperatures is greatly aided by plant hormones. The effects of these stresses may impair plant growth, yield, and general health by interfering with normal physiological processes. The interaction of many plant hormones enables plants to adapt to their environment and endure harsh conditions.

Drought Stress

The hormone most closely linked to drought stress is ABA. To minimize water loss through transpiration, it acts as a primary regulator of stomatal closure. Drought causes plants to produce more ABA, which causes the stomata to close to conserve water. Additionally, ABA encourages root development, enabling plants to probe deeper soil layers for water absorption. In addition, ABA modifies the expression of genes that respond to stress, boosting the production of defensive proteins and enzymes that counteract the harm caused by drought (Gupta et al., 2020).

Salinity Stress

The reaction to salt stress involves ethylene. It controls stomata's opening and shutting, saving water and lowering salt absorption. Additionally, ethylene encourages the expression of genes that code for ion transporters, which helps root cells exclude extra salt. Under saline conditions, auxins have an impact on the architecture of the root system. They control the development of lateral roots, which can help increase root surface area and boost the effectiveness of nutrient and water intake (Baryla and Gardeström, 2020).

Temperature Extremes

The responses to temperature stress are significantly influenced by ABA as well. ABA controls cold-responsive genes during cold stress, promoting the production of antifreeze proteins that shield plant cells from the harm caused by freezing. Gibberellins play a role in temperature reactions, particularly when heat stress is present. They control the production of heat shock proteins, which act as molecular chaperones to keep other proteins' structural and functional integrity in the face of heat stress (Benková and Hejátko, 2020).

Crosstalk and Hormonal Balance

Plant hormones frequently interact and counteract each other's effects in addition to playing their unique functions to coordinate thorough stress responses. Although these hormones are typically linked to biotic stress, they can also affect abiotic stress reactions. They interact with the ABA and ethylene pathways, influencing how the body reacts to stress in general. Jasmonates, for instance, can block ABA-regulated stomatal closure, allowing for continuous transpiration in stressful situations. In terms of stress responses, several hormones might have conflicting effects. For instance, auxins can stimulate stomatal opening while ABA promotes stomatal closure under drought stress. Plants can adjust their responses to particular stress circumstances thanks to this antagonistic relationship (Chen and Song, 2020).

Key participants in the complex web of abiotic stress reactions are plant hormones. They control stomatal closure, root development, gene expression, and the production of protective proteins, among other elements of stress tolerance. The interaction of several hormones enables plants to adapt to their surroundings and endure harsh conditions. In order to improve plant resistance to abiotic stressors, which is essential for sustainable agriculture in the face of changing climatic circumstances, it is important to understand the roles and interactions of these hormones (Ding et al., 2020).

Hydrogen Peroxide as a Secondary Messenger in Stress Signaling

One important secondary messenger in plant stress signaling pathways is hydrogen peroxide (H_2O_2). Recent study has revealed its function as a versatile and dynamic signaling molecule that coordinates responses to varied environmental stimuli. It was previously thought of as a detrimental consequence of cellular metabolism. By combining chemistry and biology in an intriguing way, H_2O_2-mediated stress signaling enables plants to efficiently monitor and respond to their environment.

Generation of Hydrogen Peroxide in Stress Responses

An imbalance between the formation of ROS and their detoxifying mechanisms is frequently caused by environmental conditions such drought, salt, severe temperatures, and pathogen infections. Due to the partial reduction of oxygen molecules, H_2O_2 is one of the main ROS produced under stressful circumstances. While high quantities of H_2O_2 can harm cells, moderate levels work as important signaling molecules that help plants activate their defense mechanisms (Das et al., 2020).

H_2O_2 as a Secondary Messenger

H_2O_2 participates in the sensing and transmission of stress, acting as a secondary messenger. Plant cells interpret variations in ROS levels, particularly H_2O_2, as an early warning mechanism when faced with stresses. A number of receptors, including protein kinases and redox-sensitive proteins, are activated during perception. Once detected, H_2O_2 is quickly transported both inside and across cells to trigger the proper reactions. The oxidation states of particular amino acid residues in proteins are modified by H_2O_2 as a redox signal, changing the structure and activity of the resulting proteins. Protein-protein interactions, enzyme activity regulation, transcription factor activation, and other biological functions are all impacted by this redox-sensitive signaling. Redox changes brought on by H_2O_2 function as molecular switches that turn on or off particular signaling cascades. By interacting with hormonal pathways, H_2O_2 makes stress signaling even more complex. It interacts with the pathways for SA, jasmonic acid (JA), and ABA, affecting hormone synthesis, perception, and subsequent reactions. Incorporating hormonal pathways with H_2O_2-mediated communication allows for the fine-tuning of stress responses in response to certain environmental signals. A crucial regulator in the activation of genes associated to defense is hydrogen peroxide. Through interactions with transcription factors, co-regulators, and epigenetic machinery, it contributes to the control of gene expression. As a result, genes that code for antioxidant enzymes, antimicrobial proteins, and other stress response elements express themselves more strongly. A phenomenon known as "priming," in which plants pre-activate defensive mechanisms in anticipation of upcoming stress, can be brought on by exposure to moderate levels of stress. The process of priming involves hydrogen peroxide-mediated communication, which improves the plant's capacity to react more quickly and successfully to future stresses. This helps people become more resilient and accustomed to stress (Gizachew et al., 2020).

Implications

The understanding of hydrogen peroxide's function as a secondary messenger in stress signaling has important ramifications for plant biotechnology, environmental protection, and agriculture. Crop stress tolerance may be improved by manipulating H_2O_2-mediated signaling pathways. Researchers may be able to create plants that are more resistant to a variety of environmental challenges by fine-tuning the production and perception of H_2O_2, which might boost agricultural productivity. Utilizing H_2O_2 signaling's potential could result in creative bioengineering techniques. Researchers may use exogenous genes involved in H_2O_2-mediated signaling or strengthening plants' natural antioxidant defense mechanisms to create crops that are more resilient to stress. H_2O_2 level fluctuations can be used as a warning sign of stress in both agricultural and natural settings. This might help with prompt treatments to lessen the effect of stressors on the health and productivity of plants.

Our understanding of plant stress responses has become more nuanced as hydrogen peroxide has come to play a secondary messenger role in stress signaling. It emphasizes the incredible plasticity and adaptability of plants in dealing with difficult situations. In the face of changing global climates, improving sustainable farming methods and producing stress-tolerant crops both stand to benefit from unraveling the complexities of H_2O_2-mediated signaling (Gupta et al., 2020).

Integrative Role of Hormones and Hydrogen Peroxide in Stress Adaptation

Hydrogen peroxide (H_2O_2) and plant hormones play an integrative role in stress adaptation, demonstrating a complex and well-rehearsed reaction to environmental stresses. Intricate signaling networks that have developed in plants enable them to recognize, transmit, and convert stress signals into appropriate responses. This process is centered on the interaction between hormones and H_2O_2, which enables plants to customize their responses based on the precise type and severity of the stresses.

Crosstalk between Hormones and H_2O_2

In stress signaling pathways, hormones such ABA, ethylene, JA, and SA frequently serve as key messengers. The cross-regulation of these hormones is intricately intertwined with hydrogen peroxide signaling. In response to stress, ABA stimulates the generation of hydrogen peroxide, and H_2O_2 in turn strengthens ABA signaling by modifying the activity of transcription factors that are receptive to ABA. This cross-regulation optimizes stress acclimatization, water stress responses, and ABA-induced stomatal closure. Hydrogen peroxide generation caused by ethylene is a key component of ethylene-mediated responses to stressors like pathogen assault and dehydration. Ethylene increases H_2O_2 generation, which can therefore affect ethylene perception and synthesis and strengthen defensive systems. SA and JA are important components of plant defense mechanisms. They may interact with hydrogen peroxide pathways, affecting the expression of genes involved in stress response. As a mediator, hydrogen peroxide may combine JA- and SA-dependent signaling for a well-balanced defensive reaction to various stressors.

Hormone-H_2O_2 Interaction in Stomatal Regulation

In order to minimize water loss under stress and maintain an adequate gas exchange for photosynthesis, stomatal control is essential. A dynamic equilibrium between these functions is ensured by the interaction of hormones and H_2O_2. During drought stress, ABA-induced ROS generation, especially H_2O_2, causes stomatal closure. The sensitivity of guard cells to ABA is increased by hydrogen peroxide, resulting in more effective stomatal closure. By coordinating efforts, it is ensured that water is conserved while minimizing any potential harm to photosynthesis. Ethylene can affect stomatal behavior by communicating with hydrogen peroxide. The mobility of guard cells can be altered by ethylene-induced ROS, which therefore affects stomatal aperture and overall water usage efficiency (Chan et al., 2019).

Redox Homeostasis and Stress Acclimation

Through its effects on redox homeostasis and stress memory, hydrogen peroxide plays a dual function in stress adaption. ROS levels rise under stress, especially H_2O_2. This oxidative burst acts as a cue to trigger defensive mechanisms. To control ROS levels and avoid cellular damage, plants also make use of antioxidant molecules and enzymes. In combination with H_2O_2, hormones like ABA and JA control the expression of antioxidant genes, preserving redox equilibrium. Furthermore, the stress memory phenomena can be brought on by hydrogen peroxide-mediated stress reactions. Mild stress exposure results in the buildup of defensive molecules and improves the plant's capacity to react to future stresses. The basis for this adaptive plasticity is the coupling of hormonal pathways with H_2O_2 signaling (Feng et al., 2018).

Hormone-H_2O_2 Interaction in Growth-Defense Trade-offs

Plants frequently have to choose between growth and defense, and focusing energies on defense might hinder growth. These compromises are mediated by hormones and hydrogen peroxide. A balanced distribution of resources under stress is ensured by the interactions between growth-promoting hormones like auxins and growth-inhibiting hormones like ABA, ethylene, and H_2O_2. Signaling mediated by hydrogen peroxide can cause growth inhibition, diverting resources to defensive systems. The intricate nature of plant responses to environmental stressors is demonstrated by the integrative involvement of hormones and hydrogen peroxide in stress adaptation. By avoiding growth trade-offs, this integration enables plants to mount proportionate and adequate defenses. Plants have the ability to fine-tune their responses depending on the kind and severity of stressors because to the complex interactions between hormones and H_2O_2 signaling networks. This ultimately increases their resilience and survival in dynamic situations. Understanding these connections can help develop tactics for increasing crops' resistance to stress and enhancing sustainable agricultural methods (Gapper et al., 2018).

APPLICATIONS AND FUTURE PROSPECTS

Future agricultural innovations now have interesting new paths to explore thanks to the dynamic interaction between plant hormones and hydrogen peroxide (H_2O_2) in stress reactions. In light of climate change and rising global food demand, harnessing these interactions offers significant promise for crop improvement and improved stress tolerance. Based on the kind and severity of stressors, the integration of hormone- H_2O_2 networks enables the fine-tuning of stress responses. Researchers may be able to create crops with precise stress tolerance by modifying these connections. This implies that plants could be engineered to react to particular conditions more effectively, increasing yield stability and lowering losses. Researchers can alter hormone signaling pathways by understanding the complex interactions between hormones and H_2O_2 (Johansson and and Mähler et al., 2018). This knowledge can be used to create crops with hormone responses that are optimal under stress. For instance, by encouraging effective water consumption and stress adaptation, designing plants with increased ABA and H_2O_2 signaling may improve drought resistance. Multiple stress responses can be enhanced simultaneously by hormone- H_2O_2 interactions. Researchers may be able to create crops that display broad-spectrum stress tolerance by altering these relationships. This all-encompassing approach may produce more resilient plants that can withstand a variety of stresses, from disease attacks and drought to salinity. Resources being limited, particularly in terms of water availability, is one of the difficulties faced by modern agriculture. Crops could be modified to react to water stress better by enhancing the interactions between hormones and H_2O_2. Resource-efficient agriculture, which conserves water while maintaining productivity, may result from improved water usage efficiency brought on by controlled stomatal closure and greater root growth. Plant defense against diseases and pests is also influenced by hormone- H_2O_2 interactions. Researchers may be able to create crops with improved pest resistance through the activation of hormone-mediated defensive responses by taking advantage of these connections. This might lessen the demand for chemical pesticides and encourage more environmentally friendly insect control techniques. Hormones, H_2O_2, and epigenetic pathways interact in interesting ways that could lead to crop improvement. Gene expression and stress memory are both influenced by epigenetic changes (Kaur et al., 2020). The creation of crops with improved stress memory and more flexibility may result from a better understanding of how hormones and H_2O_2 affect these alterations. The development of climate-resilient crops is essential as the frequency and severity of extreme weather events rise due to climate change. Hormone- H_2O_2 interactions can be altered to produce crops that are more resilient to climate-related stresses like flooding and high temperature fluctuations. A certain degree of personalization in agricultural operations is made possible by the complex nature of hormone- H_2O_2 interactions. Based on the anticipated pressures for a specific growing season, farmers may choose to modify their crop management practices. This individualized strategy may result in better crop performance and more effective resource management (Mangano et al., 2018).

Through hormone-hydrogen peroxide crosslinking, the fusion of hormone biology and oxidative chemistry has created tantalizing new possibilities for biotechnological applications. This novel strategy has enormous promise in a variety of industries. Hormone-hydrogen peroxide crosslinking in agriculture has the potential to revolutionize crop management. Researchers have been able to promote desirable growth responses, improve stress tolerance, and even induce flowering or fruiting by selectively targeting particular plant hormones with controlled amounts of hydrogen peroxide. This method encourages environmentally responsible agricultural practices by providing a sustainable substitute for conventional chemical treatments (Chen and Song, 2020). This crosslinking technique may lead to the development of inventive drug delivery systems in the medical field. The exact and triggered release of therapeutic medicines is made possible by attaching hormones to hydrogen peroxide-sensitive carriers. This strategy shows promise for diseases like diabetes or hormonal malignancies, where hormone modulation is important. Furthermore, environmental bioremediation can benefit from hormone-hydrogen peroxide crosslinking. It is feasible to improve the ability of bioengineered microbes to digest contaminants by creating hormone-induced hydrogen peroxide in them. It may be possible to effectively detoxify contaminated locations as a result of the interaction between hormone signaling and the oxidative stress response (Kaur et al., 2020).

CONCLUSION

In summary, the complex interaction between several plant hormones and hydrogen peroxide emerges as a key factor in determining plant growth and development. This chapter has highlighted the complexity and adaptability of plant signaling networks by illuminating the multiple interactions that direct many physiological and development processes. It is clear from the perspective of ABA that the interaction of hormones and hydrogen peroxide functions as a crucial regulatory mechanism in stress reactions. A prime example of the fusion of hormonal and oxidative cues is ABA-mediated stomatal closure, which is essential for water conservation under challenging conditions. The finely tuned orchestration necessary for ideal plant performance under stress is highlighted by the careful balance between ABA-induced ROS generation and detoxifying systems. Contrarily, gibberellins (GAs) draw attention to the duality of interactions between hormones and hydrogen peroxide in the promotion of growth. Increased GA production promotes elongation and developmental changes by causing hydrogen peroxide buildup. This dual role serves as an example of how these interactions are context-dependent and can have different results depending on concentration gradients, spatial distribution, and physiological settings. The interaction of auxins and hydrogen peroxide connects organogenesis to root growth during the embryonic phase. The complex interaction required for precise developmental patterning is exemplified by the interaction between auxin-induced gene expression and hydrogen peroxide signaling. This hormone-hydrogen peroxide combination presents prospective biotechnological pathways for root architecture modulation and crop production enhancement. The complex interactions between ethylene and hydrogen peroxide shed light on the crucial role ethylene plays in coordinating responses to environmental stimuli. The complexity of hormone-oxidative interaction, which affects ethylene-mediated senescence and stress adaption, is best illustrated by the hormone's dual ability to drive both antioxidant enzyme production and ROS buildup. These understandings of hormone-hydrogen peroxide interactions go beyond the scope of basic plant biology and have potential biotechnological implications. Through the manipulation of these interactions, crop management techniques may be revolutionized, stress tolerance may be improved, and regulated growth patterns may be stimulated. While this chapter has illuminated the fascinating interaction between plant hormones and hydrogen peroxide, there are still a lot of unresolved issues. Further research is required to fully understand the intricacies of these interactions, which calls for cutting-edge methods and interdisciplinary approaches.

REFERENCES

Abeles, F. B., Morgan, P. W., & Saltveit, M. E. (2012). *Ethylene in Plant Biology*. Academic Press, Cambridge, MA.

Acharya, B. R., Jeon, B. W., Zhang, W., Assmann, S. M., & Early, A. Y. (2013). Open stomata 1 (OST1) kinase regulates nutritional responses in Arabidopsis. *Plant Physiology*, *162*(2), 1020–1032.

Alhagdow, M., Mounet, F., Gilbert, L., Nunes-Nesi, A., Garcia, V., Just, D., Petit, J., Beauvoit, B., Fernie, A.R., & Rothan, C. (2017). Silencing of the tomato sugar partitioning affecting protein TPP1 results in a fruit phenotype reminiscent of cuticular waxes. *Plants*, *6*(4), 1–20. https://doi.org/10.3390/plants6040048.

An, F., Zhang, X., Zhu, Z., Ji, Y., He, W., Jiang, M., Sheng, Y., Shi, L., Capell, T., Fraser, P. D., Christou, P., & Li, J. (2020). Hydrogen peroxide generated by NADPH oxidase 5 promotes grapevine stigma petal-like tissue abscission by retarding peroxidase mediated wall lignification. *Plant Cell Reports*, *39*(1), 119–133.

Anderson, J. P., Badruzsaufari, E., Schenk, P. M., Manners, J. M., Desmond, O. J., Ehlert, C., Maclean, D. J., Ebert, P. R., & Kazan, K. (2018). Antagonistic interaction between abscisic acid and jasmonate-ethylene signaling pathways modulates defense gene expression and disease resistance in Arabidopsis. *Plant Cell*, *20*(2), 115–131.

Arshi, A., & Rastogi, R. P. (2018). Hydrogen peroxide-induced cell death and oxidative stress in Chlamydomonas reinhardtii. *Ecotoxicology and Environmental Safety*, *20*(2), 143–149.

Audenaert, K., & Höfte, M. (2015). Redox-active small molecules as hydrogen peroxide signaling equivalents in plants. *Antioxidants and Redox Signaling*, *23*(14), 1101–1116.

Bahuguna, R. N., Joshi, R., Shukla, A., & Pandey, A. (2018). Hydrogen peroxide: A versatile molecule with multi-faceted roles in plants. *Chemosphere*.

Balbi, V., Devoto, A., & Busch, W. (2010). Cell differentiation and organ initiation at the shoot apical meristem. *Plant Molecular Biology*, *9*(1), 21–27.

Bandara, N., & Gardner, M. (2015). Roles for plant peroxidases in cross-linking lignin and disease resistance. *Phytochemistry Reviews*, *14*(5), 975–985.

Baryla, A., & Gardeström, P. (2020). The integration of redox signaling with other pathways in plant development. *Environmental and Experimental Botany*, *173*, 103947.

Baud, S., Wuillème, S., Dubreucq, B., de Almeida, A., Vuagnat, C., Lepiniec, L., Miquel, M., & Rochat, C. (2017). Function of plastidial pyruvate kinase in developing embryos. *Frontiers in Plant Science*, *8*, 266. https://doi.org/10.3389/fpls.2017.00266

Bellé, C., Asensi, M., & Savassi-Coutinho, A. P. (2018). Nitric oxide modulates the degree of tolerance to salt stress in non-nodulating soybean plants. *Plant Science*, *20*(4), 595–603.

Benková, E., & Hejátko, J. (2020). Hormone interactions at the root apical meristem – Concepts, puzzles, and perspectives. *Journal of Experimental Botany*, *71*(5), 1534–1549.

Bhattacharya, A., & Sood, P. (2015). Hormonal regulation of plant development and stress tolerance. *Methods in MolecularBiology*, *1284*, 1–12.

Bielach, A., Hrtyan, M., Tognetti, V. B., Blilou, I., & Benková, E. (2019). Pericycle delimits formative cell divisions in the root meristem. *Nature Cell Biology*, *21*(3), A4–A5.

Chan, P. S., Wong, C. Y., Li, M. F., Wang, Y. F., & Jiang, L. W. (2019). Phytohormone crosstalk between brassinosteroids and polyamines in plants. *International Journal of Molecular Sciences*, *20*(3), 573.

Chehab, W. E., Kim, S., & Joshi, T. (2012). Bacterial disease resistance in Arabidopsis through flagellin perception. *Nature*, *493*(7431), 38–42.

Chen, L., & Song, Y. (2020). An overview of hydrogen peroxide in plants: Metabolism, redox signaling, and stress responses. *Frontiers in Plant Science*, *11*, 608.

Chen, X., Liu, J., & Lin, W. H. (2015). Multiple environmental stresses induce transcription and accumulation of OsHSP17. 0 in seedlings of rice. *Plant Physiology and Biochemistry*, *89*, 64–71.

Cui, J., Li, J., Liu, L., & Yu, F. (2018). Plasma membrane receptor-like kinases, BAK1 and BIR1, regulate immune signaling and plant development. *Frontiers in Plant Science*, *9*, 2.

Das, P., Kalita, M. C., & Goyal, A. K. (2020). Ethylene and hydrogen peroxide mediated upregulation of antioxidant defense mechanism enhances thermotolerance in chickpea (Cicer arietinumL.) seedlings. *Physiologia Plantarum*, *169*(1), 113–130.

Ding, X., Zhang, Y., Zhang, C., Lan, Z., Yuan, L., & He, Q. (2020). Interaction between abscisic acid and hydrogen peroxide in regulating stomatal closure in plants. *Environmental and Experimental Botany*, *171*, 103950.

Divi, U. K., & Krishna, P. (2010). Overexpression of the β subunit of ATP synthase enhances ATP synthesis and decreases reactive oxygen species production in Arabidopsis. *Plant Physiology*, *154*(2), 666–677.

Escariz, A. M., Santos, L. S., DaMatta, F. M., & Ramalho, J. C. (2020). Interactions between abscisic acid, cytokinins, and hydrogen peroxide in maize leaves exposed to water deficit. *Plant Physiology and Biochemistry*, *151*, 211–221.

Feng, H., Zhang, Q., & Yao, Y. (2018). Jasmonic acid signaling in plant defense and development: Crosstalk with other phytohormones. *Molecular Plant*, *20*(1), 19–30.

Gapper, C., Dolan, L., & Roberts, K. (2018). IAA: Chemistry in plants and beyond. *Journal of Experimental Botany*, *20*(3), 465–478.

Gifford, M. L., Dean, G. D., Gutierrez, R., Coruzzi, G. M., & Birnbaum, K. D. (2018). Cell-specific nitrogen responses mediate developmental plasticity. *Proceedings of the National Academy of Sciences of the United States of America*, *115*(5), 11573–11578.

Gill, S. S., & Tuteja, N. (2010). Reactive oxygen species and antioxidant machinery in abiotic stress tolerance in crop plants. *Plant Physiology and Biochemistry*, *48*(12), 909–930. https://doi.org/10.1016/j.plaphy.2010.08.016.

Gizachew, W., Seifu, D., & Adgaba, N. (2020). Role of plant hormones in ethylene-induced oxidative stress and its outcome on plants. *International Journal of Horticultural Science and Technology*, *7*(4), 397–410.

Gupta, D. K., Palma, J. M., Dat, J. F., & Corpas, F. J. (2020a). Reactive oxygen species and plant hormone signaling under drought stress. *Plant Signaling and Behavior*, *15*(8), 1750–1764.

Hossain, M. A., Mostofa, M. G., Fujita, M., & Tran, L. S. P. (2012). Nitric oxide mediates plant responses to dehydration and salt stress: A review. *Frontiers in Plant Science*, *3*, 251.

Johansson, H., & Mähler, R. (2018). Salicylic acid in plants: An ancient hormone with new ideas. *Plant Signaling and Behavior*, *12*(5), 6.

Kangasjärvi, S., Neukermans, J., & Li, S. (2012). A high-throughput screening system for Arabidopsis transcription factors and its application to Med25-dependent transcriptional regulation. *Molecular Plant*, *5*(4), 734–746.

Kaur, A., Pati, P. K., & Yadav, V. (2020). Role of plant hormones in plant response to biotic stress. *Plant Hormones Under Challenging Environmental Factors*, 35–46.

Key, G., Pajor, R., & Kőmíves, T. (2018). Effects of exogenous hydrogen peroxide on oxidative stress tolerance and photos ynthetic capacity in plants. *International Journal of Molecular Sciences*, *19*(5), 8.

Khorassani, R., Miransari, M., & Mohammadi, A. (2018). The effect of plant hormones on cell division and elongation in roots: Auxin and cytokinin–a review. *Netherlands Journal of Agricultural Sciences*, *20*(6), 89–106.

Li, J., Zhu, L., Hull, J. J., Liang, S., Qiao, H., & Niu, S. (2020). Crosstalk between auxin and hydrogen peroxide in plant development and stress response. *Journal of Experimental Botany*, *71*(3), 895–907.

Liu, J., Zhang, S., Zhu, S., Wang, L., & Wang, J. (2018). Interplay between hydrogen peroxide and plant hormones during plant development and stress responses. *Frontiers in Plant Science*, *9*, 1576.

Ljung, K., & Nemhauser, J. L. (2018). From seedling growth to flowering time: Crosstalk between brassinosteroids and environmental signals. *Journal of Experimental Botany*, *20*(4), 87–98.

Mangano, S., Juárez, S. P., & Moldes, C. A. (2018). Nitric oxide and hydrogen peroxide function as signaling molecules in the mediation of abiotic stress responses in plants. *Frontiers in Plant Science*, *9*, 70.

Medeiros, V. S., Sperotto, R. A., & Daloso, D. M. (2018). Hydrogen peroxide metabolism modulates plant development and stress responses. *Frontiers in Plant Science*, *9*, 10.

Müller, M., & Munné-Bosch, S. (2015). Ethylene response factors: A key regulatory hub in hormone and stress signaling. *Plant Physiology*, *169*(1), 32–41. https://doi.org/10.1104/pp.15.00677.

Nambara, E., & Marion-Poll, A. (2018). Abscisic acid biosynthesis and catabolism. *Annual Review of Plant Biology*, *20*(2), 109–145.

Navrot, N., Rouhier, N., Gelhaye, E., & Jacquot, J. P. (2018). Reactive oxygen species generation and antioxidant systems in plant mitochondria. *Physiologia Plantarum*, *20*(4), 201–214.

Niu, H., Liao, W., Gao, J., Yang, Z., Liu, W., Sun, S., Chen, F., & Xu, Y. (2019). Hydrogen peroxide mediates abscisic acid-induced stomatal closure in Vicia faba. *Biologia Plantarum*, *63*(1), 98–105.

Peleg-Grossman, S., Melamed-Book, N., Cohen, G., & Levine, A. (2018). Cytoplasmic H_2O_2 prevents translocation of NPR1 to the nucleus during the induction of systemic acquired resistance in plants. *Nature Communications*, *9*(1), 1.

Riemann, M., Müller, A., Korte, A., & Furuya, M. (2018). Plant hormone levels in seeds of different dormancy classes during dry storage: Changes in abscisic acid, gibberellins, and cytokinins. *Journal of Experimental Botany*, *20*(3), 79–91.

Rodriguez Shongo, J., Ursino, M., Galatro, A., Reinoso, H., & Casalongué, C. (2019). Crosstalk between hydrogen peroxide and abscisic acid signaling during root growth and development. *Plant Signaling and Behavior*, *14*(1), 1551941.

Sehrawat, A., Deswal, R., & Balyan, H. S. (2020). Crosstalk between plant hormones and reactive oxygen species during various stress responses. *Journal of Plant Growth Regulation*, *39*(1), 117–131.

Su, R., & Yang, Y. (2019). The emerging roles of hydrogen peroxide in plant responses to environmental stimuli. *Journal of Integrative Plant Biology*, *61*(1), 131–143.

Wang, J., Kang, Y., Hu, K., Sun, C., Qi, Y., Sun, Y., & Sun, C. (2020). Auxin-cytokinin interaction regulates aluminum tolerance by modulating antioxidant defense and organic acid anion exudation in maize. *Journal of Experimental Botany*, *271*.

Yousefi, N., Zarei, A., & Niinemets, Ü. (2020). How do jasmonates affect different traits of plant physiology? A case study on photosynthesis, water use efficiency and growth. *Plant, Cell and Environment*, *43*(2), 457–469.

Zhang, L., Yang, X., & Zhang, Y. (2020). Crosstalk between plant hormonal regulation and nitrate signaling in plant nutrient utilization. *Frontiers in Plant Science*, *11*, 1–13.

11 Plant Transcriptomics and Proteomics Study in Relation to Hydrogen Peroxide

Hamza Rafiq and Jamila Fatima

INTRODUCTION

Hydrogen peroxide (H_2O_2) in plants is a pivotal reactive oxygen species (ROS) with multifaceted roles. H_2O_2 is a chemical compound composed of two hydrogen atoms and two oxygen atoms and a simple peroxy compound containing a peroxide bond (O-O) (Ofoedu et al., 2021). It is a clear, colorless liquid with a slightly sharp odor and a slightly acidic taste. The fact that hydrogen peroxide is an ROS is one of its essential properties. ROS are oxygen-containing molecules with unpaired electrons that make them extremely reactive. Hydrogen peroxide is considered a non-radical ROS, meaning it is not a free radical itself but can give rise to highly reactive radicals in certain conditions (Sies et al., 2022).

Its significance in plant biology stems from its involvement in various physiological processes including cell signaling, development, growth and responses to environmental stimuli (Bopitiya et al., 2021; Černý et al., 2018). The production of H_2O_2 occurs in multiple cellular compartments, including chloroplasts (Ivanov et al., 2022), mitochondria (Brand, 2020), and peroxisomes (Fransen & Lismont, 2024), and is mediated by several metabolic pathways (Huchzermeyer et al., 2022). The hydrogen peroxide production in plants is not a mere consequence of metabolic activity but is also a deliberate response to abiotic stimuli (Qureshi et al., 2022; Sachdev et al., 2021). As a reaction to environmental stresses, such as drought (Yang et al., 2021), salinity (Jamshidi Goharrizi et al., 2024; Mariyam et al., 2023), or extreme temperatures (Devireddy et al., 2021; Firmansyah & Argosubekti, 2020), H_2O_2 levels in plant cells can rise significantly. These elevated levels of H_2O_2 serve as signals that cause various stress inductive pathways (Sachdev et al., 2021). Thus, H_2O_2 acts as a key mediator in the plant's adaptation to adverse environmental conditions.

Despite its beneficial roles, the accumulation of H_2O_2 can be detrimental if not properly regulated. Excessive H_2O_2 can lead to oxidative damage and deleterious impact on the growth as well as development of the plants (Claeys et al., 2014). To mitigate this risk, plants have evolved sophisticated antioxidant defense systems. These include enzymatic antioxidants, such as peroxidases as well as catalases, which directly break down hydrogen peroxide into oxygen and water, and antioxidants which are non-enzymatic, such as glutathione and ascorbate, which scavenge ROS and mitigate oxidative stress. Hydrogen peroxide (H_2O_2) in plants is a versatile molecule that plays a dual role as a signaling molecule that regulates different physiological mechanisms in living organisms and a harmful oxidizing agent that can cause damage to cells and tissues in high concentrations (Claeys et al., 2014). Its levels must therefore be strictly controlled to avoid oxidative damage, highlighting the significance of the plant's antioxidant defense mechanisms.

Understanding how plants react to different abiotic and biotic stimuli is made possible through the use of transcriptomics, the study of the whole set of RNA transcripts generated by the genome under particular conditions. Transcriptomic approaches, such as RNA sequencing (RNA-seq) and microarrays, enable scientists to measure how the levels of gene expression are altered under various stress scenarios and developmental phases (Imadi et al., 2015). These high-throughput techniques

DOI: 10.1201/9781032647692-11

enable the identification of differentially expressed genes (DEGs) under oxidative stress conditions, providing insights into the regulatory networks and pathways activated by H_2O_2 (Manimekalai et al., 2023, Sewelam et al., 2014). Transcriptomic studies have also highlighted the crosstalk between H_2O_2 signaling and other stress response pathways. This crosstalk enables plants to integrate multiple signals and mount a coordinated response to complex environmental stresses (Saxena et al., 2016). Moreover, transcriptomic data have shown that H_2O_2 can induce genes associated with cell wall modification, secondary metabolism, and protein folding, suggesting a broad impact on various cellular processes (Manimekalai et al., 2023).

The extensive study of proteins, their structures, and functions, structures as well as their interactions within the context of the plant's biological systems is known as proteomics. Proteins are fundamental to virtually every biological process, serving as enzymes, structural components, signaling molecules, and regulators of gene expression. By analyzing the plant proteome, researchers can gain comprehensive insights into how plants grow, develop, and adapt to their surroundings. As an ROS, hydrogen peroxide plays a dual part in plant cells: It functions as a signaling molecule when present in low amounts as well as a potential source of oxidative damage at higher concentrations (Claeys et al., 2014; Rahman et al., 2021). Proteomics provides detailed insights into the dynamic changes in protein abundance, modifications, and interactions that occur in response to H_2O_2, elucidating the molecular mechanisms plants employ to manage oxidative stress and harness H_2O_2 for signaling purposes (Bhardwaj et al., 2021). Proteomics encompasses a range of techniques used to study the protein complement of the genome, i.e., tandem mass spectrometry (MS/MS), mass spectrometry (MS), two-dimensional gel electrophoresis (2-DE), and liquid chromatography-mass spectrometry (LC-MS) (Li et al., 2020). These methods allow researchers to identify as well as quantify proteins involved directly in antioxidant defense mechanism, signal transduction pathways, and crosslinking with hormonal signaling, providing a comprehensive view of the proteome under oxidative stress conditions.

Identifying key proteins and pathways involved in oxidative stress tolerance can inform breeding initiatives to create crops more resistant to abiotic stresses. For instance, overexpressing genes encoding antioxidant enzymes or proteins involved in H_2O_2 signaling can improve stress tolerance in transgenic plants. This research offers a thorough summary of plant responses to hydrogen peroxide (H_2O_2), examining its dual role as both a signaling molecule and as damaging agent. It aims to detail the impact of plant transcriptomics and proteomics on H_2O_2 studies, highlighting key changes in gene expression and protein modifications. This study offers a holistic understanding of H_2O_2 and their transcriptome and proteome studies.

SOURCES AND PRODUCTION MECHANISMS

H_2O_2 is generated in plant cells through various metabolic processes, including photosynthesis, respiration, and β-oxidation in peroxisomes. Abiotic stresses such as attack of pathogen, and high light intensity, and drought can significantly elevate H_2O_2 levels. H_2O_2 production occurs in several cellular compartments, including chloroplasts, mitochondria, and peroxisomes.

Chloroplastic H_2O_2

In chloroplasts, H_2O_2 is generated during photosynthesis, under normal light conditions, the photosynthetic electron transport chain operates efficiently, transferring electrons through a series of protein complexes to ultimately produce NADPH. However, under high light conditions, the absorption of light energy can exceed the capacity of the photosynthetic machinery to process it, leading to the formation of excess electrons. These electrons can reduce molecular oxygen to superoxide ($O_2^{\bullet -}$), a precursor of H_2O_2. The dismutation of superoxide is catalyzed by the enzyme superoxide dismutase (SOD) into H_2O_2 as well as oxygen. The main ways that the H_2O_2 priming helped were by preserving high concentrations of photosynthetic pigments, high photochemical efficiency metrics

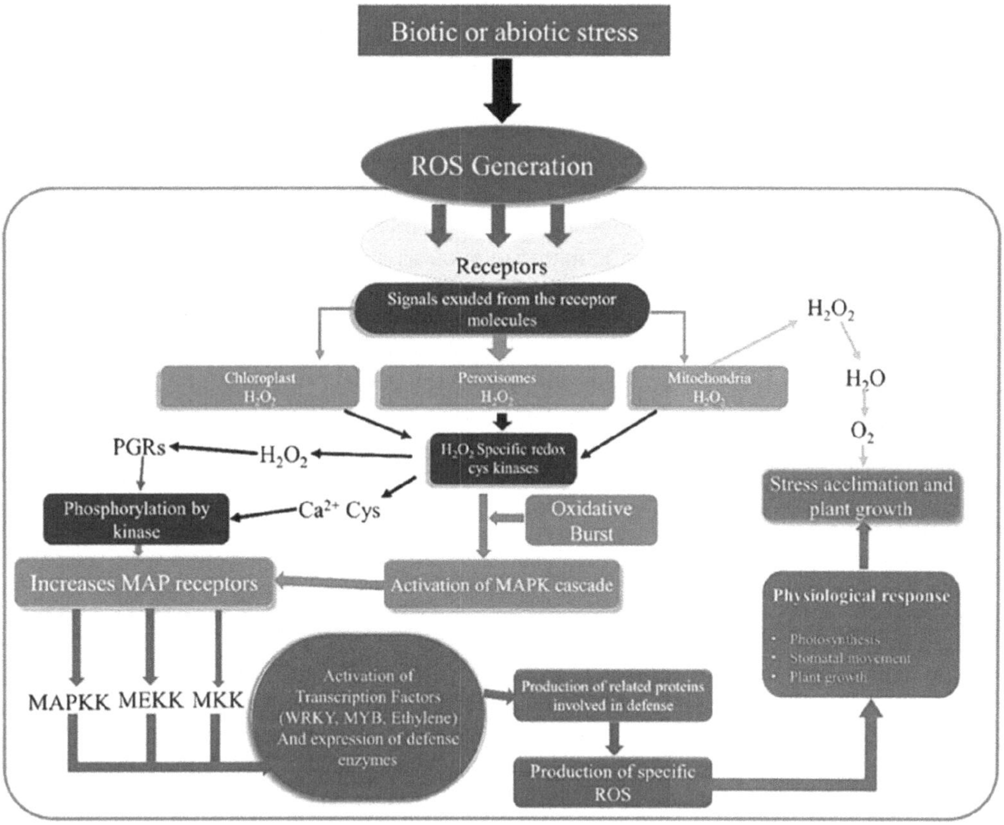

FIGURE 11.1 Production of H_2O_2 through chloroplast, peroxisomes and mitochondria and their role in activation of different signaling pathways ultimately the production physiological responses and stress acclimation against various biotic and abiotic stresses.

(Fv/Fm and NPQ), and consistent PEPcase activity (dos Santos Araújo et al., 2021). Additionally, it also preserves the ultrastructure of chloroplast by decreasing both endogenous and O_2 and H_2O_2−contents. By utilizing spin trapping electron paramagnetic resonance spectroscopy and H_2O_2-sensitive fluorescent dyes, Mubarakshina et al. (2010) also examined the generation of H_2O_2 through photosynthetic electron transport and its capacity to diffuse across the chloroplast envelope membranes. Researchers discovered that some of the H_2O_2 generated inside the chloroplasts can escape the efficient antioxidant mechanisms inside the chloroplasts, even in low light conditions.

When exposed to photoinhibitory light intensities, hydrogen peroxide (H_2O_2), which is produced during photosynthesis, is thought to both begin and transmit a retrograde signal. One possible explanation for the signaling specificity attained by using hydrogen peroxide supplied from chloroplasts for signal transduction is the frequently noted tight connection of several of these organelles with the nucleus (Breeze & Mullineaux, 2022). The photosynthetic electron transport chain in chloroplasts produces more H_2O_2 under high light stress, which may lead to photo-oxidative damage. While thylakoid and stomal membrane-bound ascorbate peroxidases (tAPX and sAPX, respectively) are important enzymes in chloroplasts that scavenge H_2O_2, their knockout mutants do not show any discernible phenotype under conditions of intense light stress. A study conducted in which they developed new mutant phenotypes and concluded that in comparison to its parental lines, the sapx tapx pgr5 triple mutant showed extraordinary sensitivity to intense light (Kameoka et al., 2021). The H_2O_2 produced by chloroplast is a best source of combating plants against environmental stresses especially light stress to maintain the normal functioning of the plant.

Mitochondria H_2O_2

Similarly, in mitochondria, H_2O_2 is generated as a byproduct of cellular respiration. The electron transport chain (ETC) in mitochondria can leak electrons, reducing oxygen to superoxide, which SOD then converts to H_2O_2. Nuclear gene expression can be altered by retrograde signals produced by changes in metabolism in the chloroplasts and mitochondria. ROS, which are mobile signals that can activate protein redox switches by oxidizing particular cysteine residues, are likely implicated (Figure 11.1). Numerous ROS-scavenging and ROS-induced damage repair pathways rely on redox buffers, including the highly reduced glutathione pool, as reservoirs of reducing power (Ugalde et al., 2021). Using genetically encoded biosensors roGFP2-Orp1 (for H_2O_2) and Grx1-roGFP2 (for EGSH) targeted to mitochondria as well as chloroplasts, researchers used the herbicide methyl viologen (MV) to produce ROS locally in the chloroplasts and mitochondria of intact Arabidopsis (Arabidopsis thaliana) seedlings and recorded dynamic changes in H_2O_2 and EGSH levels (Ugalde et al., 2021). Mitochondrion H_2O_2 is produced in association with the environmental stresses, alongside with the chloroplast H_2O_2 and activate defense response of the plant.

Peroxisome H_2O_2

Multiple enzymes that produce and degrade hydrogen peroxide (H_2O_2) are found within the functionally specialized organelles known as peroxisomes. Peroxisomes contribute to H_2O_2 production through processes such as fatty acid oxidation and photorespiration. Photorespiration, which occurs when the enzyme ribulose-1,5-bisphosphate carboxylase/oxygenase (RuBisCO) fixes oxygen instead of carbon dioxide, leads to the production of glycolate. Glycolate is subsequently oxidized in peroxisomes, producing H_2O_2 as a byproduct. These pathways underscore the centrality of peroxisomes in managing ROS levels within plant cells. As a primary redox signaling agent, this oxidant gives peroxisomes the inherent capacity to mediate and modify H_2O_2-driven activities, such as autophagy. In addition to RuBisCO, a large number of other enzymatic systems that can produce H_2O_2 are found in plant peroxisomes, including polyamine oxidase, glycolate oxidase (GOX), acyl-CoA oxidase (AOX), urate oxidase (UO), polyamine oxidase, copper amine oxidase (CuAO), sulfite oxidase (SO), sarcosine oxidase (SOX), or SOD (Figure 11.2) (Corpas et al., 2020). The endogenous metabolism of plant peroxisomes depends on these H_2O_2-generating enzymes, which are also involved in a number of metabolic processes that are crucial to the interactions of plant peroxisomes with other subcellular compartments such mitochondria, plastids, nuclei cytosols, and oil bodies. H_2O_2 itself is a key signaling molecule in crosstalk between organelles in these subcellular linkages, helping to coordinate cell function.

In order to control the generation of peroxisomes—H_2O_2, HEK-293 and HeLa—pexophagy reporter cell lines were created and confirmed in this investigation. High concentrations of peroxisomes- H_2O_2 impede autophagic flux by oxidatively inhibiting the enzymes involved in the synthesis of LC3II, and peroxisomes- H_2O_2 can oxidatively change several selective autophagy receptors and core autophagy proteins (Li et al., 2023). They came to the conclusion that peroxisomes with increased H_2O_2 release can elude pexophagy and suppress autophagic activity throughout organismal and celluar aging, which would continue the buildup of harmful and lethal cellular debris. New approaches were investigated in order to obtain a more comprehensive understanding of the contribution of peroxisomes in redox signaling. Specifically, a yeast AP1-like-based sulfenome mining strategy was employed in conjunction with an Flp-In T-REx 293 cell system, which allows for the modulation of peroxisomal H_2O_2 production, to inventory protein thiol targets of peroxisome-derived H_2O_2 in various subcellular compartments. They discovered that peroxisomes, the mitochondria, cystol, all contain more than 400 substrates for peroxisome-derived H_2O_2 (Lismont et al., 2022). Furthermore, a number of peroxisome-dependent genes were discovered in another study to be frequently regulated under various stress scenarios that affect peroxisomal ROS metabolism. Transcriptomic responses to various stresses that are abiotic share these genes (Terrón-Camero et al., 2022).

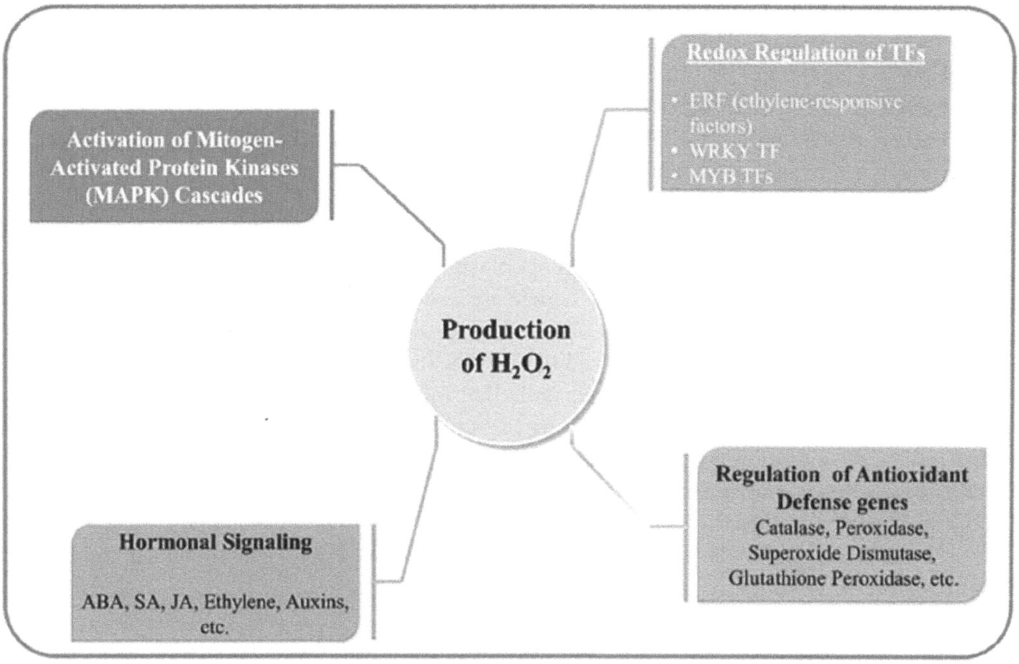

FIGURE 11.2 Production of H_2O_2 carried out through redox regulation of different transcription factors, Regulation of antioxidants, Activation oof MAPK cascades and hormonal signaling in the plant cell.

These studies suggested that the peroxisomes- H_2O_2 is beneficial for maintain the cell system by eliminating the damaged and toxic cellular debris outside the cell. Other than peroxisomes, H_2O_2 can also be produced through apoplastic area of the cell.

Apoplastic H_2O_2

The apoplast, the space outside the plant cell membrane but within the cell wall, is another important site for H_2O_2 production. Apoplastic H_2O_2 is often associated with the first line of defense against plant's response to abiotic as well as biotic stresses. For instance, during pathogen attack, plants generate an oxidative burst—an abrupt and high reactive ROS generation, including H_2O_2. This oxidative burst is mediated by plasma membrane-bound NADPH oxidases (also known as respiratory burst oxidase homologs or RBOHs), which transfer electrons from NADPH to oxygen, producing superoxide that dismutates to H_2O_2. Apoplastic H_2O_2 strengthens cell walls by crosslinking cell wall proteins and polysaccharides, thereby enhancing physical barriers against pathogen invasion. It also functions as a signaling molecule, triggering stomatal closure as well as the activation of defense-related genes. The closing of stomata is primarily caused by the release of stress-induced H_2O_2 from apoplasts and its buildup inside guard cells. It seems that H_2O_2-controlled stomatal closure is mostly dependent on the control of aquaporins and NADPH oxidases. Researchers have previously reported that RESPIRATORY BURST OXIDASE HOMOLOG (RBOH)D/F NADPH oxidases are activated by both calcium-dependent and calcium-independent pathways, which in turn cause closure of stomatal cells (Rodrigues & Shan, 2022). Under conditions of water stress, it was observed that apoplastic ROS, especially hydrogen peroxide, rose notably in the apical area of the growth zone. This rise was partially attributed to enhanced oxalate oxidase activity in this location. They assessed in comparison with the transgenic maize lines that the rise in oxalate oxidase levels increases apoplastic H_2O_2 and ultimately increasing the production OF CELLS in stressed roots (Voothuluru et al., 2020). A balanced interplay between SNO and H_2O_2 is essential for preserving thermotolerance and cellular redox homeostasis, whereas prior research also indicated that GSNOR plays a significant role in controlling the SlRBOH1-dependent apoplastic H_2O_2 synthesis when exposed to high temperatures (Song et al., 2022).

An earlier report on the Arabidopsis thaliana CuAOβ (AtCuAOβ) suggested that this enzyme plays a role in water balance by regulating xylem-dependent water supply as well as stomata-dependent losses of water under stress. This involvement in early root protoxylem differentiation and stomatal closure was triggered by the wound-signal MeJA via apoplastic H_2O_2 production (Fraudentali et al., 2023). When apoplastic peroxidases and NADPH oxidases detect bacteria, they produce apoplastic ROS, which are crucial in initiating the closure of stomatal cells. The connection between cytosolic and apoplastic dynamics of ROS was observed by Arnaud et al. (2023), who also emphasized the significance of RBOHF in plant defense. Using a genetically encoded hydrogen peroxide sensor called roGFP2-Orp1, Arnaud et al. (2021) discovered that, in contrast to the brief apoplastic ROS burst, sustained hydrogen peroxide generation occurs in the cytosol following PAMP detection, independent of cell wall peroxidases as well as NADPH oxidase. Apoplastic H_2O_2 plays a very vital part in defense mechanism of plant against pathogens attack and other than that they also play role in abiotic stress management. There are other indirect processes which are involved in the production of H_2O_2 other than its direct production through cellular compartments.

Other Process of Production of H_2O_2

Other enzymatic sources of H_2O_2 in plants include amine oxidases and xanthine oxidases. Amine oxidases catalyze the oxidative deamination of polyamines as well as other amines, producing H_2O_2 as a byproduct (Gupta et al., 2016) (Figure 11.2). Xanthine oxidases, involved in purine metabolism, generate H_2O_2 during the oxidation of xanthine and hypoxanthine to uric acid. Inadequate XOR activation increases oxidative stress-related tissue damage because XO produces hydrogen peroxide as well as superoxide, which contribute to an increase in ROS (Furuhashi, 2020). These enzymes contribute to the overall ROS pool and participate in several physiological processes, including regulation of growth and the reaction to stress.

Environmental factors such as drought, salinity, and heavy metal exposure can also induce H_2O_2 production in plants. Under these stress conditions, enhanced metabolic activity and electron leakage in chloroplasts, mitochondria, and peroxisomes lead to increased H_2O_2 levels (Mariyam et al., 2023; Yang et al., 2021). Furthermore, stress conditions can upregulate the expression and activity of ROS-producing enzymes, amplifying H_2O_2 production. This stress-induced H_2O_2 acts as a signal to turn on defense systems, which includes synthesis of osmoprotectants, stress-responsive proteins, as well as antioxidants (Nawaz & Wang, 2020; Sohag et al., 2020). In the case of salinity stress, H_2O_2 signaling can lead to the activation of ion transporters that help in maintaining ion homeostasis and preventing toxic buildup of salts in cells. Understanding the sources and production mechanisms of H_2O_2 provides valuable insights into plant physiology which offers potential strategies for enhancing stress tolerance and improving crop resilience.

THE ROLE OF HYDROGEN PEROXIDE IN PLANT SYSTEMS

Hydrogen peroxide plays an essential role in plant systems, functioning as both a signaling molecule and a mediator of stress responses. Within plant cells, a variety of metabolic processes and interactions with the environment result in the production of this ROS. Despite its potential to cause cellular damage due to its reactive nature, H_2O_2 is indispensable for regulating numerous physiological processes, including growth, development, defense, and adaptation to environmental stresses (Table 11.1).

TABLE 11.1
Plant Processes Boosted or Activated because of H_2O_2-Mediated Response with the Studies Crops

Plant Processes	H_2O_2-Mediated Effects	Crop/Plant	References
Disease resistance and Root development	Activation of H_2O_2, JA, ad phytoalexin metabolic pathways	*Oryza sativa*	Sun et al. (2023)
Phosphorylation	Regulate the production of apo plastic H_2O_2 involved in ABA signaling		Wang et al. (2023)
Lignin Accumulation	MeRAV5 interacts with MePOD and MeCAD15 and promotes their activity and influence the production of H_2O_2 and lignin	*Manihot esculenta*	Yan et al. (2021)
Root Growth and Cell elongation	Increase expression of oxalate oxidase which increase the production of apoplastic H_2O_2 resulting in root growth	*Triticum astivum*	Voothuluru et al. (2020)
Enhance activity of Antioxidant enzymes	Over accumulate ROS at 5mmol/L H_2O_2 and improve stress tolerance	*Oryza sativa*	Sohag et al. (2020)
Increase biosynthesis of Photosynthetic pigments	improvement in the activities of antioxidant enzymes and accumulation of non-enzymatic antioxidants, which effectively reduced the membrane lipid peroxidation	*Triticum astivum*	Habib et al. (2020)
Stomatal Opening	Induced by exogeneous H_2O_2 by the drop in K^+ concentration	*Triticum astivum*	Kumari and Verma (2019)
Cold tolerance	Overexpression of PtrERF109 leading to increase in production of antioxidant enzymes and ROS	*Poncirus trifoliata*	Wang et al. (2019)
Photosynthesis	Root dipping treatment of H_2O_2 improves chlorophyll content	*Solanum lycopersicum*	Nazir et al. (2019)
Physiological process	Seed soaking treatment of H_2O_2 improves the rate of photosynthesis, chlorophyll content and stomatal conductance	*Vigna radiata*	Khan et al. (2018)
Pollen development	H_2O_2 and other ROS play a significant function in pollen functionality and gametophyte integration	*Propagating plants*	Černý et al. (2018)
Root growth	Elevated level inhibited root elongation, however a lower concentration enhanced cell expansion and root diameter	*Oryza sativa*	Xiong et al. (2015)
Seed dormancy and germination	Down regulation of ABA catabolism and increase in GA biosynthesis		Liu et al. (2010)

Signaling Molecule

A hallmark of H_2O_2 signaling in plants is its role in regulating the way antioxidant enzyme-encoding genes are expressed and proteins involved in ROS detoxification. Increased H_2O_2 levels stimulate the expression of genes encoding catalase, peroxidases, SOD, glutathione peroxidase, and other ROS-scavenging enzymes. Transcriptomic studies have elucidated complex networks of antioxidant genes that are coordinately regulated in response to H_2O_2, underscoring the importance of maintaining redox homeostasis in plant cells under stress conditions. It acts as a secondary messenger that conveys stress signals by modulating the activity of redox-sensitive proteins and initiating signaling cascades and influences several areas of plant physiology. Jiang et al. (2012) reported that H_2O_2 activates photosynthetic genes, implying that H_2O_2 regulates the metabolism of sugar and Calvin cycle via redox signaling, hence increasing efficacy of photosynthesis. Furthermore, seeds treated with a high concentration of H_2O_2 at 120 mM accelerated chlorophyll breakdown and reduced efficiency of photosynthesis in the seedlings (Zheng et al., 2018). Transcriptomics and proteomics studies collectively help in finding the role of H_2O_2 in maintaining plant health. Understanding the fundamentals of plant transcriptomics, illuminates how plants regulate gene expression and produce specific proteins involved to cope with oxidative stress and modulate physiological processes.

A study demonstrated that supplementing H_2O_2 had boosted content of chlorophyll, photosynthetic rate of photosynthesis, and buildup of CO_2 in *Vigna radiata* (Khan et al., 2015). H_2O_2 can activate other signaling molecules that regulate gene expression for stomatal opening, growth, and development. In wheat, foliar spraying with H_2O_2 affected opening of stomata, which was associated with development and size of stomata (Kumari & Verma, 2019). It has been demonstrated that a decrease in K+ concentration and an increase in K+ discharge at the guard cells' plasma membrane via K+ channels cause the stomata to close in response to exogenous H_2O_2 (An et al., 2016). This suggests that the effects of H_2O_2 is due to the signaling cascade activation mediated by protein kinases and phosphatases, whereas the effects of H_2O_2 may be brought on by changes in membrane stability at high levels. These enzymes phosphorylate or dephosphorylate target proteins in response to H_2O_2, thereby transmitting signals that regulate the expression of ions in plant cells.

H_2O_2 also acts as a stress signaling molecule, such as during drought stress, H_2O_2 levels rise, activating genes that respond to stress and benefit the plant cope with adverse conditions. A study attributed that the enhanced resistance to water stress seedlings of rice at 5mmol/L H_2O_2 might be as a result of H_2O_2-mediated reduction of the excessive ROS buildup imposed by the drought, potentially through boosting antioxidant enzyme activity. They advised farmers to use these hydrogen peroxide dosages as a practical way to improve the establishment of early rice seedlings in drought-prone areas (Sohag et al., 2020). According to a hypothesis proposed by another study, MeRAV5 physically interacts with MePOD and MeCAD15, encouraging their processes and affecting the buildup of lignin and H_2O_2, which are critical for cassava's resilience to drought stress. Additionally, they emphasized the molecular pathways via which MeRAV5 influences lignin buildup and H_2O_2 to enhance drought stress tolerance in cassava (Yan et al., 2021). The development and grain yield of water-stressed wheat plants were also found to be significantly enhanced by seed priming with hydrogen peroxide and SNP alone or in combination. This improvement was linked to an increase in the production of photosynthetic pigments, a rise in the activities of enzymes that inhibit oxidation, and a buildup of non-enzymatic antioxidants, which effectively decreased its peroxidation. Better cellular membrane function followed, and as a consequence, water-stressed wheat plants grew and produced more. (Habib et al., 2020).

In *A. compressus*, drought stress increases ROS generation, peroxidation of lipid osmolytes buildup, and activity of antioxidants. Moreover, the exogenous administration of GB and ABA increased the accumulation of osmolytes, the activity of antioxidants, and the activity of photosynthesis while lowering ROS as well as lipid peroxidation. Improving the mechanisms for drought resistance was most successfully achieved by using GB alone (Nawaz & Wang, 2020). A research revealed that the first ABA-induced apoplastic H_2O_2 burst is caused by SAPK8/9/10-mediated phosphorylation of OsRbohB Ser-140, which controls the generation of apoplastic H_2O_2 and activates OsDMI3 in signaling of ABA (Wang et al., 2023). These genes may encode for protective proteins, osmo-protectants production enzymes, or components of antioxidant defense systems. Study of the specific genes and proteins involved in proper functioning of H_2O_2 is most important and will provide the deeper insights in to the plant defense mechanism.

Defensive Compound

Genes involved in antioxidant defense pathways have their expression influenced by H_2O_2. Transcriptomic studies have revealed intricate networks of antioxidant genes that are coordinately regulated in response to H_2O_2, highlighting the complexity of plant antioxidant systems and their function in preserving the redox equilibrium of cells. When a pathogen attack is detected by plants, they frequently undergo an oxidative burst, characterized by the rapid and transient production of ROS (Figure 11.3). This oxidative burst triggers additional defense responses: it strengthens cell walls by crosslinking cell wall proteins which makes it harder for pathogens to penetrate; it directly inhibits pathogen growth due to its toxic effects; and production of pathogenesis-related (PR) proteins and secondary metabolites with antimicrobial properties. Sun et al. (2023) have verified that OsGLP3–7 positively modulates rice resistance to bacterial blight, panicle, and leaf blast, through transgenic techniques. *OsGLP3–7* promotes disease resistance by activating the H_2O_2, phytoalexin, and jasmonic acid (JA) metabolic pathways in rice. These hormones such as JA, salicylic acid and ethylene modulate stress response, plant growth and development. OsGLP3–7 is extensively expressed in growing roots and root tissues during the booting stage, according to GUS staining study.

Other than plant resistance H_2O_2 also have detrimental effect on microorganisms. The aerosolized H_2O_2 causes greater sensitivity in the Gram-positive bacteria. Reactive species either damage intracellular components or react with the envelope of cell. Gram-negative bacteria such as E. coli were inactivated by cell envelope damage-induced leakage, while Gram-positive bacteria were primarily inactivated by intracellular damage. Reactive species like H_2O_2 and ozone have antibacterial properties because they cause lipoproteins as well as phospholipids in the inner membranes of both Gram-negative as well as Gram-positive cells to peroxide (Song & Fan, 2020). Only 10.5% H_2O_2 and above was able to kill above 90% of the total microorganisms. H_2O_2 serves as a highly effective disinfectant, demonstrating efficacy against a broad spectrum of microorganisms. It is typically used in aqueous solutions with concentrations ranging from 3% to 9% (w/w) (Tao et al., 2024). H_2O_2 interacts with the outer most layer of the pathogens and react with it or cause intracellular damage

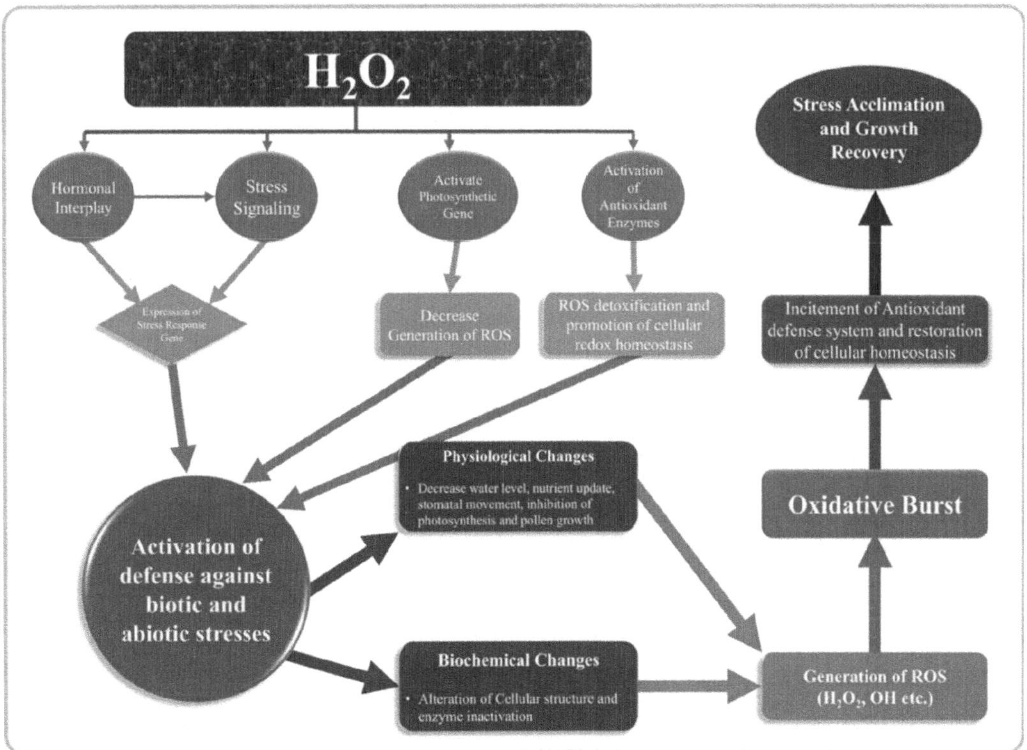

FIGURE 11.3 The process of activation of plant defense through H_2O_2 in case of any biotic or abiotic factors and the process of activation of oxidative burst, until plant recovery against stress.

by oxidative burst. This oxidative burst plays a vital part in protecting plants, by reinforcing cell walls, thereby inhibiting pathogen penetration and spread. Additionally, the synthesis of antibacterial substances and the expression of genes linked to defense are both stimulated by H_2O_2. The oxidative burst is an essential part of innate immune response of plant, facilitating rapid deployment of defenses at the site of infection against biotic and abiotic infections.

Abiotic Defense Compound

In plants, H_2O_2 is produced in response to environmental stresses such as drought, high light intensity, pollutants, and temperature fluctuations. It acts as a secondary messenger that conveys stress signals by modulating the activity of redox-sensitive proteins and initiating signaling cascades. The perception of H_2O_2 involves a complex interplay of receptors, kinases, phosphatases, and transcription factors, which collectively regulate gene expression to promote adaptive responses. A recent study states that the chloroplast APXs function as 'antioxidant enzymes' and is compensated by ΔpH-dependent mechanisms, especially photosynthetic control, which prevents the plant with light stress. When photosynthesis is excessively reduced as a result the rate of production of H_2O_2 is considerably enhanced. Thus, chloroplasts produce extracellular NADPH oxidases to minimize H_2O_2 levels (Kameoka et al., 2021). TFs play a crucial role in controlling the expression of stress-responsive genes by binding to specific DNA sequences in the promoters of target genes. Upon exposure to H_2O_2, cysteine residues within TFs undergo oxidative post-translational modifications, altering their DNA-binding affinity and transcriptional activity (Zhou et al., 2023). For example, members of the AP2/ERF (ethylene-responsive factor) and WRKY TF families are known to possess redox-sensitive cysteine residues that allow them to integrate H_2O_2 signals and regulate the downstream target genes' expression in antioxidant defense, pathogen resistance, as well as other responses due to stress (Wu et al., 2022).

Apart from its involvement in biotic stress responses, H_2O_2 is integral to plant adaptation to abiotic stressors include heavy metals, salt, and extremely high or low temperatures. Under these conditions, H_2O_2 functions as a signal that initiates the activation of antioxidant defenses and other protective mechanisms. For instance, in response to salt stress, H_2O_2 can induce expression of genes encoding for ion transporters, osmoprotectants, and detoxifying enzymes, helping the plant maintain cellular homeostasis and reduce damage from excess salts. Together with other essential signaling elements like Ca^{2+}, TFs, NO, kinases as well as other metabolites, ROS and phytohormones plan the physiological, molecular, and metabolic that control acclimation responses to high- or low-temperature extremities' (Devireddy et al., 2021). PtrERF109 from *Poncirus trifoliata*, known for enhancing cold tolerance, acts as a positive regulator. Its overexpression induces widespread changes in the global transcriptome, reprogramming numerous genes involved in antioxidant mechanisms and metabolic pathways (Wang et al., 2019).

Role in Growth and Development

Signaling from in plants also intersects with developmental processes, hormone signaling pathways, and metabolic regulation. By comparing transcriptomic profiles of plants exposed to H_2O_2 with those of control plants, researchers can pinpoint specific genes and regulatory elements involved in H_2O_2 perception, signal transduction, and stress responses. The interplay between H_2O_2 production and scavenging is crucial for maintaining cellular homeostasis and ensuring that H_2O_2 fulfill its signaling functions without causing undue harm. This balance is dynamically regulated and context-dependent, varying with developmental stages, tissue types, and environmental conditions. For example, during seed germination, H_2O_2 plays a part in breaking dormancy of seed as well as promoting growth, while in mature leaves, it may be involved in regulation of closing of stomata as a reaction to the drought stress. Seed dormancy and germination are important developmental processes that are regulated by several signaling pathways. Plant hormones like gibberellin (GA) and ABA are pivotal in controlling seed dormancy induction, maintenance, as well as release and seed germination. Additionally, signal compounds such as ROS, including H_2O_2, also influence these developmental processes by adjusting the balance between ABA and GA. However, excessive ROS production can cause oxidative stress, which adversely impacts growth and developmental processes of plant (Koramutla et al., 2021).

H_2O_2 function in growth and development often involves the activation of MAPK cascades, which serve as central hubs in stress-responsive signaling pathways. Target proteins are phosphorylated by serine/threonine protein kinases, or MAPKs, in response to extracellular stimuli, including H_2O_2. A study conducted on the gene silencing of H_2O_2 producing genes concluded that following the silencing of WHITEFLY INDUCED 1 (WFI1) and tomato RESPIRATORY BURST OXIDASE HOMOLOG 1 (RBOH1), more bud expansion, a decrease in the FZY's transcript, a restricting gene in IAA production, and IAA buildup in the apex (Chen et al., 2016). These effects were fully abolished by the application of exogenous H_2O_2. In a recent study, scientists suggested that the Lepidium seeds germination, which necessitates the rupture of the endosperm, involves the breakdown of cell wall polymers by ROS. They demonstrated that hydrogen peroxide counteracted the abscisic acid's inhibiting impact on endosperm rupture, highlighting the interaction between these two compounds (El-Maarouf-Bouteau & Bailly, 2008). Research on transcriptomes showed that H_2O_2 controlled seed germination and dormancy by upregulating GA production and downregulating ABA catabolism during seed imbibition (Liu et al., 2010). In addition to stimulating GA production, H_2O_2 causes the up-regulation of ABA catabolism. Elevated ABA levels prevent GA production; nonetheless, an equilibrium between two of these hormones can regulate the germination as well as dormancy of Arabidopsis seeds. By manipulating H_2O_2-associated pathways, we can develop crops that are more resilient to environmental stresses, ultimately contributing to food security and sustainable agricultural practices.

Hydrogen Peroxide as a Signaling Molecule as Transportation of Nutrients

Hydrogen peroxide (H_2O_2) serves as a critical signaling molecule in plants, playing a multifaceted role in regulating various physiological processes, including the transportation of nutrients. For instance, hydrogen peroxide (H_2O_2) can promote the development of root hairs, which are extensions of root epidermal cells that significantly increase the surface area of roots for nutrient absorption. Certain transporter proteins found in root cell plasma membranes facilitate nutrient uptake in plants. H_2O_2 can modulate the activity and expression of these transporters through redox signaling pathways. For example, studies have shown that H_2O_2 can regulate the expression of nitrate transporters, thereby influencing nitrate uptake and assimilation. Based on root oxidation ability and phenyl-propanoid metabolic processes that are H_2O_2-dependent, the study found that nitrate uptake-induced rhizosphere alkalization caused an Iron (Fe) shortage in rice (Chen et al., 2018). By influencing transporter activity, H_2O_2 ensures that plants can efficiently absorb and utilize nutrients under varying environmental conditions.

In addition to its role in roots, H_2O_2 is involved in the long-distance transport of nutrients through the vascular system, which includes xylem and phloem. H_2O_2 can affect the functionality of these vascular tissues, particularly during stress conditions. For instance, under drought stress, H_2O_2 is produced in root cells and can act as a signal to induce the expression of aquaporins— Water channel proteins that make it easier for water to pass through cell membranes (Shivaraj et al., 2021). This helps maintain water and nutrient flow even under adverse conditions. The ability of plants to balance H_2O_2 production and scavenging is essential for harnessing its signaling capabilities while preventing oxidative stress, underscoring the complex yet vital role of H_2O_2 in plant nutrient management.

Role of H_2O_2 in Metabolism

This H_2O_2 plays a role in regulating lipid metabolism and energy homeostasis. During photorespiration, which takes place when ribulose-1,5-bisphosphate carboxylase/ oxygenase (RuBisCO), an enzyme which fixes O_2 rather than CO_2, glycolate is produced and oxidized to glyoxylate in the peroxisomes, generating H_2O_2 (Busch, 2020). This process is crucial for recycling carbon and nitrogen, and the H_2O_2 produced can signal adjustments in photo-respiratory activity, impacting overall carbon metabolism and plant growth. In addition to its involvement in primary metabolism, H_2O_2 influences secondary metabolism, which includes the production of various secondary metabolites like flavonoids, alkaloids, and terpenoids (Fan et al., 2024). These metabolites provide crucial role in defense of plants, pigmentation, as well as signaling. H_2O_2 can modulate the activity of key enzymes in the biosynthetic pathways of these secondary metabolites. For instance, a study reported that in case of salinity stress, all the other observed parameters remains the same only the amount of H_2O_2 production increases along with the production of the enzyme phenylalanine ammonia-lyase (PAL), which is essential to the phenylpropanoid route, leading to the production of flavonoids and other phenolic compounds (Mrázová et al., 2017). These compounds not only protect the plant against UV radiation and pathogens but also contribute to plant structural integrity and communication.

Moreover, H_2O_2 plays a crucial part in the control of hormonal metabolism. Phytohormones, or plant hormones, such as JA, ethylene, gibberellins, abscisic acid (ABA), auxins, gibberellins, are essential regulators of plant growth and development (Chen et al., 2016; Ishibashi et al., 2012; Mhamdi et al., 2010). H_2O_2 can influence the biosynthesis, signaling, and action of these hormones. For instance, during abiotic stress conditions like drought, H_2O_2 can enhance ABA synthesis and signaling (Liu et al., 2010), which leads to stomatal closure to reduce water loss and induce stress-responsive gene expression. Similarly, H_2O_2 can interact with JA signaling to modulate defense responses against herbivores and pathogens (Sun et al., 2023). Its role as a signaling molecule is central to coordinating complex physiological processes, making it a key player in plant biology. By influencing these diverse metabolic processes, H_2O_2 helps plants maintain homeostasis, adapt to environmental changes, and optimize growth and development.

NEGATIVE IMPACT

Despite its beneficial roles, excessive H_2O_2 can be detrimental, causing oxidative harm to the constituents of cells such as nucleic acids, lipids as well as proteins. In order to reduce this potential risk, plants have evolved sophisticated antioxidant defense mechanisms that tightly regulate H_2O_2 levels. These systems include enzymatic antioxidants like catalase, which directly degrades H_2O_2 into water and oxygen, and peroxidases, which utilize H_2O_2 to oxidize various substrates. Non-enzymatic antioxidants, such as ascorbate (vitamin C) and glutathione, are also essential for scavenging ROS and maintaining redox balance within cells. Research indicates that the buildup of ROS can weaken organisms by causing oxidative damage. This damage impacts mitochondrial efficiency, leading to an increase in ROS production. The effect of oxidative damage and its influence on aging vary depending on the type of tissue affected (Firmansyah & Argosubekti, 2020). Understanding the intricate roles of H_2O_2 in plant processes is essential for advancing agricultural practices and improving crop resilience and productivity.

TECHNOLOGICAL ADVANCES AND FUTURE DIRECTIONS

Transcriptomics and proteomics approaches are transforming agricultural research by offering detailed insights into how plants respond to environmental stresses, including the production of hydrogen peroxide (H_2O_2). These omics technologies provide comprehensive snapshots of gene expression patterns and protein dynamics, enabling scientists to unravel the molecular mechanisms underlying plant responses to oxidative stress and to explore their practical applications in agriculture. By profiling gene expression changes and protein modifications in response to H_2O_2, researchers can identify key genes and proteins involved in stress tolerance mechanisms. By elucidating these pathways, researchers can manipulate gene expression through genetic engineering or molecular breeding techniques to enhance antioxidant defenses or optimize stress response mechanisms.

Proteomic analyses complement transcriptomics by providing insights into post-translational modifications and protein interactions that are crucial for adapting to oxidative stress conditions. This integrative approach not only enhances our fundamental understanding of plant biology but also translates into practical strategies for sustainable agriculture.

Future directions in transcriptomics and proteomics with respect to H_2O_2 production in plant systems focus on advancing technological platforms, expanding multi-omics integrations, and applying findings in agricultural contexts. Technological advancements and further studies in high-throughput sequencing and mass spectrometry continue to improve sensitivity, accuracy, and throughput, allowing for more comprehensive profiling of the plant transcriptome and proteome under oxidative stress conditions. These advancements facilitate the discovery of novel biomarkers, regulatory networks, and candidate genes for targeted crop improvement strategies. Integrating transcriptomic and proteomic data with metabolomics and phenomics datasets will enable a holistic understanding of plant responses to oxidative stress. Moreover, applying omics technologies in precision agriculture holds promise for optimizing resource use efficiency and sustainable crop production. By monitoring gene expression and protein dynamics in real-time, farmers are able to make knowledgeable irrigation decisions, nutrient application, as well as pest management strategies tailored to specific environmental conditions and crop growth stages. This personalized approach minimizes input costs, reduces environmental impact, and maximizes yield potential in a changing climate.

CONCLUSION

The study on the hydrogen peroxide's function in the biology of plant as well as its crosstalk with signaling networks provides valuable insights into the multifaceted functions of this molecule in plant growth and development. The study emphasizes the role that hydrogen peroxide plays as a signaling molecule in the regulation of many physiological processes in plants, which includes stress responses, hormonal pathways, and metabolic regulation. Understanding the molecular basis and studying the proteins involved in these processes provide powerful tools to obtain better knowledge of H_2O_2 production and its impact on plant systems in agriculture. These technologies offer unprecedented insights into reactions to oxidative stress, facilitating development of those crop varieties that are resistant to stress and sustainable agricultural practices. By leveraging omics data, researchers can accelerate breeding efforts, optimize crop management strategies, and contribute to global food security challenges posed by climate change and environmental variability. Continued advancements in omics technologies and their integration with agronomic practices will drive innovation in agricultural research and enable transformative solutions to meet future agricultural demands.

REFERENCES

An, Y., Feng, X., Liu, L., Xiong, L., & Wang, L. (2016). ALA-induced flavonols accumulation in guard cells is involved in scavenging H_2O_2 and inhibiting stomatal closure in *arabidopsis* cotyledons. *Frontiers in Plant Science*, *7*, 1713.

Arnaud, D., Deeks, M. J., & Smirnoff, N. (2021). Differences between apoplastic and cytosolic reactive oxygen species production in Arabidopsis during pattern-triggered immunity. *BioRxiv*. https://doi.org/10.1101/2021.04.20.440614

Arnaud, D., Deeks, M. J., & Smirnoff, N. (2023). RBOHF activates stomatal immunity by modulating both reactive oxygen species and apoplastic pH dynamics in Arabidopsis. *Plant Journal*, *116*(2), 404–415.

Bhardwaj, R. D., Singh, N., Sharma, A., Joshi, R., & Srivastava, P. (2021). Hydrogen peroxide regulates antioxidant responses and redox related proteins in drought stressed wheat seedlings. *Physiology and Molecular Biology of Plants*, *27*(1), 151–163.

Bopitiya, D., Christensen, D., Martin, M., Zhang, J., & Bennett, L. E. (2021). Production of hydrogen peroxide in formulated beverages is associated with the presence of ascorbic acid combined with selected redox-active functional ingredients. *Food Chemistry*, *338*, 127947.

Brand, M. D. (2020). Riding the tiger – physiological and pathological effects of superoxide and hydrogen peroxide generated in the mitochondrial matrix. *Critical Reviews in Biochemistry and Molecular Biology*, *55*(6), 592–661.

Breeze, E., & Mullineaux, P. M. (2022). The passage of H_2O_2 from chloroplasts to their associated nucleus during retrograde signalling: reflections on the role of the nuclear envelope. *Plants*, *11*(4), 552. https://doi.org/10.3390/plants11040552

Busch, F. A. (2020). Photorespiration in the context of Rubisco biochemistry, CO_2 diffusion and metabolism. *The Plant Journal*, *101*(4), 919–939. https://doi.org/10.1111/tpj.14674

Černý, M., Habánová, H., Berka, M., Luklová, M., & Brzobohatý, B. (2018). Hydrogen peroxide: Its role in plant biology and crosstalk with signalling networks. *International Journal of Molecular Sciences*, *19*(9), 2812. https://doi.org/10.3390/ijms19092812.

Chen, H., Zhang, Q., Cai, H., Zhou, W., & Xu, F. (2018). H_2O_2 mediates nitrate-induced iron chlorosis by regulating iron homeostasis in rice. *Plant, Cell & Environment*, *41*(4), 767–781. https://doi.org/10.1111/pce.13145

Chen, X., Xia, X., Guo, X., Zhou, Y., Shi, K., Zhou, J., & Yu, J. (2016). Apoplastic H_2O_2 plays a critical role in axillary bud outgrowth by altering auxin and cytokinin homeostasis in tomato plants. *New Phytologist*, *211*(4), 1266–1278. https://doi.org/10.1111/nph.14015

Claeys, H., Van Landeghem, S., Dubois, M., Maleux, K., & Inzé, D. (2014). What is stress? Dose-response effects in commonly used in vitro stress assays. *Plant Physiology*, *165*(2), 519–527. https://doi.org/10.1104/pp.113.234641

Corpas, F. J., González-Gordo, S., & Palma, J. M. (2020). Plant peroxisomes: A factory of reactive species. *Frontiers in Plant Science*, *11*. https://doi.org/10.3389/fpls.2020.00853

Devireddy, A. R., Tschaplinski, T. J., Tuskan, G. A., Muchero, W., & Chen, J.-G. (2021). Role of reactive oxygen species and hormones in plant responses to temperature changes. *International Journal of Molecular Sciences*, *22*(16), 8843. https://doi.org/10.3390/ijms22168843

dos Santos Araújo, G., de Oliveira Paula-Marinho, S., de Paiva Pinheiro, S. K., de Castro Miguel, E., de Sousa Lopes, L., Camelo Marques, E., de Carvalho, H. H., & Gomes-Filho, E. (2021). H_2O_2 priming promotes salt tolerance in maize by protecting chloroplasts ultrastructure and primary metabolites modulation. *Plant Science*, *303*, 110774.

El-Maarouf-Bouteau, H., & Bailly, C. (2008). Oxidative signaling in seed germination and dormancy. *Plant Signaling and Behavior*, *3*(3), 175–182. https://doi.org/10.4161/psb.3.3.5539

Fan, J., Hu, H., Lu, Y., Tao, L., Shen, X., Sun, C., & Shen, Y. (2024). Mechanism by which exogenous H_2O_2 improves main secondary metabolites contents in post-harvest fresh Ginkgo leaves through induced physiological responses mimicking stress. *Industrial Crops and Products*, *218*, 118908. https://doi.org/10.1016/j.indcrop.2024.118908

Firmansyah, & Argosubekti, N. (2020). A review of heat stress signaling in plants. *IOP Conference Series: Earth and Environmental Science*, *484*(1), 012041. https://doi.org/10.1088/1755-1315/484/1/012041

Fransen, M., & Lismont, C. (2024). Peroxisomal hydrogen peroxide signaling: A new chapter in intracellular communication research. *Current Opinion in Chemical Biology*, *78*, 102426. https://doi.org/10.1016/j.cbpa.2024.102426

Fraudentali, I., Pedalino, C., D'Incà, R., Tavladoraki, P., Angelini, R., & Cona, A. (2023). Distinct role of AtCuAOβ- and RBOHD-driven H_2O_2 production in wound-induced local and systemic leaf-to-leaf and root-to-leaf stomatal closure. *Frontiers in Plant Science*, *14*. https://doi.org/10.3389/fpls.2023.1154431

Furuhashi, M. (2020). New insights into purine metabolism in metabolic diseases: role of xanthine oxidoreductase activity. *American Journal of Physiology-Endocrinology and Metabolism*, *319*(5), E827–E834. https://doi.org/10.1152/ajpendo.00378.2020

Gupta, K., Sengupta, A., Chakraborty, M., & Gupta, B. (2016). Hydrogen peroxide and polyamines act as double edged swords in plant abiotic stress responses. *Frontiers in Plant Science*, *7*. https://doi.org/10.3389/fpls.2016.01343

Habib, N., Ali, Q., Ali, S., Javed, M. T., Zulqurnain Haider, M., Perveen, R., Shahid, M. R., Rizwan, M., Abdel-Daim, M. M., Elkelish, A., & Bin-Jumah, M. (2020). Use of nitric oxide and hydrogen peroxide for better yield of wheat (*Triticum aestivum* L.) under water deficit conditions: Growth, osmoregulation, and antioxidative defense mechanism. *Plants*, *9*(2), 285. https://doi.org/10.3390/plants9020285

Huchzermeyer, B., Menghani, E., Khardia, P., & Shilu, A. (2022). Metabolic pathway of natural antioxidants, antioxidant enzymes and ROS providence. *Antioxidants*, *11*(4), 761. https://doi.org/10.3390/antiox11040761

Imadi, S. R., Kazi, A. G., Ahanger, M. A., Gucel, S., & Ahmad, P. (2015). Plant transcriptomics and responses to environmental stress: An overview. *Journal of Genetics*, *94*(3), 525–537. https://doi.org/10.1007/s12041-015-0545-6

Ishibashi, Y., Tawaratsumida, T., Kondo, K., Kasa, S., Sakamoto, M., Aoki, N., Zheng, S.-H., Yuasa, T., & Iwaya-Inoue, M. (2012). Reactive oxygen species are involved in gibberellin/abscisic acid signaling in barley aleurone cells. *Plant Physiology*, *158*(4), 1705–1714. https://doi.org/10.1104/pp.111.192740

Ivanov, B., Borisova-Mubarakshina, M., Vilyanen, D., Vetoshkina, D., & Kozuleva, M. (2022). Cooperative pathway of O_2 reduction to H_2O_2 in chloroplast thylakoid membrane: New insight into the Mehler reaction. *Biophysical Reviews*, *14*(4), 857–869. https://doi.org/10.1007/s12551-022-00980-4

Jamshidi Goharrizi, K., Karami, S., & Ghanaei, S. (2024). Hydrogen peroxide priming promotes salinity tolerance in plants—A comprehensive review. *Agronomy Journal*, *116*(2), 612–629. https://doi.org/10.1002/agj2.21534

Jiang, Y., Cheng, F., Zhou, Y., Xia, X., Mao, W., Shi, K., Chen, Z., & Yu, J. (2012). Hydrogen peroxide functions as a secondary messenger for brassinosteroids-induced CO_2 assimilation and carbohydrate metabolism in Cucumis sativus. *Journal of Zhejiang University Science B*, *13*(10), 811–823. https://doi.org/10.1631/jzus.B1200130

Kameoka, T., Okayasu, T., Kikuraku, K., Ogawa, T., Sawa, Y., Yamamoto, H., Ishikawa, T., & Maruta, T. (2021). Cooperation of chloroplast ascorbate peroxidases and proton gradient regulation 5 is critical for protecting Arabidopsis plants from photo-oxidative stress. *The Plant Journal*, *107*(3), 876–892. https://doi.org/10.1111/tpj.15352

Khan, T. A., Yusuf, M., & Fariduddin, Q. (2015). Seed treatment with H_2O_2 modifies net photosynthetic rate and antioxidant system in mung bean (*Vigna radiata* L. Wilczek) plants. *Israel Journal of Plant Sciences*, *62*(3), 167–175. https://doi.org/10.1080/07929978.2015.1060806

Khan, T. A., Qazi Fariduddin, Q. F., & Mohammad Yusuf, M. Y. (2018). Effect of exogenously sourced hydrogen peroxide treatments on growth, photosynthesis and antioxidant traits in two contrasting cultivars of tomato: A mode and concentration dependent study. https://doi.org/10.15413/ajar.2018.1201

Koramutla, M. K., Negi, M., & Ayele, B. T. (2021). Roles of glutathione in mediating abscisic acid signaling and its regulation of seed dormancy and drought tolerance. *Genes*, *12*(10), 1620. https://doi.org/10.3390/genes12101620

Kumari, S., & Verma, V. K. (2019). Hydrogen peroxide mediates suberization, root thickness and stomatal movement in wheats. *International Journal of Current Microbiology and Applied Sciences*, *8*(04), 2448–2454.

Li, H., Lismont, C., Costa, C. F., Hussein, M. A. F., Baes, M., & Fransen, M. (2023). Enhanced levels of peroxisome-derived H_2O_2 do not induce pexophagy but impair autophagic flux in HEK-293 and HeLa cells. *Antioxidants*, *12*(3), 613. https://doi.org/10.3390/antiox12030613

Li, L.-Q., Lyu, C.-C., Li, J.-H., Wan, C.-Y., Liu, L., Xie, M.-Q., Zuo, R.-J., Ni, S., Liu, F., Zeng, F.-C., Lu, Y.-F., Yu, L.-P., Huang, X.-L., Wang, X.-Y., & Lu, L.-M. (2020). Quantitative proteomic analysis of alligator weed leaves reveals that cationic peroxidase 1 plays vital roles in the potassium deficiency stress response. *International Journal of Molecular Sciences*, *21*(7), 2537. https://doi.org/10.3390/ijms21072537

Lismont, C., Revenco, I., Li, H., Costa, C. F., Lenaerts, L., Hussein, M. A. F., De Bie, J., Knoops, B., Van Veldhoven, P. P., Derua, R., & Fransen, M. (2022). Peroxisome-derived hydrogen peroxide modulates the sulfenylation profiles of key redox signaling proteins in Flp-In T-REx 293 cells. *Frontiers in Cell and Developmental Biology*, *10*. https://doi.org/10.3389/fcell.2022.888873

Liu, Y., Ye, N., Liu, R., Chen, M., & Zhang, J. (2010). H_2O_2 mediates the regulation of ABA catabolism and GA biosynthesis in Arabidopsis seed dormancy and germination. *Journal of Experimental Botany*, *61*(11), 2979–2990. https://doi.org/10.1093/jxb/erq125

Manimekalai, R., Selvi, A., Narayanan, J., Vannish, R., Shalini, R., Gayathri, S., & Rabisha, V. (2023). Comparative physiological and transcriptome analysis in cultivated and wild sugarcane species in response to hydrogen peroxide-induced oxidative stress. *BMC Genomics*, *24*(1), 155. https://doi.org/10.1186/s12864-023-09218-3

Mariyam, S., Bhardwaj, R., Khan, N. A., Sahi, S. V., & Seth, C. S. (2023). Review on nitric oxide at the forefront of rapid systemic signaling in mitigation of salinity stress in plants: Crosstalk with calcium and hydrogen peroxide. *Plant Science*, *336*, 111835. https://doi.org/10.1016/j.plantsci.2023.111835

Mhamdi, A., Hager, J., Chaouch, S., Queval, G., Han, Y., Taconnat, L., Saindrenan, P., Gouia, H., Issakidis-Bourguet, E., Renou, J.-P., & Noctor, G. (2010). Arabidopsis GLUTATHIONE REDUCTASE1 plays a crucial role in leaf responses to intracellular hydrogen peroxide and in ensuring appropriate gene expression through both salicylic acid and jasmonic acid signaling pathways. *Plant Physiology*, *153*(3), 1144–1160. https://doi.org/10.1104/pp.110.153767

Mrázová, A., Belay, S. A., Eliášová, A., Perez-Delgado, C., Kaducová, M., Betti, M., Vega, J.M., & Paľove-Balang, P. (2017). Expression, activity of phenylalanine-ammonia-lyase and accumulation of phenolic compounds in Lotus japonicus under salt stress. *Biologia*, *72*(1), 36–42. https://doi.org/10.1515/biolog-2017-0001

Mubarakshina, M. M., Ivanov, B. N., Naydov, I. A., Hillier, W., Badger, M. R., & KriegerLiszkay, A. (2010). Production and diffusion of chloroplastic H_2O_2 and its implication to signalling. *Journal of Experimental Botany*, *61*(13), 3577–3587. https://doi.org/10.1093/jxb/erq171

Nawaz, M., & Wang, Z. (2020). Abscisic acid and glycine betaine mediated tolerance mechanisms under drought stress and recovery in axonopus compressus: A new insight. *Scientific Reports*, *10*(1), 6942. https://doi.org/10.1038/s41598-020-63447-0

Nazir, F., Hussain, A., & Fariduddin, Q. (2019). Hydrogen peroxide modulate photosynthesis and antioxidant systems in tomato (Solanum lycopersicum L.) plants under copper stress. *Chemosphere*, *230*, 544–558. https://doi.org/10.1016/j.chemosphere.2019.05.001

Ofoedu, C. E., You, L., Osuji, C. M., Iwouno, J. O., Kabuo, N. O., Ojukwu, M., Agunwah, I. M., Chacha, J. S., Muobike, O. P., Agunbiade, A. O., Sardo, G., Bono, G., Okpala, C. O. R., & Korzeniowska, M. (2021). Hydrogen peroxide effects on natural-sourced polysacchrides: Free radical formation/production, degradation process, and reaction mechanism—A critical synopsis. *Foods*, *10*(4), 699. https://doi.org/10.3390/foods10040699

Qureshi, M. K., Gawroński, P., Munir, S., Jindal, S., & Kerchev, P. (2022). Hydrogen peroxideinduced stress acclimation in plants. *Cellular and Molecular Life Sciences*, *79*(2), 129. https://doi.org/10.1007/s00018-022-04156-x

Rahman, M. A., Alam, I., Sharmin, S. A., Kabir, A. H., Kim, Y.-G., Liu, G., & Lee, B.-H. (2021). Physiological and proteomic analyses reveal the protective roles of exogenous hydrogen peroxide in alleviating drought stress in soybean plants. *Plant Biotechnology Reports*, *15*(6), 805–818. https://doi.org/10.1007/s11816-021-00719-9

Rodrigues, O., & Shan, L. (2022). Stomata in a state of emergency: H_2O_2 is the target locked. *Trends in Plant Science*, *27*(3), 274–286. https://doi.org/10.1016/j.tplants.2021.10.002

Sachdev, S., Ansari, S. A., Ansari, M. I., Fujita, M., & Hasanuzzaman, M. (2021). Abiotic stress and reactive oxygen species: Generation, signaling, and defense mechanisms. *Antioxidants*, *10*(2), 277. https://doi.org/10.3390/antiox10020277

Saxena, I., Srikanth, S., & Chen, Z. (2016). Cross talk between H_2O_2 and interacting signal molecules under plant stress response. *Frontiers in Plant Science*, *7*. https://doi.org/10.3389/fpls.2016.00570

Sewelam, N., Jaspert, N., Van Der Kelen, K., Tognetti, V. B., Schmitz, J., Frerigmann, H., Stahl, E., Zeier, J., Van Breusegem, F., & Maurino, V. G. (2014). Spatial H_2O_2 signaling specificity: H_2O_2 from chloroplasts and peroxisomes modulates the plant transcriptome differentially. *Molecular Plant*, *7*(7), 1191–1210. https://doi.org/10.1093/mp/ssu070

Shivaraj, S. M., Sharma, Y., Chaudhary, J., Rajora, N., Sharma, S., Thakral, V., Ram, H., Sonah, H., Singla-Pareek, S. L., Sharma, T. R., & Deshmukh, R. (2021). Dynamic role of aquaporin transport system under drought stress in plants. *Environmental and Experimental Botany*, *184*, 104367. https://doi.org/10.1016/j.envexpbot.2020.104367

Sies, H., Belousov, V. V., Chandel, N. S., Davies, M. J., Jones, D. P., Mann, G. E., Murphy, M. P., Yamamoto, M., & Winterbourn, C. (2022). Defining roles of specific reactive oxygen species (ROS) in cell biology and physiology. *Nature Reviews Molecular Cell Biology*, *23*(7), 499–515. https://doi.org/10.1038/s41580-022-00456-z

Sohag, A. A. M., Tahjib-Ul-Arif, M., Brestič, M., Afrin, S., Sakil, M. A., Hossain, M. T., Hossain, M. A., & Hossain, M. A. (2020). Exogenous salicylic acid and hydrogen peroxide attenuate drought stress in rice. *Plant, Soil and Environment*, *66*(1), 7–13. https://doi.org/10.17221/472/2019-PSE

Song, X., Wang, T., Zhang, Y., Yu, J.-Q., & Xia, X.-J. (2022). S-nitrosoglutathione reductase contributes to thermotolerance by modulating high temperature-induced apoplastic H_2O_2 in solanum lycopersicum. *Frontiers in Plant Science*, *13*. https://doi.org/10.3389/fpls.2022.862649

Song, Y., & Fan, X. (2020). Cold plasma enhances the efficacy of aerosolized hydrogen peroxide in reducing populations of Salmonella Typhimurium and Listeria innocua on grape tomatoes, apples, cantaloupe and romaine lettuce. *Food Microbiology*, *87*, 103391. https://doi.org/10.1016/j.fm.2019.103391

Sun, B., Li, W., Ma, Y., Yu, T., Huang, W., Ding, J., Yu, H., Jiang, L., Zhang, J., Lv, S., Yang, J., Yan, S., Liu, B., & Liu, Q. (2023). OsGLP3-7 positively regulates rice immune response by activating hydrogen peroxide, jasmonic acid, and phytoalexin metabolic pathways. *Molecular Plant Pathology*, *24*(3), 248–261. https://doi.org/10.1111/mpp.13294

Tao, C., Tang, X., Gan, Y., Qin, Y., Yang, S., & Huang, F. (2024). Investigation of the disinfection efficiency of commercial hydrogen peroxide, chlorine dioxide, and chlorine disinfectant on different surfaces. *American Journal of Veterinary Research*, *85*, 1–6. https://doi.org/10.2460/ajvr.24.03.0079

Terrón-Camero, L. C., Peláez-Vico, M. Á., Rodríguez-González, A., del Val, C., Sandalio, L. M., & Romero-Puertas, M. C. (2022). Gene network downstream plant stress response modulated by peroxisomal H_2O_2. *Frontiers in Plant Science*, *13*. https://doi.org/10.3389/fpls.2022.930721

Ugalde, J. M., Fuchs, P., Nietzel, T., Cutolo, E. A., Homagk, M., Vothknecht, U. C., Holuigue, L., Schwarzländer, M., Müller-Schüssele, S. J., & Meyer, A. J. (2021). Chloroplast-derived photo-oxidative stress causes changes in H_2O_2 and E GSH in other subcellular compartments. *Plant Physiology*, *186*(1), 125–141. https://doi.org/10.1093/plphys/kiaa095

Voothuluru, P., Mäkelä, P., Zhu, J., Yamaguchi, M., Cho, I.-J., Oliver, M. J., Simmonds, J., & Sharp, R. E. (2020). Apoplastic hydrogen peroxide in the growth zone of the maize primary root. Increased levels differentially modulate root elongation under wellwatered and water-stressed conditions. *Frontiers in Plant Science*, *11*, 392.

Wang, M., Dai, W., Du, J., Ming, R., Dahro, B., & Liu, J. H. (2019). ERF109 of trifoliate orange (Poncirus trifoliata (L.) Raf.) contributes to cold tolerance by directly regulating expression of Prx1 involved in antioxidative process. *Plant Biotechnology Journal*, *17*(7), 1316–1332.

Wang, Q., Shen, T., Ni, L., Chen, C., Jiang, J., Cui, Z., Wang, S., Xu, F., Yan, R., & Jiang, M. (2023). Phosphorylation of OsRbohB by the protein kinase OsDMI3 promotes H_2O_2 production to potentiate ABA responses in rice. *Molecular Plant*, *16*(5), 882–902.

Wu, Y., Li, X., Zhang, J., Zhao, H., Tan, S., Xu, W., Pan, J., Yang, F., & Pi, E. (2022). ERF subfamily transcription factors and their function in plant responses to abiotic stresses. *Frontiers in Plant Science*, 13. https://doi.org/10.3389/fpls.2022.1042084

Xiong, J., Yang, Y., Fu, G., & Tao, L. (2015). Novel roles of hydrogen peroxide (H_2O_2) in regulating pectin synthesis and demethylesterification in the cell wall of rice (*Oryza sativa*) root tips. *New Phytologist*, *206*(1), 118–126. https://doi.org/10.1111/nph.13285

Yan, Y., Wang, P., Lu, Y., Bai, Y., Wei, Y., Liu, G., & Shi, H. (2021). MeRAV5 promotes drought stress resistance in cassava by modulating hydrogen peroxide and lignin accumulation. *The Plant Journal*, *107*(3), 847–860. https://doi.org/10.1111/tpj.15350

Yang, X., Lu, M., Wang, Y., Wang, Y., Liu, Z., & Chen, S. (2021). Response mechanism of plants to drought stress. *Horticulturae*, *7*(3), 50. https://doi.org/10.3390/horticulturae7030050

Zheng, Y., Yin, X., & Ma, H. (2018). Effects of hydrogen peroxide on seed germination, seedling growth and physiological characteristics of Bombax ceiba after heat shock. *Pakistan Journal of Botany*, *50*(4), 1327–1333.

Zhou, H., Huang, J., Willems, P., Van Breusegem, F., & Xie, Y. (2023). Cysteine thiol-based post-translational modification: What do we know about transcription factors? *Trends in Plant Science*, *28*(4), 415–428. https://doi.org/10.1016/j.tplants.2022.11.007

12 Hydrogen Peroxide's Function and Interactions with Other Phytohormones in Plants' Responses to Heavy Metal Stress

Bushra Ahmad, Ahamad Faiz Khan, Farina Mujeeb, Mohd Sajid Khan, Rafeeya Shams, and Vinay Kumar Pandey

INTRODUCTION

The fundamental area of study in the life sciences today is how living things interpret oxidative signals. H_2O_2 levels within plant cells rise as a direct result of environmental stressors like light exposure, drought conditions, and the presence of the plant hormone abscisic acid (ABA). This has become a crucial area of research in regards to how cells respond to outside stimuli. Understanding a plant's biochemical response to external stimuli is important for agriculture as it helps reveal the plant's basic biological characteristics. This is because the potential crop yield is limited by these stressors. Initially, H_2O_2 was considered a potentially harmful reactive oxygen species that could adversely affect various cell types, particularly since these stresses decrease crop productivity. Consequently, comprehending these mechanisms is imperative for agriculture. While H_2O_2 has long been recognised as an antioxidant, its role as a potent signalling molecule, actively participating in a wide range of physiological processes, has only become apparent in the past decade. It is now widely accepted that oxidative stress is induced or involved in nearly all challenges posed by biotic and abiotic factors. The ability of plants to effectively control oxygen levels is strongly associated with their capacity to endure high-stress thresholds. Whether the explanation for tolerance primarily relies on antioxidants, over time, remains to be determined. It is unclear whether this explanation will prove more effective than other factors, such as the plant's ability to store suitable osmotic agents for drought tolerance or its capability to prevent sodium buildup in shoots while maintaining optimal potassium to sodium ratios for salt tolerance. However, it is evident that a multitude of genes can be upregulated or downregulated in response to significant environmental changes caused by various stressors. Proteins related to oxidative metabolism, characterised by high stability and low turnover rates, play a crucial role in these processes. This complex web of responses extends to components like the cell wall, encompassing numerous metabolites and proteins, many of which may have implications for oxidative metabolism, despite being previously considered of minor biological significance.

It is generally known that oxidative metabolism, in particular H_2O_2, is involved in several reactionary processes and signalling mechanism those are essential for the integration of all processes involved in plant growth, from the formation of individual root hairs through the differentiation and lignification of the xylem, to root/shoot coordination, wall cross-linking, wall slackening, and stomatal regulation. Since H_2O_2 is involved in "normal" growth and metabolism as well as stress responses, it is important to take these activities into account when discussing the role of H_2O_2 in

DOI: 10.1201/9781032647692-12

stress responses. Overall, applying a stress abruptly after a period of unstressed growth when plants are being grown under controlled conditions is the current "fashion" in plant stress studies, and then compare one or more features of the reaction at a later set time, ranging from entire genome transcript networks to the activity of a single enzyme. Such environmental disturbance doesn't typically experience in the real world, fortunately for comparing some component or features of response, ranging from activity of a single enzyme to entire genome transcript networks, at a defined time later may reveal that plants and models based on them are, in the words of Manfred Eigen, "right, but irrelevant.". It is much more challenging to grasp the daily or minute-by-minute fluctuations that prevent the need for dramatic response and to design appropriate models that incorporate them. In this chapter, I will focus on how H_2O_2 functions physiologically in relation to plant reactions to stress, and to act in such a way that acknowledges "normal" actions. To achieve this, I will give more weight to overarching patterns than to individual details of responses, and I will give credit to as many of the "tools" plants have for dealing with H_2O_2 as I can, not just the ones with the most elaborate genetic models.

In addition to other soil contaminants, heavy metal contamination is a major global environmental issue that has captivated experts' attention who are working to ensure the long-term viability of ecological systems. Heavy metals (HMs) can't be broken down by living organisms, so they'll always be around in soil ecosystems. When HMs build up in soil, they have a negative effect on dynamism in the soil and on the structure and function of soil microbes (Emamverdian et al., 2015), causing a natural decline in crop yields due to diminished soil fertility (Asati et al., 2016). Crop failure can be attributed to HM stress since it disrupts plant functions like growth, photosynthetic effectiveness, and water potential (Choudhary et al., 2012a, b, Nazir et al., 2019a, b). Plant responses to HM stress are very nuanced and dependent on a wide range of factors, includes solute type and concentration, plant growth stage and genetic potential, and environmental effects, to name a few. A loss in crop profitability can result from the negative effects of HM stress on the enzymatic activity, photosynthetic rate, and membrane stability, plant development, distress of ion absorption, and especially ion dripping in leaves (Nazir et al., 2019a, b; Saxena et al., 2016; Stambulska et al., 2018). Minimising the harmful consequences of HM stress and increasing the plant's resilience to HM stress are accomplished through a variety of agricultural and physiological practices. One effective method of boosting crop production under stressful conditions is the use of plant growth regulators (PGRs) (Nazir et al., 2019a, b; Sondhi et al., 2008). It is also well known that under normal or stressful conditions, PGRs cause reactions involving biochemistry, metabolism, and/or cells in plants by intervening either close to or transporting from the site of their synthesis to other places. Hydrogen peroxide (H_2O_2) has gained attention in recent years as a signalling molecule among plant scientists. A chemical molecule discovered by Louis Jacques Thenard a century ago shows enough promise as a plant growth regulator to warrant further study (Černý et al., 2018). At high doses, it can cause cell death by causing an oxidative burst in organic molecules. The diagrammatic representation of heavy metal-indued generation of reactive oxygen species and role of H_2O_2 as a signalling molecule that triggers activation of the antioxidant system and thereby decreases the concentration of reaction oxygen species and protects plants from damage by heavy metal stress is shown in Figure 12.1.

However, H_2O_2 acts as a messenger molecule at nanomolar concentrations, and its function in regulating development and growth of plants has received more attention (Htet et al., 2019). The stress transducing effects of H_2O_2 can mitigate or ameliorate plant stress responses (Nazir al., 2019a, b). H_2O_2 has been found to mediate between a plant's stress defence system and its secondary metabolic processes (Ellouzi et al., 2017). Plant morphophysiological and biochemical responses may be significantly regulated by H_2O_2 via the proteome model (Khan et al., 2019). It is well-recognised that H_2O_2 plays a vital role in how plants respond to stress from things like HMs in the environment, low-temperature stress (Khan et al., 2019), drought, salt (Jahan et al., 2023). The main mechanism through which H_2O_2 is thought to play a significant role in plant stress reactions is as a transducer in the metabolically or environmentally regulated activation of different tolerance genes (Qi et al., 2018). Specifically, H_2O_2 is linked to other PGRs in stress resistance (Tian et al., 2018).

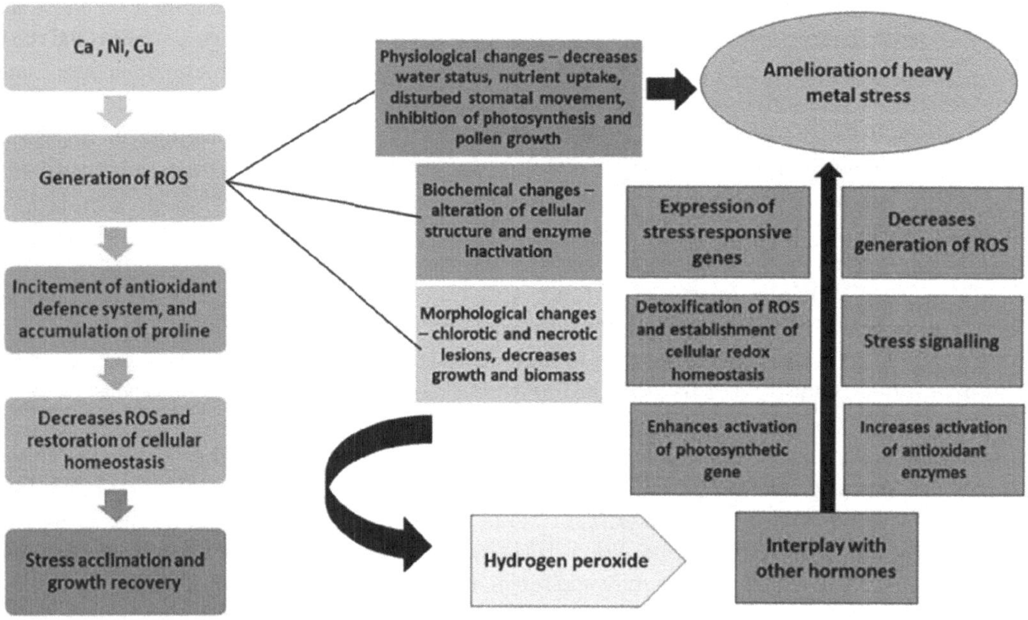

FIGURE 12.1 Representation of heavy metal-indued generation of reactive oxygen species and role of H_2O_2 to activate antioxidant system and decreases the reaction oxygen species concentration and protects plants from damage by heavy metal stress.

Several investigations showed that H_2O_2 promoted under HM stress, plant growth and development, however, the information on the mechanism by which H_2O_2 initiated augmentation of HM stress and its modulation by other important PGRs is scarce this article reviews the current research on the role and underlying processes of H_2O_2- altered the control of plant transcription factors (TFs) and mitogen-activated protein kinases (MAPKs) during morphophysiological growth and signalling under HM stress conditions, shows how H_2O_2 communicates with other PGRs, how it interacts with them in both positive and negative ways, and why this may be an important topic for future research. H_2O_2 is clearly a chemical that can have a dramatic effect on how normally functioning plant cells operate. Its function in plants has been extensively studied, although much remains to be discovered. However, in this chapter, we zero in on the current level of our understanding of how H_2O_2 works and how it interacts with other phytohormones.

WHAT IS H_2O_2?

O_2's two-electron reduction result is hydrogen peroxide. It is oxygen, which can be very reactive, but not a free radical. Unlike superoxide, $O_2^{\bullet}$, and H_2O_2, the hydroxyl radical $^{\bullet}OH$, is comparatively "safe"; it is stable and unreactive, even though quantities far greater compared to a biological system could ever produce, provided that no transition metals are present. The increased mobility throughout tissues and the possibility of its usage when used as a substrate in various processes and as a ROS-related signalling molecule are both practical benefits. However, H_2O_2 can be very reactive with Fe_2+- or other transition metal-containing compounds via Fenton reaction. H_2O_2 homolysis to 2 $^{\bullet}OH$ is the "evil" result of this reaction, and H_2O_2 toxicity is typically linked with this activity. When Rubisco is blocked by exogenous H_2O_2, for instance, the LSU (Large Subunit) is shattered at a glycine in the catalytic site, reliant on the enzyme's degree of activation. When oxidant scavenging systems were inhibited, fragmentation was also observed in otherwise whole chloroplasts. Chloroplast fragmentation was observed even in otherwise undamaged chloroplasts when oxidant scavenging mechanisms were blocked, because chloroplasts store as much as 80% of a plant's Fe and are an excellent source of reactive oxygen species (ROS). The Fenton reaction mechanism, analogous to that of catalase, has been related to FeSOD's sensitivity to H_2O_2 (or, technically, $^{\bullet}OH$), and glutamine synthase, among other enzymes. However, it has been theorised that the -SH groups on fructose bisphosphatase are what H_2O_2 reacts with, rendering the enzyme inactive. There are a number of techniques to reduce the toxicity of hydrogen peroxide, including enzymatic removal (throughthe enzyme ascorbate peroxidase or catalase) and complexing iron (III) and iron (II) with substances like tannic acid and proanthocyanins, which blocks the formation of $^{\bullet}OH$.

WHAT H_2O_2 LEVELS IN THE TISSUE ARE IMPORTANT?

The fundamental question is, what amount of H_2O_2 is present in the tissue of plant, and against what context might alterations serve as meaningful signals? H_2O_2 in the solution can be tested in a variety of ways, some of which are fairly specific, it is remarkable that there is no agreement on how great a change due to stress or treatment needs to be in order to be considered physiologically relevant, and estimates for plant tissues span approximately seven orders of magnitude. At the low end, Pea leaf apoplast levels were reported to be anything from 10 to 150 pmol/ gFW, with a difference (caused by salt) being enough to trigger lesions with oxidation. Concentrations to a maximum of 1.3% of the dry weight were found by, He et al. (2005) in the leaves of Poa pratensis, which, according to their data, is approx60 mol/gFW or 100 mM on the basis of leaf water. N deficit increases maize concentrations from 20 mol/gFW, as observed by Tewari et al. (2004), to 75 mol/ gFW. As per the study done by Ben Amor et al. (2006), *Cakile maritima*, a coastal halophyte, was the subject of an intriguing investigation due to its high tissue H_2O_2 values (up to 45 mol/gFW). Leaf levels must normally be lesser than 0.1 mol/gFW, and they were the first to notice potential interferences by plant ingredients with H_2O_2 measurement procedures. However, when looking at field-grown plants, ranges between 1 and 5 mol/gFW were deemed to be plausible after taking into consideration probable interferences and the continuing metabolism of H_2O_2 after harvesting (Cheeseman, 2006). What it means to have an "oxidative burst" in terms of H_2O_2 concentrations in tissues is also unclear. The rapid regeneration of H_2O_2 in plants occurs both under normal physiological conditions and in response to tissue damage (Cheeseman, 2006). H_2O_2 turnover is quite high in plants and after tissue harvesting found a four fold increase in H_2O_2 levels in the leaves of tobacco (cv. 'Xanthi') plants (in the range of 100 nmol/gFW, using plants grown in low ozone and irradiance environments), but grew four to five folds after 24 hours of recuperation, the leaf content of Arabidopsis grew from around 5 mol/gFW to less than 7 mol/ gFW when exposed to excess irradiance, which was ten times higher than the 200 mol m^2/s growth irradiance. It's not just a matter of scale or percentage shift here. However, the more important question is, given the practical limitations of measuring H_2O_2 in bulk tissue, what baseline conditions and modifications are required to support the role of H_2O_2 in signalling?

WHAT PRODUCES H_2O_2?

In an aerobic setting, H_2O_2 and other ROS are essential to survival (Moller, 2001). H_2O_2 can be generated and destroyed in a wide variety of ways, thus models that account for all of these processes are necessary for studying its function in plant development and stress responses, Moreover, ROS generated from one source may serve as substrates or drivers for a different source. Misfires in chloroplast and mitochondrial electron transport chains, the Mehler reaction, and numerous limited substrate oxidases all contribute significantly, to NAD(P)H oxidases and type III peroxidases. Others, however, can only generate H_2O_2 after passing through higher reactivity intermediates, such as 1 O or $O_2^{\bullet}$. Although these processes are natural and inherent to many stages of plant development, they are amplified by stressors.

IS H_2O_2 MOBILE AT ALL? WHICH WAY IS IT?

Nevertheless, it was once thought that intracellularly produced H_2O_2 reaches additional cells by diffusion where it can be used by POX and other defensive enzymes, recent evidence suggests that this is not the case. Extracellular metabolism appears to utilise H_2O_2 produced extracellularly, but it now seems more likely that quickly and locally, intracellularly generated H_2O_2 is used. Movements of hydrogen peroxide across membranes (such as from the apoplast to the cytosol) likely require regulated transport via aquaporins (Bienert et al., 2007). When two Arabidopsis aquaporins were expressed in yeast and H_2O_2 was applied to the cells, growth and survival were inhibited; blocking the channels restored normal cellular behaviour. Cucumber roots exposed to low temperatures had extracellular H_2O_2 accumulation in the millimolar range, which reduced hydraulic conductivity, and this impact interacted with other stresses, according to subsequent accounts of the results of using H_2O_2 (Ye and Steudle, 2006). In contrast, it is possible that aquaporin-mediated transport has a role in eliciting intracellular reactions at lower H_2O_2 concentrations. Toxic effects of extracellular oxidants and H_2O_2-dependent signalling may depend on this. Even though H_2O_2 diffusion that does not include membrane transit is limited, it is significantly long compared to the movements of other ROS, which are considerably more limited due to their higher reactivity. Root hairs can be a site of H_2O_2 localisation inside tissues (Carol and Dolan, 2006), injury, or stomatal opening and closing in epidermal cells, shows how well plants are able to regulate their own internal environments, and so does leaf tissue compartmentation, in vascular tissues, and in areas of regeneration (rhizogenesis). Studies comparing the effects of ozone suggest that resistance may depend on a plant's ability to keep hydrogen peroxide (H_2O_2) out of its cells and chloroplasts. Particularly well established is localisation regarding differentiation of the xylem and lignification during "normal" development and growth, H_2O_2 synthesis at the cell surface is poorly understood, in part because it varies from species to species and even within individual plants in response to varied environmental stressors. Using tissue printing methods on several plant species, Schopfer (1994) showed hypocotyl development and radial elongation gradients in response to light-mediated elongation inhibition, lignification, ethylene, and wounding, during auxin-induced rooting of grapevine cuttings, H_2O_2 localisation varies with time, as demonstrated. When Arabidopsis plants were deprived of potassium, ROS accumulated in their root areas involved in potassium (K) absorption and transport, this accumulation was inhibited, mutation of the atrbohC NADPH oxidase gene does not induce high-affinity transporters. Surprisingly, such transporters were triggered by H_2O_2 administration even in K-sufficient circumstances. When N and P were removed from mutants of the Arabidopsis root hair, the same authors looked at the effect on ROS levels (Shin et al., 2005). Increased ROS patterns revealed the importance of root hairs in K and N responses, while responses from P found only in the cortex. H_2O_2 generation can be precisely targeted to attachment sites in bacterial cell walls, which is true even in the context of bacterial infection.

SIGNALLING AND RESPONSE NETWORKS

In most cases, H_2O_2 generation is only the beginning of the narrative; in fact, it is almost always linked to additional reactions and is essential for signalling. While significant progress has been made in understanding the role of H_2O_2 in signaling and downstream processes, there is less clarity regarding the exact significance of the signal itself, how plants perceive specific threats, or how the decoded signal is processed to respond to specific metabolic demands or stress conditions. This is anowing in part to the difficulty of accurately modelling H_2O_2 concentrations in tissues, where estimates span a reported seven-orders-of-magnitude range. The space limitations at the cellular and tissue levels also play a role. It's not too dissimilar to the calcium scenario to think about the relationship between tissue H_2O_2 concentration and defence or stress. The leaf tissues have calcium concentrations that are orders of magnitude higher than the intracellular levels that can support normal signalling and metabolic regulation. It is fitting to wrap up this review by briefly discussing the results of recent promising advancements in integrative plant biology aimed at comprehending sensing, signalling, and response networks. Recognising how plants avoid problems caused by such complexity, however, continues to be an interesting yet difficult challenge because there are numerous parallels in the crosstalk between the response networks to various environmental stimuli, developmental stages, and appears to be certain.

SENSING H_2O_2

For the time being, let's assume that the "normal" range for H_2O_2 concentrations in tissues like leaves is between 0.1 and 5 mol/gFW (Cheeseman, 2006), that H_2O_2 is rapidly recycled, additionally, that ROS play crucial roles in development and physiology even when stress is absent, therefore, sense is obviously a difficult issue. Cells would need to be able to detect movement or even the quality of motion rather than just the presence or absence of a stimulus. One such model, incorporating protein phosphorylation, was presented by Chandra and Low (1995). They found that adding kinase inhibitors to cultivated soybean cells prevented the oxidative burst and, if added later, stopped the burst altogether. However, it was activated independently of external cues by phosphatase inhibitors. From this, they deduced that the kinases in question might be permanently active and that stabilisation of their phosphorylated forms signalled the onset of the burst. As an alternative, Hancock et al. (2006), lately pointed out that the H_2O_2 molecule's size precludes the involvement of specialised in its sensing, receptor proteins. They offered an alternate hypothesis, arguing that proteins serving diverse functions yet sharing the property of having thiol groups that are active and serve as redox targets acted as a moderator of ROS sensing in general. To rephrase, H_2O_2 and other ROS, such as NO, oxidised the thiol groups, resulting in sensing, which may have involved a competition between the two oxidants for the identical modification sites. Histidine kinase receptor ETR1, for instance, is required for sensing H_2O_2 and triggering stomatal closure. Not the kinase domain itself, although just one cysteine (Cys65) was needed for this Kolbe et al. (2006) have shown outcomes from proteomic and genomic studies bolstering this concept. Analysis of heat shock reactions has led to yet another integrative theory, which proposes interactions of heat shock promoter elements and H_2O_2 as sensors. ROS molecular sensors, according to Miller and Mittler's (2006) theory, are heat shock TFs, and that these factors regulate the expression of numerous genes involved in coping with stress, not just heat shock.

SIGNALLING

Multiple stress responses, such as those triggered by ABA-associated stomatal closure, pathogen elicitors, insect feeding, wounds, and extreme temperatures, have been shown to involve a connection between signalling networks and the H_2O_2. Instead, then attempting to review each of these separately, given their shared properties (such as the connection between H_2O_2 and Ca), My plan is to zero in on that connection first, then examine the beginnings of the chain of responses, and finally examine several of the most difficult and promising aspects of the problem. Pei et al. (2000) used patch clamp methods to conclusively demonstrate the relationship between H_2O_2, Ca, and stomatal closure, demonstrating that the calcium channel-mediated signalling pathway linking stomatal closure to ABA involves H_2O_2. Ca-sensitive fluorescent dye allowed researchers to demonstrate that the Ca current increased Cacyt. Based on the sensitivity of DPI to ABA-induced stomatal closure, it was hypothesised that elevated Cacyt enhanced activity of NAD(P)H oxidase, resulting in the O_2• released extracellularly, which was then dismuted to H_2O_2. Later, Chen and Gallie (2004) demonstrated that guard cell redox status, as well as diurnal fluctuations in ascorbate levels and H_2O_2 generation, are reflected in stomatal response.

RESPONSE NETWORKS

The signalling cascades have been found to have many similarities, and over the past five years, this response network has been continually enlarged and summarised, not just for stomatal responses but also for responses to various biotic and abiotic stimuli. Both the upstream TFs that regulate scavenging enzyme and NAD(P)H oxidase transcription and the early processes involve calcium or phosphatidic acid triggered serine/threonine protein kinase (OXI1) are conserved. This means that H_2O_2's initial reaction might either dampen or boost the oxidative signal, enabling a more nuanced or selective reaction to elicitation (Suzuki and Mittler, 2006).

With the advent of whole-genome microarrays, notably for *Arabidopsis thaliana*, this response network has been able to expand even more. In 2004, *Arabidopsis thaliana* has 152 genes identified as being contributed to ROS management, while this was increased to at least 400 by Hancock et al. (2006). Because ROS are crucial for metabolism, defence, cell and tissue-specific development, and because of the importance of maintaining a steady state that allow for the registration of essential signals, the network is both redundant and dynamic. By pooling their Arabidopsis microarray data, researchers have been able to create more intricate and thorough models, such as the examination of gene response clusters.

Ozone response gives a convenient example of both the breadth of the networks and their dependency regarding the experiment's settings, even though relevant research has included a wide variety of experimental situations. Despite ozone's toxicity being known for over fifty years, its method of action and the molecular response of organisms are poorly understood. Ascorbic acid's importance in ozone detoxification was recognised early on, in addition to ascorbate deficient or increased mutant research, this has been proposed as a potential contributor to ozone defence. However, this approach has not been entirely unchallenged (D'haese et al., 2005). Initial studies of the response network revealed considerable parallels to networks for various oxidative stresses at the molecular level. However, recent research by Li et al. (2006), this just goes to show how much variation there can be within a species and across closely related species in terms of how they handle stress in general. The authors compared the reaction of Arabidopsis and its resilient to stress cousin *Thellungiella halophila* using microarrays to investigate the effects of ozone. The study was unique because it was conducted in the real world. Free air concentration enrichment (FACE) conditions were used to modify ozone levels, and the effects on plants were compared at both baseline and 1.2-times baseline concentrations. Wassilewskija (WS), the Cape Verde Islands (Cvi-0), and Columbia-0 (Col-0) ecotypes of Arabidopsis were included. There was a wide variety of physiological and transcriptional responses even within this limited set of genotypes. Physiological and transcriptional responses were highly variable even among this limited set of genotypes. The WS ecotype of Arabidopsis was also the most impacted, exhibited the highest levels of real damage to the leaves and transcriptome responses comprised genes involved in photosynthetic light reactions, phenylpropanoid biosynthesis and metabolism, scavenging ROS, photosynthesis, respiration, and the reductive pentose phosphate pathway). Contrary to what was discovered with acute exposure, reversing the relative ozone resistance of the Arabidopsis genotypes under FACE circumstances, the necessity of undertaking field trial rather than depending on extremely artificial controlled environments. Additionally, the authors highlighted how the experiment's results were influenced by the local climate, the timing of plant growth and harvest, and any potential biotic stressors. The "take home lesson" based on this, the extreme caution that should be exercised when accepting any "definitive" model of a stress reaction.

PLANTS' REACTIONS TO HEAVY METAL STRESS AND H_2O_2

These are listing the functions and potential that is inherent of H_2O_2-driven plant defence systems in opposition to various HMs such as Cu, Ni, and Cd. The following is a brief overview of the research' conclusions on the effects of H_2O_2 on Copper, Nickel, Cadmium, Aluminium, Crimium, Lead, and Boron.

CU TOXICITY

Plants need Cu for a few metabolic activities, but it becomes a problem when it gets increased in excessive level. Plants' resistance to Cu stress was increased in *Vigna radiata* by a reduced concentration of H_2O_2 from outside sources (Fariduddin et al., 2014). Leaf water potential, pigments involved in photosynthetic reactions, stomatal activity, Leaf water potential, pigments involved in photosynthetic reactions, stomatal activity, the osmoprotectants and the antioxidant system against 100 mg/kg H_2O_2 root dipping treatment had a significant and profound effect on growth and yield characteristics copper stress (Nazir et al., 2019b). In instance, it has been showed that 0.1 mM H_2O_2 was most effective in lowering Cu toxicity, while 0.5 mM$H_2O_2$was less effective (Nazir et al., 2019b). H_2O_2 reduced the adverse effect of Cu stress in *Zea mays* by strengthening endogenous cellular structures via increased mineral content and osmotic solute concentration in the tissues of leaves (Guzel and Terzi, 2013). Interaction between H_2O_2 and the production of metallothioneins has also been seen, most notably in copper-fed plants of Oryza sativa (Liu et al., 2015). By increasing the activities of protein-metabolising enzymes, including arginase, glutamate dehydrogenase, ornithine aminotransferase (AOT), and D1-pyrroline-5-carboxylate synthetase (P5CS), H_2O_2 pretreatment reduced the oxidative burst in sweet corn seedlings that have been exposed to copper. In addition, H_2O_2 inhibits the production of ROS and ascorbate deposits caused by Cu. Reversal of H_2O_2, there was a decrease in hydrogen peroxide content and the functions of ascorbate peroxidases and guaiacol in *Matricaria chamomilla* when it was treated with Cu. According to González et al. (2012), *Ulva compressa* exposed to too much Cu, Genes encoding protein kinases that are calcium-dependent and calmodulins were activated in response to H_2O_2. To reduce the oxidative burst induced by ROS, hydrogen peroxide (H_2O_2) upregulated antioxidative enzymes in *Oryza sativa*' smetallothionein. An important method of Cu-resilience involves H_2O_2-modulated antioxidant activity (Naziret al., 2019a). Proline accumulation in detached rice leaves during Cu stress is likewise regulated by H_2O_2. In addition, Fariduddin et al. (2014) showed the effects of foliar spraying H_2O_2 (2.5 mM) on the antioxidant metabolism under Cu stress in *Vigna radiata* and demonstrated a reduction in H_2O_2, superoxide anion, and lipid peroxidation production. Exogenously sourced H_2O_2 was found to boost plant resistance to Cu stress mostly through promoting the activity of antioxidant enzymes and the proline content.

NI PHYTOTOXICITY

H_2O_2 has been shown to have beneficial regulatory roles in plants, particularly in regard to tolerance for nickel (Nazir et al., 2019b). Ni resilience in *Solanum lycopersicum* was improved by dipping the roots in 0.1 mMH_2O_2, which enhanced photosynthetic efficiency, growth parameters, and a variety of biochemical traits, as well as restored root shape and stomatal mobility (Nazir et al., 2019b). By increasing PS II activity and consequently the photosynthetic nitrogen-use efficiency (NUE), H_2O_2 treatment (50 mM) reduced toxicity of Ni (200 mg Ni kg1 soil), lowered production, lipid peroxidation and ion leakage in mustard plants due to increased SUE (sulphur usage efficiency) and GSH (glutathione).

CD PHYTOTOXICITY

H_2O_2's protective effect against Cd stress varies with the variety of plant species subjected to it and the concentration of H_2O_2 applied (Hasanuzzaman et al., 2017). H_2O_2 at lowest concentrations can mitigate Cd toxicity by promoting the uptake of osmoprotectants and the production of antioxidants. Additionally, through increasing AsA, GSH, CAT, APX, and NADH peroxidase activity levels, 0.1mMH_2O_2 restored MAPK (MPK1/2) activation and crop development. The oxidative stress caused by Cd and Cu in rice has been shown to be decreased by H_2O_2 and increased by metallothionein activity. Oxalate oxidase (OxO), peroxidase gene (PRX8), and OxO-like protein

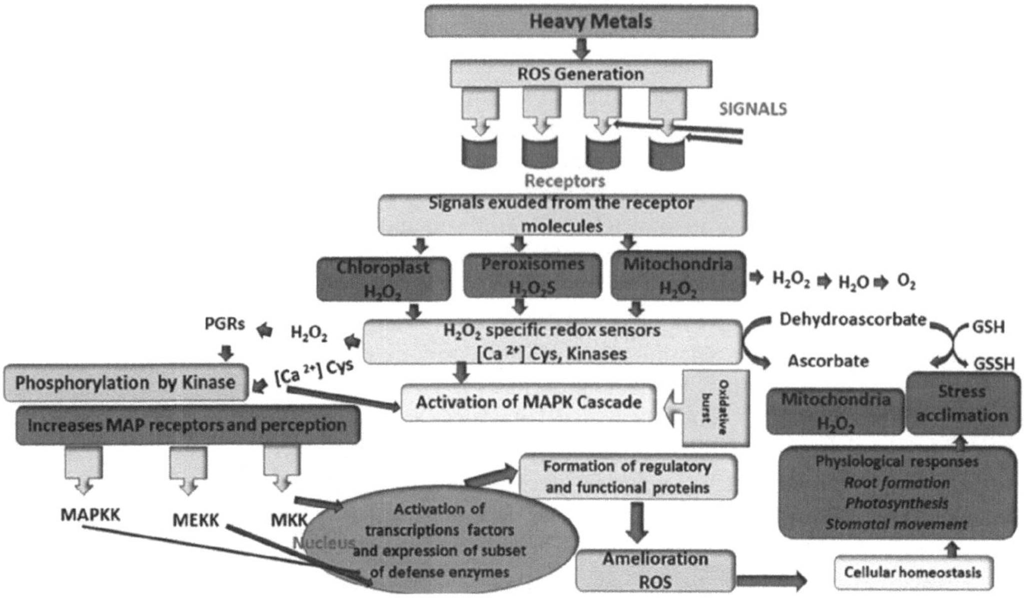

FIGURE 12.2 Detailed representation of H_2O_2 signal transduction to ameliorate heavy metal-induced ROS generation.

(OLP) expression were seen in barley root trip exposed to Cd, Hg, and oxidative stress was maximised by H_2O_2 supplementation. The ability of H_2O_2 to activate MAPK4, MAPK12, MPK1/2, MPK6, and nucleotide diphosphate kinase 2 (NDPK2), which explicitly associates with MPK3 and MPK6, thus maintaining the redox homeostasis is shown in Figure 12.2. H_2O_2 (100 m mol/L) Cd-exposed rice seedlings exhibited enhanced plant development characteristics and higher levels of glutathione (GSH), phytochelation, and cysteine (Cys). Another study found that when applied to rice seedlings, H_2O_2 (100 mM) significantly decreased Cd absorption while increasing activity of ascorbic acid (APX), SOD, CAT, GST, GPX, and ascorbic acid and reduced glutathione (GSH) levels. Additionally, H_2O_2 levels were maintained via exogenous sources, which boosted ROS detoxification by increasing enzymatic and nonenzymatic antioxidant activities. The increased GSH concentration and activity of Gly I and Gly II after H_2O_2 pretreatment are indicative of H_2O_2's participation in the methylglyoxal scavenging mechanism (Hasanuzzaman et al., 2017). H_2O_2 pretreatment increased GSH levels and triggered Gly I and Gly II production. Increased GSH synthesis was shown to regulate rice seedlings' defence against Cd stress after it was initiated by hydrogen peroxide. Cd-exposed roots of two rice genotypes (N07-6 and N07-63) showed improvements in activity of the GST, PCs, NPT, and PCs when treated with H_2O_2. Ascorbate peroxidase (APX) and glutathione reductase (GR) activation in rice seedling leaves by heat shock and Cd has also been linked to H_2O_2 (Chou et al., 2012).

AL, CR, PB, HG, AND B PHYTOTOXICITY

Both the *Triticum aestivum* L. winter wheat strains Jian-864, which is Al-tolerant, and Yangmai-5, which is Al-sensitive, genotypes were protected from Al's deleterious effects by applying H_2O_2 exogenously, which altered H_2O_2 metabolism and ROS levels (Xu et al., 2011). They additionally found that pretreatment with H_2O_2 increased levels of GSH and AsA, and the content of both genotypes of AsA/(AsA DHA) and GSH/GSSG, which may be important in causing a comparatively bigger Al adjustment in the sensitive variety rather the tolerant one(Xu et al., 2011). Lipid peroxidation was reduced, Al uptake was slowed, nutritional content was preserved, and Al toxicity in wheat was reduced when H_2O_2 was administered exogenously. H_2O_2 application also increased thiol concentration, chlorophyll content, and antioxidant enzyme activity, activation of metallothionein protein (BnMP1), and reduced lipid peroxidation in *Brassica napus*, which reduced the absorption of Cr and provided tolerance to Cr (Yıldız et al., 2013). It was hypothesised that the antioxidative enzymes APX, SOD, and CAT, which were induced by exposure to Pb (0.05, 0.15, 0.25 mmol/L), would protect Oryza sativa against Pb-toxicity. *Cajanus cajan*, when subjected to boron stress, has its H_2O_2 and glutathione levels fluctuate, affecting the plant's morphophysiological characteristics, ascorbate (AsA) and glutathione (GSH) metabolism, and mineral nutrition. Root water transport disruption due to external challenges is likely regulated in part by the plant's high H_2O_2concentration.In addition, H_2O_2 is critical for the activation of antioxidative mechanisms and the induction of physio-morphological changes in reaction of stress.

H_2O_2 INTERACTIONS WITH OTHER PLANT GROWTH REGULATORS WHEN EXPOSED TO HEAVY METALS

Phytohormones are thought to have a role in a diverse range of plant processes. Many PGRs have been implicated in plant stress resistance, including nitric oxide (NO), calcium (Ca^{2+}), polyhydroxylated steroidal PGRs (BRs), auxins, GAs, CKs, ABA, JA, ET, and SA. PGRs control a variety of physiological activities and are integral to many physiological and ecological parameters (Nazir et al., 2019a, b). Crosstalk among various PGRs has been a major stumbling block for plant researchers trying to strike a balance among growth and safety systems in plants'-controlled frameworks. PGRs can act independently, or their signalling processes might be additive or subtractive, reducing the strength of each other's biogenesis or reactions (Niu and Liao, 2016). In particular, intermodulation of hormones may influence the synthesis, signalling, and transport of additional hormones, with synergistic and antagonistic effects (Zou et al., 2015). H_2O_2, a fundamental signalling molecule, does not act alone in many signalling pathways; rather, it operates in dynamic systems that crosstalk to a number of additional phytohormonal metabolic processes. Crosstalk between phytohormones also helps plants become more resilient by reconfiguring their genetic systems, modulating their defence responses, and reducing the negative impacts of stress (Table 12.1).

TABLE 12.1
Overview of How Hydrogen Peroxide Interacts with Plant Growth Regulators to Control a Variety of Physiological Functions

Phytohormones	H_2O_2 Induced Effects	Plant Species
Auxin	The interaction between AUX and H_2O_2 promotes lateral root development	*Arabidopsis thaliana*
	By regulating the level of endogenous auxin, H_2O_2 treatment promotes the formation of adventitious roots	*Linum usitatissimum*
	In the wild-type root tip, indole-acetic acid concurrently increases endogenous H_2O_2 levels	*Solanum lycopersicum*
Gibberellin	GA-induced pheripheral endosperm programmed cell death was slowed by H_2O_2	*Hordeum vulgare*
	NtRBOH (*Nicotiana tobacum* respiratory burst oxidase homolog), which encodes the genes for NADPH oxidase production, is activated by exogenous GA whereas its expression is suppressed by H_2O_2	*Nicotiana tobaccum*
	H_2O_2 and GA work together synergistically to promote seed germination	*Nicotiana tobaccum*
Cytokinin	In order to stimulate proteins involved in plant signalling and growth during germination, H_2O_2 interacts well with CKs	*Pisum sativum*
	H_2O_2 and CK worked together synergistically to control the senescence caused by shadow in leaves	*Triticum aestivum*

(Continued)

TABLE 12.1 (*Continued*)
Overview of How Hydrogen Peroxide Interacts with Plant Growth Regulators to Control a Variety of Physiological Functions

Phytohormones	H_2O_2 Induced Effects	Plant Species
Abscisic acid	Water stress tolerance is provided by the interaction between H_2O_2 and ABA Plasma membrane calcium channels needed for opening and closing of stomata caused by ABA were activated by H_2O_2 accumulation The production of H_2O_2 was induced by the accumulation of ABA, which increased the activity of antioxidant enzymes and the expression of the genes that code for them	*Zea mays* *Vicia faba* *Cucumis sativs*
Ethylene	Assembling the wall labyrinth in transfer cells, which is essential for promoting high levels of nutrient transport in plants, requires the formation of sterol-enriched domain, which is influenced by ethylene and H_2O_2. H_2O_2 is a significant contributor to ethylene's role in mediating the mortality caused by Cd. Ethylene and H_2O_2 were once considered to act in a synergistic manner	*Vicia faba* *Solanum lycopersicum*
Brassinosteroid	H_2O_2 play a crucial role in controlling the movement of stomata in response to different BR dosages. Brassinosteroids cause CO_2 assimilation and glucose metabolism, while H_2O_2 serves as a secondary messenger for these processes	*Solanum lycopersicum* *Cucumis sativus*
Jasmonate	H_2O_2 acts as a messenger molecule to activate the antioxidant defence mechanism induced by MeJA. Application of MJ causes H_2O_2 to accumulate and is thought to be a factor in jasmonate-induced leaf senescence.	*Triticum aestivum* *Oryza sativa*
Nitric oxide	Anthocyanins biosynthesis in radish sprouts' hypocotyls is controlled by H_2O_2-dependent NO production The guard cells and other cells of mung bean leaf are likewise stimulated to produce NO by exogenous H_2O_2	*Raphanus sativus* *Vigna radiata*
Salicylic acid	SA encourages adventitious roots through H_2O_2 acquisition An SA and H_2O_2 seed soaking treatment could synergistically activate hormone metabolism and signal transduction, as well as increase energy supply and antioxidant enzyme activity	*Vigna radiate* *Zea mays*
Calcium ions	Ca^{2+} signalling was involved in the H_2O_2-induced adventitious rooting. H_2O_2 signalling promoted root development by controlling Ca^{2+} inward movement in the root cells' plasma membrane	*Arabidopsis thaliana* *Arabidopsis thaliana*

H_2O_2 AND AUXINS

Among the many growth and development processes that auxin is known to regulate are division and divergence of cells, cell expansion, apical dominance, and others. Furthermore, H_2O_2 applied exogenously stimulates adventitious root development in *Linum usitatissimum* via controlling endogenous levels of auxin. Exogenous indole-acetic acid (IAA) has been shown to simultaneously enhance endogenous levels of H_2O_2 in the wild-type *Solanum lycopersicum*'s root tip, showing that H_2O_2 is auxin-regulated in the developing root. NAA (naphthalene acetic acid) and IBA (indole-3-butyric acid) supplements boost H_2O_2 generation in Azalea micro shoots in an in vitro assay (Elmongy et al., 2018). Red light stimulates phyB-induced auxin production in tomato shoot apex, leading to H_2O_2 production in systemic leaves. Consequently, elevated H_2O_2 levels boost ATP synthesis and cyclic electron flow, this caused quick CO_2 uptake during photosynthesis in the cellular leaves, helping plants respond better to fluctuating light levels. Arabidopsis's AUXH_2O_2 interaction was observed to mediate the formation of lateral roots. Consistent with H_2O_2-AUX crosstalk,

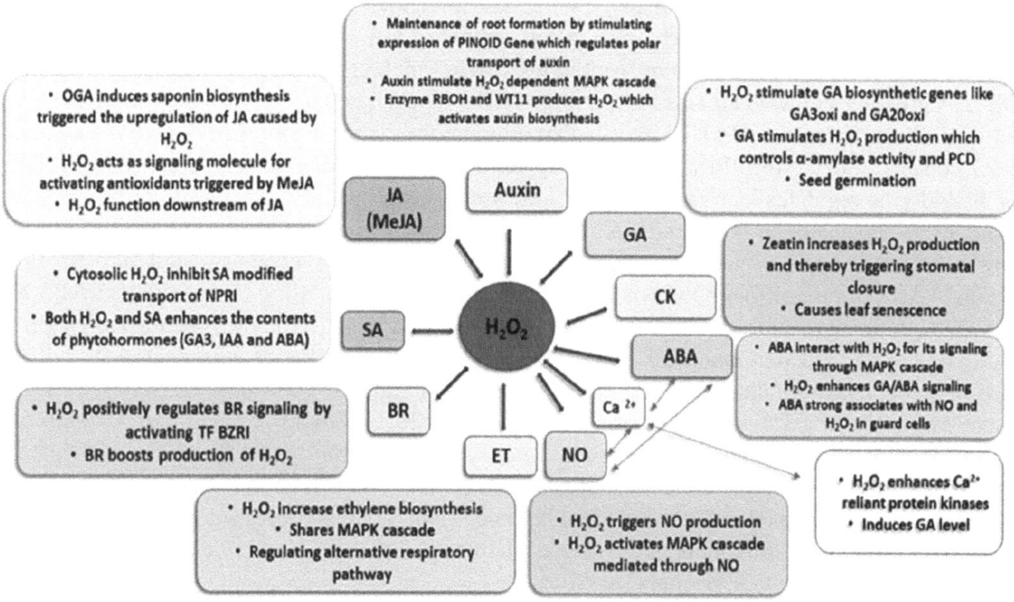

FIGURE 12.3 Representation of crosstalk of hydrogen peroxide with other plant growth regulators.

it has been found that primary root gravitropism in Arabidopsis thaliana drastically decreases. H_2O_2-auxin crosstalk can be seen in a number of developmental processes, such as when H_2O_2 controls polar transport and redistributes auxin to preserve root shape (Su et al., 2016). To learn how these two signal molecules work, we need to investigate the molecular-level interaction between H_2O_2 and AUX. Analysis of transcripts revealed that apoplastic ROS affect several aspects of auxin stability and signalling. To learn how these two signal molecules work, we need to investigate the molecular-level interaction between H_2O_2 and AUX. Analysis of transcripts revealed that apoplastic ROS affect several aspects of auxin stability and signalling. Increases in H_2O_2 have been shown to modify auxin metabolism, and structural changes in plants induced by ROS impact on auxin signalling have been detected. Changes in auxin homeostasis may result from PINOID gene expression modifications induced by H_2O_2, which regulate auxin polar transportation. In addition, genes implicated in auxin conjugation and biosynthesis mediated by Trpare activated in response to H_2O_2-related peroxidases, contributing to auxin oxidative degradation and so altering auxin homeostasis. Respiratory burst oxidase homolog 1 (RBOH1) and whitefly-induced 1 (WFI1) genes produce apoplastic H_2O_2, which stimulates auxin biosynthesis at the tomato plant apex and reduces the stem's CKs biosynthesis, which in turn reduces axillary bud expansion. These results provided novel insights into the role of H_2O_2 in lateral bud development in tomatoes via interaction with auxin signalling. H_2O_2-can trigger alterations to PINOID gene expression that influence the polar transport of auxin might alter the auxin homeostasis (Figure 12.3).

H_2O_2 AND GIBBERELLINS

Gibberellins have a pivotal function in promoting germination of seeds, growth of stems, flowers, and fruits (Huang et al., 2017). Many different morphophysiological properties (such as de-etiolation, dormancy breaking, sprouting, and the formation of the hypocotyl and primary root) are reportedly influenced by the interaction of H_2O_2 and GA (Hejazi et al., 2018). Also, ROS, especially hydrogen peroxide (H_2O_2), promote GA-stimulated eamylases in barley aleurone cells. Consequently, H_2O_2 appears to be involved in GA and ABA signalling and affects a-amylase production in barley aleurone cells. Not surprisingly, many reports also indicated a detrimental interaction between H_2O_2 and DELLA proteins (which are negative agents of GA signalling) which are regulated by GAs in modern times. Furthermore, Arabidopsis gal–3 mutants lacking GA had higher levels of DELLA proteins and were far less susceptible to ROS-induced cell death. Stress (drought or excess salt) conditions increase ABA/GA indices that promote the synthesis of DELLA proteins and raise ROS concentrations. And through regulating ROS generation, the GA-activated gene GASA14 in Arabidopsis regulates leaf growth and abiotic stress resistance. Tobacco seed germination was found to be dependent on H_2O_2 and GA's mutual induction, principally due to the ABA/GA ratio's downregulation and forcing reserve composition to mobilise. It has been postulated that H_2O_2 lowers the ABA/GA ratio through influencing the metabolism and signalling transduction of ABA and GA through the activation or silencing of genes. Exogenous H_2O_2 promotes seed germination by increasing GA production and ABA catabolism in sunflower seed, maize, and Arabidopsis. It was also shown that GA regulates the ROS-antioxidant activity and the concentration of ABA, causing latent caryopses to germinate (CembrowskaLech et al., 2015). GA in peripheral endosperm also causes apoptosis by regulating H_2O_2 generation. Interestingly, the NADPH oxidase biosynthesis genes are repressed by H_2O_2, but activated by the exogenous administration of GA. This is because *Nicotiana tabacum* respiratory burst oxidase homolog (NtRBOH) is expressed by these genes. H_2O_2 enhances biosynthesis in Arabidopsis by upregulating seed germination-critical genes involved in biosynthesis of GA like GA3ox and GAw20ox.

H_2O_2 AND CYTOKININ

Leaf senescence, cell division, and the stimulation of light-controlled plant growth are just a few of the many physiology, cognition, and metabolism processes in plants that are commonly associated with cytokinins (CKs) (Chang et al., 2016). Maintaining an optimal level of CK for plant growth, H_2O_2 can also boost cytokinin oxidase and gene expression. Shade-induced leaf senescence in wheat is regulated by a CK–H_2O_2 axis. Stimulating proteins involved in plant development and signalling required a close relationship between H_2O_2 and phytohormones such as zeatin riboside (ZR) and ABA. Peas may have better germination rates because H_2O_2 from the outside the plant causes a drop in the amount of zeatin (zeatin riboside) that is more active. These investigations show that H_2O_2, along with other PGRs, plays a crucial role as a messenger in the germination of seeds. These findings suggest that CK and H_2O_2 interact with one another, possibly in an antagonistic fashion. Root and shoot CK concentrations are reduced and CK production is inhibited by H_2O_2. The CK content and the H_2O_2 content of the roots of the transgenic line (S41) of creeping bentgrass were both significantly greater than those of the wild-type (WT) roots. Overexpression of the isopentyltransferase (IPT) gene and RNAi silencing of the HvCKX1 gene encoding the cytokinin oxidase (COX) that catalyses CK cleavage improved the development and fibrous root systems of creeping bentgrass and monocot species like barley (Hordeum vulgare). Increased buildup of hydrogen peroxide (H_2O_2) in guard cells and activation of histidine kinase 3 (AHK3) and response regulator 2 (ARR2) receptors have been linked to stomatal closure induced by exogenously supplied -zeatin in Arabidopsis thaliana (Arnaud et al., 2017). CKs are crucial in cellular signal transduction mechanisms. Phosphorelay activity is controlled by cytokinin, and it also initiates transcription and affects protein standardisation and stability. H_2O_2 has been shown to play a more significant role in stress and activating programmed cell death responses by upregulating the C_2H_2 zinc finger ZAT11 gene.

H_2O_2 AND ABSCISIC ACID

Through its specific impact on aquaporins, ABA controls water transport, stomatal behaviour, bud and seed dormancy, and leaf growth and development. Furthermore, ABA is crucial in conditions of drought and salinity. When it comes to signalling through MAPK cascades, ABA also interacts with H_2O_2. H_2O_2 is an important secondary transmitter involved in ABA transduction. Accumulation of

H_2O_2 in *Vicia faba* causes ABA-mediated stomatal behaviour via activating plasma membrane calcium channels. The genes Atrboh D and Atrboh F, which encode NADPH oxidase catalytic subunits and are embedded in the plasma membrane, have been shown to be required for ABA-induced stomatal ROS generatio. Recently, it has come to light that NADPH oxidase can activate the SA, JA, and ethylene hormone signal attenuation pathways by stimulating Ca^{2+} and MAPK signalling routes. The production of DELLA protein, which in turn decreases H_2O_2 concentration, is regulated by a ratio of ABA to gibberellic acid (GA) that is affected by abiotic stress factors (Considine and Foyer, 2014). Involved in ABA signalling is the calcium/calmodulin-dependent protein kinase (CCaMK) superfamily, which is activated by free Ca^{2+} and coupled Ca^{2+} in the form of CaMvia interfering with ABA's upstream antioxidant system. During exposure to abiotic stress, MAPK is activated, and ABA is signalled. Plants have been shown to adapt to water stress by a combination of ABA and hydrogen peroxide (Li et al., 2016). Drought-stressed leaves of *Zea mays* showed a significant increase in ABA content. H_2O_2 concentrations rose in a same fashion. However, in rice seedlings growing in soils rich in K, exogenous treatment with ABA boosted expression of OsRab16 A (an ABA sensitive gene) and effectively accelerated the synthesis of H_2O_2 in the leaves. In addition, NADPH oxidase had been identified as the enzyme responsible for H_2O_2 synthesis in rice seedling roots and leaves. The accumulation of ABA in pumpkin-grafted cucumber leaves causes H_2O_2 generation, which in turn activates the genes for the plant's antioxidant defence system (SOD, POD, and cAPX) by upregulating their expression (Shu et al., 2016). Increased levels of H_2O_2 and nitric oxide (NO) in tobacco's guard and mesophyll cells are a result of ABA enrichment, which also activates a number of antioxidant enzymes. Stress tolerance is maintained for longer periods of time when H_2O_2 levels are high because the stimulation of NADPH oxidase causes more ABA to be released. Wheat plants' SA-induced freezing tolerance may be regulated by a positive feedback system involving hydrogen peroxide and ABA. Eventually, Elevated H_2O_2 and ABA independently activate their defence mechanisms, enabling activation of expression of cold-responsive gene and activity of antioxidant enzyme to stop ions from leaking out and lipid peroxidation, ultimately improving the photosynthesis efficiency (Fv/Fm and ETR) of wheat plants under freezing stress. Ishibashi et al. (2017) focused on how H_2O_2 and ABA affect inactive (quiescent) and active (non-dormant) barley seeds. After being soaked, the percentage of H_2O_2 in embryos of barley seed was discovered to be more in seeds that were not sleeping than in seeds that were sleeping. An enzyme called ABA-80-hydroxylase that breaks down ABA was controlled by H_2O_2 in the embryos. Also, the activity of the enzyme NADPH oxidase that makes ROS was lower in dormant seeds than in non-dormant seeds, while the activity of the H_2O_2-detoxifying enzyme catalase was higher because it was controlled by an ABA-responsive transcription factor (HvABI5) that was turned on by ABA. These results show that a change in the balance between ABA and ROS is effective in the embryo of soaked barley seeds and controls whether the seeds stay dormant or start to grow. Exogenous inhibitors ofH_2O_2 and NADPH oxidase enhanced the breakdown of ABA by increase the expression of the ABA 80-hydroxylase-coding CYP707A gene. They also improved the activity of genes for synthesis of GA in Arabidopsis seeds that were dormant. But H_2O_2 works with the ABA signalling system in sunflower seeds at the transcriptional level. H_2O_2 was discovered to be involved in the ABA-controlled growth of new roots in Cucumis sativus. It did this through increasing the amount of chlorophyll, relative water content, quantum yield, carbohydrates, and nitrogen, and by making some enzymes work better (Li et al., 2016).

H_2O_2 AND ETHYLENE

A gaseous phytohormone is ethylene that helps plants with things like leaf, flower, and fruit development and ageing, as well as making them more resistant to abiotic stress. Productivity or stress response can be influenced by H_2O_2 and ethylene in an antagonistic or synergistic way. Most of the interactions between hydrogen peroxide and ethylene had been linked to ETHYLENE INSENSITIVE3 (EIN3), a transcription factor that is turned on by ethylene (ET), and its close relative EIN3-Like1 (EIL1). Ethylene response factors (ERFs) are important for ethylene communication and help plants be less tolerant of abiotic stress. Nicotiana tabacum's TERF1 is turned on by ROS-making substrates that come from outside the plant. It is also controlled by upstream EIN3-like regulators, LeEIL3 and LeEIL4. When TERF1 is turned on in tobacco plants, it affects a number of stress-responsive genes and makes them more resistant to H_2O_2. It does this by connecting their promoters, such as GCC-box, DRE/CRT, CE1, JERE, and CT-rich, to the cis-elements. It had been suggested that ethylene and H_2O_2 work together in tomato, where hydrogen peroxide is a major source of ethylene formation that helps Cd kill cells. H_2O_2 and ethylene might be involved in controlling the alternative respiratory route (AP) when the plant is under salt stress. Zhu et al. (2016) showed that ethylene and H_2O_2 play important roles in salt stress-induced BR-reliant endurance in tomato plants. When plants don't get enough oxygen, their energy system changes from aerobic to anaerobic, which changes how much water and nutrients they can take in hypoxia signals then causes the production of both H_2O_2 and ethylene. In the initial stages of hypoxia signalling in Arabidopsis, the amounts of gene expression for AtERF73/HRE1 and ADH1 are affected by how H_2O_2 and ethylene work together (Yang, 2014). Hypoxia signalling makes ethylene enzymes like ACC synthase (ACS) and ACC oxidase (ACO) work faster. Ethylene and H_2O_2 also affect the formation of sterol-rich domains in developing *Vicia faba* cotyledons. These domains are crucial for the building of wall labyrinths in transfer cells, which helps move nutrients into and around the plant body. ROS and ethylene are also part of the signalling chain set off by Nod factor, which is needed for legumes to make nodules. A collection of receptors that act as ethylene signalling negative regulators also have a role in ethylene perception. The transcription factor EIN3 (a direct transmitter of ERFs) that links the GCC-box located in the ethylene-triggered genes promoters is activated by a cascade of MAPKs regulated by the negative regulator CTR1. The ethylene receptor ETR1 was discovered to be a key player in the interplay between ethylene and hydrogen peroxide signals in stomatal guard cells. It is also shown that the MAPK cascade is involved in both ethylene and H_2O_2 signaling. Increased resistance to ROS was linked to changes in the ethylene receptor ETR1 and transcription factor EIN3 in the plant species Arabidopsis. Song et al. (2007) found that mung bean hypocotyls react negatively to exogenous H_2O_2 and ethylene. Increased ethylene accumulation was seen in H_2O_2-treated hypocotyls due to the activation of ethylene biosynthesis enzymes, although the application of exogenous ACC (a readily convertible intermediary to ethylene) had no effect on ROS production. Soybean seeds that have been imbibed produce H_2O_2in the embryonic axis, which signals the body to produce ethylene, hence promoting seed germination and root tip cell elongation. In particular, H_2O_2 acts as a primary phytohormone in the production of 1-aminocyclopropane-1-carboxylic acid (ACC, a significant precursor of ethylene) in the context of ACS synthase.

H_2O_2 AND BRASSINOSTEROIDS

There is a strong hormonal interplay between H_2O_2 and BRs, which contributes to plant development and physiological responses. Plant steroid hormones, or BRs, perform a crucial role in plant development and environmental regulation. The major promoter of BR signalling, the transcription factor BRASSINAZOLE-RESISTANT1 (BZR1), is turned on by H_2O_2, hence favourably regulating brassinosteroid signalling. Exogenously applied BR also increases H_2O_2 synthesis in the niche of the root stem cell, possibly contributing to BR-influenced QC division and elasticity of cell (Tian et al., 2018). Cucumber plants that have had BR applied to them have collected H_2O_2 due to an increase in NADPH oxidase activity, while increased RBOH1 activity achieves a similar result in tomatoes. Similarly, knocking down RBOH in *Nicotiana benthamiana* reduced the detoxification of ROS, making the plant more susceptible to abiotic stresses. Jiang et al., (2012) found that *Cucumis sativus*' photosynthetic efficiency and productivity are enhanced by H_2O_2 via redox signalling, which is initiated by brassinosteroids for carbohydrate metabolism and CO_2 assimilation. In *Solanum lycopersicum*, the solubility and dosages of H_2O_2 are crucial for controlling stomatal movement in response to varied BR doses. The ABA-responsive pathway is negatively affected by a transient increase in H_2O_2 generated by a low dosage of BR, which acts as a catalyst to promote the regeneration and/or manufacture of GSH. The ABA-responsive pathway is negatively affected by a transient increase in hydrogen peroxide generated by a low dosage of BR, which acts as a catalyst to promote the regeneration and/or manufacture of GSH.

H_2O_2 AND JASMONATE

Another type of plant growth regulator, methyl jasmonate (MJ) promotes growth, photosynthesis, fruit maturity, and senescence in response to varying environmental conditions. There is mounting evidence that H_2O_2 and JA work together synergistically in plants (Hung et al., 2006).H_2O_2 and JA were determined to be the primary signalling agents during the cellular processes responsible for oligo galacturonic acid (OGA)-induced saponin production. *Oryza sativa's* detached leaves accumulated H_2O_2 after being treated with MJ, a compound linked to the jasmonate-induced senescence of the plant (Hung et al., 2006). The acquisition of H_2O_2 by MJ-treated parsley suspension-cultured cells is another area that has been investigated, whereas other research has linked it to luring hidden tomato genes out of hiding. Antioxidant stilbenoids (polyphenolic phytoalexins) were differentially activated after being supplemented with methyl jasmonate (MJ) and hydrogen peroxide (H_2O_2) in the hairy roots of Vitis rotundifolia. H_2O_2 acts as a signalling molecule in the wheat callus to activate the antioxidant defence mechanism induced by MeJA. In addition to its involvement in redox equilibrium, glutathione production is boosted by JA. H_2O_2 reduces ROS production by activating nuclear factor erythroid 2-related factor2 (NrF2), which was found to be activated and JA-induced H_2O_2 triggered MEK1/2 phosphorylation and stimulation, which activated metabolic enzymes (AsA and GSH), which in turn improved the AsA and GSH redox state. DPI and DMTU, two compounds that inhibit and scavenge H_2O_2, were utilised to drastically reduce the amount of H_2O_2 produced by MJ. Both wounding and the application of JA have been shown to significantly increase the activities of phenylalanine ammonia-lyase and plasma membrane NADPH oxidase in the leaves of pea (*Pisum sativum* L.) plants. In addition, the addition of an H_2O_2 scavenger inhibited this increase, suggesting that H_2O_2 plays a role downstream of JA.

H_2O_2 AND NITRIC OXIDE

Multiple studies have shown that the NO signalling route and the H_2O_2 signalling pathway interact with one another in a parallel fashion. Numerous studies have demonstrated that H_2O_2 therapy is the origin of NO, which plays an important role in boosting the effectiveness of alternate oxidase (AOX) and also alleviating the damage to the photosystem under stress (Arfan et al., 2019). In addition, the manufacturing of anthocyanins in radish sprout hypocotyls is controlled by NO generation that is H_2O_2-dependent (Wu et al., 2016). The guard cells and other cells of the mung bean leaf are also stimulated to produce NO when exposed to exogenous H_2O_2. After being exposed to H_2O_2, NO levels in the roots of wild-type Arabidopsis plants increased by a factor of eight. The inhibition of NO synthesis in H_2O_2-stimulated guard cells of *Vicia faba* suggests that ROS play a role in regulating NO production. It is interesting to note that a calcium channel blocker can likewise reduce NO production in response to H_2O_2, suggesting that Ca_2 ions may modify the effect of H_2O_2 on NO formation. ABA's involvement in the production and regulation of ROS and nitric oxide (NO) in Arabidopsis guard cells has been well established. However, ABA-controlled NO generation is dependent on H_2O_2 production. In contrast, exogenously administered NO did not induce H_2O_2 synthesis in guard cells, and H_2O_2 synthesis was not necessary for NO-induced stomatal closure. Further, NO prevents H_2O_2 accumulation in leaf senescence by blocking the activation of numerous senescence-related genes (SAGs) and down-regulating photosynthetic genes, the primary regulators in H_2O_2-mediated photosynthetic efficiency. H_2O_2 also induced MAP Kinase 6 activation, which mediated NO generation and signal transduction in developing Arabidopsis roots. These results provide novel insights into how existing processes at the molecular level in signalling cascades related to stress and development might be used to regulate gene expression in plants.

H_2O_2 AND SALICYLIC ACID

A key player in numerous physiological pathways, salicylic acid (SA) is a major intracellular growth regulator as well as phenolic signalling molecule (Garg and Bharti, 2018). Depending on the situation, H_2O_2 and SA might be positively or adversely correlated (Cocetta and Ferrante, 2018). SA is probably an effective elicitor that causes the buildup of rosmarinic acid (RA), which can react either singly or in concert with H_2O_2 in *Salvia miltiorrhiza* cell cultures. The intracellular H_2O_2 contributes to RA buildup and serves as a signal molecule for SA stimulants. An SA and H_2O_2 seed soaking treatment have been shown to synergistically promote hormone metabolism and signal transmission, as well as increase energy supply and activate antioxidant enzymes in maize seeds. Benzoic acid, a complex precursor to SA, interacts with H_2O_2 in a process that results in the synthesis of H_2O_2-responsive benzoic acid 2-hydroxylase. The corresponding pathways of this situation show that H_2O_2 stimulation in cell organelles (peroxisome and chloroplast) causes SA synthesis, which helps the plant's defence system by closing stomata and triggering programmed cell death. SA also plays a role in making H_2O_2. Tomato plants cultivated in sand-ponic culture and given H_2O_2 and SA had a positive effect on growth. PGR concentration (GA3, IAA, and ABA), antioxidant enzyme activity (POD and PPO), and fruit yield characteristics. SA-mediated H_2O_2 generation increased the transcription of heat shock protein (Hsp) genes via AtHsfA2, a crucial component in providing thermal resistance in Arabidopsis. Additionally, SA and an alternate respiratory pathway may support PSII alteration brought on by H_2O_2 in the presence of excess light, according to Hou et al. (2018). According to several studies, the signalling pathways of SA and H_2O_2 interact negatively. Yang et al. (2013) demonstrated that SA plays a specific role in encouraging adventitious roots by acquiring H_2O_2. It was discovered that SA-induced adventitious root formation in mung bean seedlings might change the status of the antioxidant system, which could lead to the acquisition of free H_2O_2. One of the key factors that modify SA and H_2O_2 signalling is nuclear translocation of the SAR-linked transcription activator NPR1 (non-expressor of pathogenesis-related gene 1). It was demonstrated in this case that cytoplasmic H_2O_2 prevents the SA-triggered transfer of NPR1 to the nucleus, impairing the activity of pathogenesis-related (PR) genes. Glutaredoxins (GRXC9, also known as GRX480) may also have an impact on the conflict between SA- and ROS-governed signalling. In a wide range of biological functions, GRXs, which are small, common glutathione (GSH)-relientoxidoreductases that facilitate the breakdown of protein GSH disulfide bridges (S-glutathionylated proteins), are involved. NPR1 was necessary for the activation of GRXC9 through SA and TGA TFs, which in turn activated PR genes that are responsive to SA. On the other hand, elevated GRXC9 levels reduce the activation of its own momentary SA-dependent transcriptional induction.

H_2O_2 AND CALCIUM IONS

Ca^{2+} ion is one of the key participants in the abiotic stress signalling that controls varied responses to stress among a variety of secondary messengers. Ca^{2+} homeostasis regulation is the main goal of H_2O_2 signalling. Ca^{2+} is pumped into the cell through cation channels in the tonoplast, ER, and plasma membrane. This results in the generation of cytoplasmic Ca^{2+}, which sets off a variety of physiological and developmental cues. The opening of stomata, one of the reactions, is a thoroughly researched mechanism that involves H_2O_2 Ca^{2+} signalling But Ca^{2+} is also a key signalling molecule in the control of guard cell pore opening and closure under the effect of ABA, NO, and ROS, a study that examined how Ca^{2+} and H_2O_2 signals interacted to control plant defence. CAT enzyme activity was shown in certain studies to be activated by Ca^{2+} to reduce cellular H_2O_2 levels and function as an upregulating element in H_2O_2 stress signals in plants. González et al. (2012) found that in *Ulva compressa*, Cu-triggered increase of calcium, NO, and H_2O_2 concentrations and the interaction between these signalling molecules led to up- or downregulation of the gene activation cascade through calmodulins and Ca^{2+}-dependent protein kinases. Treatment with H_2O_2 improved the responsiveness of eight Ca^{2+}-dependent protein kinases in wheat plants. H_2O_2 also detected fast gene activation in tomato leaves. Rentel and Knight (2004) hypothesised that the change in gene expression (GSTI) caused by H_2O_2 and Ca^{2+} may be the result of a redox balance disruption. Gao et al. (2018) according to research, cytoplasmic ROS and Ca^{2+} concentrations significantly influence *Ganoderma lucidum*'shyphal development and induce GA levels in response to Cu stress. Han et al. (2015) investigated the hypothesis that H_2O_2 signalling controls the inward flow of Ca^{2+} in root cell plasma membranes and may cause root elongation in Arabidopsis seedlings. Additionally, Anexin1 (Ca^{2+} transporter) may control H_2O_2 and Ca^{2+} signalling to support growth and development in Arabidopsis thaliana root cells, according to Richards et al. (2014). It was discovered that Ca_2+ signalling inhibitors may prevent the growth of adventitious roots in marigolds, which is a response to H_2O_2-induced adventitious rooting.

CONCLUSION

H_2O_2 is a very promising chemical that plays a key role in enhancing a wide variety of plant functions, but it has thus far proven difficult to study. Furthermore, it is becoming increasingly clear that H_2O_2 functions as a universal molecule of signalling in plants. Improved understanding of how H_2O_2 significantly counters HM caused alterations in processes of development like plant development and growth, photosynthesis, stomatal behaviour, pollen formation, blooming, fruit development and growth, and wilting of leaf is presented. Multiple complex cues that react positively or negatively have influenced H_2O_2 reactions. H_2O_2 interacts with additional PGRs to control a variety of responses to physiological and ecological variables. The complex crosstalk between H_2O_2 and other phytohormones (such as auxins, GAs, CKs, ABA, ET, and SA), NO and Ca^{2+} (as signalling molecules), and BRs (polyhydroxylated steroidal PGRs) makes it imperative to evaluate the control of a variety of physiological functions. Intriguingly, recent molecular studies have linked H_2O_2 to the regulation of signal transduction by TFs and MAPKs, a crucial mechanism for adapting to external conditions and allowing moderated intracellular changes necessary for cellular homeostasis. As far as we can tell, H_2O_2 greatly aids plant development and metabolism. The knowledge of physiological and molecular aspects of plant tolerance to heavy metal stress appears to be crucial for developing new plant varieties that would be more successful in coping with global climate changes. In the near future, it will be important to investigate the precise roles of H_2O_2 signalling in the conflict between crops and environmental conditions. While our understanding of the physiology affected by H_2O_2 has advanced significantly in recent years, the signalling system underlying this effect still contains many unknowns. These results have realistic significance in addition to being of basic significance.

REFERENCES

Amor, N. B., Jiménez, A., Megdiche, W., Lundqvist, M., Sevilla, F., and Abdelly, C. (2006). Response of antioxidant systems to NaCl stress in the halophyte Cakile maritima. *Physiologia Plantarum*, *126*(3), 446–457.

Arfan, M., Zhang, D. W., Zou, L. J., Luo, S. S., Tan, W. R., Zhu, T., and Lin, H. H. (2019). Hydrogen peroxide and nitric oxide crosstalk mediates brassinosteroids induced cold stress tolerance in *Medicagotruncatula*. *International Journal of Molecular Sciences*, *20*(1), 144.

Arnaud, D., Lee, S., Takebayashi, Y., Choi, D., Choi, J., Sakakibara, H., and Hwang, I. (2017). Cytokinin-mediated regulation of reactive oxygen species homeostasis modulates stomatal immunity in Arabidopsis. *Plant Cell*, *29*(3), 543–559.

Asati, A., Pichhode, M., and Nikhil, K. (2016). Effect of heavy metals on plants: An overview. *International Journal of Application or Innovation in Engineering & Management*, *5*(3), 56–66.

Bienert, G. P., Møller, A. L., Kristiansen, K. A., Schulz, A., Møller, I. M., Schjoerring, J. K., and Jahn, T. P. (2007). Specific aquaporins facilitate the diffusion of hydrogen peroxide across membranes. *Journal of Biological Chemistry*, *282*(2), 1183–1192.

Carol, R. J., and Dolan, L. (2006). The role of reactive oxygen species in cell growth: lessons from root hairs. *Journal of Experimental Botany*, *57*(8), 1829–1834.

Cembrowska-Lech, D., Koprowski, M., and Kępczyński, J. (2015). Germination induction of dormant Avena fatua caryopses by KAR(1) and GA(3) involving the control of reactive oxygen species (H_2O_2 and $O_2(-)$) and enzymatic antioxidants (superoxide dismutase and catalase) both in the embryo and the aleurone layers. *Journal of Plant Physiology*, *176*, 169–179.

Černý, M., Habánová, H., Berka, M., Luklová, M., and Brzobohatý, B. (2018). Hydrogen peroxide: Its role in plant biology and crosstalk with signalling networks. *International Journal of Molecular Sciences*, *19*(9), 2812.

Chandra, S., and Low, P. S. (1995). Role of phosphorylation in elicitation of the oxidative burst in cultured soybean cells. *Proceedings of the National Academy of Sciences*, *92*(10), 4120–4123.

Chang, Z., Liu, Y., Dong, H., Teng, K., Han, L., and Zhang, X. (2016). Effects of cytokinin and nitrogen on drought tolerance of creeping bentgrass. *PLoS One*, *11*(4), e0154005.

Cheeseman, J. M. (2006). Hydrogen peroxide concentrations in leaves under natural conditions. *Journal of Experimental Botany*, *57*(10), 2435–2444.

Chen, Z., and Gallie, D. R. (2004). The ascorbic acid redox state controls guard cell signalling and stomatal movement. *The Plant Cell*, *16*(5), 1143–1162.

Chou, T. S., Chao, Y. Y., and Kao, C. H. (2012). Involvement of hydrogen peroxide in heat shock- and cadmium-induced expression of ascorbate peroxidase and glutathione reductase in leaves of rice seedlings. *Journal of Plant Physiology*, *169*(5), 478–486.

Choudhary, S. P., Kanwar, M., Bhardwaj, R., Yu, J. Q., and Tran, L. S. P. (2012a). Chromium stress mitigation by polyamine-brassinosteroid application involves phytohormonal and physiological strategies in Raphanus sativus L. *PLoS One*, *7*(3), e33210.

Choudhary, S. P., Oral, H. V., Bhardwaj, R., Yu, J. Q., and Tran, L. S. P. (2012b). Interaction of brassinosteroids and polyamines enhances copper stress tolerance in Raphanus sativus. *Journal of Experimental Botany*, *63*(15), 5659–5675.

Cocetta, G., and Ferrante, A. (2018). Postharvest application of hydrogen peroxide and salicylic acid differently affects the quality and vase life of cut rose (Rosa hybrida L.) petals and leaves. *Advances in Horticultural Science*, *32* (3), 371–378.

Considine, M. J., and Foyer, C. H. (2014). Redox regulation of plant development. *Antioxidants & Redox Signaling*, *21*(9), 1305–1326.

D'haese, D., Vandermeiren, K., Asard, H., and Horemans, N. (2005). Other factors than apoplasticascorbate contribute to the differential ozone tolerance of two clones of *Trifolium repens* L. *Plant, Cell & Environment*, *28*(5), 623–632.

Ellouzi, H., Sghayar, S., and Abdelly, C. (2017). H_2O_2 seed priming improves tolerance to salinity; drought and their combined effect more than mannitol in Cakilemaritima when compared to Eutremasalsugineum. *Journal of Plant Physiology*, *210*, 38–50.

Elmongy, M. S., Cao, Y., Zhou, H., and Xia, Y. (2018). Root development enhanced by using indole-3-butyric acid and naphthalene acetic acid and associated biochemical changes of in vitro Azalea microshoots. *Journal of Plant Growth Regulation*, *37*, 813–825.

Emamverdian, A., Ding, Y., Mokhberdoran, F., and Xie, Y. (2015). Heavy metal stress and some mechanisms of plant defense response. *The Scientific World Journal*, 2015, 756120.

Fariduddin, Q., Khan, T. A., and Yusuf, M., (2014). Hydrogen peroxide mediated tolerance to copper stress in the presence of 28-homobrassinolide in Vignaradiata. *ActaPhysiologiae Plantarum*, *36*(10), 2767–2778.

Gao, T., Shi, L., Zhang, T., Ren, A., Jiang, A., Yu, H., and Zhao, M. (2018). Cross talk between calcium and reactive oxygen species regulates hyphal branching and ganoderic acid biosynthesis in ganodermalucidum under copper stress. *Applied and Environmental Microbiology*, *84*(13), e00438.

Garg, N., and Bharti, A. (2018). Salicylic acid improves *arbuscular mycorrhizal* symbiosis, and chickpea growth and yield by modulating carbohydrate metabolism under salt stress. *Mycorrhiza*, *28*(8), 727–746.

González, A., Cabrera, M. de L., Henríquez, M. J., Contreras, R. A., Morales, B., and Moenne, A. (2012). Cross talk among calcium, hydrogen peroxide, and nitric oxide and activation of gene expression involving calmodulins and calcium-dependent protein kinases in Ulva compressa exposed to copper excess. *Plant Physiology*, *158*(3), 1451–1462.

Guzel, S., and Terzi, R. (2013). Exogenous hydrogen peroxide increases dry matter production, mineral content and level of osmotic solutes in young maize leaves and alleviates deleterious effects of copper stress. *Botanical Studies*, *54*(1), 1–10.

Han, S., Fang, L., Ren, X., Wang, W., and Jiang, J. (2015). MPK6 controls H_2O_2-induced root elongation by mediating Ca^{2+} influx across the plasma membrane of root cells in Arabidopsis seedlings. *The New Phytologist*, *205*(2), 695–706.

Hancock, J. T., Henson, D., Nyirenda, M., Desikan, R., Harrison, J., Lewis, M., and Neill, S. J. (2005). Proteomic identification of glyceraldehyde 3-phosphate dehydrogenase as an inhibitory target of hydrogen peroxide in Arabidopsis. *Plant Physiology and Biochemistry*, *43*(9), 828–835.

Hancock, J., Desikan, R., Harrison, J., Bright, J., Hooley, R., and Neill, S. (2006). Doing the unexpected: proteins involved in hydrogen peroxide perception. *Journal of Experimental Botany*, *57*(8), 1711–1718.

Hasanuzzaman, M., Nahar, K., Gill, S. S., Alharby, H. F., Razafindrabe, B. H., and Fujita, M. (2017). Hydrogen peroxide pretreatment mitigates cadmium-induced oxidative stress in Brassica napus L.: An intrinsic study on antioxidant defense and glyoxalase systems. *Frontiers in Plant Science*, *8*, 115.

He, J. M., Xu, H., She, X. P., Song, X. G., and Zhao, W. M. (2005). The role and the interrelationship of hydrogen peroxide and nitric oxide in the UV-B-induced stomatal closure in broad bean. *Functional Plant Biology: FPB*, *32*(3), 237–247.

Hejazi, Z., Sadat, M., Raghib, M., Zerak, A., and Fitrat, K. (2018). Effect of different concentrations of GA3, H_2O_2 and bleach solutions on seed germination of guava (Psidium guajava L). *International Journal of Chemical Studies, New Delhi*, *6*(5), 2679–2681.

Hou, Q. Z., Wang, Y. P., Liang, J. Y., Jia, L. Y., Feng, H. Q., Wen, J., and Bai, J. Y. (2018). H_2O_2-induced acclimation of photosystem II to excess light is mediated by alternative respiratory pathway and salicylic acid. *Photosynthetica*, *56*(4), 1154–1160.

Htet, Y., Lu, Z., Trauger, S. A., and Tennyson, A. G. (2019). Hydrogen peroxide as a hydride donor and reductant under biologically relevant conditions. *Chemical Science*, *10*(7), 2025–2033.

Huang, Y., Lin, C., He, F., Li, Z., Guan, Y., Hu, Q., and Hu, J. (2017). Exogenous spermidine improves seed germination of sweet corn via involvement in phytohormone interactions, H_2O_2 and relevant gene expression. *BMC Plant Biology*, *17*(1), 1.

Hung, K. T., Hsu, Y. T., and Kao, C. H. (2006). Hydrogen peroxide is involved in methyl jasmonate-induced senescence of rice leaves. *Physiologia Plantarum*, *127*(2), 293–303.

Ishibashi, Y., Aoki, N., Kasa, S., Sakamoto, M., Kai, K., Tomokiyo, R., and Iwaya-Inoue, M. (2017). The interrelationship between abscisic acid and reactive oxygen species plays a key role in barley seed dormancy and germination. *Frontiers in Plant Science*, *8*, 275.

Jahan, A., Khan, M. M. A., Ahmad, B., Ahmed, K. B. M., Pandey, R. P., and Gulfishan, M. (2023). Hydrogen peroxide: Regulator of plant development and abiotic stress response. In: Faizan, M., Hayat, S., and Ahmed, S. M. (eds.) *Reactive Oxygen Species: Prospects in Plant Metabolism* (pp. 213–228). Singapore: Springer Nature Singapore.

Jiang, Y. P., Cheng, F., Zhou, Y. H., Xia, X. J., Mao, W. H., Shi, K., Chen, Z. X., and Yu, J. Q. (2012). Hydrogen peroxide functions as a secondary messenger for brassinosteroids-induced CO_2 assimilation and carbohydrate metabolism in Cucumis sativus. *Journal of Zhejiang University Science B*, *13*(10), 811–823.

Khan, T. A., Yusuf, M., Ahmad, A., Bashir, Z., Saeed, T., Fariduddin, Q., and Wu, T. (2019). Proteomic and physiological assessment of stress sensitive and tolerant variety of tomato treated with brassinosteroids and hydrogen peroxide under low-temperature stress. *Food Chemistry*, *289*, 500–511.

Kolbe, A., Oliver, S. N., Fernie, A. R., Stitt, M., van Dongen, J. T., and Geigenberger, P. (2006). Combined transcript and metabolite profiling of Arabidopsis leaves reveals fundamental effects of the thiol-disulfide status on plant metabolism. *Plant Physiology*, *141*(2), 412–422.

Li, P. H., Sioson, A., Mane, S. P., Ulanov, A., Grothaus, G., Heath, L. S., Murali, T. M., Bohnert, H. J., and Grene, R. (2006) Response diversity of Arabidopsis thaliana ecotypes in elevated [CO_2] in the field. *Plant Molecular Biology*, *62*, 593–609.

Li, X. P., Xu, Q. Q., Liao, W. B., Ma, Z. J., Xu, X. T., Wang, M., ... and Zhu, Y. C. (2016). Hydrogen peroxide is involved in abscisic acid-induced adventitious rooting in cucumber (Cucumis sativus L.) under drought stress. *Journal of Plant Biology*, *59*, 536–548.

Liu, J., Shi, X., Qian, M., Zheng, L., Lian, C., Xia, Y., and Shen, Z. (2015). Copper-induced hydrogen peroxide upregulation of a metallothionein gene, OsMT2c, from Oryzasativa L. confers copper tolerance in Arabidopsis thaliana. *Journal of Hazardous Materials*, *294*, 99–108.

Miller, G. A. D., and Mittler, R. O. N. (2006). Could heat shock transcription factors function as hydrogen peroxide sensors in plants? *Annals of Botany*, *98*(2), 279–288.

Moller, I. M. (2001). Plant mitochondria and oxidative stress: Electron transport, NADPH turnover, and metabolism of reactive oxygen species. *Annual Review of Plant Biology*, *52*(1), 561–591.

Nazir, F., Hussain, A., and Fariduddin, Q. (2019a). Interactive role of epibrassinolide and hydrogen peroxide in regulating stomatal physiology, root morphology, photosynthetic and growth traits in Solanumlycopersicum L. under nickel stress. *Environmental and Experimental Botany*, *162*, 479–495.

Nazir, F., Hussain, A., and Fariduddin, Q. (2019b). Hydrogen peroxide modulate photosynthesis and antioxidant systems in tomato (Solanumlycopersicum L.) plants under copper stress. *Chemosphere*, *230*, 544–558.

Niu, L., and Liao, W. (2016). Hydrogen peroxide signaling in plant development and abiotic responses: Crosstalk with nitric oxide and calcium. *Frontiers in Plant Science*, *7*, 230.

Pei, Z., Murata, Y., Benning, G., Thomine, S., Kluesener, B., Allen, G. J., Grill, E., and Schroeder, J. I. (2000) Calcium channels activated by hydrogen peroxide mediate abscisic acid signalling in guard cells. *Nature*, *406*, 731–734.

Qi, J., Song, C. P., Wang, B., Zhou, J., Kangasjärvi, J., Zhu, J. K., and Gong, Z. (2018). Reactive oxygen species signaling and stomatal movement in plant responses to drought stress and pathogen attack. *Journal of Integrative Plant Biology*, *60*(9), 805–826.

Rentel, M. C., and Knight, M. R. (2004). Oxidative stress-induced calcium signaling in Arabidopsis. *Plant Physiology*, *135*(3), 1471–1479.

Richards, S. L., Laohavisit, A., Mortimer, J. C., Shabala, L., Swarbreck, S. M., Shabala, S., and Davies, J. M. (2014). Annexin 1 regulates the H_2O_2-induced calcium signature in Arabidopsis thaliana roots. *The Plant Journal: For Cell and Molecular Biology*, *77*(1), 136–145.

Saxena, I., Srikanth, S., and Chen, Z. (2016). Cross talk between H_2O_2 and interacting signal molecules under plant stress response. *Frontiers in Plant Science*, *7*, 570.

Schopfer, P. (1994). Histochemical demonstration and localization of H_2O_2 in organs of higher plants by tissue printing on nitrocellulose paper. *Plant Physiology*, *104*(4), 1269–1275.

Shan, C., Sun, H., Zhou, Y., and Wang, W. (2019). Jasmonic acid-induced hydrogen sulfide activates MEK1/2 in regulating the redox state of ascorbate in *Arabidopsis thaliana* leaves. *Plant Signaling & Behavior*, *14*(8), 1629265.

Shin, R., Berg, R. H., and Schachtman, D. P. (2005). Reactive oxygen species and root hairs in Arabidopsis root response to nitrogen, phosphorus and potassium deficiency. *Plant and Cell Physiology*, *46*(8), 1350–1357.

Shu, S., Gao, P., Li, L., Yuan, Y., Sun, J., and Guo, S. (2016). Abscisic acid-induced H_2O_2 accumulation enhances antioxidant capacity in pumpkin-grafted cucumber leaves under Ca $(NO_3)_2$ stress. *Frontiers in Plant Science*, *7*, 1489.

Sondhi, N., Bhardwaj, R., Kaur, S., Kumar, N., and Singh, B. (2008). Isolation of 24-epibrassinolide from leaves of Aeglemarmelos and evaluation of its antigenotoxicity employing Allium cepa chromosomal aberration assay. *Plant Growth Regulation*, *54*, 217–224.

Song, Y. J., Joo, J. H., Ryu, H. Y., Lee, J. S., Bae, Y. S., and Nam, K. H. (2007). Reactive oxygen species mediate IAA-Induced ethylene production in mungbean (Vignaradiata L.) hypocotyls. *Journal of Plant Biology*, *50*(1), 18–23.

Stambulska, U. Y., Bayliak, M. M., and Lushchak, V. I. (2018). Chromium (VI) toxicity in legume plants: modulation effects of rhizobial symbiosis. *BioMed Research International*, 2018, 8031213.

Su, C., Liu, L., Liu, H., Ferguson, B. J., Zou, Y., Zhao, Y., and Li, X. (2016). H_2O_2 regulates root system architecture by modulating the polar transport and redistribution of auxin. *Journal of Plant Biology*, *59*, 260–270.

Suzuki, N., and Mittler, R. (2006). Reactive oxygen species and temperature stresses: a delicate balance between signaling and destruction. *Physiologiaplantarum*, *126*(1), 45–51.

Sylvester, D., and Minati, S. K. (2018) Levels and health risk assessment of heavy metals in soil water and vegetables of dar es salaam tanzania. *Journal of Chemistry*, 2018, 1–9.

Tewari, R. K., Kumar, P., Tewari, N., Srivastava, S., and Sharma, P. N. (2004). Macronutrient deficiencies and differential antioxidant responses—influence on the activity and expression of superoxide dismutase in maize. *Plant Science*, *166*(3), 687–694.

Tian, Y., Fan, M., Qin, Z., Lv, H., Wang, M., Zhang, Z., and Bai, M. Y. (2018). Hydrogen peroxide positively regulates brassinosteroid signaling through oxidation of the BRASSINAZOLE-RESISTANT1 transcription factor. *Nature Communications*, *9*(1), 1063.

Wu, Q., Su, N., Zhang, X., Liu, Y., Cui, J., and Liang, Y. (2016). Hydrogen peroxide, nitric oxide and UV RESISTANCE LOCUS8 interact to mediate UV-B-induced anthocyanin biosynthesis in radish sprouts. *Scientific Reports*, *6*, 29164.

Xu, F. J., Jin, C. W., Liu, W. J., Zhang, Y. S., and Lin, X. Y. (2011). Pretreatment with H_2O_2 alleviates aluminum -induced oxidative stress in wheat seedlings. *Journal of Integrative Plant Biology*, *53*(1), 44–53.

Yang, C. Y. (2014). Ethylene and hydrogen peroxide are involved in hypoxia signaling that modulates AtERF73/HRE1 expression. *Plant Signaling & Behavior*, *9*(3), 28583.

Yang, W., Zhu, C., Ma, X., Li, G., Gan, L., Ng, D., and Xia, K. (2013). Hydrogen peroxide is a second messenger in the salicylic acid-triggered adventitious rooting process in mung bean seedlings. *PloS One*, *8*(12), e84580.

Ye, Q., and Steudle, E. (2006). Oxidative gating of water channels (aquaporins) in corn roots. *Plant, Cell & Environment*, *29*(4), 459–470.

Yıldız, M., Terzi, H., and Bingül, N. (2013). Protective role of hydrogen peroxide pretreatment on defense systems and BnMP1 gene expression in Cr (VI)-stressed canola seedlings. *Ecotoxicology*, *22*, 1303–1312.

Zhu, T., Deng, X., Zhou, X., Zhu, L., Zou, L., Li, P., Zhang, D., and Lin, H. (2016). Ethylene and hydrogen peroxide are involved in brassinosteroid-induced salt tolerance in tomato. *Scientific Reports*, *6*, 35392.

Zou, J. J., Li, X. D., Ratnasekera, D., Wang, C., Liu, W. X., Song, L. F., Zhang, W. Z., and Wu, W. H. (2015). Arabidopsis CALCIUM-DEPENDENT PROTEIN KINASE8 and CATALASE3 function in abscisic acid-mediated signaling and H_2O_2 homeostasis in stomatal guard cells under drought stress. *The Plant Cell*, *27*(5), 1445–1460.

13 Role of Hydrogen Peroxide on Horticultural Crops

Nilofer and Abdul Mazeed

INTRODUCTION

Horticulture is a subfield of agriculture which deals with garden crops usually fruits, vegetables, and ornamental plants. The word horticulture is derived from two Latin words *hortus*, which means "garden or orchard" and *cultura*, which means "to cultivate". Although it alludes to heavy commercial production, in everyday usage it applies to all sorts of garden maintenance. Though all forms of production inherently have tight ties to one another, horticulture is positioned between field agriculture and residential gardening in terms of scale. The two types of plants grown in horticulture are plants for food (pomology and olericulture) and plants for adornment (floriculture and landscape horticulture). Fruit and nut crops are the focus of pomology; edible herbaceous plants, moreover such as carrots (edible root), asparagus (edible stem), lettuce (edible leaf), cauliflower (edible flower buds), tomatoes (edible fruit), and peas (edible seed), are the subject of olericulture. Culturing flowers and decorative plants, such as cut flowers, potted plants, and greenery, is known as floriculture. Plants for the landscape, such as grass turf, and particular interest in nursery crops including shrubs, trees, and vines are included in the wide area of landscape horticulture. Numerous elements, including topography, climate, and other geographical variances, affect both the crop's success and the horticulturist's area of expertise. Climatic changes such as altering soil and irrigation water, increasing the buildup of salts in agricultural soil, and decreasing the quantity and quality of irrigation water in agricultural areas globally, particularly in arid regions will have a detrimental effect on plant production; this will result in soil salinization and poor irrigation water quality (Corwin, 2021). One of the main abiotic stresses affecting horticultural crops, particularly vegetables, is the salinity of the soil and water. Light, temperature, water, humidity, and nutrition are additional environmental factors that impact plant growth. It's critical to comprehend how these elements impact horticulture crops' growth and development.

Oxidative stress causes stressed plants to produce excess hydrogen peroxide, a non-radical reactive oxygen species (ROS) that builds up. Although hydrogen peroxide (H_2O_2) is known to be toxic and to cause oxidative damage to plant tissues, some recent research has demonstrated that, when applied at low concentrations, H_2O_2 can improve plant growth and mitigate the negative effects of stress on plants by increasing the activity of antioxidant enzymes, the stability of plant membranes, and the metabolism of lipids, fatty acids, and carbohydrates (Sun et al., 2016; Gohari et al., 2020). H_2O_2 also alters physiological processes, which boost a plant's resistance to abiotic stress. Applying H_2O_2 is an easy, practical, economical, and effective strategy to support plants in their defense against abiotic stress. It improves plant growth and yield, relative water content, membrane stability index, and photosynthetic efficiency (Hajivar and Zare-Bavani, 2019).

Low-concentration H_2O_2 application increases accumulation of inorganic and organic solutes, fresh herb production, and essential oil *ocimum* spp. while Na^+ and Cl^- content decreases under salt stress (Silva et al., 2019); in tomato improved growth traits and increased fruit count was observed by Hajivar and Zare-Bavani (2019); under salt stress, growth and development of wheat root increased (Li et al., 2011); improved plant resistance to salt stress by decreased accumulation of H_2O and MDA and increased antioxidant enzyme activity (Li et al., 2011;Liu et al., 2022); and H_2O_2 spray improved the growth and production of soursop plants while reducing the detrimental effects of salt stress.

DOI: 10.1201/9781032647692-13

Additionally, it improved the water and mineral condition of stressed maize plants (Guzel and Terzi, 2013). However, the concentration, application technique, plant type, and developmental stage all affect the effectiveness of the application of H_2O_2 (Capitulino et al., 2023), while both seed priming and foliar spraying with H_2O_2 mitigate the negative effects of drought stress, foliar treatment is the most effective method that significantly increased yield and yield components of rice stressed plants (Jira-anunkul and Pattanagul, 2021). As documented in wheat, high concentrations of H_2O_2 hindered plant development, produced substantial oxidative damage, and eventually resulted in plant cell death and low concentrations of H_2O_2 strengthened the antioxidative defense system and raised crop yields (Asghar et al., 2021).

INDUSTRIAL HISTORY OF HYDROGEN PEROXIDE

Louis Jacques Thenard made the initial discovery of H_2O_2 in 1818 when he observed the interaction between barium peroxide and nitric acid (McDonnell, 2014). Up to the development of three methods-electrolytic, anthraquinone, and isopropyl-alcohol oxidation in the middle of the 20th century, variations of this reaction were use and nowadays, the bulk of industrial manufacturing uses the anthraquinone auto-oxidation process, which was initially created by DuPont Chemical (Wernimont et al., 1998). As the formula indicates, H_2O_2 is composed of two hydrogen and two oxygen atoms, with one oxygen atom bonded to both hydrogen and the additional oxygen atom as a valence tetrad. At industrial quantities above 10%, it exists as a pale blue liquid; however, when diluted to the 3% concentration often seen in pharmacies, it loses all color. H_2O_2 is a moderately unstable substance that may break down spontaneously or with the help of a catalyst under a range of environmental circumstances. The spectrum of catalysts includes naturally occurring enzymes like catalase (Bosmans et al., 2016; Van Haute et al., 2015), inorganic metals like manganese or iron oxides (McDonnell, 2014), and UV radiation from the sun (Sichel et al., 2009). This reaction can be regulated in a way that is particular to the application by using catalysts, whether used alone or in conjunction with catalysts. H_2O_2 is used in a wide range of global sectors, including pulp and paper, detergent production, wastewater purification, and fertigation systems.

Water treatment processes and industrial applications have long employed hydrogen peroxide (H_2O_2) as a potent oxidant. When H_2O_2 is catalyzed in water, it can generate a variety of free radicals and other reactive species. These organisms possess the capacity to alter or decompose organic substances (Petri et al., 2003). It is a chemical that is safe for the environment and is largely produced in plant cells during respiration, photosynthesis, and photorespiration. A small quantity is also produced during photosynthesis (Slesak et al., 2007). Concern over hydrogen peroxide's impact on fruit quality and plant growth and development is growing on a global scale. During photosynthesis, photorespiration, and respiration processes, plant cells are the primary source of hydrogen peroxide production. Being the most stable of the so-called ROS, it is essential for physiological processes as a signaling molecule. Increases in intra- and intercellular H_2O_2 levels when applied at the ideal concentration positively influenced fruit growth, development, and quality.

Carbon dioxide (CO_2) and water (H_2O) are used by plants during photosynthesis to make carbohydrate ($C_6H_{12}O_6$) and release oxygen (O_2). The carbohydrate is then turned into energy during respiration and utilized to construct new tissues. While respiration may happen at any time, in both light and dark conditions, photosynthesis requires light. The presence of O_2 in the earth's atmosphere is believed originates from photosynthesis activity (Slesak et al., 2007). A molecule with two unpaired electrons, each in a distinct π^* anti-bonding, is oxygen in its ground state. O_2's orbital and divalent reduction is a difficult procedure that calls for the creation of univalent intermediates. When a lone electron is compiled, O_2 is reduced to the superoxide anion radical, which demands energy input.

H_2O_2 is a molecule that has no net charge and is a member of the non-radical ROS (Halliwell, 2006). Hydrogen peroxide is much preferred to be a long-range signaling element than superoxide for this reason as well as the fact that H_2O_2 has a longer half-life than the radical superoxide anion (Vranova et al., 2002). Neill et al. (2002) provide support for this theory, stating that superoxide produces H_2O_2 during electron transport activities including photosynthesis and mitochondrial respiration, most likely uncontrollably.

According to Slesak et al. (2007), hydrogen peroxide is the most stable ROS, and as such, H_2O_2 is essential as a signaling molecule in many physiological processes, including as respiration, photosynthesis, translocation, and transpiration. As a result, these procedures will increase agricultural production and output. In both large- and small-scale farming, farmers are increasingly using H_2O_2 as a chemical to promote plant development. According to Quan et al. (2008), it is thought to be a signaling molecule and a regulator of several genes' expression in cells. Additionally, it was revealed that H_2O_2 regulates fruit shape and size elaboration, production, and quality.

EFFECT OF H_2O_2 ON THE SEED EMERGENCE AND EARLY SEEDLING OF FRUIT TREES

H_2O_2 can be applied as a treatment to increase the germination percentage of seeds and the growth of seedlings in a concentration-dependent manner, which can help with the horticultural problem of mild growth of seedlings under spontaneous conditions (Barba-Espin et al., 2010). A large amount of hydrogen peroxide, an extracellular ROS, is created during seed germination, early seedling growth, and seed aging (Kranner et al., 2010). Many recent studies have demonstrated that ROS would be essential for signaling the success of significant seed life events including germination and dormancy release. In reality, it has been demonstrated that H_2O_2 builds up during seed germination in a number of species. Exogenous employment of H_2O_2 has been shown in several studies to enhance seed germination in a variety of plants, including *Hordeum vulgare* (Fontaine et al., 1994), and spinach (Katzman et al., 2001). The commencement of seed germination is coordinated by H_2O_2, which acts as a main factor involving particular modifications at the transcriptome, proteome, and hormone levels. This accelerates the germination process, most likely because it invigorates the seeds (Barba-Espín et al., 2011). H_2O_2 would cause the abscisic acid content of the seed to decrease in a manner that is dependent on MAPK. It would also cause the carbonylation of seed storage proteins, which would favor their mobilization, and some glycolytic enzymes that could activate the phosphate pentose pathway (ox PPP), which in turn supplies NADPH for the thioredoxin system, which is involved in seed germination. It is now well known that H_2O_2 has a useful role as a major hub integrating signaling network during developmental processes and in response to biotic and abiotic stress. The most important stage of the plant life cycle, seed germination determines plant development and production. Seed germination is initiated by the intake of water, which also results in the production of ROS, particularly H_2O_2. The physiology of seeds has been shown to accumulate H_2O_2 and other ROS during imbibition and the early stages of germination, mostly due to a considerable increase in their intracellular and extracellular synthesis (Kubala et al., 2015). While H_2O_2 is also produced in dry seeds, it fulfills its function as cellular messenger or toxic molecule mainly when seeds become saturated such as imbibition and emergence. H_2O_2 is a long-lived ROS that is known to be a crucial signaling molecule because it can diffuse efficiently through membranes and reach targets that are far from the generation site (Møller et al., 2007). A decrease in the antioxidant system combined with a buildup of H_2O_2 and related oxidative damages might be considered as a form of stress that could impede the complete completion of germination. To achieve a balance between oxidative signaling, which promotes germination, and oxidative damage, which prevents or delays germination, cellular antioxidant machinery must precisely regulate H_2O_2 accumulation. H_2O_2 is also thought to be a signaling hub for the regulation of seed dormancy and germination (Bailly et al., 2008). It was reported that the moment seed submerged in water, it started to absorb and expand. Within 40 hours, it had grown around 40% of its original weight, and for the next 60 hours, not much more was absorbed (Alexander, 1966). The optimal time to add H_2O_2 should be now, by combining it with irrigation water. Ascorbate peroxidase (APX), peroxidase (POX), and ascorbate oxidase (AAO) all increased following the H_2O_2-pretreatment (Barba-Espin et al., 2010). Both the rise in pea seedling growth and the fall in ascorbate's redox state were connected with the increases in these ascorbate-oxidizing enzymes. Conversely, H_2O_2 had no effect on ABA suppression on the emergence percentage and coleoptile elongation, but it somewhat lessened ABA inhibitions on radicle elongations and fresh weights. Applying H_2O_2 to plants in their early stages also greatly benefited them by removing the impacts of temperature rises that hindered and delayed germination (Cavusoglu and Kabar, 2010). The pretreatment with hydrogen peroxide proved to be highly effective in mitigating the germination-delaying and preventing effects of elevated temperatures and salt levels. At all temperatures, it also significantly reduced the inhibitory effects of salt on seedling development. The proteomic analysis revealed that H_2O_2 activated proteins involved in plant signaling and development, cell cycle control, and cell elongation and division (Barba-Espin et al., 2010). Thus, H_2O_2 mediates the relationship between plant hormones and the redox state in the early phases of seedling development. Plant growth and signaling proteins are activated as a result of this interaction.

EFFECT OF H_2O_2 ON FLOWERING AND BUD FORMATION

Flower initiation marks the transition from vegetative to reproductive growth in seed plants (Zhou et al., 2012). As a result, it is a pivotal moment in the life cycle of these plants because it affects the typical relationship between vegetative and reproductive growth in seed plants, resulting from the flower's morphology. Modified shoot meristems, or flower primordia, produce transformed shoots, which are flowers. But if a meristem is identified as a floral primordium, it may be the very first steps toward returning to vegetative growth, as in seed plants, vegetative growth and reproductive development are rather mutually incompatible. According to Lang and Nitsch (1965), flower initiation denotes the conclusion of a specific meristem. During normal metabolism, H_2O_2 is regularly generated from several sources and is a signaling compound that interstitial responses to biotic and abiotic stresses (Neil et al., 2002), also accumulates in the buds during endodormancy. Further, H_2O_2 has been suggested to play a pivotal role in the control of endodormancy and bud break (Beauvieux et al., 2018). Since the plant antioxidant system typically eliminates H_2O_2 effectively, a decrease in antioxidant enzyme activity may moderate the oxidative impact of numerous stimuli as opposed to an increase in H_2O_2 production (Neil et al., 2002). Plants include a family of ubiquitous enzymes called peroxidase class III (POD), whose primary function is to oxidize molecules at the expense of H_2O_2. POD is therefore, generally considered an H_2O_2 detoxifying enzyme; however, in some fruits (grapevine and peach) POD can also produces H_2O_2. Most studies using chemical and physical dormancy breakdown stimuli such as hydrogen cyanamide (HC), sodium azide (AZ) and heat shock (HS) (Pérez and Burgos, 2004; Pérez and Lira, 2005; Pérez et al., 2008; Beauvieux et al., 2018) show a transient increase in H_2O_2 level within the grapevine bud which has been interpreted as a positive signal for the release of bud from the endodormancy. They were in fierce competition with one another during the development of panicles in fruit blossoming and bud formation, while nitric oxide (NO) and hydrogen peroxide (H_2O_2) play significant roles in the competition to stimulate reproductive growth by ameliorating the expression of the flower-related gene, *LcLFY* by preventing the growth of rudimentary leaves and chilling-occurred flowering increased H_2O_2 and NO contents in the mixed buds of litchi (Zhou et al., 2012). Once bud is produced, it normally goes into a latent state for a while. But early bud drop poses a significant challenge to fruit production. By treating the trees with 20 mM of H_2O_2, the bud drop of wax apple fruit (*Syzygium samarangense*) can be decreased (Khandaker et al.,2012).

EFFECT OF H_2O_2 ON FRUIT GROWTH AND DEVELOPMENT

Fruit is produced when one or more flowers reach maturity; the entire fruit is made up of the gynoecium of the flowers (Mauseth, 2003). Applying an H_2O_2 treatment to the crop can improve fruit development. Abscisic acid (ABA)-induced stomatal closure is mediated by H_2O_2, a signal that is crucial for the activation of calcium permeable channels in the plasma membrane.

Stomatal conductance, which is positively associated with photosynthesis, controls CO_2 fixation in leaf mesophyll tissue, which in turn impacts the photosynthetic rate. The H_2O_2-treated plants exhibited the lowest frequency of bud drop, a sign of more buds and superior quality. H_2O_2 increases the receptivity of flowers, which in turn increases fruit set (Souza et al., 2004). H_2O_2 can play a very important role in fruit set, preventing initial fruit drop and enhance fruit yield (Ozaki et al., 2009). Several previous studies also suggest that H_2O_2 treatment can improve fruit size as well as quality and quantity of fruit juice and biomass (Bryce et al., 1982). Applying a weekly 5 mM H_2O_2 spray to wax apple trees will improve fruit development. The photosynthetic rates, transpiration, stomatal conductance, chlorophyll, and dry matter content of the leaves, as well as the total soluble solids and total sugar content of the fruits of the wax apple (*Syzygium samarangense*), all shown a considerable increase (Khandaker et al., 2012). According to Geros et al. (2012), hydrogen peroxide has the ability to relax cell walls, which can either modify cell expansion at phase II or improve cellular growth during early cell division at phase I. In comparison to the control, the treated wax apple fruit with 5 mM H_2O_2 displayed greater fruit size, improved fruit set, fruit number, biomass, and yield (Khandaker et al., 2012). However, the fruit's natural H_2O_2 content may have aided in the ripening process (Geros et al., 2012).

EFFECT OF H_2O_2 ON FRUIT QUALITY DURING PREHARVEST

Public interest in the effects of food quality on human health has been growing. The goal of the agricultural sector in the past was to produce as many fruits as possible for commercial markets. But in order to improve their health, today's customers are more concerned with maximizing the nutritional content of food (Wang, 2010). As a result, a lot of focus is currently being paid to agricultural techniques that will improve the nutritional value of horticultural products grown today. Research has demonstrated that fruits are rich in antioxidant substances, including phenols, carotenoids, vitamins, flavonoids, dietary glutathione, and endogenous metabolites. According to Wang (2010), these antioxidants have the ability to function as enzyme inhibitors, free radical scavengers, peroxide decomposers, singlet and triplet oxygen quenchers, and synergists. In addition to many other health advantages, the many antioxidant components included in fruits, including bananas, may offer protection against heart disease and cancer (Agoreyo, 2012; Wang, 2010).

Preharvest factors include soil type, temperature, light intensity, composting, mulching, fertilizing, applying naturally occurring chemicals, and rising atmospheric carbon dioxide concentration will all have an impact on the antioxidant activity and content. The treated wax apple fruits with a concentration of 20 mM of H_2O_2 showed increased K (+), anthocyanin and carotene contents, flavonoid, phenol and soluble protein content, Sucrose Phosphate Synthase (SPS), Phenylalanine Ammonia Lyase (PAL), and antioxidant activities in relation to these factors (Khandaker et al., 2012). In fruits treated with H_2O_2, there was a positive association between peel color (hue) and Total Soluble Solids (TSS), between net photosynthesis and SPS activity, and between phenol and flavonoid content and antioxidant activity (Khandaker et al., 2012). Sweetness is a significant determinant of fruit quality. Standard handheld refractometers, hydrometers, electronic tongues, and High-Pressure Liquid Chromatography (HPLC) with various detectors can all be used to quantify sweetness and sugar content (Magwaza and Opara, 2015). H_2O_2 treatment improved melon fruit sweetening, according to a study by Ozaki et al. (2009), fruit will become more valuable because of improved quality, which will grow the fruit sector. When assessing if some fruit is susceptible to a certain illness, the harvesting date or time is crucial for "Fuji" apples, late harvest fruits showed the lowest levels of H_2O_2, malondialdehyde (MDA), and damage incidence and severity compared to early harvest fruits (Lu et al., 2014). Meanwhile, golden delicious apples (*Malus domestica*) is best to be picked seven days before the commercial harvest time as the H_2O_2 level is significantly increased with the increasing of superoxide dismutase activity (SOD) (Torres et al., 2003). For preservation, the picked fruits also require the ideal temperature. After being warmed, all "Fuji" apple fruits exhibited a sharp rise in both the severity and frequency of injuries, particularly during the first four days, during which polyphenol oxidase (PPO) activities as well as H_2O_2 and MDA levels increased. PPO activity was constant throughout storage at 0°C, and they demonstrated a relative resistance to harm. The development of injury symptoms was shown to be strongly correlated with the accumulations of H_2O_2 and MDA (Lu et al., 2014).

EFFECT OF H_2O_2 ON POSTHARVEST FRUIT QUALITY

Fruit quality can only be maintained once it has been harvested; it cannot be enhanced beyond the point of production. Mango quality features and the development of postharvest diseases have been found to be influenced by several factors, including genetics, geographic location, environmental conditions, and preharvest cultural methods such as canopy control, nutrition, and irrigation management (Rehman et al., 2015). The harvested fruits must be handled carefully to preserve their freshness, overall look, extended shelf life, and some nutritional content. Proper handling also minimizes the development of rot during storage. According to Aida et al. (2011), no discernible variations in microbe counts, physiochemical values, or sensory qualities were found between samples left untreated or treated with 1% or 3% of H_2O_2, hydrogen peroxide can be used as a sanitizing treatment on fresh-cut pineapple held at 5°C. Meanwhile, fresh-cut pineapple treated with 3% H_2O_2 had the highest lightness value and maintained flesh firmness better than 1% H_2O_2 (Aida et al., 2011).

HYDROGEN PEROXIDE FOR DISINFECTION

H_2O_2, a powerful oxidant, is mostly used to stop plant diseases from growing in fertigation water, however it also works effectively against pathogens that have already grown. H_2O_2 has a greater oxidizing potential (1.7 V) than typical disinfectants such free chlorine (1.36 V) and chlorine dioxide (1.57 V), although it is still less than ozone (2.07 V) (Newman, 2007). According to Linley et al. (2012), H_2O_2 breaks down into superoxide anions, hydroxyl radicals, and other very reactive chemicals that weaken cell membranes, cellular proteins, and occasionally DNA. Since neither disinfectant produces the dangerous disinfection byproducts (DPB) created when chlorine is used, their modes of action are comparable to those of ozone (Ronen et al., 2010). Two frequently occurring DPBs are the chlorates (ClO^{3-}) that are created when chlorine reacts with different organic compounds in fertigation water, as well as trihalomethanes (THMs) such dibromochloromethane and chloroform (Singer, 1994). These are recognized class B carcinogens that build up in recirculated fertigation systems' central return tank (Hua and Reckhow, 2007) and within plant tissues (Lopez-Galvez et al., 2018). Because of this, recirculated water becomes phytotoxic to crops and may cause environmental harm when it leaks out during discharge events (Singer, 1994). Any negative impact that a particular substance usually a chemical or an insect has on a plant's appearance, development, or germination is known as phytotoxicity (Raviv and Lieth, 2008). Regardless of the intermediate chemical pathways, employing H_2O_2 prevents environmental harm by ultimately breaking down into oxygen and water as byproducts (Olmez and Kretzschmar, 2009). This makes H_2O_2 a perfect fit for organic farming and green chemistry, where disinfectants must not create any DPBss regardless of the components of the water matrix (Carrasco and Urrestarazu, 2010). Decomposition of H_2O_2 will also raise the fertigation water's dissolved oxygen concentration, which might improve crop development by providing more oxygen in the root zone (Butcher,2017). Furthermore, because H_2O_2 is inexpensive and simple to employ in recirculated nutrient solutions, greenhouse producers may easily obtain it.

HYDROGEN PEROXIDE AND PLANT GROWTH

Hydrogen peroxide may promote plant development even if it can cause a phytotoxic reaction at doses lower than those advised for disease control. It has been demonstrated through research that H_2O_2 applied at concentrations lower than 100 mg/L can enhance shoot weight (Ahmad et al., 2014;

Hasan et al., 2016), induce root elongation (Lin and Kao, 2001; Li and Jai, 2013), boost root hair density (Foreman et al., 2003), enhance leaf chlorophyll content (Hasan et al., 2016), intensify leaf growth (Deng et al., 2012; Guzel and Terzi, 2013), and regulate overall metabolism (Carol and Dolan, 2006; Gechev and Hille, 2005). Although plant species, development stages, treatment frequency, and methods varied between experiments, phytotoxic effects were not recorded in our literature sample. There is scant evidence in the literature indicating that H_2O_2 treatments result in beneficial plant growth responses, despite variations in experimental techniques. It needs to be examined if this tendency holds true under regular or daily irrigation events utilizing H_2O_2-containing fertigation water.

Kato and Shimizu (1987) showed that deterioration of the subcellular environment engendered by leaf senescence accelerates chlorophyll bleaching by a peroxidase- H_2O_2 catalyzed antiphon. Nurnaeimah et al. (2020) indicates that applications of exogenous H_2O_2 significantly improved chlorophyll content in *F. deltoidea* plants. Hydrogen peroxide, a molecule that promotes growth, may shield the chloroplast ultrastructure, preserving photosynthetic pigments and increasing photosynthesis. Furthermore, *F. deltoidea* plants' stomatal conductance and photosynthetic rate increased in response to H_2O_2 treatments. This might be because of the H_2O_2 application increasing the concentration of CO_2, stomatal opening, and accumulated photosynthetic pigments. H_2O_2 was produced in plant cells by a variety of activities, including electron transport during photosynthesis, and it was beneficial to the growth and development of the plant. When a plant grows and develops, a variety of physiological activities, including stomatal movement, photosynthesis, photorespiration, senescence, the cell cycle, and biochemical reactions, can be mediated by H_2O_2 (Deng et al., 2012). Research also indicates that the photosynthetic production of *F. deltoidea* plants is considerably impacted by the application of H_2O_2.

CONCLUSIONS

In conclusion, fruit development, germination, bud formation, blossoming, and fruit quality preservation throughout the pre- and postharvest stages are all greatly impacted by hydrogen peroxide. By clearing the obstruction of abscisic acid, hydrogen peroxide aids in the germination delay and contributes to a reduction in pathogen virulence as well as the development of illness symptoms. The application of H_2O_2 is thus a practical and advantageous tool that may be used in the horticulture business.

REFERENCES

Ahmad, I., Basra, S.M.A. and Wahid, A. 2014. Exogenous application of ascorbic acid, salicylic acid and hydrogen peroxide improves the productivity of hybrid maize at low temperature stress. *Intl. J. Agr. Biol.* 16: 825–830.

Agoreyo, B.O., 2012. Oxidative stress and antioxidant enzyme response in ripening bananas (Musa sapientum). *Acta Hortic.*, 945: 119–124.

Aida, M.P.N., Hairiyah, M., Reza, W.H.W.M. and Ilida, M.N. 2011. Effect of hydrogen peroxide on quality of fresh cut pineapple stored at 50°C. Proceedings of the 7th International Pineapple Symposium, Volume 1, July 31, 2011, Johor Bahru, Malaysia.

Asghar, N., Akram, N.A., Ameer, A., Shahid, H., Kausar, S., Ashghar, A., Idrees, T., Mumtaz, S., Asfhan, H.M., Sultan, M. and Jahangir, I. 2021. Foliar applied hydrogen peroxide and proline modulates growth, yield and biochemical attributes of wheat (Triticum aestivum L.) under varied N and P levels. *Fresenius Environ. Bull.* 30(5): 5445–5465.

Bailly, C., El-Maarouf-Bouteau, H., and Corbineau, F. 2008. From intracellular signalling networks to cell death: The dual role of reactive oxygen species in seed physiology. *C. R.Biol.* 331, 806–814. https://doi.org/10.1016/j.crvi.2008.07.022

Barba-Espin, G., Diaz-Vivancos, P., Albacete, A. and Faize, L. 2010. Interaction between hydrogen peroxide and plant hormones during germination and the early growth of pea seedlings. *Plant Cell Environ.* 33: 981–994.

Barba-Espín, G., Diaz-Vivancos, P., Job, D., Belghazi, M., Job, C. and Hernández, J.A. 2011. Understanding the role of H_2O_2 during pea seed germination: A combined proteomic and hormone profiling approach. *Plant Cell Environ.* 34: 1907–1919. https://doi.org/10.1111/j.1365-3040.2011.02386.x

Beauvieux, R., Wenden, B. and Dirlewanger, E. 2018. Bud dormancy in perennial fruit tree species: A pivotal role for oxidative cues. *Front. Plant Sci.* 9: 657.

Bosmans, L., van Calenberge, B., Paeleman, A., Moerkens, R., Wittemans, L., van Kerckhove, S., de Mot, R., Lievens, B. and Rediers, H. 2016. Efficacy of hydrogen peroxide treatment for control of hairy root disease caused by rhizogenic agrobacteria. *J. Appl. Microbiol.* 121: 519–527.

Bryce, J.H., Focht, D.D. and Stolzy, L.H. 1982. Soil aeration and plant growth response to urea peroxide fertilization. *Soil Sci.* 134: 111–116.

Butcher, J.D., Charles, P., Laubscher, C.C. and Johannes, C.C. 2017. A study of oxygenation techniques and the chlorophyll responses of Pelargonium tomentosum grown in deep water culture hydroponics. *HosrtScience* 52: 952–957.

Capitulino, J.D., de Lima, G.S., Azivedo, C.A.V., da Silva, A.A.R., Arruda, T.F., Soares, L.A., dos, A., Gheyi, H.R., Fernandes, P.D., de Farias, M.S.S., de Assis da Silva, F. and Dias, M.D.S. 2023. Influence of foliar application of hydrogen peroxide on gas exchange, photochemical efficiency and growth of soursop under slat stress. *Plants* 12: 599. https://doi.org/10.3390/plants12030599

Carol, R.J. and Dolan, L. 2006. The role of reactive oxygen species in cell growth: Lessons from root hairs. *J. Exp. Bot.* 57: 1829–1834.

Carrasco, G. and Urrestarazu, M. 2010. Green chemistry in protected horticulture: The use of peroxyacetic acid as a sustainable strategy. *Intl. J. Mol. Sci.* 11: 1999–2009.

Cavusoglu, K. and Kabar, K. 2010. Effects of hydrogen peroxide on the germination and early seedling growth of barley under NaCl and high temperature stresses. *EurAsian J. BioSci.*, 4: 70–79.

Corwin D.L. 2021. Climate change impacts on soil salinity in agricultural areas. *Eur. J. Soil Sci.* 72(2): 842–862. https://doi.org/10.1111/ejss.13010

Deng, X.-P., Cheng, Y.-J., Wu, X.-B., Kwak, S.S., Chen, W. and Eneji, A.E. 2012. Exogenous hydrogen peroxide positively influences root growth and metabolism in leaves of sweet potato seedlings. *Aust. J. Crop Sci.* 6(11): 1572–1578.

Fontaine, O., Huault, C., Pavis, N. and Billard, J.P. 1994. Dormancy breakage of Hordeum vulgare seeds: Effect of hydrogen peroxide and stratification on glutathione level and glutathione reductase activity. *Plant Physiol. Biochem.* 32: 677–683.

Foreman, J., Demidchik, V., Bothwell, J.H.F., Mylona, P., Miedema, H., Torres, M.A., Linstead, P., Costa, S., Brownlee, C., Jones, J.D.G., Davies, J.M. and Dolan, L. 2003. Reactive oxygen species produced by NADPH oxidase regulate plant cell growth. *Nature* 422: 1–5.

Gechev, T.S. and Hille, J. 2005. Hydrogen Peroxide as a signal controlling programmed cell death. *J. Cell Biol.* 168: 17–20.

Geros, H., Chaves, M. and Delrot, S. 2012. *The Biochemistry of the Grapefruit*. Bentham Science Publishers, USA, p. 304. ISBN: 9781608053605.

Gohari, G., Alavi, Z., Esfandiari, E., Panahirad, S., Hajihoseinlou, S. and Eotopoulos, V. 2020. Interaction between hydrogen peroxide and sodium nitroprusside following chemical priming of *Ocimum basilicum* L. against salt stress. *Physiol. Plant.* 168(2): 361–373. https://doi.org/10.1111/ppl.13020

Guzel, S. and Terzi, R. 2013. Exogenous hydrogen peroxide increases dry matter production, mineral content and level of osmotic solutes in young maize leaves and alleviates deleterious effects of copper stress. *Bot. Sci.* 54: 1–10.

Hajivar, B. and Zare-Bavani, M.R. 2019. Alleviation of salinity stress by hydrogen peroxide and nitric oxide in tomato plants. *Adv. Hortic. Sci.* 33(3): 409–416. https://doi.org/10.13128/ahs-24335

Halliwal, B. 2006. Reactive species and antioxidants. Redox biology is a fundamental theme of aerobic life. *Plant Physiol.* 141(2); 312-322. https://doi.org/10.1104/pp.106.077073

Hasan, S.A., Irfan, M., Masrahi, Y.S., Khalaf, M.A. and Hayat, S. 2016. Growth, photosynthesis and antioxidant responses of Vigna unguiculata L. treated with hydrogen peroxide. *Cogent Foods Agri.* 2: 1–13.

Hua, G. and Reckhow, D.A. 2007. Comparison of disinfection byproduct formation from chlorine and alternative disinfectants. *Water Res.* 41: 1667–1678.

Jira-Anunkul, W. and Pattanagul, W. 2021. Effect of hydrogen peroxide application on agronomic traits of rice (*Oryza sativa* L.) under draught stress. *Plant Soil Environ.* https://doi.org/10.17221/628/2020-PSE

Kato, M. and Shimizu, S. 1987. Chlorophyll metabolism in higher plants. VII. Chlorophyll degradation in senescing tobacco leaves: Phenolic-dependent peroxidative degradation. *Can. J. Bot.* 65: 729–735.

Katzman, L.S., Taylor, A.G. and Langhans, R.W. 2001. Seed enhancements to improve spinach germination. *Hort Sci.* 36: 979–981.

Khandaker, M.M., Boyce, A.N. and Osman, N. 2012. The influence of hydrogen peroxide on the growth, development and quality of wax apple (*Syzygium samarangense*, [Blume] Merrill and L.M. Perry var. jambu madu) fruits. *Plant Physiol. Biochem.* 53: 101–110.

Kranner, I., Roach, T., Beckett, R.P., Whitaker, C. and Minibayeva, F.V. 2010. Extracellular production of reactive oxygen species during seed germination and early seedling growth in Pisum sativum. *J. Plant Physiol.* 167: 805–811.

Kubala, S., Wajtyla, L., Quinet, M., Lechowska, K., Lutts, S. and Garnczarska, M. 2015. Enhanced expression of proline synthesis gene P5CSA in relation to seed osmopriming improvement of Brassica napus germination under salinity stress. *J. Plant Physiol.* 183: 1–12. https://dx. doi.org/10.1016/j.jplph.2015.04.009

Lang, A. and Nitsch, J.P. 1965. Differentiation and development part IV(I). In: *Flower and Fruit Formation*, Lang, A. and J.P. Nitsch (Eds.). Springer, New York, pp. 1380–1647.

Li, A., Zhang, R., Pan, L., Tang, L., Zhao, G., Zhu, M., Chu, J., Sun, X., Wei, B., Zhang, X., Jia, J. and Mao, L. 2011. Transcriptome athenard nalysis of H_2O_2 treated wheat seedlings reveals a H_2O_2 responsive fatty acid desaturase gene participating in powdery mildew resistance. *PLoS One*. 6(12): e28810. https://doi.org/10.1371/journal.pone.0028810

Li, J., and Jia, H. 2013. Hydrogen peroxide is involved in cGMP modulating the lateral root development of Arabidopsis thaliana. *Plant Signal. Behav.* 8: 1–4.

Lin, C.C. and Kao, C.H. 2001. Cell wall peroxidase activity, hydrogen peroxide level and NaCl-inhibited root growth of rice seedlings. *Plant Soil* 230: 135–143.

Linley, E., Denyer, S.P., McDonnell, G., Simons, C. and Maillard, J.Y. 2012. Use of hydrogen peroxide as a biocide: new consideration of its mechanisms of biocidal action. *J. Antimicrob. Chemother.* 67: 1589–1596.

Liu, T., Pan, Z., Vequizo, J.J.M., Kato, K., Wu, B., Yamakata, A., Katayama, K., Chen, B., Chu, C. and Domen, K. 2022. Overall photosynthesis of H_2O_2 by an inorganic semiconductor. *Nat Commun.* https://doi.org/10.1038/s41467-022-28686-x

Lopez-Galvez, F., Andujar, S., Marin, A., Tudela, J.A., Allende, A. and Gil, M.I. 2018. Disinfection by-products in baby lettuce irrigated with electrolysed water. *J. Sci. Food Agric.* 98: 2981–2988.

Lu, X., Ma, Y. and Liu, X. 2014. Effects of hydrogen peroxide accumulation, lipid peroxidation and polyphenol oxidation during superficial scald development in 'Fuji' apples. *Hortic. Environ. Biotechnol.* 55: 299–307.

Magwaza, L.S. and Opara, U.L. 2015. Analytical methods for determination of sugars and sweetness of horticultural products-A review. *Sci. Hortic.* 184: 179–192.

Mauseth, J.D., 2003. *Botany: An Introduction to Plant Biology*. Jones and Bartlett, Burlington, MA, pp. 271–272.

McDonnell, G. 2014. The use of hydrogen peroxide for disinfection and sterilization applications. In: *PATAI's Chemistry of Functional Groups*, pp. 1–34. https://doi.org/10.1002/9780470682531.pat0885

Møller, I.M., Jensen, P.E. and Hansson, A. 2007. Oxidative modifications to cellular components in plants. *Annu. Rev. Plant Biol.* 58: 459–481.

Neill, S.J., Desikan, R., Clarke, A. and Hancock, J.T. 2002. Nitric oxide is a novel component of abscisic acid signalling in stomatal guard cells. *Plant Physiol.* 128: 13–16.

Neill, S.J., Desikan, R., Clarke, A. Hurst, R.D. and Hancock, J.T. 2002. Hydrogen Peroxide and Nitric oxide as signalling molecule in plants. J.Exp. Bot. 53(372); 1237-1247. https://doi.org/10.1093/jexbot/53.372.1237.

Newman, S.F., Buratti, B.J., Jaumann, R., Bauer, J.M. and Momary, T.W. 2007. Hydrogen peroxide on Enceladus. *Astrophys. J.* 670: L143–L146.

Nurnaeimah, N., Mat, N., Mohd, K.S., Baduluddin, N.A., Yusoff, N., Sajili, M.H., Mahmud, K, Adnan, A.F.M. and Khandaker, M.M. 2020. The effect of hydrogen peroxide on plant growth, mineral accumulation as well as biological and chemical properties of *Ficus deltoidea*. *Agronomy* 10, 599. https://doi.org/10.3390/agronomy10040599

Olmez, H. and Kretzschmar, U. 2009. Potential alternative disinfection methods for organic fresh-cut industry for minimizing water consumption and environmental impact. *LWT- Food Sci. Tech.* 42: 686–693.

Ozaki, K., Uchida, A., Takabe, T., Shinagawa, F., Tanaka, Y., et al., 2009. Enrichment of sugar content in melon fruits by hydrogen peroxide treatment. *J. Plant Physiol.* 166: 569–578.

Pérez, F.J. and Burgos, B. 2004. Alterations in the pattern of peroxidase isoenzymes and transient increases in its activity and in H_2O_2 levels take place during the dormancy cycle of grapevine buds: The effect of hydrogen cyanamide. *Plant Growth Regul.* 43: 213–220.

Pérez, F.J. and Lira, W. 2005. Possible role of catalase in post-dormancy bud break in grapevines. *J. Plant Physiol.* 162: 301–308.

Pérez, F.J., Vergara, R. and Rubio, S. 2008. H_2O_2 is involved in the dormancy-breaking effect of hydrogen cyanamide in grapevine buds. *Plant Growth Regul.* 55: 149–155.

Petri, B.G., Watts, R.J., Teel, A.L., Huling, S.G. and Brown, R.A. 2003. Fundamentals of ISCO using hydrogen peroxide. In: *In Situ Chemical Oxidation for Groundwater Remediation*, Siegrist, R.L., M. Crimi and T.J. Simpkin (Eds.). Springer, New York.

Quan, L.J., Zhang, B., Shi, W.W. and Li, H.Y. 2008. Hydrogen peroxide in plants: A versatile molecule of the reactive oxygen species network. *J. Integr. Plant Biol.* 50: 2–18.

Raviv, M., and Lieth, J.H. 2008. *Soilless Culture: Theory and Practice*. 2nd ed. Elsevier Science, Amsterdam.

Rehman, A., Malik, A.U., Ali, H., Alam, M.W. and Sarfraz, B. 2015. Preharvest factors influencing the postharvest disease development and fruit quality of mango. *J. Environ. Agric. Sci.* 3: 42–47.

Ronen, Z., Guerrero, A., and Gross, A. 2010. Greywater disinfection with the environmentally friendly Hydrogen Peroxide Plus (HPP). *Chemosphere* 78: 61–65.

Sichel, C., Fernandez-Ibanez, P., de Cara, M. and Tello, J. 2009. Lethal synergy of solar UV-radiation and H_2O_2 on wild Fusarium solani spores in distilled and natural well water. *Water Res.* 43: 1841–1850.

Silva, H., Neto, A.Z., Menezes, R., Silva, P. and Gheyi, H. 2019. Use of hydrogen peroxide in acclimation of basil (Ocimum basilicum L.) to salt stress. *Turk. J. Bot.* 43(2): 208–217. https://doi.org/10.3906/bot-1807-80

Singer, P.C. 1994. Control of disinfection by-products in drinking water. *J. Environ. Eng.* 120: 727–744.

Slesak, I., Libik, M., Karpinska, B., Karpinski, S. and Miszalski, Z. 2007. The role of hydrogen peroxide in regulation of plant metabolism and cellular signaling in response to environmental stresses. *Acta Biochim. Pol.* 54: 39–50.

Souza, M.M., Pereira, T.N. and Martins, E.R. 2004. Flower receptivity and fruit characteristics associated to time of pollination in the yellow passion fruit *Passiflora edulis* Sims f. *flavicarpa Degener (Passifloraceae). Sci. Hort.* 101: 373–385.

Sun, Y., Wang, H., Liu, S. and Peng, X. 2016. Exogenous application of hydrogen peroxide alleviates drought stress in cucumber seedlings. *South African J. Bot.* 106: 23–28.

Torres, R., Valentines, M.C., Usall, J., Vinas, I. and Larrigaudiere, C. 2003. Possible involvement of hydrogen peroxide in the development of resistance mechanisms in 'Golden Delicious' apple fruit. *Postharvest Biol. Technol.* 27: 235–242.

Vranova, E., Inze, D. and Van Breusegem, F. 2002. Signal transduction during oxidative stress. *J. Exp. Bot.* 53: 1227–1236.

Van Haute, S., Tryland, I., Veys, A. and Sampers, I. 2015. Wash water disinfection of a full-scale leafy vegetables washing process with hydrogen peroxide and the use of a commercial metal ion mixture to improve disinfection efficiency. *Food Control* 50: 173–183.

Wang, S.Y. 2010. Antioxidants in fruits and their possible anticancer property. *Acta Hortic.* 864: 463–469.

Wernimont, E., Ventura, M., Garboden, G. and Mullen, P. 1998. *Past and Present Uses of Rocket Grade Hydrogen Peroxide*. General Kinetics LLC, Aliso Viejo, CA.

Zhou, B., Li, N., Zhang, Z., Huang, X., Chen, H., et al., 2012. Hydrogen peroxide and nitric oxide promote reproductive growth in Litchi chinensis. *Biol. Plant.* 56: 321–329.

14 Protecting Effects of Hydrogen Peroxide in Contrast to Herbicides

Fasiha Qurashi, Alina Mansoor, Aiman Fatima, Mahnoor Fatima, and Hamda Hussain Qureshi

INTRODUCTION

Hydrogen peroxide (H_2O_2) works as a green shield against agricultural wields. Instead of killing weeds, its role focuses on the protection of crops. With its incredible potency and high reactivity, H_2O_2 is here to challenge traditionally used herbicides (Winterbourn, 2013). In this chapter, the effectiveness and mechanism of action of conventional herbicides will be discussed and compared with hydrogen peroxide and other alternative herbicides. The detrimental effects of chemical herbicides on soil, microorganisms, animals, aquatic life, and human beings will be explored. The strategies to manage residues of chemical herbicides will also be discussed. The ability of hydrogen peroxide to target invasive weeds while nourishing and strengthening agricultural staples selectively will also be explored. Unlike herbicides, which often leave lingering chemical residues and disrupt delicate ecosystems, hydrogen peroxide offers a clean and sustainable solution. The discussion on the chemistry of hydrogen peroxide and its reactive nature with different compounds will help the readers better understand the mode of action of this molecule. Its impacts on the environment in contrast to herbicides will be investigated. Furthermore, the limitations of herbicides will be addressed using case studies from a vast library of literature and how in the present age H_2O_2 plays a better role as a protectant of crops. The literature regarding the effectiveness of some natural products as herbicides will also be reviewed (Figure 14.1).

OVERVIEW OF H_2O_2

Hydrogen peroxide, a multifaceted molecule with a unifying theme of redox signaling, has a simplistic nature with two hydrogen and two oxygen molecules. Defying its humble composition, H_2O_2 is a biologically reactive species of oxygen. It is generated by respiration processes and the end product of metabolic pathways, especially peroxidase oxidation pathways (Figure 14.2).

There are several examples of exogenous H_2O_2 starting stress responses or redox signals, and receptor-mediated redox signaling which extensively involves endogenously produced H_2O_2 (Methods in Enzymology, Christian C.) This overview delves into the unique properties and intricate pathways through which H_2O_2 exerts its effects, focusing on its potential for both weed control and crop protection (Abdollahi & Hosseini, 2014) (Figure 14.3).

EXOGENOUS AND ENDOGENOUS H_2O_2

Two categories of hydrogen peroxide are discussed as; exogenous H_2O_2 is hydrogen peroxide that comes from external sources, such as environmental factors, pollutants, or chemical agents. Endogenous H_2O_2 is hydrogen peroxide that is produced within the cells as a byproduct of metabolic

 DOI: 10.1201/9781032647692-14

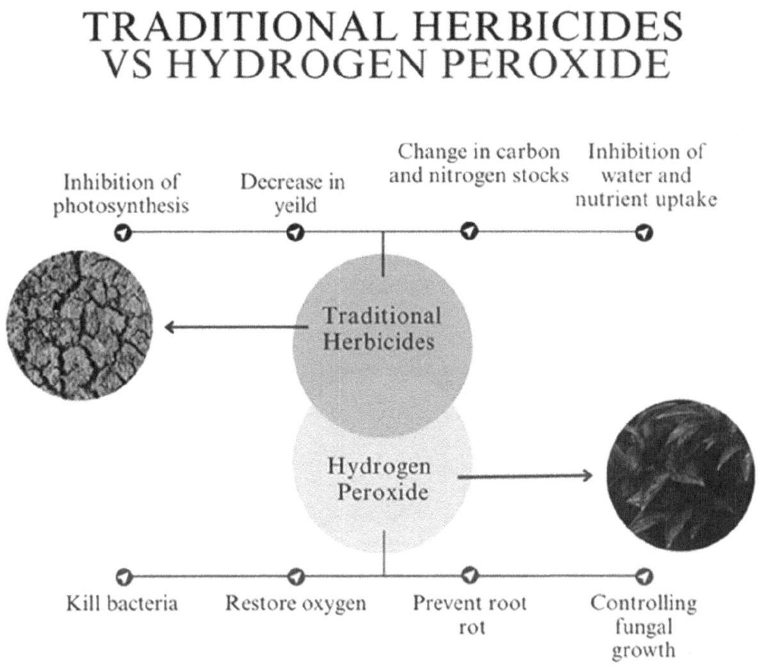

FIGURE 14.1 Comparison of traditional herbicides and hydrogen peroxide.

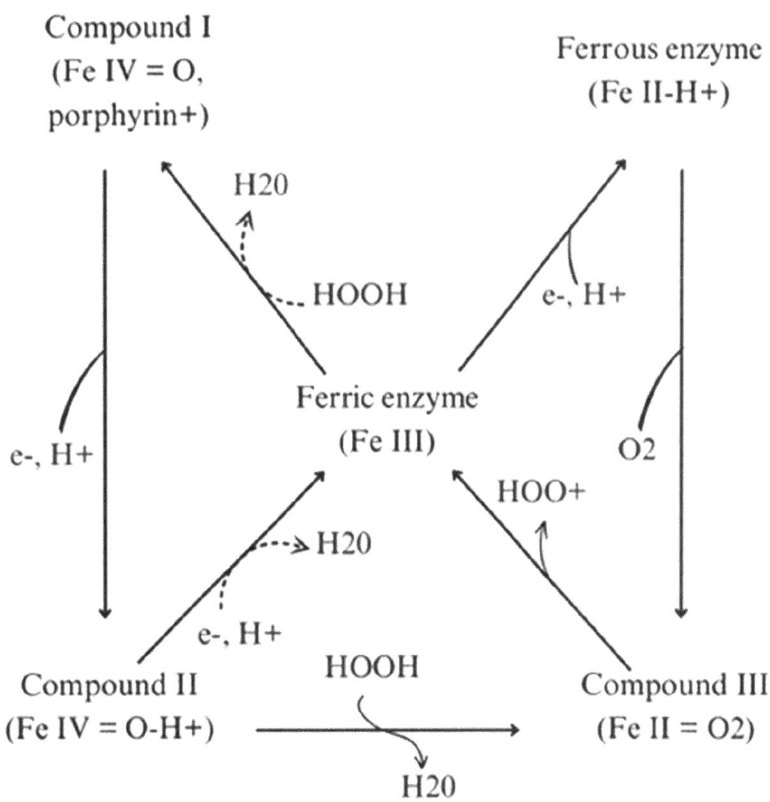

FIGURE 14.2 Peroxidase pathway.

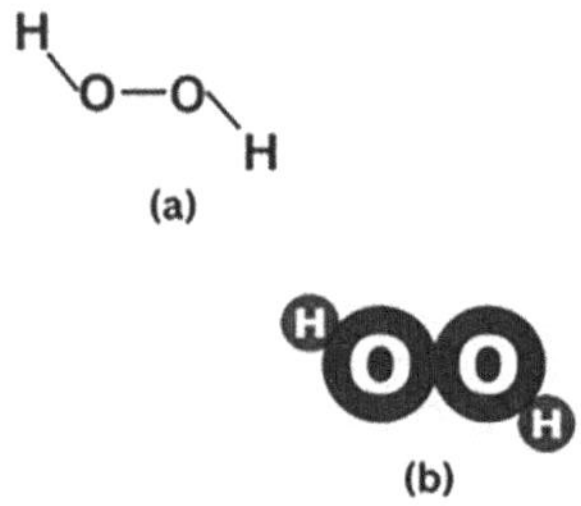

FIGURE 14.3 Molecule of H_2O_2.

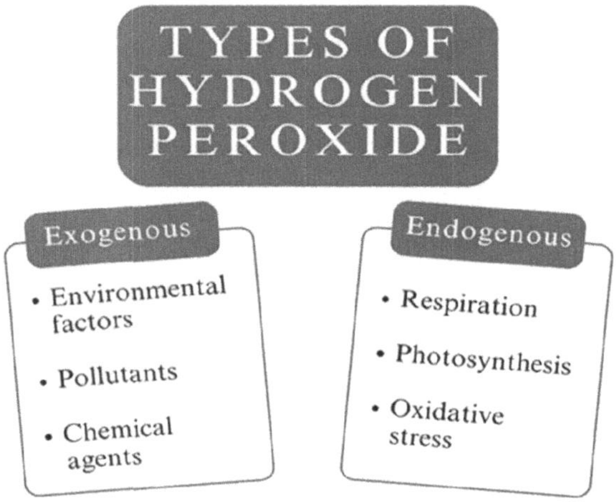

FIGURE 14.4 Types of Hydrogen peroxide; H_2O_2

processes including oxidative stress, respiration, and photosynthesis. Both exogenous and endogenous H_2O_2 can function as signaling molecules that regulate a range of cellular functions, like development, defense, growth, and death. However, excessive or unbalanced levels of H_2O_2 can also cause damage to components of the cell, such as proteins, DNA, and lipids. Therefore, plants have developed various mechanisms to scavenge or detoxify H_2O_2, such as catalase, peroxidase, and glutathione (Winterbourn, 2013) (Figure 14.4).

CHEMICAL NATURE OF H_2O_2

Hydrogen peroxide has a reduction potential of 1.32 V at pH 7.0 and it is a vigorous two-electron oxidant that is even more oxidizing than peroxynitrite or hypochlorous acid. However, since H_2O_2 has a barrier of high activation energy that needs to be broken through to release its oxidizing activity, it does not properly react with the majority of biomolecules, such as antioxidants having low molecular weight. Though the hydroxyl radical, the one-electron reduction product, is one of the strongest oxidants known, H_2O_2 itself is a modest one-electron oxidant. At physiological pH, hydrogen peroxide has a pKa of 11.6 and is largely uncharged (Winterbourn, 2013).

SYNTHESIS OF H_2O_2

H_2O_2 is generated through the anthraquinone process. In an organic working solution, anthraquinone gets reduced with H_2 and reoxidized with O_2 and a heterogeneous palladium catalyst catalyzes this reaction. In the next step of this reaction, H_2O_2 is released and then extracted with water.

However, this process of H_2O_2 synthesis is highly costly and is usable for large-scale production only. Moreover, high transportation and storage costs also limit using this process for small-scale production (Goor et al., 2012; Nishimi et al, 2011). For small-scale production of H_2O_2, two alternative methods are being used.

DIRECT SYNTHESIS OF H_2O_2

During direct synthesis, oxygen, and hydrogen are utilized to generate H_2O_2. This procedure can be carried out by mainly using heterogeneous catalysts applying platinum, gold, and palladium as active metals. Particularly, bimetallic catalysts made of gold and palladium showed excellent selectivity and activity (Ntainjua et al., 2009). To manage the decomposition of H_2O_2, this process is carried out under mild conditions at low pressure (20 bar) and room temperature. Polar reaction media include small alcohols and water to carry out this reaction (Salmi et al., 2015). A quick deactivation of the catalyst occurs in the presence of large quantities of a polar solvent. When small alcohols are compared with water, it has been observed that a higher amount of H_2O_2 is produced when the reaction occurs in alcohol than in water. This is due to the fact that gasses are highly soluble in alcohol. Moreover, it has been observed that the order of reaction of hydrogen is different in each medium (Paunovic et al., 2015).

OXIDATION OF SUGARS TO SUGAR ACIDS

Sugar acids have a significant role in pharmaceutical, alimentary, and fine chemical industries. Presently, sugars such as arabinose and glucose are oxidized to produce sugar acids (Chheda et al., 2007). In this reaction, oxygen is used as the oxidant and the procedure normally occurs in a semi-batch reactor. The catalysts used in this reaction are metal complexes, supported metal nanoparticles, and enzymes. The commonly used metals are platinum, gold, and palladium. Gold is particularly efficient in activating molecular oxygen to selectively produce sugar acid (Delidovich et al., 2010). However, the use of gaseous oxygen results in a low reaction rate, which is a cause of concern. Additionally, gold is highly structure-activity dependent, so more advanced methods are required for catalyst synthesis. The by-products of this reaction are performic acid and formic acid, depending on the conditions. Most of the previous studies suggest that nanoparticles of less than 3 nm were used and these are active for classical oxidation reactions with gold (Omri et al., 2016).

H_2O_2 AS A GREEN OXIDANT

Although chemical fumigants and herbicides applied in soil produce desired effects, they also have harmful effects such as environmental pollution and accumulation of toxic residues. So, it is crucial to find an efficient and environmentally friendly alternative agent. It is a reliable, safe, and simple agent with the byproduct, water, also known as *Green Oxidant.* But due to this researchers have been concerned about the use of hydrogen peroxide for a long time. At present H_2O_2 is commonly used in food safety, agriculture, medicine, environmental, and clinical applications (Xu et al., 2023). Hydrogen peroxide is used to kill and inhibit the growth of microbes on food containers and packaging bags. Its vapors are highly effective for that purpose and are commonly used in the field of agriculture. These H_2O_2 vapors enter the microbial cell by reacting with the lipid double bonds in the cell wall and changing the permeability of the cell membrane by reacting with lipids, polysaccharides, and proteins in the membrane, ultimately leading to cell lysis and death of the microbe. Moreover, conventional chemical fumigation requires film covering and time consuming. Conversely, H_2O_2 is a simpler and eco-friendly measure to kill pathogens in replanted soils (Salazar et al., 2017).

H_2O_2 ROLE IN PLANTS

Hydrogen Peroxide (H_2O_2) is part of the reactive oxygen species; ROS network in plants. It is generated in various compartments of cell as a result of normal metabolic processes or environmental stresses. It has two roles in plants: it acts as a signal molecule involved in growth at low concentrations, development, and stress responses; at high concentrations, it causes cell death due to oxidative damage. Plants have developed anticipated antioxidant systems to regulate H_2O_2 and maintain redox homeostasis. Hydrogen peroxide is a versatile molecule that also has several significant roles in plant physiology and adaptation (Ślesak et al., 2007). It also regulates the expression of many genes and interacts with other signaling molecules including salicylic acid; SA, nitric oxide; NO and ethylene.

COMPARATIVE METHOD

The comparative study analyzes the different variables and then combines them to highlight the similarities and differences between the applications of herbicides and hydrogen peroxide (H_2O_2) in agriculture (Mokhtarianpour, 2016). This method is helpful in determining the weaknesses and strengths of research designs, as it provides an idea about the pros and cons of the research (Holt, 1970). Comparative analysis explains the homologies and analogies of different kinds of situations between higher scales of social units, including cultures, nations, societies, and regions. This definition encompasses traditions such as cross-societal analysis in sociology, cross-national analysis in political science, cross-cultural analysis in anthropology, psychological analysis, and comparative historical analysis in history. Comparative analysis includes both quantitative and qualitative comparisons, making it a broad term. Social entities can be based on several lines, including political or geographical ones, in the form of regional or cross-national comparisons. There are different types of this kind of analysis: quantitative, qualitative, and fuzzy. Quantitative analysis is a variable-oriented analysis. Qualitative comparative analysis is a type of case-oriented analysis, whereas fuzzy comparative analysis is collection-oriented.

This research methodology examines different aspects of social science and social life globally. Researchers use case-study analysis to determine the similarities and differences between different groups through qualitative analysis. In this method, theoretical concepts are combined with data collection. Each comparative analysis has its own outline and precise goals, which the scholars explain in the introduction to their studies. It is useful to consider the intended audience of the study to locate the goals more broadly. To conduct a detailed analysis and an in-depth study of the different characteristics of two particular samples without neglecting any detail, a comparative study can be carried out. While it appears helpful to isolate the two general objectives scientifically, they can complement each other (Freiberger, 2019).

COMPARATIVE AGRICULTURE

One of the most important issue that mankind will have to face in the coming decades is the global agriculture and food issue. There are several questions regarding these issues, some of which are as follows; how to produce enough food to feed everyone? what are the most effective food distribution methods? There is need to solve all these problems and explore new farming techniques which can enhance food production. With the current approach our goal is to not only address the current agricultural challenges but find solutions that are safe for the ecosystem. This directly contributes to filling the gap in agricultural advancements and living standards had been increasing for the past years. The aim is to target cultivation regions worldwide and construct a strategy to overcome food shortages. Research has been conducted on both theoretical and practical levels. It is crucial to explore the practices done in the past and find the reasons for the depletion of resources, including habitat destruction, and how to control it. This is the role of a comparative agriculturist (Cochet, 2015).

Comparative agriculture attempts to explain the origin, transformations, and responsibilities of agriculturists in the evolution of life and humanity. It tracks the record of activities in different eras of history and in various regions of the world. This body of knowledge can be integrated into the general culture and provide a conceptual, theoretical, and methodological basis for those involved in agricultural, economic, and social development (Mazoyer & Roudart, 2017).

It is the science of transformations and adaptations of agricultural development processes and deals with the alterations and modifications of these processes. Establishing a meaningful comparative approach is essential for a precise understanding and interpretation of each specific type of agriculture, as well as the modes and consequences of their interlinking under the growing influence of markets. The aims of comparative agriculture are: (i) to identify common, rare, essential and optional aspects in the organization of agriculture and their changes; (ii) to understand and explain these differences by situating each specific situation within the broader framework of the differential evolutions of agriculture on a global scale (Dumont, 1954); (iii) to emphasize continuities or breaks, progressive series, and relationships, and to identify one or several overall changes; (iv) to preserve from the agrarian legacy of humanity useful mechanisms, machines, ideas, plant and animal materials, approaches, skills, and devices that can illuminate and favor agricultural development in a manner that aligns with the common good (Cochet, 2015).

OVERVIEW OF HERBICIDES

Herbicides are chemical agents that control unwanted vegetation and play a crucial role in modern agriculture. They boost crop yields by eliminating weed competition, facilitate land management across various sectors, and even contribute to protecting ecosystem health by combating invasive species. They represent a broad array of chemical classes and thus play an undeniable role in energy transfer and several metabolic pathways in plants. However, their widespread use necessitates a critical examination of their potential drawbacks. Herbicides can influence not only target organisms (plants) but also non-target species, thus directly affecting plant-pathogen interactions and often promoting symbiotic relations (Duke et al., 2006).

CHEMICAL NATURE OF HERBICIDES

The chemical diversity among herbicides parallels the spectrum of targeted weeds they control. Glyphosate, a potent chelator, inhibits the 5-enolpyruvylshikimate-3-phosphate; EPSPS enzyme, disrupting amino acid synthesis and inducing metabolic disarray. Auxin mimics, such as 2,4-D, infiltrate hormonal pathways, provoking uncontrolled cell division and damaging cellular structures. Atrazine engages in a molecular assault on photosynthesis, capturing electrons in photosystem II and inducing energy starvation in plants. Paraquat, seemingly inert, triggers the Fenton reaction, generating free radicals that damage cellular membranes (Singh & Tiwari, 2020). This intricate chemical landscape governs selectivity, with varying affinities for specific targets distinguishing crops from undesirable plants. An in-depth analysis of the nuanced interplay between herbicide chemistry and plant biochemistry allows us to harness their weed-controlling efficacy precisely, laying the groundwork for sustainable weed management. Mode of action of herbicides is discussing further (Figure 14.5).

IMPORTANT ADVANTAGES OF HERBICIDES

Herbicides play a crucial role in enhancing productivity. By mitigating weed competition, which ensure crops receive vital resources, maximizing yield potential and quality. Translates to improved food security, economic benefits for farmers and also promotes efficient land management. It enables effective management of unwanted vegetation in diverse settings, from vast agricultural fields to meticulously maintained landscapes like parks and golf courses. Furthermore, it is important to

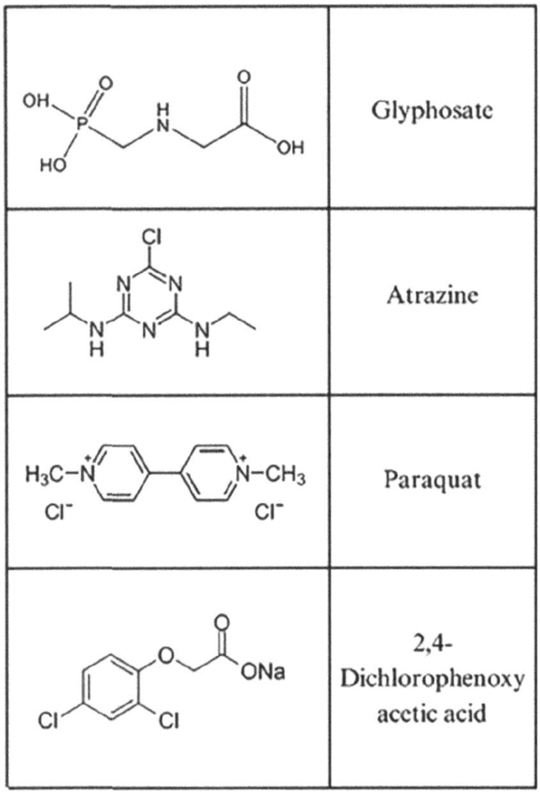

FIGURE 14.5 Structure of glyphosate, atrazine, paraquat, and 2, 4-dichlorophenoxy acetic acid.

highlight their role in ecosystem protection. Invasive species and detrimental weeds can disrupt ecological balance. Herbicides, when used strategically, can be valuable tools in combating these threats, safeguarding biodiversity and the health of ecosystems (Qasem, 2011).

MODE OF ACTION OF HERBICIDES

Herbicides affect several biochemical processes of invasive plants. Initially, it can obstruct a single biochemical reaction or multiple reactions at the same time, ultimately leading to plant death. These biochemical reactions include metabolic processes of plants such as respiration, cell wall metabolism, photosynthesis, or metabolism of protein, carbohydrate, nucleic acid, lipid, and amino acid. Some herbicides do not inhibit the biochemical processes but directly affect the membranes (DiTomaso, 2005).

GROWTH REGULATORS

Herbicides that imitate naturally occurring auxins are known as growth regulator herbicides. There are several groups of growth regulators including phenoxy carboxylic acids, phenolic acid derivatives, indole acid, and benzoic acid. Indole acids have a growth regulator known as indole acetic acid and are not available as herbicides commercially. Auxins as herbicides control plant cell differentiation and growth, but their mode of action is not well known, when plants are treated with auxins, their initial response is categorized into two phases. The first response occurs within minutes and is regarded as a fast response. During this response, quick acidification and loosening of the cell wall occurs. The second phase starts 35–40 minutes after the initial treatment and involves nucleic acid synthesis. Several metabolic and biochemical changes occur in plants in response to the application of auxinic herbicides. These changes include alteration of stomatal function, root initiation in tissues of stem, proliferation of stem tissues, leaf chlorosis, abnormal apical growth, and breakdown of root tissues. Most of these effects are secondary. Primarily, low levels of these herbicides stimulate RNA polymerase, resulting in the synthesis of RNA and proteins. However, high auxin levels restrict RNA synthesis and growth in meristematic tissues. Synthetic auxin treatment results in quick cell wall loosening and abnormal stimulation of cell division, resulting in uncontrolled growth and subsequent collapse of the vascular tissues (DiTomaso, 2005).

PHOTOSYNTHETIC INHIBITORS

Photosynthesis is a complex process that involves the production of organic compounds, oxygen from water and carbon dioxide. This process mainly occurs in the chloroplasts of leaf cells. The transformation of light energy to chemical energy (ATP and NADPHA) drives the process of photosynthesis. So, light is essential for photosynthesis and the performance of photosynthesis inhibiting herbicides. These inhibitors are divided into two groups. One of these groups restricts the electron transfer in photosystem II, while the other group accepts electrons from photosystem I (DiTomaso, 2005).

BLEACHING AGENTS

Pigment inhibitors restrict the production of carotenoids, with subsequent bleaching of foliar tissues and stems. This absence of protective carotenoids makes chlorophyll susceptible to photooxidation. Norflurazon, clomazone, fluridone, and amitrole restrict an enzyme of the carotenoid biosynthetic pathway. Norflurazon and fluridone restrict the performance of phytoene desaturase in the thylakoids of chloroplast. Amitrole inhibits the activity of zetacarotene desaturase, an enzyme of the carotenoid synthesis pathway. Clomazone is also known to restrict carotenoid synthesis; however, its particular site of action is not yet identified. The diphenyl ether herbicides also affect the biosynthesis pathways of pigments. But, they act by rapidly producing phytotoxic free radicals. These herbicides act by rapidly destroying the cell membranes (DiTomaso, 2005).

LIPID SYNTHESIS INHIBITORS

The cyclohexanedione and aryloxy phenoxy propionate herbicides work by inhibiting lipid synthesis. Their primary target is a plastid-localized enzyme, A carboxylase, that catalyzes the process of formation of malonyl Co-A from acetyl-CoA. This conversion is one of the early steps in fatty acid formation. Broadleaf species possess a different binding site that is not affected by these herbicides. When fatty acid synthesis is inhibited in grasses, the production of phospholipids that are used in the formation of new membranes is blocked. Synthesis of cell membrane becomes unable to keep up with the process of cell enlargement under rapid growing circumstances. It causes the destruction of cells in the growing points of the greatest cell expansion. Several biotypes of weeds have become resistant to cyclohexanedione and aryloxy phenoxy propionate probably due to the improved metabolism of the herbicide (DiTomaso, 2005).

CELL WALL SYNTHESIS INHIBITORS

Cell walls are an intricate matrix predominantly comprising cellulose, but also pectin, structural glycoproteins, and hemicelluloses. Cell wall synthesis inhibitors such as isoxaben and dichlobenil restrict the synthesis of cellulose in vulnerable plants. Quinclorac also blocks cell wall synthesis but it inhibits both hemicelluloses and cellulose. It indicates that the site of action of quinclorac differs from isoxaben and dichlobenil. All of these three herbicides cause injury to the cell wall by first forming necrotic bands close to the areas of elongation in roots and shoots (DiTomaso, 2005).

ENERGY PRODUCTION INHIBITORS

Energy production inhibitors act by restricting the formation of energy required to carry out metabolic functions. These herbicides such as chloracetamide and thiocarbamate inhibit cell elongation, decrease wax production, and cause abnormal foliar development in susceptible plants. Both of these herbicides induce the same symptoms, which mean that they have a similar mode of action. It is evident that these inhibitors do not restrict a single metabolic process or enzyme. It has been documented that these herbicides restrict lipid synthesis, leaf waxes, flavonoids, fatty acids, proteins, gibberellins, and terpenes. Moreover, these herbicides also affect hormone regulation and cell division. The main sites of action involve an acyl carrier protein, coenzyme A, and lipoic acid. As these inhibitors do not have a definite site of action, weed resistance is unlikely to be developed due to a different site of action. The mode of action of bensulide is not well known, but it appears to inhibit cell division, root elongation, and different metabolic processes. Similarly, the definite site of action of ethofumesate is also not known, but it is thought to restrict the synthesis of long-chain fatty acids (DiTomaso, 2005).

CELL MEMBRANE DESTRUCTION

Herbicides can harm cell membranes by restricting glutamine synthetase, producing more free radicals, or exerting a direct effect on membrane integrity, such as oxadiazon and diphenylethers cause damage to the membrane by accumulating toxic oxygen molecules. The primary site of action of these herbicides is the protoporphyrinogen oxidase enzyme. This enzyme is involved in the synthesis of heme and chlorophyll. When this enzyme is inhibited by herbicides, an accumulation of protoporphyrinogen IX occurs. This compound gets diffused out of the membranes of thylakoids and is oxidized, leading to the formation of protoporphyrin IX, when present outside the regular environment, not metabolized anymore. Protoporphyrin IX absorbs light, releasing electrons to oxygen molecules, leading to the formation of singlet oxygen and other free radicals. It initiates a chain reaction that quickly causes lipid peroxidation and destruction of the membrane (DiTomaso, 2005).

GENERAL CELL TOXICANTS

Acrolein is a general cell toxicant that reacts with sulfhydral group related to many macromolecules and proteins. In plants, sodium chlorate is a strong oxidizing agent and it inhibits protein sulfation. Due to this effect, these compounds interfere with the activity of several enzymes and disrupt many metabolic processes of plants (DiTomaso, 2005).

HARMFUL EFFECTS OF HERBICIDES

Although effective in suppressing undesirable vegetation, herbicides can have negative effects on the environment and human health. Prolonged exposure to these chemicals can lead to soil erosion, water pollution and biodiversity loss. Furthermore, some herbicides have been linked to serious health issues, including respiratory problems and an increased risk of cancer. Understanding these side effects is important to make informed decisions about the use of herbicide.

EFFECT ON SOIL QUALITY

Soils retain herbicides to various degrees, and it depends upon the interaction between the herbicides and the properties of the soil. The most important factor is the organic content of the soil. The adsorption of herbicides increases with an increase in the organic matter content. So, the amount of molecules of herbicide absorbed by the soil becomes high (Stevenson et al., 1972) (Table 14.1).

TABLE 14.1
Active Persistence Period of Different Herbicides

Herbicide	Active Persistence Period (Months)	References
Alachor	1.5–3	Borner et al. (1979), Monaco et al. (2002)
Acetachlor	3–12	
Atrazine	4–18	
Bromacil	9–18	
Cholorxuron	3–6	
Chlordbiamid	12–24	
Dichlobenil	12–24	
Isoxaben	6–12	
Metsulfuron	2–12	
Oxadiazon	9–18	
Terbacil	9–18	
Trifluralin	5–10	

Herbicides negatively affect the factors that provide nutrition to the plants such as soil structure, organization of organic substances, and transformation of soil. Many herbicides can inhibit the growth of soil microorganisms. As a result of this microbial disruption, more toxic and permanent metabolites are formed in the soil. Although microbial complexes present in the soil have adaptations to changes in environmental conditions, it has been reported that herbicides induce significant changes in their populations (Sitaramaraju et al., 2014). Triclopyr restricts the growth of microorganisms that convert ammonia to nitrite. Nitrogen-fixing bacteria living in the roots of bean plants are negatively affected by the presence of 2,4- dichlorophenoxyacetic acid as it inhibits their nitrogen-fixing ability (Fabra et al., 1997). Glyphosate destroys the communities of arbuscular mycorrhizal fungi in the soil. It negatively affects the growth of soil microbes that depend on nutrition from this fungus. It has been reported that glyphosate is harmful for beneficial soil bacteria, including the Bacillus family. Furthermore, it inhibits soil minerals such as iron and magnesium and disrupts the growth of nitrogen-stabilizing free-living bacteria that are essential for plants (Zaller et al., 2014). Earthworms are an essential component of the soil ecosystem, as they act as bioindicators and show deterioration in the quality of the soil. According to a study conducted in Argentina, glyphosate has negative effects on the population of earthworms (Casabé et al., 2007). Various studies have suggested that the contamination level of soil due to herbicides depends on different factors such as frequency of application, quantity used, toxicity of the herbicide, and persistence in the soil. However, these herbicides are intended to persist in the soil for a longer time to have the desired impact on the weeds (Greer et al., 1990). Two main processes, transfer, and degradation, govern the fate of a herbicide applied in the soil. The process of transfer includes penetration, overflow, uptake by flora and fauna, absorption, and desorption. In this way, the applied herbicide remains intact in the soil. The process of degradation involves plant detoxification, photodecomposition, microbial decomposition, and chemical breakdown. The processes of transfer and degradation determine the efficacy of herbicide for weeds, its persistency, and the possibility of groundwater and soil contamination (Subhani et al., 2000). So, it is essential to recognize the factors that affect the processes of degradation of herbicide to adopt appropriate strategies to minimize its persistence in the soil environment (Baran et al., 2007) (Figure 14.6).

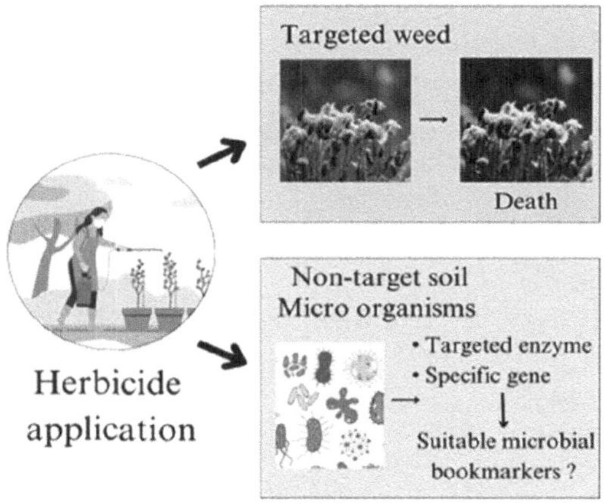

FIGURE 14.6 Effect of herbicides on soil microorganisms.

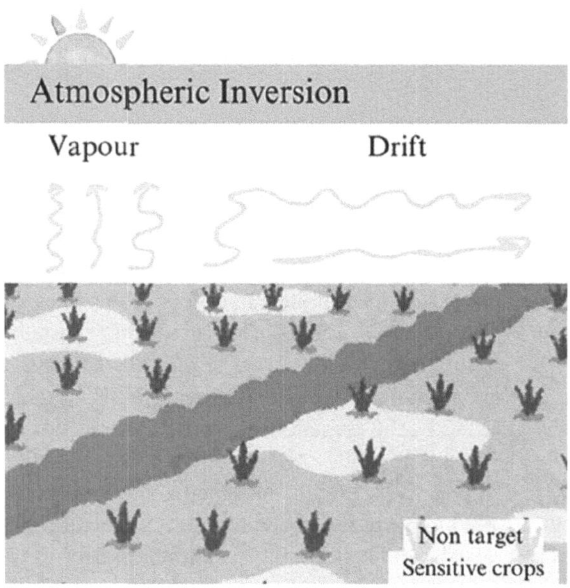

FIGURE 14.7 Transfer of herbicides from target species to non-target species.

EFFECT ON NON-TARGET SPECIES

Herbicides are useful for farmers as they provide a tool for the management of harmful weed species for better crop production. However, these can also negatively affect the non-target plants. These plants are not included in the target area, and the movement of herbicide away from the target area will disturb them. This non-target movement of herbicides can occur as a result of vapor movement, droplet drift, inappropriate disposal, as well as leaching, and erosion. An abundance of plant species is at risk of being negatively exaggerated by such movement (Hewitt, 2000). The direct negative effects of non-target movement include the death of a plant, reduced flowering and minor effects on a plant or a population, also effective against regenerative plant cycle or seed production. The effects on early recruitment and germination of plants are thought to be very significant during a growth stage that is predominantly susceptible to pesticides (Breeze et al., 1999) (Figure 14.7).

Non-target effects can have a slight impact on the composition of the plant community, mediated by effects on the chemical or water environment in the rhizosphere or by plant competition. The importance of the non-target effects of herbicides is unclear for example, it is not known if low doses of mixtures of herbicides or recurring drift events can affect the recruitment of plants. Droplet movement is the most common and significant form of non-target movement of herbicides. If the concentrations of pesticides are very high, pesticides can go through secondary redistribution after application (Marshall, 2001).

EFFECTS ON HUMAN HEALTH

Herbicides have deleterious effects on human health, and many of these have been banned due to causing serious diseases in humans. However, it does not suggest that the other herbicides that are allowed to be used do not cause any damage, as their indirect and direct effects are hard to identify due to high cost and complexity. Some of the banned herbicides such as paraquat can cause serious diseases and even death (Kortekamp, 2011). In 1990, pesticide poisoning caused the death of around one million individuals around the world. According to the report of the World Health Organization; WHO poisoning cases reported due to the harmful effects of pesticides are 3 million individuals per year generally. There are two types of toxicity caused by herbicides, chronic and acute. Most of the herbicides have weak acute toxicity. It is predicted that the chronic effects of herbicides may be very significant and can be a threat to human life. A study was conducted in Syria to investigate the role of herbicides in causing cancer, positive results were obtained in the hamster embryo assay (Maire et al., 2007). Moreover, it has been observed that the use of pesticides is linked to prostate cancer in farmers and causes a decrease in sperm count, an increase in sperm abnormalities, and endocrine disruption.

A lot of cases are reported the risk of cancer due to exposure to 2,4- dichlorophenoxyacetic acid (2,4-D) and dioxin pollution. It also caused in chromosomal aberrations in mice, human cells, and plants, including lymphocytes of humans who were occupationally exposed to pesticides while working with them. The 3,6-dichloro-2- methoxy benzoic acid and 2,4- dichlorophenoxyacetic acid belong to the auxin group of herbicides. The most commonly used post-emergence herbicides in the world are those for grain crops to inhibit broadleaf weeds. According to WHO, both of these herbicides belong to Class II, as they are moderately dangerous. It has been reported that herbicides such as dicamba and 2,4- dichlorophenoxyacetic acid cause DNA damage in human lymphocytes and Chinese hamster ovary cells when the DNA and chromosomal levels were analyzed. Both of these herbicides have been observed to cause sister chromatid exchange in mammals, which has a clastogenic activity (Soloneski & Larramendy, 2011). According to US Environmental Protection Agency (EPA), dicamba; 2-methoxy-3,6-dichlorobenzoic acid is an increased toxin and it is a cholinesterase inhibitor. Moreover, it negatively affects the reproductive system.

In human history, glyphosate is the best-selling herbicide and comprises 60% of the total sales of broad-spectrum herbicides. In laboratory animals, this chemical substance causes chronic health effects. Nutritional experiments using glyphosate suggest that it causes damage to the eye lens, results in weight reduction, and damages the kidney and liver. These effects were obtained only at higher doses of glyphosate. Clinical studies suggest that workers who used to handle glyphosate had a higher frequency of developing non-Hodgkin lymphoma than those who did not use this herbicide. Non-Hodgkin lymphoma is a rare form of cancer in the lymphatic system. Most of the toxic tests have been performed on glyphosate, while a little testing is done on formulated products. It has been observed that glyphosate damages DNA in the kidneys and liver of mice. An ingredient known as polyoxymethylene amine is used in many formulations and is known to have more acute toxicity than glyphosate itself (Navarro & Martinez, 2014) (Figure 14.8).

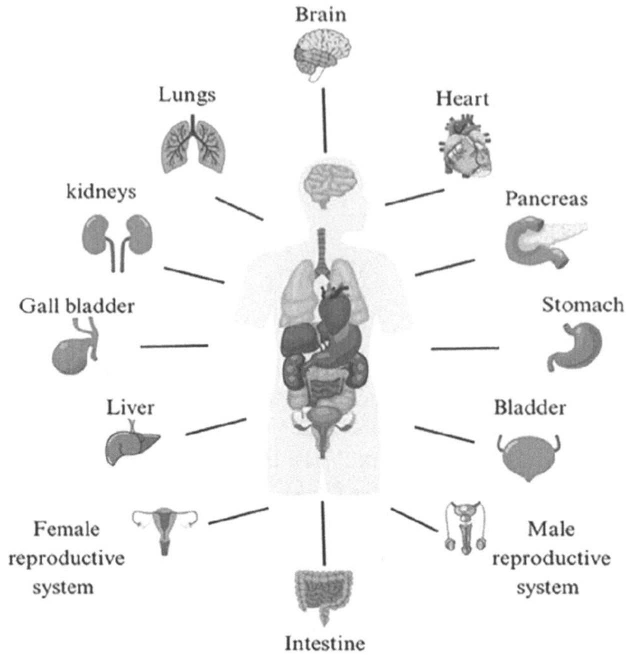

FIGURE 14.8 Potential side effects on human body.

EFFECTS ON ANIMALS

Herbicides have both direct and indirect effects on animals. A decline in the number of species of some endangered birds and rare animals has been observed due to the harmful effects of fungicides and herbicides. Herbicides can affect the dynamics of the arthropod community and also the plant community. Moreover, in agroecosystems, these chemicals have an impact on the biological control factors (Evans et al., 2010). When the effects of glyphosate were studied on frogs, a high level of death was observed. It has been reported that herbicide application negatively affects birds and chicken species. Moreover, it has been suggested that atrazine causes damage to the genetics of frogs. When chicken eggs were exposed to 2,4- dichlorophenoxyacetic acid, a reduction in incubation time was observed which led to infertility and feminization in pheasant chicks (Duffard et al., 1981).

EFFECTS ON AQUATIC ENVIRONMENT

Herbicides have several negative effects on water bodies. It can reach aquatic ecosystems by direct and indirect ways. These are used directly in water bodies to inhibit the growth of harmful aquatic weeds. Indirect ways of introduction of herbicides in aquatic ecosystems include leaching processes and agricultural runoff. Herbicides reduce the quality of the aquatic environment and affect essential functions of the ecosystem as they modify food chains, reduce species diversity, change the resilience and stability of the ecosystem, alter the patterns of nutrient cycling and energy flow, alter community structures, and modify food chains (Pérez et al., 2011). Herbicides reach the aquatic bodies in different ways and cause negative effects on living organisms. It has been reported that herbicides applied in agricultural areas to kill the weeds pollute the aquatic environment and this has been observed in both field and laboratory (Daam et al., 2009). It is estimated that around 1% of the used herbicide leaks in the surrounding environment and this rate can increase up to 3% due to rain precipitation (Bengtson-Nash et al., 2004).

It has been observed that herbicides are being used in amounts above legal standards in the Netherlands. In Australia, pollution caused by herbicides has been reported in sediment and water samples (Knauer et al., 2010; Magnusson et al., 2008). Moreover, it has been reported that herbicides including simazine, metolachlor, trifluralin, atrazine, and alachlor are found in surface and groundwater near agricultural areas (Anasco et al., 2010). Aquatic plants are crucial for the survival of aquatic organisms as they provide approximately 80% of the dissolved oxygen. Herbicides have been observed to kill aquatic plants, thus lowering oxygen levels and eventually killing aquatic animals (Helfrich et al., 2009). Herbicides disrupt the food chains in aquatic ecosystems. Alachlor and Atrazine have lethal effects on diatoms and algae, which play a crucial role in the food chain of aquatic environments. Even low concentrations of herbicides have been reported to prevent photosynthesis, degrade compounds in cells, and inhibit plant growth.

As herbicides are applied close to the fish environment, the reproductive potential of fish and many other aquatic organisms has decreased. Trifluralin is very toxic to fish and causes the formation of spinal abnormalities in fish. It has also been reported to be very harmful to marine organisms including mussels and shrimps. In Sockeye salmon, 2,4-D herbicides have been observed to induce a physiological stress response. Additionally, these herbicides reduce the food-collecting ability of rainbow trout and are toxic to shellfish (Cheney et al., 1997). It has been reported that glyphosate causes irregular swimming in fish and a lack of oxygen (Liong et al., 1988).

RESIDUES OF HERBICIDES

A herbicide is considered to be persistent if, even after its mission has been accomplished, it remains present in the soil in its original or any other closely related but phytotoxic forms, and the quantity that remains in the soil is referred to as residue (Sondhia, 2014). Actually, the potential of the herbicides to persist in the soil is often varied from one another. There are different families of herbicides that have persistent members, i.e., triazines, sulfonylureas, uracils, phenylureas, dinitroanilines, imidazolinones, isoxazolidinones, and also some other plant growth regulators that belong to the pyridine family of heterocyclic organic compounds. The persistence of the herbicides is also affected by the chemical properties of these herbicides. Some of the other factors also include vapor pressure, solubility of these herbicides within water, and vulnerability to chemical and microbial alteration or sometimes degradation. In addition to all these factors, the climatic conditions of the area and physicochemical properties of the soil is also responsible for the persistence of that type of herbicide in the soil and its prevalence. A comprehensive review on the degradation and prevalence of different types of herbicides has already been published which is also important for health issues (Sharma & Angiras, 2004).

MANAGEMENT OF HERBICIDE RESIDUES IN SOIL

There are different ways to circumvent herbicide persistence and to overcome the complications created by herbicide persistence. Works of literature are also available on utilizing various management techniques to minimize the persistence and residue risks in soil. For ease of understanding, the problems caused by herbicide persistence are categorized into five different categories (Janaki et al., 2015).

CULTURAL AND MECHANICAL MANAGEMENT PRACTICES

Among all the management practices, mechanical methods stand as one of the oldest management practices for weeds. Although chemical management is widely used because of its ease and widespread availability, nonchemical weed management practices are opted when dealing with organic crop productions (Melander et al, 2017).

INTEGRATED WEED MANAGEMENT

Integrated weed management (IWM) is a type of weed control that includes the administration of a wide variety of different kinds of management practices to manage weeds. Weed killers are utilized only when the population of weed is very high or simply exceeds an economic threshold level that allows their use. Field scouting is required at regular intervals to monitor the level of weed populations. There are also some nonchemical weed control methods used as management practices, which may minimize the use of herbicides. These methods include crop rotation, rotary hoeing, competitive hybrids, and altered planting. The efficacy of integrated weed management utilizing different mechanical and chemical methods in various crops has been studied widely in Pakistan and globally. Scientists worked on different types of crops to check the pre-emergence use of pendimethalin 1.0 kg/ha or pre-plant incorporation of fluchloralin 1.0 kg without phytotoxicity on that particular crop (Vaishya et al., 2003). Researchers have also reported the pre-emergence use of metribuzin 0.7 kg/ha three days after planting (DAP) of turmeric and hand weeding on 45 as well as 75 DAP for efficient weed killing and better yield with no phytotoxicity to crop and some other issues (Sathiyavani, 2014). In a familiar fashion, it was also observed that the use of 2,4-D at three levels of 2.0, 1.0, and 0.5 kg/ha in the wheat crops at 35 days after sowing remained in soil up to 15, 45, and 75 days, respectively, but, residues of 2,4-D were found lower than the detectable level (0.02 ppm) in wheat straw as well as wheat grains (Sharma et al., 2013). It is also observed that the pre-emergence use of pendimethalin (38.7%) at 2.0 kg/ha 3 days after sowing; DAS and hand weeding on 45 DAS exhibited efficient weed management in cotton and left no residues in the soil at harvesting and carryover issues to the successive crops, cowpea sunflower and pearl millet, grown in sequence (Nalini et al., 2010). Pre-emergence use of pendimethalin leaves no residues in the soil ahead of the harvest of the garlic at any of the used doses (Sharma et al., 2013). Researchers observed that the residue of pendimethalin 1.5 kg/ha used in wheat and mustard intercropping system was merely 0.001 ppm in the soil post-harvest of the crops (Sharma & Angiras, 1997).

PLOWING OR CULTIVATING THE LAND

Different kinds of plowing and cultivating operations in the field help to decontaminate by volatilization in a way by taking herbicide residues that are present deep in the soil and bringing them to the surface. By using the disc plow or inter-cultivators, there is a considerable cutback in the herbicide toxicology, as in this way, the herbicide is diluted by mixing with a considerable volume of soil. Researchers reported that the loss of atrazine was less in the chisel-disk system with incorporation than in no-till and ridge-till systems (Olson et al., 1998). Rapid fluometuron degradation was observed in conventional tillage; CT than to no-tillage; NT soils in the 2–10 cm depth as well as observed that the ryegrass systems of cover crop, incorporated under CT, initiated microbiological properties of soil and supported the degradation of herbicide in surface soils. The herbicide residue and prevalence in the soil depends chiefly on the soil environment. Most transformations of herbicides in soil occur by the metabolism of microbes, changes of the soil environment and populations of microbes by less tillage or cover crops can impact the fate of the herbicides (Zablotowicz et al., 2007). Herbicide degradation was faster in the upper layers of the soil than in the deeper layers, and it was also faster at conventional tillage than no tillage at all (Gaston & Locke, 2000). In compliance with a study on the behavior of atrazine in the soil at no-till, only in a few isolated cases, the faster herbicide degradation was in deeper layers of the soil layers than in the surface layers (Hang et al., 2010). Unbalanced degradation of metribuzin and S-ethyl metribuzin due to slow herbicide release fixated in the crop residues. Straw affinity can reduce the efficiency of herbicides that were activated in the soil if they were applied to the soil surface at no-till (Dao, 1991). The degradation of molecules can be affected by the presence of mulch but in contrasted ways depending on the nature of crop residues. The impact of tillage practices on herbicides' sorption, degradation, and movement was also discussed. In no-tillage, vetch residues accelerated the degradation of metolachlor by 1.5–3 times but did not affect the degradation of atrazine. Vetch residues seemed to slow down fluometuron degradation compared with soil samples without vetch residues or other types of residues such as wheat residues, rye residues. The abundance of nitrogen in legumes could be the reason for this slowdown (Alletto et al., 2010).

INCORPORATION OF HERBICIDES

Mechanical amalgamation of an herbicide helps to reduce runoff loss by placing it below the mixing zone (e.g., atrazine), which takes place through the solution or water phase and not much with soil particles. A few of the herbicides are reactive to sunlight and should be mixed into the soil to limit the misfortunes. On the other hand, some of the herbicides are volatile and evaporation can result in the loss of such kinds of herbicides, mainly when applied to wet soil. Because the sowing operation removes a certain amount of herbicide away from the seed row, applying the pre-emergent herbicides as pre-sowing and introducing these into the seed bed during the sowing process will frequently increase crop safety. In this case, because of the fact that the chemicals are moved out of the seed row, this can reduce weed control. That is why it is rational to include a water-soluble herbicide in the mix to incorporate a portion of it into the seed furrow. Moreover, utilization of mulch and crop residue from the preceding year can lessen the loss. Subsurface herbicide application can provide an introduction with minimal disturbance of surface residue, with the development of special tools. The subsurface herbicide application with the mulch master and soil incorporation can reduce herbicide losses through surface runoff when compared to surface application with no-till (Mickelson et al., 2001).

CROP ROTATION

There are different kinds of herbicide-resistant plants that are used for the removal of herbicide residues from the contaminated soil. Crop rotation during the season reduces the risk of herbicide contamination. Some kinds of crops should be avoided in the rotation as succeeding crops because these are more sensitive to herbicides, such as rape seed and sugar beet are more sensitive to the imidazolinones (imazethapyr+ imazamox), and when the previous crop was applied with these herbicides, the successive crops may be affected later. It is also reported that some of the crops; maize, barley and winter wheat were unaffected by herbicides while used in crop rotation, so these can be used for crop rotation (Süzer & Büyük, 2010). If the sunflower received imidazolinones, the re-cropping with canola and sugar beet should be ignored in the growing season to eliminate the risk of carryover problems (Bresnahan et al., 2000). The sensitivity of rotational crops is also different from one another; acifluorfen residues were in decreasing order from sorghum to corn, rice, and then millet and in the case of imazamox and fomesafenis from sorghum to corn, millet, and then rice. There are also different kinds of conditions, such as low moisture content of the soil and high content of clay and organic matter, that are responsible for injury under certain environmental conditions, but it also appears to be low with high precipitation conditions, as in the case of corn, rice and millet (Cobucci et al., 1998). The application of herbicides can also be avoided by the use of appropriate crop rotation with legume and non-legume crops, and growing of cover crops also helps to suppress the weed population (Verma et al., 2014).

LIGHT IRRIGATION AFTER APPLICATION

The aqueous medium is helpful for microbial activity; moist soils provide many favorable conditions for microbial activity and often result in a more rapid breakdown of herbicides. Controlled irrigations reduce the risk, while heavy irrigations leach the herbicide's harmful chemicals out of the crop's root zone. The saturated soil favored forming soil-bound residues and deleting metolachlor, significantly greater quantities of a de-chlorinated metabolite were measured in the saturated surface soil compared to the unsaturated soil (Rice et al., 2002). The dissipation of isoxaflutole was faster in soil maintained at −100 or −1,500 kPa compared to that in air-dry soil. At 25°C, the half-lives for isoxaflutole were 9.6, 2.4, and 1.5 days in air-dry, −1,500 and −100 kPa moisture regimes, respectively (Taylor-Lovell et al., 2002).

SITE-SPECIFIC APPLICATION USING VARIABLE RATE APPLICATOR

Weed control efficacy and the potential for crop injury are very much influenced by the interrelationship between soil properties and the herbicide chemistry applied to the particular crop. The fields that have a significant variability between the soil properties and herbicide chemistry, these kinds of crops are useful for the application of herbicides at a variable rate. The combination of automatic tractor steering and variable rate technology is well suited for site-specific application of pre-emergence herbicides. The farmers can enhance the efficiency of chemical application by applying optimum rates according to the soil texture with the help of tractor guidance control and variable rate control systems. These different kinds of technologies have primarily been adopted by farmers of major crops such as corn, wheat, and soybeans (Koch & Khosla, 2003). The recommended application rates of pre-emergence soil-applied herbicides can vary as much as 50% in a given field due to varying soil textures (Bauer & Schefcik, 1994). There was notably less crop injury and remarkably more marketable yield as compared to uniform application. It was also observed that the effectiveness of benefin herbicides in controlling weeds on vegetable crops using the variable rate technology in different textured soils is also increased. In the portions of the lettuce field with loamy sand-textured soils, 35% less herbicide was applied, and up to 40% more heads were harvested. Furthermore, there was no significant difference found in weed control efficacy between the two application methods examined. The type of the soil is also responsible for the amount of herbicide that is used for that particular crop because of the fact that different kinds of soils; with a high or low clay content. Sandy soils require fewer amounts of herbicides for weed control as compared to the soils with high clay content. On the other hand, higher rates of herbicides are required for good weed control in high clay content soils as compared to the sandy soils, but it was also noted that the excessive herbicide concentrations in soil solution can cause injury to lettuce seedling roots (Nolte et al., 2011; Kurt et al., 2011).

BIOSTIMULATION

The term "biostimulation" is frequently used to describe the addition of electron acceptors, electron donors, or nutrients to stimulate naturally occurring microbial populations (Scow & Hicks, 2005). In its broadest sense, in order to enhance the activity of indigenous microbial degraders or to promote co-metabolism, biostimulation can be regarded as the introduction of an adequate amount of water, nutrients, and oxygen into the soil (Lorenzo, 2008). The idea of biostimulation has been used for a wide variety of xenobiotics to stimulate the intrinsic degradation potential of a polluted matrix through the accumulation of nutrients or other limiting factors (Kadian et al., 2008). The less amount of sufficient decomposable organic matter present in the soil leaves the herbicides unmanageable in the soil and for several years without decomposition because there is a less amount of the substrate present in the soil to stimulate microorganisms for the decomposition of these herbicides (Felsot & Dzantor, 1990). Microbial activity or biodegradation can be initiated in the soil by introducing different kinds of organic matter, bioprocessed materials, or compost. All the components can naturally initiate microbial activity in the soil and could also be utilized to treat contaminated soils. These materials contain different kinds of organic compounds such as nitrogen, carbon, and other nutrients, providing an excellent material for microbial activity and growth. The herbicide concentrations in the soil can be reduced to a much lower level by the use of biodegradable materials within short spans of time. Enhanced the degradation of butachlor, pretilachlor, and 2,4-D in the soil, was also controlled by the continuous application of farmyard manure; FYM to the rice crop through enhanced microbial activity.

NUTRIENT ADDITION

Mostly, the nutrients the microorganisms require for microbial activity are in fewer amounts than the microbes need for degradation. Supplementing required nutrients to the soil stimulates the biodegradation of the pollutants and other chemicals effectively. All of these nutrients are required by the microbes for the synthesis of the essential enzymes for their growth and also for the degradation of the contaminants. In a similar research, the effect of the addition of inorganic nutrients into the soil was studied on the breakdown of atrazine in the soil (Hance, 1973). After that, various researchers also studied the concept of nutrient supplementation in the soil for the purpose of enhanced decomposition of contaminants, and the prospects of microbial biostimulation through the manipulation of organic and inorganic nutrient status in the soil have since been investigated. In some cases, the limited availability of some of the inorganic ions; nitrogen, is more effective in promoting degradation and has been reported for atrazine and other heterocyclic compounds. The reduction of such an ion in the soil is carried out by the enhanced supply of some other ion into the soil (Bichat et al., 1999). In the case of nitrogen, this can potentially be carried out by supplying excess carbon to the soil, and in this way, nitrogen becomes limited. Recently, research confirmed that phosphorus; P was a limiting nutrient during the degradation of coexisting dichlobenil and atrazine by bacterial degraders in soil. Dichlobenil was degraded entirely in 60 hours in the P-supplemented soil extract, in comparison to less than 40% degradation without P supplement. It was also noted that the degree of enhancement of atrazine degradation was even greater, in that it was completely degraded in P-supplemented extract within 40 hours, compared to less than 10% degradation without the P supplement (Qiu et al., 2009).

BIOAUGMENTATION

The introduction of a particular indigenous and non-indigenous microbe for the enhancement of biodegradation of a target compound or serving for the catabolic gene is referred to as the process of bioaugmentation. Usually, this is also used with the term biostimulation (Kanissery & Sims, 2011). Different kinds of microbes present in the soil are capable of degrading many of the compounds present in the soil. Herbicide compounds can also be degraded by these microorganisms. The microorganisms used these compounds as a source of nutrients and energy. When there is an increase in a particular herbicide in the soil, a pure culture of bacteria can be applied to the soil by artificial means to solve the problem. Instead of pure cultures, mixing pure cultures restored atrazine-mineralizing activity and also observed increased rates of atrazine metabolism with the repeated transfer of the mixed cultures even at the elevated concentrations. An atrazine-degrading culture of bacteria contained the organisms that were capable of using atrazine as the single source of carbon and nitrogen in an aerobic environment. The concerned culture of microbes was interacted by herbicide spills earlier and then the same culture was used to amplify the degradation in soil (Mandelbaum et al., 1993). It was reported that a fungal isolate known as *Rhizopus orzea* was used in a bioremedial process of alachlor. It was revealed that the half-life values of the sterile and non-sterile soil brooded with the concerned fungal culture was found to be different from one another; 7.2 and 8.6 days, respectively. It was also found in another research conducted with lactofen that lactophan consumption was faster in the soil solarization technique and was increased by up to 90% by the straw amendment technique. Besides this, it was also suggested that this technique is beneficial to remove herbicide residues from the concerned soil.

USE OF NON-PHYTOTOXIC OIL, ADJUVANTS AND SURFACTANTS

Non-phytotoxic oil, adjuvants, and surfactants are beneficial for weed killing. Because these reduce the residual hazard and enhance the weed-killing capability. The adjuvants change different physical characteristics, such as the surface tension of the chemicals and their wetting ability and hence may modify the spray solution's response to move in the soil. Adjuvants and especially surfactants influence the herbicide residues in soil by affecting the amount of water available for evaporation from the soil surface. In both laboratory and in pot-field experiments, the rate of herbicide degradation was increased when olejan was added to the trifluralin applications, causing a significant increase (Mandelbaum et al., 1993). When cationic adjuvants are applied to the soil, then they form neutral compounds by reacting with the anionic molecules present in the soil. The resultant complex may have dissolved the herbicide rendering it less mobile in soil. Surfactants are an important small group of chemicals among adjuvants. They behave as wetters and emulsifiers as well as spreaders (Hall et al., 1993).

IMPORTANCE OF PROTECTING EFFECTS IN AGRICULTURE

Protective effects are crucial for vibrant yields in agriculture, playing a role as vital as weed control. These effects, encompassing disease resistance, stress tolerance, and nutrient enhancement, safeguard productivity and resilience. Like nature's arsenal, these multifaceted protecting mechanisms shield plants from unseen threats and optimize resource utilization. Research explains these protective effects in the critical areas discussed below.

ENHANCED DISEASE RESISTANCE

It is observed that 20% reduction in fungal infections in tomatoes with overexpressed chitinase activity stands as a potent defense against microbial foes (O'Connell et al., 2008). Similarly, in natural plants demonstrates drought resilience, where maize engineered with aquaporin genes exhibited a 35% improvement in water-use efficiency, potentially boosting yields by 10% in underwater stress (Shi et al., 2020). Finally, nutrient-harvesting prowess, as showcased in plant cell who equipped rice with silicon transporters, leading to a 15% yield increase and enhanced lodging resistance, highlights the potential for optimizing resource acquisition (Sher Khan et al., 2019, Kim et al., 2012).

DROUGHT TOLERANCE

Shitao et al. (2020) unveiled remarkable feat engineering maize to improve its water-use efficiency by 35%. This triumph utilized overexpression of aquaporin genes, integral membrane proteins facilitating water transport across cell membranes. This enhanced permeability allowed maize to utilize water more effectively, translating to a potential 10% yield increase under water-stressed conditions. Essentially, these maize plants became adept at "sipping" and utilizing every drop of available water, a crucial adaptation for regions facing drought challenges (Hu & Xiong, 2014).

NUTRITIONAL ENHANCEMENT

Rice, a staple for billions, received a significant boost courtesy (Govindaraju et al., 2015). Research has been focused on silicon, a micronutrient often deficient in rice-growing soils. By equipping rice with genes for silicon transporters, they increased silicon uptake by a remarkable 15%. This not only led to a 15% increase in grain yield but also enhanced lodging resistance the ability of rice plants to stand tall amidst strong winds and heavy rains. This silicon-infused upgrade essentially strengthened the rice stems, reducing the risk of crop loss due to lodging and ultimately contributing to greater food security (Tucker, 2003).

ALTERNATIVE METHODS FOR WEED CONTROL

Management of crops and vegetation demands to use the methods that are both efficient in controlling weeds and cost-effective. Managers of vegetation rely mainly on chemical, cultural, and mechanical methods to eliminate weeds as these methods are efficient, quick, and effective for reducing weeds. The concern about chemical methods is being raised due to their hazardous effects. To solve this problem, the interest in using residue-free products is growing in the field of agriculture. Farmers are looking for some alternative chemical methods that are less toxic and efficient as well. Natural herbicides are weed-controlling products that consist of naturally occurring ingredients such as soaps, oils, and acids (Young, 2004).

These products work through contact action and destroy the tissues of plants while acting very fast on weeds. The extent of damage these natural products cause to the plants depends on various factors such as coverage, time of application, and concentration of the product. The amount of product that must be used to kill weeds is not usually mentioned on the label. It is recommended that these products should be used as needed to control the weeds. Particular information, including the timing of repeat application, product formulation, and stage of plant growth for applications, is usually missing (Neal, 1998). Due to regulatory actions and environmental concerns, homeowners and green industry managers are seeking effective alternatives to glyphosate. Organic, nonchemical, and natural herbicides are available as an alternative to glyphosate. These products act by imitating the natural chemicals of plants and are generally less toxic and break down in the environment rapidly. These natural products can be used against weeds by painting, drenching, and spraying. However, they do not have residual activity, vary in quality from batch to batch, and are more expensive.

Alternative natural products for weed control do not usually provide 100% results. These products work more efficiently on broadleaf weeds than grassy weeds. To obtain the best results, these products should be used in combination with proper cultural practices such as enhancing plant nutrition, improving soil health, over seeding, irrigation, plant establishment, proper moving, and cultivar selection. To increase their effectiveness, conditions that are suitable for weed growth, such as excessive nitrogen application, overwatering, and compacted soils, should be avoided. Moreover, adjusting the pH of the soil that promotes the growth of plants over weeds can also help. Most of these products are not based on chemicals, so they do not require permission from the environmental protection agency; EPA. Available data suggests that these products have little or no toxicity on exposure to human beings. Some of these can have negative effects on humans during application and mixing due to eye or skin contact, inhalation, or spray mist. Others can have allergic effects or unpleasant smells. Corn gluten meal, one of the herbicide alternatives, should not be used by a person who is allergic to corn. There are seven categories of alternative herbicides including herbicidal soaps, natural acids, salt-based herbicides, natural acids, corn gluten, iron-based herbicides, phytotoxic oils, combination products; mixture of other categories (Smith-Fiola & Gill, 2022)

ACETIC ACID

Acetic acid, also known as ethanoic acid and vinegar, is used as a weed killer. It works by affecting the cell membranes of the cells of plants and causes the breakdown of foliage tissues rapidly on contact. The concentration of acetic acid is 10%–20% for herbicidal vinegar and 5% for household vinegar. The use of acetic acid with less than 8% concentration does not require registration from the environmental protection agency and is regarded as a minimum risk pesticide. However, registration for using acetic acid is required in many states. When applied in sufficient amounts, acetic acid causes rapid burn down of the plant, particularly in the presence of bright sunlight. Products based on acetic acid are non-selective, and foliar sprays eliminate most broad-leaved weeds. The application of acetic acid usually requires a surfactant. A natural sticking agent such as yucca extract is sometimes added. Acetic acid is very efficient in controlling annual small broadleaf weeds. Its kill rate is rapid, and it can eliminate 90% of the treated weeds within 24 hours of application. Products of acetic acid tend to break down in the environment very quickly. Acetic acid is primarily valuable for controlling weeds on sidewalks and gravel. It is non-selective but mainly kills broadleaf weeds. It can temporarily burn back the grasses (Smith-Fiola & Gill, 2022).

SALT

Salt is another alternative to chemical herbicides. It is an efficient and non-toxic herbicide, not all types of salts can be used as a herbicide. Regular non-iodized or iodized table salt can be used effectively. Make sure to check the package for the presence of sodium chloride and not rock salt, sea salt, or magnesium sulfate; Epson salt. Salt should be applied carefully when it is being used as a herbicide as it can harm other plants besides weeds as well and negatively affect the soil health by leeching into it. Therefore, salt is mainly used to manage the weeds that are present away from other plants, for example, weeds present in the cracks in pavement or asphalt or between patio stones. As far as its mechanism of action is concerned, salt works to eliminate weeds by disturbing the balance of internal water of the cells and making the plants dehydrated. Salt is easily soluble in water so it is applied by mixing in water and gets easily absorbed by the weeds. The recommended concentration of the salt in water depends on the type of herbicide and site of application. If the goal is to use salt to control garden weeds and protect the other plants present around, a weaker mixture of salt and water, such as 1:2, is recommended.

However, if salt is used in areas such as between cracks, patio stones, and in driveways, a stronger mixture such as 2:1 and 3:1 can be used. But if salt is applied in areas where the health of the soil is concerned, a weaker mixture is a better choice, as a stronger mixture can affect the pH of the soil making it sterile. Solutions of saltwater must be applied to the foliage of the weed directly. Soaking the roots with the solution must be avoided to protect other plants and soil. The solution can be applied directly through a container or by using a spray bottle. To ensure that the other plants do not get affected by the salt, water them after application of the saltwater to the weeds so that they can flush out if present. The treatment of saltwater is not recommended in tight-spaced veggie gardens and flowerbeds. As sodium chloride is harmful to all plants, caution must be taken while using it for weed control. Generally, salt is most effectively used for the management of weeds on a small scale (Sears, 2022).

HERBICIDAL SOAP

Herbicidal soap is an effective alternative to conventional herbicides. The acids used to produce a herbicidal soap include caprylic acid, undecylenic acid, sorbic acid, and pelargonic acid. The active ingredients of these soaps are generally sodium, ammonium, or potassium soaps of the acids and the ratio of soap to free acid in the active ingredient ranges from 3:4 to 9:1. Herbicidal soaps can effectively work against a broad range of liverworts, algae, higher plants, and mosses. These are usually effective for use on hard substrates such as stairs, walks, patios, and walls. Herbicidal soaps are advantageous over traditional herbicides because they do not leave behind any long-lasting residue on the surface to which it is applied. This is a desirable characteristic, as several naturally occurring herbicides are not suitable for use on certain substrates due to the residues and damage they leave behind. Additional substances such as anti-foaming agents, solvents, gums, and emulsifiers can also be added as a component in a herbicidal soap to enhance its processing and shelf-life. These additional substances do not have any herbicidal activity, non-staining herbicidal soap.

HYDROGEN PEROXIDE AS A PROTECTANT

Hydrogen peroxide not only provides essential benefits in daily life, but it also has a crucial role in the kingdom of flora. Its remarkable and multifaceted nature allows it to perform several protective roles in animals and plants. A summarized list of defensive measures played by hydrogen peroxide in plants is mentioned below.

PATHOGEN FIGHTER

H_2O_2 directly attacks invading bacteria and fungi by disrupting their cell membranes, hindering their growth and reproduction. This oxidative burst creates an inhospitable environment for pathogens, often allowing the plant to mount further defenses (Bozso et al., 2005).

WALL REINFORCEMENT

H_2O_2 strengthens plant cell walls by triggering the crosslinking of polymers, making them more rigid and resistant to pathogen penetration. This fortified barrier acts like a suit of armor, further deterring microbial invaders (Dempsey & Klessig, 1995).

CHEMICAL WARFARE

H_2O_2 acts as a precursor for producing phytotoxins, potent antimicrobial compounds are synthesized by plants to target specific pathogens, adding another layer of defense to their arsenal (Quan et al., 2008).

STRESS SIGNALING

Beyond direct defense, H_2O_2 acts as a signaling molecule, alerting the plant to various stresses like drought, wounding, and even UV radiation. This triggers the activation of stress-responsive genes, helping the plant adapt and survive in challenging environments (Desikan et al., 2004).

A DELICATE BALANCE

It's important to note that H_2O_2 is a double-edged sword. At high concentrations, it can damage plant cells. However, plants have evolved to regulate their production and scavenging tightly, acting as protectors instead of enemies (Quan et al., 2008) (Figure 14.9).

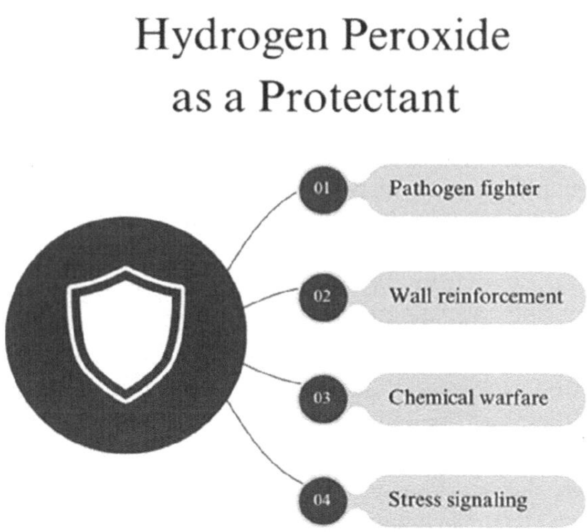

FIGURE 14.9 Role of hydrogen peroxide as a protectant in plants.

MODE OF ACTION OF HYDROGEN PEROXIDE

In the plant kingdom, hydrogen peroxide; H_2O_2 transitions from an antimicrobial agent to a signaling molecule, responding to biotic and abiotic stressors such as pathogen intrusion or environmental fluctuations. In controlled concentrations, it acts as a molecular signal, accumulating rapidly in specific cellular compartments to initiate a cascade of biochemical responses. This includes the upregulation of antioxidant enzymes and the synthesis of defensive secondary metabolites. Antioxidant molecules, such as glutathione and superoxide dismutase, counteract stress-induced free radicals, mitigating potential cellular damage. Simultaneously, it activates defense pathways, promoting the production of antimicrobial compounds like phytoalexins to impede pathogen growth (Wojtyla et al., 2016). Beyond direct defense, it contributes to cell wall reinforcement and strengthens physical barriers, enhancing protection against external threats. The orchestrated interplay between controlled oxidative stress and defense activation highlights the multifaceted role of H_2O_2 in safeguarding plant life.

INDUCTION OF PLANT DEFENSE MECHANISMS

H_2O_2 has the highest stability among all of the ROS. When transition metal ions are absent, they are poorly reactive with most organic molecules. It is capable of diffusing across cell membranes and reaches locations present away from its site of production. H_2O_2 plays a crucial role in defending plants against pathogens. It directly restricts the growth of the pathogens to reduce their viability which inhibits the spread of infection. Studies suggest that micromolar concentrations of H_2O_2 restrict the in vitro germination of harmful fungi. Moreover, *Erwinia carotovora* is observed to be highly sensitive toward micromolar concentrations of H_2O_2 (Wu et al., 1995).

Plant cells exhibit more tolerance to H_2O_2 than pathogens. Additionally, increased endogenous levels of H_2O_2 result in more resistance to pathogens. The glucose oxidase encoding gene oxidizes glucose and generates H_2O_2. It is observed that potato plants having this gene are more resistant to Phytophthora infestans and *E. carotovora*. Catalase is an enzyme that removes H_2O_2 from plants. Transgenic tobacco plants lacking this enzyme exhibited more resistance to *Pseudomonas syringae*. It suggests that increased levels of endogenous H_2O_2 enhanced the tolerance of transgenic plants to fungal and bacterial pathogens. When a pathogen attacks a plant, H_2O_2 is produced during this phase and contributes to the strengthening processes of the cell wall, such as lignifications, papillae formation, and crosslinking of cell wall polymers.

Cytological studies suggest that H_2O_2 contributes toward the development of effective papillae in cells. Fungal pathogens are unable to penetrate these papillae successfully (Hückelhoven et al., 1999). It has been observed that H_2O_2 acts as a substrate in lignification. It has been documented that the levels of H_2O_2 and the activity of peroxidases that oxidize the phenylpropanoid alcohols involved in the process of lignification increase during the interactions of plants and pathogens (Gajewska & Urbanek, 1999).

The highest concentration of H_2O_2 matched with the highest activity of peroxidase when examined with syringaldazine in tomato leaves treated with Fusarium toxin. This high concentration is an indicator of the process of lignifications (Kuźniak et al., 1999). These reinforcing processes of the cell wall occur in the tissue around the intrusion site and restrict the transport of pathogenic toxins to the plant, increase the resistance of cell wall polymers to the enzymes secreted by pathogens, limit the availability of nutrients to pathogens, and eventually inhibit the spread of infection before the development of defense mechanisms dependent on transcription. In many species of plants, H_2O_2 was found to be essential for phytoalexin production. These are low molecular weight, hydrophobic compounds produced in response to infection and possess antimicrobial activity.

It is reported that when potato, soybean, and parsley were treated with scavengers of reactive oxygen species, the induction of phytoalexins was prevented (Kombrink & Hahlbrock, 1986). It suggests that there is a correlation between the production of H_2O_2 and the synthesis of phytoalexin. However, some studies indicate that the synthesis of phytoalexin in the absence of formation of reactive oxygen species; ROS was independent. Hypersensitive response is a resistance mechanism that causes rapid and localized death of plant cells at the infection site. In plants, the hypersensitive response is initiated when an avirulent pathogenic or a non-pathogenic strain attacks and limits the growth of the pathogen. Studies have suggested the involvement of ROS in hypersensitive response. It has been observed that H_2O_2 produced during oxidative burst plays an essential role in hypersensitive response. In soybean cell cultures, an increase in the concentration of H_2O_2 increased the amount of dead plant cells. It suggests that H_2O_2 plays a key role in cell death associated with hypersensitive response (Lu & Higgins, 1999). Transgenic tobacco plants possess yeast catalase genes to produce peroxisomal catalase. These plants were observed to begin a hypersensitive response after an attack of the tobacco mosaic virus. However, the necrotic lesions were bigger in size than those present on infected control leaves. It indicates that an enhanced catalase activity and subsequently reduced levels of H_2O_2 affected the intensity of hypersensitive response upon pathogen attack (Krzymowska et al., 1997). However, some other studies suggest that bacterial mutants that could not induce a hypersensitive response in tobacco suspension culture still induced the oxidative burst (Glazener et al., 1996). Recently, a new function of H_2O_2 in disease resistance has been postulated. It is named as an endogenous signal for systemic acquired resistance. This process is initiated by pathogens that cause local cell death, and it provides broad-spectrum and long-lasting resistance against subsequent infections. Signals that trigger systemic acquired resistance travel from the site of the primary inoculation to the uninfected or healthy parts of the plant. The induction of systemic acquired resistance occurs when a set of systemic acquired resistance genes gets activated (Ryals et al., 1996).

INDUCTION OF STRESS TOLERANCE IN PLANTS

Several environmental constraints, including flooding, chilling, SO_2, drought, salt highlight, wounding, ozone, UV radiation, and herbicides, cause enhanced production of ROS and lead to severe oxidative damage to tissues. ROS are generally involved in abiotic stresses. To improve the oxidative stress tolerance of plants, various antioxidative enzymes that eliminate damaging ROS were genetically engineered. However, it is suggested that H_2O_2 produced during abiotic stress carries out signal transduction and protects against this stress. This process is similar to the role of H_2O_2 in plant-pathogen interactions. It has been reported that exogenous H_2O_2 induces stress tolerance in plants (Iturbe-Ormaetxe et al., 1998).

Studies have shown that tolerance to chilling stress was enhanced in maize seedlings after the application of H_2O_2. It occurred because the temporary increase in H_2O_2 initiated acclamatory mechanisms. A dual effect of H_2O_2 has been observed in maize. It acts as a signal to initiate the production of ROS-scavenging enzymes during chilling acclimation and in non-acclimated seedlings; it plays a role as a destructive agent by accumulating in higher levels. H_2O_2 induces thermo tolerance in potato microplants grown in vitro using standard methods and improves the survival of plantlets during the thermotherapy for virus eradication (Lopez-Delgado et al., 1998). In maize seedlings, salicylic acid pretreatment led to increased levels of H_2O_2 which increased the activity of the antioxidant enzyme subsequently causing enhanced chilling tolerance.

In the same way, H_2O_2 was found to be involved in signal transduction, causing acclimation when under heat stress (Dat et al., 1998). It is suggested that the initiation of tolerance to particular environmental stresses that involve oxidative stress also results in enhanced tolerance to other biotic and abiotic stresses. This phenomenon is known as cross-tolerance and is observed in many plants. In cotton plants, paraquat tolerance was induced due to water deficiency, whereas resistance to tobacco mosaic virus in tobacco plants was caused by exposure to UV radiations. Exposure to ozone activated a set of proteins related to pathogens and initiated resistance to *P. syringae* in Arabidopsis thaliana (Sharma et al., 1996).

SAFETY CONSIDERATIONS OF HYDROGEN PEROXIDE

There is a growing interest in using hydrogen peroxide in agriculture and horticulture. Although H_2O_2 is less harmful than conventional herbicides, some points need to be considered while using it as a herbicide. H_2O_2 is unstable, and one of its oxygen atoms is always trying to escape the molecule. Once this oxygen atom leaves, it becomes available for reactions. When this oxygen atom comes in contact with any organic matter, it reacts with it and burns it releasing carbon dioxide. All microorganisms are composed of organic molecules. Hydrogen peroxide can kill harmful microorganisms but it also harms the microorganisms beneficial for soil. The use of hydrogen peroxide beyond recommended concentrations is very harmful to the organic matter of the soil and decomposes it within a few days. This process occurs in two phases. The materials containing nitrogen and carbon are oxidized in the first phase. In the second phase, the materials containing nitrogen-free carbon compounds are oxidized. The most important fungi in the soil, the mycorrhiza, are also negatively affected by the use of peroxides. The hyphae and mycelia are killed, which destroys the association of plants with mycorrhiza, especially the plantation crops and perennial crops such as fruits. The absence of mycorrhiza destroys the structure of the soil. It also negatively affects the glomaline, which is the binding material (Ganeshamurthy, 2022).

TRADITIONAL HERBICIDES VS. HYDROGEN PEROXIDE

There are some glaring differences between traditional herbicides, which have been in circulation for decades, and hydrogen peroxide, an emerging solution to combat weeds while considering soil health. The main differences are discussed further.

CHEMICAL COMPOSITION

Conventional herbicides are composed of synthetic chemicals, while hydrogen peroxide is composed of molecules of oxygen and hydrogen.

MODE OF ACTION

Traditional herbicides work against weeds work by disrupting specific biological processes such as cell division, photosynthesis, and protein synthesis. Hydrogen peroxide oxidizes plant tissues and causes damage to cells and cell death.

METHOD OF APPLICATION

Various methods are used to apply traditional herbicides. These include spraying, incorporation in soil, and granule spreading. Special care and equipment are required for the effective and safe application of chemical herbicides. Hydrogen peroxide is easy to use and is usually directly applied to the target plants. Moreover, it rarely requires special equipment for its application.

SELECTIVITY

Conventional chemical herbicides are both selective and non-selective in their target. Some of them can act on selective weeds, while others can act on a wide range of weeds. Hydrogen peroxide is usually non-selective in its target and can affect almost every plant that comes in contact.

EFFECTIVENESS

Chemical herbicides are very effective against a wide range of weeds and can work both selectively and non-selectively. They are specifically designed to take into account the types of weeds and the challenges to control them. Whereas, hydrogen peroxide is primarily effective against small weeds and shows less efficiency against mature weeds with extensive root systems. The effectiveness of hydrogen peroxide against weeds also depends on environmental conditions.

IMPACT ON ENVIRONMENT

Traditional herbicides negatively impact the environment. They can cause significant damage to other non-target plants. They persist in water and soil and disrupt their chemical composition. They can harm aquatic life and beneficial organisms in the soil. On the other hand, hydrogen peroxide breaks down into non-toxic products, water and oxygen. Moreover, it is less toxic to other plants and animals.

IMPACT ON HUMAN BEINGS

Herbicides based on chemicals can cause serious negative impacts on human beings. They can affect the skin and respiratory systems of individuals who come in contact with them. Moreover, certain traditional herbicides, such as glyphosate, are carcinogenic.

IMPACT ON SOIL

Conventional herbicides have a negative impact on soil health as they disturb the natural microorganisms of the soil that are required for maintaining a proper soil environment. The chemical residues of these herbicides are accumulated in the soil. These herbicides can affect soil fertility and can also cause its degradation. On the other hand, hydrogen peroxide does not affect the soil health significantly. It breaks down into water and oxygen, which are harmless to the soil. Additionally, it does not impact the microbiota of the soil, and no harmful residues accumulate in the soil.

RESIDUES

Traditional herbicides tend to persist in soil for a long period of time. They can control weeds for several months after application. It suggests the less frequent application of these herbicides for weed control. However, the harmful residues of these herbicides can affect other plants and animals and must be managed properly. Hydrogen peroxide does not reside in the soil for a long time. Therefore, it is applied more frequently to manage weeds. Moreover, it does not have harmful residues that can affect other plants and animals.

SAFETY CONSIDERATIONS

Chemical herbicides are generally toxic so special care is required for their handling. Protective equipment must be used while applying these herbicides to avoid any damage to other plants and workers. Alternatively, hydrogen peroxide does not require special care and equipment for use. Its process of application is simple and is suitable for both, professionals and home gardeners.

REGULATORY RESTRICTIONS

As traditional herbicides have negative impacts on humans, other plants, animals, and the environment, their use is strictly regulated. Health and environmental agencies tend to regularly review these herbicides.

CASE STUDIES

Due to the harmful effects of traditional herbicides on living organisms and the environment, researchers are looking for alternative methods for weed control that are less toxic and environmentally friendly. For this purpose, research work is being conducted at various levels to find out the effectiveness of these alternative methods against weeds. The effectiveness of some natural products for the management of weeds was compared. The natural products used in this study include citric acid (10%), acetic acid (30%), clove oil (45.6%), corn gluten meal, garlic (0.2%)+citric acid (5%), and garlic (0.2%)+citric acid (10%). These herbicides were applied to broadleaf weeds. The results of this study suggest that these natural products are effective against weeds (Abouziena et al., 2009). A study was conducted to determine the efficacy of an acetic acid-based formula against broadleaf weeds in medicinal crop fields. The experiment was conducted on small plots and large plots. A control group was treated with a synthetic herbicide known as 2,4-D amine. At the end of this study, the scientists concluded that acetic acid-based formula was effective against broadleaf weeds. The mode of action of this formula is a contact poison. It was found the effectiveness of this formula was the same as that of 2,4-D amine. This formulation requires repeated application for prolonged effect (Rahayuningsih & Supriadi, 2016).

A unique study was conducted to determine the effectiveness of *Asteraceae* essential oils as a herbicide. Twenty species of *Asteraceae* were analyzed for their quality of essential oils, and half of them were found to have sufficient production of essential oils. Further analysis suggested that essential oils of *Xanthium strumarium* and two species of *Artemisia* can inhibit germination completely. The weeds were sprayed with essential oils of these species at different concentrations. The essential oils of *X. strumarium* and *Artemisia annua* were found to be the most effective in weed control. So, the researchers concluded that essential oils of both of these species can act as herbicides for weed management (Benvenuti et al., 2017).

In a similar study, scientists conducted an experiment to investigate the effectiveness of methanol extracts from eight species of plants for weed control. The species of plants that were used are *Alhagi maurorum*, *Haloxylon salicornicum*, *Cyperus conglomerates*, *Ochradenus bacctus*, *Lactuca virosa*, *Tamarix mannifera*, and *Neurada procumbens.* The effect of these plant species was examined on the germination and growth of *Echinochloa crusgalli*, *Lactuca sativa*, *Portulaca oleracea*, and *Phalaris minor.* The results revealed that *Lactuca virosa* was the most effective species for controlling seed germination and the growth of seedlings in selected species. Among weed species, *Phalaris minor* was the most sensitive toward plant extracts. *Lactuca virosa* and *Tamarix mannifera* can be used for the development of new natural herbicides (El-Mergawi & Al-Humaid, 2019). To figure out the effectiveness of essential oils and their compounds for weed management, Korean plants were used to extract a total of 18 essential oils. Moreover, 64 commercial essential oils were also examined for their phytotoxic activity. Among these 82 samples, 11 commercial oils, which include clove, geranium, lemongrass, spearmint, jasmine, cumin seed, cinnamon, rose otto, pimento, citronella, and palmarosa, effectively disrupted the seedling growth. The individual components of these essential oils were also identified. In conclusion, individual components of essential oils, citral, and geraniol can be used for weed management as natural herbicides. In a similar study, the herbicidal potential of essential oils from rosemary, white cedar, lawson cypress, and eucalypt was investigated, all of these essential oils have a strong herbicidal effect on the germination of the selected weed species. The most efficient of these essential oils was found to be eucalyptus oil, when used at 300 ppm. Results indicate that essential oils can be applied for weed management and can reduce the use of chemical herbicides (Ramezani et al., 2008).

The effectiveness of mulching and some natural products against weed control also has been examined including citric acid, acetic acid, oxyflurfen, oxadiargyl, and rice straw mulch. The experiment was conducted on sandy soils and examines weed control, onion productivity after the application of the products and all of these methods were effective for weed control. So, the researchers suggested that farmers can rely on these natural methods of weed control in onion fields instead of chemical herbicides (El-Metwally & Shalaby, 2019). To determine the herbicidal effects of plant essentials, acetic acid, and pine oil on vegetation along the roadside, the effectiveness of these natural products was compared with the effectiveness of a chemical herbicide known as glyphosate for weed killing. Glyphosate was more efficient and cost-effective than natural products. Moreover, the frequency of application was greater for natural products than for glyphosate (Young, 2004). The effect of natural products as weed killers are also used for tree establishment, the natural herbicides used citronella oil and bilanaphos. The efficacy of these products was compared to the effectiveness of glufosinate-ammonium. All of these products were sprayed on weeds in different seasons, and both of these natural products were efficient for controlling weeds and tree establishment (Clay et al., 2005).

FUTURE PERSPECTIVES

To avoid the harmful effects of chemical herbicides, there is a need to move toward natural products for the management of weeds. As described in the case studies included in this chapter, research work is being conducted for the development of more sustainable and environmentally friendly herbicides. Essential oils, hydrogen peroxide, acetic acid, salt, herbicidal soaps, and many other products are being tested for their herbicidal potential. The use of these herbicides will help in avoiding the undesirable effects of traditional herbicides on the environment and other organisms.

CHALLENGES AND OPPORTUNITIES

The significant challenges in the field of agriculture regarding weed management are weed resistance against herbicides and the harmful effects of chemical herbicides. Weeds become resistant to herbicides over time. It is very challenging for the farmers and demands the development of stronger herbicides. It can be managed by developing herbicide formulations that are difficult to be resisted by the weeds. If a weed becomes resistant to a certain herbicide, the use of an alternative herbicide can help. The use of integrated weed management strategies can minimize the development of resistance. The mixture of two herbicides can also be used to manage herbicide-resistant weeds. The harmful effects of chemical herbicides can be managed by developing herbicides based on natural products.

RECOMMENDATIONS AND PREDICTIONS

After reading this chapter and analyzing the case studies mentioned, it is recommended that the comparative method is effective for analyzing the efficacy of various herbicides. The comparison of chemical herbicides with natural products suggests that these products will be more effective and environmentally friendly in the future. If proper research is conducted in this area of research, more beneficial and less hazardous herbicides can be developed for weed management.

CONCLUSION

The use of herbicides for weed control is inevitable in the field of agriculture. Conventional herbicides are effective against weeds, but they cause many harmful effects on soil, microorganisms, animals, human beings, and aquatic life. The use of alternative herbicides such as essential oils, acetic acid, hydrogen peroxide, salt, and herbicidal soap can help to avoid these undesirable effects. Literature is available regarding the effectiveness of the use of these products for weed management. It is a simple compound that breaks down into water and oxygen and leaves no toxic residues in the soil. It plays very important roles in plants such as induction of defense mechanisms and stress tolerance. The use of hydrogen peroxide is a safe method for weed control with minimum side effects on other organisms. Further research is required to investigate the herbicidal potential of hydrogen peroxide and various other products.

REFERENCES

Abdollahi, M., & Hosseini, A. (2014). Hydrogen peroxide. *Encyclopedia of Toxicology, 3*, 967–970.

Abouziena, H. F., Omar, A. A., Sharma, S. D., & Singh, M. (2009). Efficacy comparison of some new natural-product herbicides for weed control at two growth stages. *Weed Technology, 23*(3), 431–437.

Alletto, L., Coquet, Y., Benoit, P., Heddadj, D., & Barriuso, E. (2010). Tillage management effects on pesticide fate in soils. A review. *Agronomy for Sustainable Development, 30*(2), 367–400.

Anasco, N., Uno, S., Koyama, J., Matsuoka, T., & Kuwahara, N. (2010). Assessment of pesticide residues in freshwater areas affected by rice paddy effluents in Southern Japan. *Environmental Monitoring and Assessment, 160*, 371–383.

Baran, N., Mouvet, C., & Negrel, P. (2007). Hydrodynamic and geochemical constraints on pesticide concentrations in the groundwater of an agricultural catchment (Brevilles, France). *Environmental Pollution, 148*(3), 729–738.

Bauer, W. D., & Schefcik, M. (1994). Using differential GPS to improve crop yields. *GPS World, 5*(2), 38–41.

Bengtson-Nash, S., Eaglesham, G., & Mueller, J. (2004). *Application of a Novel Phytotoxicity Test for the Detection of Herbicides in Hervey Bay and the Great Sandy Stait*, CRC Reef, Townsville.

Benvenuti, S., Cioni, P. L., Flamini, G., & Pardossi, A. (2017). Weeds for weed control: Asteraceae essential oils as natural herbicides. *Weed Research, 57*(5), 342–353.

Bichat, F., Mulvaney, R., & Sims, G. (1999). Microbial utilization of heterocyclic nitrogen from atrazine. *Soil Science Society of America Journal, 63*(1), 100–110.

Borner, T., Manteuffel, R., & Wellburn, A. (1979). Enzymes of plastid ribosome-deficient mutants chloroplast ATPase (CF 1). *Protoplasma, 98*, 153–161.

Bozso, Z., Ott, P., Szatmari, A., Czelleng, A., Varga, G., Besenyei, E., Sardi, E., Banyani, E., & Klement, Z. (2005). Early detection of Bacterium-induced basal resistance in tobacco leaves with diaminobenzidine and dichlorofluorescein diacetate. *Journal of Phytopathology, 153*(10), 596–607.

Breeze, V., Marshall, E., Hart, A., Vickery, J., Crocker, J., Walters, K., Packer J., Kendall, D., Fowbert, J., & Hodkinson, D. (1999). Assessing pesticide risks to non-target terrestrial plants. *A desk study. Commission PN0923. MAFF Pesticides Safety Directorate.*

Bresnahan, G. A., Koskinen, W. C., Dexter, A. G., & Lueschen, W. E. (2000). Influence of soil pH^- sorption interactions on imazethapyr carry-over. *Journal of Agricultural and Food Chemistry,* 48(5), 1929–1934.

Casabé, N., Piola, L., Fuchs, J., Oneto, M. L., Pamparato, L., Basack, S., Kesten, E. (2007). Ecotoxicological assessment of the effects of glyphosate and chlorpyrifos in an Argentine soya field. *Journal of Soils and Sediments, 7*, 232–239.

Cheney, M. A., Fiorillo, R., & Criddle, R. S. (1997). Herbicide and estrogen effects on the metabolic activity of Elliptio complanata measured by calorespirometry. *Comparative Biochemistry and Physiology Part C: Pharmacology, Toxicology and Endocrinology, 118*(2), 159–164.

Chheda, J. N., Huber, G. W., & Dumesic, J. A. (2007). Liquid-phase catalytic processing of biomass-derived oxygenated hydrocarbons to fuels and chemicals. *Angewandte Chemie International Edition, 46*(38), 7164–7183.

Clay, D., Dixon, F., & Willoughby, I. (2005). Natural products as herbicides for tree establishment. *Forestry, 78*(1), 1–9.

Cobucci, T., Prates, H. T., Falcão, C. L., & Rezende, M. M. (1998). Effect of imazamox, fomesafen, and acifluorfen soil residue on rotational crops. *Weed Science, 46*(2), 258–263.

Cochet, H. (2015). *Comparative Agriculture*. Springer, Cham.

Daam, M. A., Rodrigues, A. M., Van den Brink, P. J., & Nogueira, A. J. (2009). Ecological effects of the herbicide linuron in tropical freshwater microcosms. *Ecotoxicology and Environmental Safety, 72*(2), 410–423.

Dao, T. H. (1991). *Field decay of wheat straw and its effects on metribuzin and S-ethyl metribuzin sorption and elution from crop residues*. (Vol. 20, No. 1, pp. 203-208). American Society of Agronomy, Crop Science Society of America, and Soil Science Society of America.

Dat, J. F., Lopez-Delgado, H., Foyer, C. H., & Scott, I. M. (1998). Parallel changes in H_2O_2 and catalase during thermotolerance induced by salicylic acid or heat acclimation in mustard seedlings. *Plant Physiology, 116*(4), 1351–1357.

Lorenzo, D. V. (2008). Systems biology approaches to bioremediation. *Current Opinion in Biotechnology, 19*(6), 579–589.

Delidovich, I. V., Taran, O. P., Matvienko, L. G., Simonov, A. N., Simakova, I. L., Bobrovskaya, A. N., & Parmon, V. N. (2010). Selective oxidation of glucose over carbon-supported Pd and Pt catalysts. *Catalysis Letters, 140*, 14–21.

Dempsey, D., & Klessig, D. (1995). Signals in plant disease resistance. *Bulletin de l'Institut Pasteur, 93*(3), 167–186.
Desikan, R., Hancock, J. T., & Neill, S. J. (2004). Oxidative stress signalling. In: *Plant Responses to Abiotic Stress*, 121–149. https://doi.org/10.1007/978-3-540-39402-0_6.
DiTomaso, J. M. (2005). Understanding Herbicides: What They are and How They Work. *University of California. Division of Agriculture and Natural Resources.*
Duffard, R., Traini, L., & de Duffard, A. (1981). Embryotoxic and teratogenic effects of phenoxy herbicides. *Acta Physiologica Latino Americana, 31*(1), 35–38.
Duke, S. O., Cedergreen, N., Velini, E. D., & Belz, R. G. (2006). Hormesis: Is it an important factor in herbicide use and allelopathy? *Outlooks on Pest Management, 17*(1), 29–33.
Dumont, R. (1954). *Economie Agricole Dans le Monde*. Dalloz, Paris.
El-Mergawi, R. A., & Al-Humaid, A. I. (2019). Searching for natural herbicides in methanol extracts of eight plant species. *Bulletin of the National Research Centre, 43*, 1–6.
El-Metwally, I., & Shalaby, S. (2019). Herbicidal efficacy of some natural products and mulching compared to herbicides for weed control in onion fields. *Journal of Plant Protection Research, 59*(4), 479–486.
Evans, S. C., Shaw, E. M., & Rypstra, A. L. (2010). Exposure to a glyphosate-based herbicide affects agrobiont predatory arthropod behaviour and long-term survival. *Ecotoxicology, 19*, 1249–1257.
Fabra, A., Duffard, R., & De Duffard, A. E. (1997). Toxicity of 2, 4-dichlorophenoxyacetic acid to Rhizobium sp in pure culture. *Bulletin of Environmental Contamination and Toxicology, 59*(4), 645–652.
Felsot, A., & Dzantor, E. (1990). Enhancing biodegradation for detoxification of herbicide waste in soil. ACS Publications.
Freiberger, O. (2019). The comparative method in the study of religion. In: *Considering Comparison: A Method for Religious Studies*, pp. 7–44. https://doi.org/10.1093/oso/9780199965007.003.0002.
Gajewska, E., & Urbanek, H. (1999). Hydogen peroxide and ezymes metabolizing it in tomato leaves after infection with fungus. *Acta Physiologiae Plantarum, 21*(3 suppl.), 41.
Ganeshamurthy, A.N. (2022). Should hydrogen peroxide be used in agriculture and horticulture? Available at https://www.soulofkrishi.com/post/should-hydrogen-peroxide-be-used-in-agriculture-and-horticulture.
Gaston, L., & Locke, M. (2000). Acifluorfen sorption, degradation, and mobility in a Mississippi delta soil. *Soil Science Society of America Journal, 64*(1), 112–121.
Glazener, J. A., Orlandi, E. W., & Baker, C. J. (1996). The active oxygen response of cell suspensions to incompatible bacteria is not sufficient to cause hypersensitive cell death. *Plant Physiology, 110*(3), 759–763.
Goor, G., Glenneberg, J., Jacobi, S., Dadabhoy, J., & Candido, E. (2012). *Ullmann's Encyclopedia of Industrial Chemistry. American Cancer Society*, New York, 1–40.
Greer, C. W., Hawari, J., & Samson, R. (1990). Influence of environmental factors on 2, 4-dichlorophenoxyacetic acid degradation by Pseudomonas cepacia isolated from peat. *Archives of Microbiology, 154*, 317–322.
Hall, F. R., Chapple, A. C., Downer, R. A., Kirchner, L. M., & Thacker, J. R. M. (1993). Pesticide application as affected by spray modifiers. *Pesticide Science, 38*(2–3), 123–133.
Hance, R. (1973). The effect of nutrients on the decomposition of the herbicides atrazine and linuron incubated with soil. *Pesticide Science, 4*(6), 817–822.
Hang, S., Andriulo, A., Sasal, C., Nassetta, M., Portela, S., & Cañas, A. I. (2010). Integral study of atrazine behavior in field lysimeters in Argentinean humid pampas soils. *Chilean Journal of Agricultural Research, 70*(1), 104–112.
Helfrich, L. A., Weigmann, D. L., Hipkins, P. A., & Stinson, E. R. (2009). Pesticides and aquatic animals: a guide to reducing impacts on aquatic systems. http://hdl.handle.net/10919/48060.
Hewitt, A. (2000). Spray drift: Impact of requirements to protect the environment. *Crop Protection, 19*(8–10), 623–627.
Holt, R. T. (1970). *The Methodology of Comparative Research*. The Free Press, New York.
Hu, H., & Xiong, L. (2014). Genetic engineering and breeding of drought-resistant crops. *Annual Review of Plant Biology, 65*, 715–741.
Hückelhoven, R., Fodor, J., Preis, C., & Kogel, K.-H. (1999). Hypersensitive cell death and papilla formation in barley attacked by the powdery mildew fungus are associated with hydrogen peroxide but not with salicylic acid accumulation. *Plant Physiology, 119*(4), 1251–1260.
Iturbe-Ormaetxe, I., Escuredo, P. R., Arrese-Igor, C., & Becana, M. (1998). Oxidative damage in pea plants exposed to water deficit or paraquat. *Plant Physiology, 116*(1), 173–181.
Janaki, P., Sundaram, K. M., Chinnusamy, C., & Sakthivel, N. (2015). Determination of residues of metribuzin in soil and sugarcane by QuEChERS. https://doi.org/10.14233/ajchem.2015.18926.

Kadian, N., Gupta, A., Satya, S., Mehta, R. K., & Malik, A. (2008). Biodegradation of herbicide (atrazine) in contaminated soil using various bioprocessed materials. *Bioresource Technology, 99*(11), 4642–4647.

Kanissery, R. G., & Sims, G. K. (2011). Biostimulation for the enhanced degradation of herbicides in soil. *Applied and Environmental Soil Science,* 2011, 1–10.

Kim, Y. H., Khan, A. L., Shinwari, Z. K., Kim, D. H., Waqas, M. U. H. A. M. M. A. D., Kamran, M. U. H. A. M. M. A. D., & Lee, I. J. (2012). Silicon treatment to rice (Oryza sativa L. cv.'Gopumbyeo') plants during different growth periods and its effects on growth and grain yield. *Pakistan Journal of Botany, 44*(3), 891–897.

Knauer, K., Leimgruber, A., Hommen, U., & Knauert, S. (2010). Co-tolerance of phytoplankton communities to photosynthesis II inhibitors. *Aquatic Toxicology, 96*(4), 256–263.

Koch, B., & Khosla, R. (2003). The role of precision agriculture in cropping systems. *Journal of Crop Production, 9*(1–2), 361–381.

Kombrink, E., & Hahlbrock, K. (1986). Responses of cultured parsley cells to elicitors from phytopathogenic fungi: Timing and dose dependency of elicitor-induced reactions. *Plant Physiology, 81*(1), 216–221.

Kortekamp, A. (2011). *Herbicides and Environment.* BoD–Books on Demand, Norderstedt.

Krzymowska, M., Talarczyk, A., & Hennig, J. (1997). Is tobacco response to TMV infection modulated by catalase activity? *Acta Physiologiae Plantarum, 19*(4), 577–579.

Kucharski, M., & Sadowski, J. (2009). Influence of adjuvants on behavior of phenmedipham in plant and soil. *Polish Journal of Agronomy, 1,* 32–36.

Kuźniak, E., Patykowski, J., & Urbanek, H. (1999). Involvement of the antioxidative system in tomato response to fusaric acid treatment. *Journal of Phytopathology, 147*(7–8), 385–390.

Liong, P., Wan Portiah Hamzah, W. P. H., & Velu Murugan, V. M. (1988). Toxicity of some pesticides towards freshwater fishes. http://myagric.upm.edu.my/id/eprint/10973.

Lopez-Delgado, H., Dat, J. F., Foyer, C. H., & Scott, I. M. (1998). Induction of thermotolerance in potato microplants by acetylsalicylic acid and H_2O_2. *Journal of Experimental Botany, 49*(321), 713–720.

Lu, H., & Higgins, V. J. (1999). The effect of hydrogen peroxide on the viability of tomato cells and of the fungal pathogen Cladosporium fulvum. *Physiological and Molecular Plant Pathology, 54*(5–6), 131–143.

Magnusson, M., Heimann, K., & Negri, A. P. (2008). Comparative effects of herbicides on photosynthesis and growth of tropical estuarine microalgae. *Marine Pollution Bulletin, 56*(9), 1545–1552.

Maire, M., Rast, C., Landkocz, Y., & Vasseur, P. (2007). 2, 4–Dichlorophenoxyacetic acid: Effects on Syrian hamster embryo (SHE) cell transformation, c-Myc expression, DNA damage and apoptosis. *Mutation Research/Genetic Toxicology and Environmental Mutagenesis, 631*(2), 124–136.

Mandelbaum, R., Wackett, L. P., & Allan, D. L. (1993). Mineralization of the s-triazine ring of atrazine by stable bacterial mixed cultures. *Applied and Environmental Microbiology, 59*(6), 1695–1701.

Marshall, E. (2001). Biodiversity, herbicides and non-target plants. *Paper presented at the Brighton Crop Protection Conference Weeds.* (Vol. 2, pp. 855-862) BCPC, Farnham, UK.

Mazoyer, M., & Roudart, L. (2017). *Histoire des agricultures du monde. Du néolithique à la crise contemporaine.* Média Diffusion, Paris.

Melander, B., Liebman, M., Davis, A. S., Gallandt, E. R., Bàrberi, P., Moonen, A. C., Rasmussen, J., van der Weide, R., & Vidotto, F. (2017). Non-chemical weed management. In: *Weed Research: Expanding Horizons*, 245–270. https://doi.org/10.1002/9781119380702.ch9.

Mickelson, S. K., Anton, S., Baker, J. L., & Ahmed, S. I. (2001). Subsurface herbicide application with the modified John Deere mulch master. *Transactions of the ASAE, 44*(4), 807.

Mokhtarianpour, M. (2016). Islamic model of Iranian pattern development process model. *The Pattern of Islamic Development of Iran, 4*(8), 9–30.

Monaco, T. J., Weller, S. C., & Ashton, F. M. (2002). *Weed Science: Principles and Practices.* John Wiley & Sons, Hoboken, NJ.

Nalini, K., Muthukrishan, P., Chinnusamy, C., & Janaki, P. (2010). Pre-emergence herbicide pendimethalin residue in cotton cultivated soil. *Paper presented at the Proceedings of the Biennial Conference of Indian Society of Weed Science on Recent Advances in Weed Science Research.* (pp. 25-26) IGKV, Raipur, Chhattisgarh.

Navarro, C. D., & Martinez, C. B. (2014). Effects of the surfactant polyoxyethylene amine (POEA) on genotoxic, biochemical and physiological parameters of the freshwater teleost Prochilodus lineatus. *Comparative Biochemistry and Physiology Part C: Toxicology & Pharmacology, 165*, 83–90.

Neal, J. (1998). Postemergence, Non-selective Herbicides for Landscapes and Nurseries. *Horticulture Information Leaflets.* North Carolina State University, Raleigh, NC.

Nishimi, T., Kamachi, T., Kato, K., Kato, T., & Yoshizawa, K. (2011). Mechanistic study on the production of hydrogen peroxide in the anthraquinone process. Wiley Online Library. https://doi.org/10.1002/ejoc.201100300.

Nolte, K., Siemens, M. C., & Andrade-Sanchez, P. (2011). Integrating variable rate technologies for soil-applied herbicides in Arizona vegetable production. The University of Arizona Cooperative Extension, AZ1538, 1-5.

Non-staining herbicidal soap. (n.d.) Available at https://patents.google.com/patent/EP0868849A1/en.

Ntainjua, N. E., Piccinini, M., Pritchard, J. C., Edwards, J. K., Carley, A. F., Moulijn, J. A., & Hutchings, G. J. (2009). Effect of halide and acid additives on the direct synthesis of hydrogen peroxide using supported gold–palladium catalysts. *ChemSusChem: Chemistry & Sustainability Energy & Materials, 2*(6), 575–580.

O'Connell, S., Rivard, C., Peet, M. M., Harlow, C., & Louws, F. (2012). High tunnel and field production of organic heirloom tomatoes: Yield, fruit quality, disease, and microclimate. *HortScience, 47*(9), 1283-1290.

Olson, B. L., Regehr, D. L., Janssen, K. A., & Barnes, P. L. (1998). Tillage system effects on atrazine loss in surface water runoff. *Weed Technology, 12*(4), 646–651.

Omri, M., Pourceau, G., Becuwe, M., & Wadouachi, A. (2016). Improvement of gold-catalyzed oxidation of free carbohydrates to corresponding aldonates using microwaves. *ACS Sustainable Chemistry & Engineering, 4*(4), 2432–2438.

Paunovic, V., Ordomsky, V. V., Sushkevich, V. L., Schouten, J. C., & Nijhuis, T. A. (2015). Direct synthesis of hydrogen peroxide over Au-Pd catalyst—the effect of co-solvent addition. *ChemCatChem, 7*(7), 1161–1176.

Pérez, G. L., Vera, M. S., & Miranda, L. A. (2011). Effects of herbicide glyphosate and glyphosate-based formulations on aquatic ecosystems. *Herbicides and Environment, 16*, 343–368.

Qasem, J. R. (2011). Herbicides applications: Problems and considerations. In: *Herbicides and Environment.* IntechOpen, University of Jordan, Jordan, pp. 643–664.

Qiu, Y., Pang, H., Zhou, Z., Zhang, P., Feng, Y., & Sheng, G. D. (2009). Competitive biodegradation of dichlobenil and atrazine coexisting in soil amended with a char and citrate. *Environmental Pollution, 157*(11), 2964–2969.

Quan, L. J., Zhang, B., Shi, W. W., & Li, H. Y. (2008). Hydrogen peroxide in plants: A versatile molecule of the reactive oxygen species network. *Journal of Integrative Plant Biology, 50*(1), 2–18.

Rahayuningsih, S., & Supriadi, S. (2016). Herbicidal efficacy of acetic acid and citric acid base on broad leaf weeds of medicinal crops fields. *Buletin Penelitian Tanaman Rempah dan Obat, 25*(2), 137–143.

Ramezani, S., Saharkhiz, M. J., Ramezani, F., & Fotokian, M. H. (2008). Use of essential oils as bioherbicides. *Journal of Essential Oil Bearing Plants, 11*(3), 319–327.

Rice, P. J., Anderson, T. A., & Coats, J. R. (2002). Degradation and persistence of metolachlor in soil: Effects of concentration, soil moisture, soil depth, and sterilization. *Environmental Toxicology and Chemistry: An International Journal, 21*(12), 2640–2648.

Ryals, J. A., Neuenschwander, U. H., Willits, M. G., Molina, A., Steiner, H.-Y., & Hunt, M. D. (1996). Systemic acquired resistance. *The Plant Cell, 8*(10), 1809.

Salazar, P., Rico, V., & González-Elipe, A. R. (2017). Non-enzymatic hydrogen peroxide detection at NiO nanoporous thin film-electrodes prepared by physical vapor deposition at oblique angles. *Electrochimica Acta,* 235, 534–542.

Salmi, T., Gemo, N., Biasi, P., & Serna, J. G. (2015). Product distribution analysis of the hydrogen peroxide direct synthesis in an isothermal batch reactor. *Catalysis Today, 248*, 108–114.

Sathiyavani, E. (2014). *Evaluation of integrated weed management with pre and post emergence herbicides in turmeric*. Ph. D. thesis submitted to Dept. of Agronomy, TNAU, Coimbatore.

Scow, K. M., & Hicks, K. A. (2005). Natural attenuation and enhanced bioremediation of organic contaminants in groundwater. *Current Opinion in Biotechnology, 16*(3), 246–253.

Sears, C. (2022). How to Kill Weeds with Salt. Available at https://www.thespruce.com/killing-weeds-with-salt-5080243.

Sharma, N., & Angiras, N. (1997). Studies on standardization of bioassay technique and its use in estimation of pendimethalin residues in soil under wheat+ sarson intercropping system. *Indian Journal of Weed Science, 28*(1 and 2), 82–84.

Sharma, N., & Angiras, N. (2004). Effect of organic matier content on persistence of atrazine. *Agricultural Science Digest, 24*(3), 227–228.

Sharma, N., Sharma, S., Kumar, S., & Joshi, R. (2013). Dissipation and harvest time residue of 2, 4-D in soil and wheat crop. *Indian Journal of Weed Science, 45*(1), 68–70.

Sharma, Y. K., León, J., Raskin, I., & Davis, K. R. (1996). Ozone-induced responses in Arabidopsis thaliana: the role of salicylic acid in the accumulation of defense-related transcripts and induced resistance. *Proceedings of the National Academy of Sciences, 93*(10), 5099–5104.

Sher Khan, R., Iqbal, A., Malak, R., Shehryar, K., Attia, S., Ahmed, T., Khan, M. A., Arif, M., & Mii, M. (2019). Plant defensins: types, mechanism of action and prospects of genetic engineering for enhanced disease resistance in plants. *3 Biotech, 9*, 1–12.

Singh, S., & Tiwari, S. (2020). Responses of plants to herbicides: Recent advances and future prospectives. In: *Plant Life Under Changing Environment*, 237–250. https://doi.org/10.1016/B978-0-12-818204-8.00011-4.

Sitaramaraju, S., Prasad, N., Reddy, V., & Narayana, E. (2014). Impact of pesticides used for crop production on the environment. *Journal of Chemical and Pharmaceutical Sciences, 3*, 75–79.

Ślesak, I., Libik, M., Karpinska, B., Karpinski, S., & Miszalski, Z. (2007). The role of hydrogen peroxide in regulation of plant metabolism and cellular signalling in response to environmental stresses. *Acta Biochimica Polonica, 54*(1), 39–50.

Smith-Fiola, D. & Gill, S. (2022). Vinegar: An Alternative to Glyphosate? Available at https://extension.umd.edu/resource/vinegar-alternative-glyphosate/.

Soloneski, S., & Larramendy, M. L. (2011). *Herbicides in Argentina. Comparative evaluation of the genotoxic and cytotoxic effects on mammalian cells exerted by auxinic members*. IntechOpen. https://doi.org/10.5772/13452.

Sondhia, S. (2014). Herbicides residues in soil, water, plants and non-targeted organisms and human health implications: an Indian perspective. *Indian Journal of Weed Science, 46*(1), 66–85.

Stevenson, F. J. (1972). Organic matter reactions involving herbicides in soil. *Journal of Environmental Quality, 1*(4), 333–343.

Subhani, A., El-ghamry, A. M., Changyong, H., & Jianming, X. (2000). Effects of pesticides (herbicides) on soil microbial biomass-a review. *Pakistan Journal of Biological Sciences, 3*(5), 705–709.

Süzer, S., & Büyük, H. (2010). Residual effects of spraying imidazolinone-family herbicides on Clearfield® sunflower production from the point of view of crop rotation. *Helia, 33*(52), 25–36.

Taylor-Lovell, S., Sims, G. K., & Wax, L. M. (2002). Effects of moisture, temperature, and biological activity on the degradation of isoxaflutole in soil. *Journal of Agricultural and Food Chemistry, 50*(20), 5626–5633.

Tucker, G. (2003). Nutritional enhancement of plants. *Current Opinion in Biotechnology, 14*(2), 221–225.

Vaishya, R., Srivastava, B., & Ghanshyam Singh, G. S. (2003). Integrated weed management in summer urdbean. *Indian Journal of Pulses Research, 16* (2), 161–162.

Verma, S., Dubey, L., & Verma, A. (2014). Management of weeds under organic farming through conservation agriculture. In *Proceedings of the Biennial Conference of Indian Society of Weed Science on Emerging Challenges in Weed Management*, Jabalpur, Madhya Pradesh, India, 15–17.

Winterbourn, C. C. (2013). The biological chemistry of hydrogen peroxide. *Methods in Enzymology, 528*, 3–25.

Wojtyla, Ł., Lechowska, K., Kubala, S., & Garnczarska, M. (2016). Different modes of hydrogen peroxide action during seed germination. *Frontiers in Plant Science, 7*, 175649.

Wu, G., Shortt, B. J., Lawrence, E. B., Levine, E. B., Fitzsimmons, K. C., & Shah, D. M. (1995). Disease resistance conferred by expression of a gene encoding H_2O_2-generating glucose oxidase in transgenic potato plants. *The Plant Cell, 7*(9), 1357–1368.

Xu, X., Zhou, Y., Wang, X., Jiang, W., Qin, L., Wang, J., Yu, H., Chen, X., Shen, X., Yin, C., & Mao, Z. (2023). Effect of hydrogen peroxide on the soil microbial community structure and growth of malus hupehensis rehd. Seedlings under replant conditions. *ACS Omega, 8*(7), 6411–6422.

Young, S. L. (2004). Natural product herbicides for control of annual vegetation along roadsides. *Weed Technology, 18*(3), 580–587.

Zablotowicz, R. M., Locke, M. A., & Gaston, L. A. (2007). Tillage and cover effects on soil microbial properties and fluometuron degradation. *Biology and Fertility of Soils, 44*, 27–35.

Zaller, J. G., Heigl, F., Ruess, L., & Grabmaier, A. (2014). Glyphosate herbicide affects belowground interactions between earthworms and symbiotic mycorrhizal fungi in a model ecosystem. *Scientific Reports, 4*(1), 5634.

15 Potential Applications of Hydrogen Peroxide in Boosting Crop Yield

Mohd Saleem, Qazi Fariduddin, Taruba Ansari, Barket Hussain, Harpreet Bhatia, and Sugandha Mahajan

INTRODUCTION

In recent years, the global agricultural landscape has faced unprecedented challenges in meeting the escalating demands for food production due to a burgeoning population and changing climate conditions (Thornton and Herrero, 2014; Kleinman et al., 2018). This necessitates innovative and sustainable approaches to enhance crop yields while minimizing the environmental impact. One such promising avenue is the utilization of H_2O_2 as a potential agent to boost crop productivity (Cui et al., 2011; Nazir et al., 2020). H_2O_2 plays multifaceted role in agriculture, exploring its impact on plant growth, stress mitigation, and overall yield enhancement (Fariduddin et al., 2014; Yu et al., 2015; Farouk and Qados, 2018). Hydrogen peroxide, a simple and environmentally friendly compound composed of two hydrogen and two oxygen atoms, has garnered attention for its diverse roles in cellular processes (Melchionna et al., 2019). In plants, endogenous hydrogen peroxide functions as a signaling molecule involved in various physiological responses, including cell growth, development, and defense mechanisms against biotic and abiotic stresses (Quan et al., 2008; Gupta et al., 2016; Černý et al., 2018). Understanding and harnessing these natural functions form the basis for exploring hydrogen peroxide's applications in agriculture.

One pivotal aspect of H_2O_2 effect on crop yield is its involvement in regulating plant growth and development (Niu and Liao, 2016; Talaat, 2019; Nazir et al., 2019b). Studies have shown that controlled application of exogenous hydrogen peroxide can stimulate seed germination, root development, and overall plant growth (Nazir et al., 2019b). By acting as a signaling molecule, H_2O_2 participates in intricate signaling pathways that modulate gene expression and hormonal regulation, thereby influencing key processes such as photosynthesis and nutrient uptake (Hossain et al., 2015; Saxena et al., 2016; Niu and Liao, 2016). Unraveling these molecular mechanisms provides insights into how hydrogen peroxide can be strategically employed to optimize crop performance. Furthermore, hydrogen peroxide emerges as a potent tool in ameliorating the adverse effects of environmental stressors on plants (Raza et al., 2022). Abiotic stresses, including drought, salinity, and extreme temperatures, pose significant threats to crop production worldwide (Saleem et al., 2021a; Saleem and Fariduddin, 2022; Khan et al., 2022). Hydrogen peroxide as a stress-responsive signaling molecule equips plants with the ability to activate defense mechanisms and enhance their resilience to adverse conditions (Petrov and Van Breusegem, 2012; Saleem and Fariduddin, 2022). Consequently, the application of hydrogen peroxide presents a promising avenue for bolstering crop tolerance to environmental stresses, contributing to increased agricultural sustainability.

This chapter also delves into the interaction between hydrogen peroxide and plant-microbe relationships. Beneficial microorganisms in the rhizosphere play a crucial role in promoting plant health and nutrient acquisition (Deb et al., 2021). Recent research suggests that hydrogen peroxide-mediated signaling pathways may influence the establishment of symbiotic relationships between plants and beneficial microbes (Puppo et al., 2013; Mukherjee and Corpas, 2023). Understanding

DOI: 10.1201/9781032647692-15

these interactions opens up possibilities for harnessing hydrogen peroxide to enhance soil health and promote sustainable agricultural practices. As the agricultural sector grapples with the imperative to reduce reliance on conventional chemical inputs, hydrogen peroxide stands out as an eco-friendly alternative for crop management. Its inherent properties, such as rapid decomposition into water and oxygen, align with the principles of sustainable agriculture (Hasan et al., 2016; Chattopadhyay et al., 2021; Jira-anunkul and Pattanagul, 2021). By exploring the potential applications of hydrogen peroxide in boosting crop yield, this chapter contributes to the ongoing discourse on sustainable and innovative agricultural practices, paving the way for a more resilient and productive crop management practices.

WHAT IS HYDROGEN PEROXIDE?

H_2O_2 is a chemical compound consisting of two hydrogen atoms and two oxygen atoms, arranged in a diatomic molecule. It is a pale blue liquid at room temperature and is characterized by the presence of a highly reactive peroxide bond. H_2O_2 has a bent molecular geometry as depicted in Figure 15.1, with an angle of approximately 101.9° between the two O-H bonds (Litorja and Ruscic, 1998). The oxygen-oxygen bond in hydrogen peroxide is relatively weak, and the molecule is thermodynamically unstable, leading to its decompositon over time (Ellingson et al., 2007). H_2O_2 can be produced through various methods, including the anthraquinone process, autoxidation of hydroquinone, and electrolysis. The primary industrial method involves the anthraquinone process, which utilizes anthraquinone as a precursor in a multistep process (Campos-Martin et al., 2006; Gill and Tuteja, 2010; Garcia-Munoz et al., 2023). In biological systems, hydrogen peroxide is generated as a natural byproduct of various metabolic processes, particularly in peroxisomes and mitochondria (Winterbourn, 2013, 2018; Yan et al., 2020; Saleem et al., 2021a, b). It functions as a reactive oxygen species (ROS), participating in redox signaling and oxidative stress responses (Schieber and Chandel, 2014; Sies, 2017; Sies and Jones, 2020). The dual role of hydrogen peroxide as a signaling molecule and a potential oxidative stress inducer is well documented in plant biology, influencing processes such as defense responses, cell growth, and programmed cell death (Gadjev et al., 2008; Habibi, 2014; Liu et al., 2021). In the context of plant-microbe interactions, hydrogen peroxide plays a crucial role. It serves as a signaling molecule in plant defense mechanisms against pathogens and in the establishment of symbiotic relationships, such as those with mycorrhizal fungi and nitrogen-fixing bacteria (Li et al., 2013; Zipfel and Oldroyd, 2017; Mittler, 2017). The modulation of hydrogen peroxide levels influences the activation of defense-related genes and systemic acquired resistance (SAR), contributing to the plant's overall ability to respond to environmental challenges (Fu and Dong, 2013; Kamle et al., 2020; Maithani et al., 2021). However, high concentrations of hydrogen peroxide can be harmful and lead to oxidative stress, damaging cells and tissues (Prasad et al., 1994; Torres and Dangl, 2005; Cho and Seo, 2005; Anjum et al., 2022). Overall, its impact on living organisms depends on the concentration and context in which it is present.

HOW IS HYDROGEN PEROXIDE SYNTHESIZED IN PLANTS?

H_2O_2 is an ROS that serves various physiological functions in plant cells, including signaling, defense mechanisms, and regulation of cellular processes. The sites of H_2O_2 production are depicted in Figure 15.1b. The biosynthesis of hydrogen peroxide in plant cells involves complex enzymatic and non-enzymatic processes.

Enzymatic Biosynthesis: it involves synthesis by NADPH Oxidase (Respiratory Burst Oxidase Homologs, RBOHs) and polyamine oxidases enzymes;

a. **NADPH Oxidase (Respiratory Burst Oxidase Homologs, RBOHs):** One of the major enzymatic sources of hydrogen peroxide in plants is the NADPH oxidase family, specifically the Respiratory Burst Oxidase Homologs (RBOHs) (Torres et al., 2002). RBOHs are

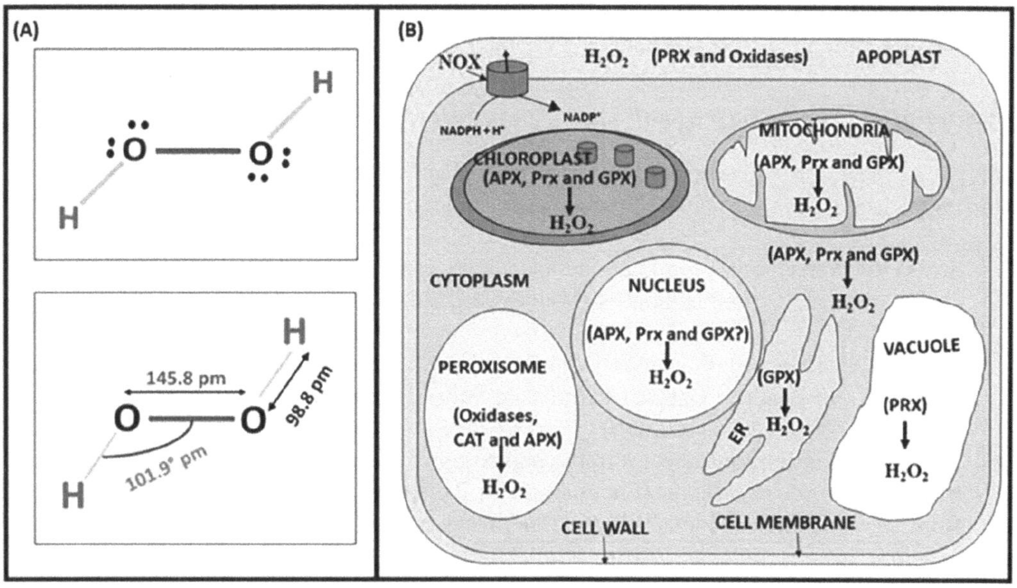

FIGURE 15.1 (a) Chemical structure of hydrogen peroxide. (b) Sites of hydrogen peroxide (H_2O_2) generation. The diagram shows the main sites of H_2O_2 production typical plant cell include aploplast, chloroplast, mitochondria, peroxisome, endoplasmic reticulum, vacuole and cytoplasm. H_2O_2 transport from chloroplast to nucleus is shown via the endoplasmic reticulum (ER), but could be more direct. APX, ascorbate peroxidase; GPX, glutathione peroxidase; NOX, NADPH oxidase; PRX, type III peroxidase; Prx, peroxiredoxin.

integral membrane proteins localized in the plasma membrane and are responsible for catalyzing the production of superoxide radicals ($O_2^{\bullet -}$) from molecular oxygen (O_2) using NADPH as an electron donor (Marino et al., 2012). Subsequently, dismutation of superoxide radicals, facilitated by superoxide dismutase (SOD), leads to the formation of hydrogen peroxide (Wang et al., 2018).

b. **Polyamine Oxidases (PAOs):** Polyamine oxidases are enzymes involved in the oxidation of polyamines, leading to the generation of hydrogen peroxide as a byproduct (Moschou et al., 2008). This enzymatic process occurs in various cellular compartments, contributing to the overall cellular pool of hydrogen peroxide (Wang et al., 2019).

Non-Enzymatic Biosynthesis: it involves synthesis by either photochemical or mitochondrial respiratory chain;

c. **Photochemical Processes:** Hydrogen peroxide can also be generated through non-enzymatic processes involving photochemical reactions (Mittler, 2017). The interaction of light with cellular components, such as flavonoids and other photosensitizers, may lead to the production of hydrogen peroxide as a result of the transfer of electrons (Smirnoff and Arnaud, 2019).
d. **Mitochondrial Respiratory Chain:** Mitochondria, in addition to being energy-producing organelles, are potential sources of ROS, including hydrogen peroxide (Møller et al., 2007). The electron transport chain in mitochondria can leak electrons, leading to the formation of superoxide radicals that are subsequently converted to hydrogen peroxide (Tiwari et al., 2002; Juszczuk et al., 2012).

In summary, the biosynthesis of hydrogen peroxide in plant cells involves a combination of enzymatic and non-enzymatic processes, with key enzymes such as NADPH oxidases and polyamine oxidases playing pivotal roles. These processes are tightly regulated and contribute to the dynamic redox balance within plant cells, influencing various physiological and developmental processes.

WHICH PROCESSES ARE AFFECTED BY HYDROGEN PEROXIDE IN PLANTS?

H_2O_2 plays crucial roles in various cellular and physiological processes (Figure 15.2). It serves as a signaling molecule in various cellular processes, including:

i. **Stress Response:** Acts as a signaling molecule in response to environmental stresses, such as drought, salinity, and pathogens (Malik et al., 2014; Caverzan et al., 2016; Khan et al., 2022, 2023).
ii. **Defense Mechanism:** Functions in plant defense against pathogens by initiating programmed cell death (apoptosis) in infected cells (Hammond-Kosack and Jones, 1996; Shirasu and Schulze-Lefert, 2000).
iii. **Seed Germination:** Participates in the regulation of seed germination by influencing the breakdown of stored reserves (Liu et al., 2010; Wojtyla et al., 2016).
iv. **Root Development:** Modulates root development and architecture, especially in response to abiotic stresses (Su et al., 2016).
v. **Cell Expansion:** Regulates cell expansion by influencing cell wall loosening through the activation of certain enzymes (Rhee, 2006; Xiong et al., 2015).
vi. **Regulation of Hormone Signaling:** Interacts with plant hormones, such as abscisic acid (ABA), gibberellins, and auxins, influencing their signaling pathways (Barba-Espin et al., 2010; Saxena et al., 2016; Černý et al., 2018).
vii. **Photosynthesis:** Acts as a signal in response to changes in light intensity, affecting photosynthetic processes (Exposito-Rodriguez et al., 2017; Liu et al., 2022).

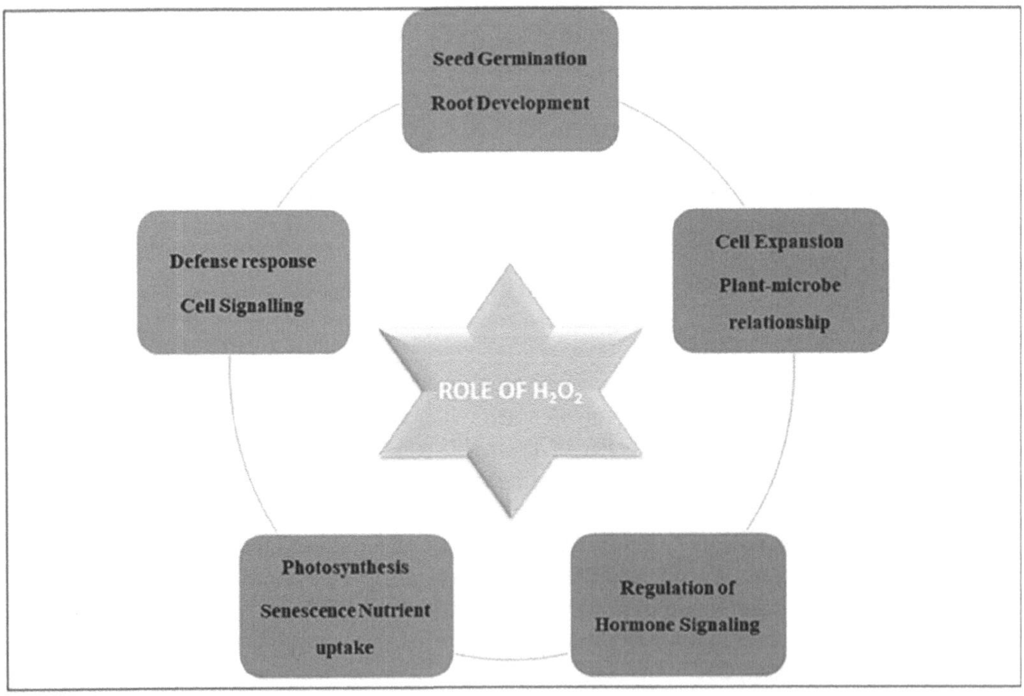

FIGURE 15.2 Generalized role of hydrogen peroxide (H_2O_2) in plants.

viii. **Regulation of Senescence**: Participates in the regulation of senescence, influencing the timing of leaf aging and nutrient remobilization (Chen et al., 1998; Zimmermann and Zentgraf, 2005; Balazadeh et al., 2011).
ix. **Redox Homeostasis:** Contributes to redox homeostasis by participating in both oxidative and antioxidative processes (Dietz, 2003; Foyer and Noctor, 2005; Sies, 2014).

Understanding these roles highlights the versatile and complex involvement of hydrogen peroxide in plant growth, development, and adaptation to changing environmental conditions.

HOW DOES HYDROGEN PEROXIDE AFFECT PLANT-MICROBE RELATIONSHIPS?

H_2O_2 plays a significant role in plant-microbe relationships, acting as a signaling molecule and participating in various physiological processes (Cheeseman, 2007; Mishra et al., 2015; Singh et al., 2021). Both plants and microbes can produce hydrogen peroxide, and its concentration in the rhizosphere (the soil region in close proximity to plant roots) can influence interactions between plants and microorganisms (Singh et al., 2013; Motamedi et al., 2022). Here are some aspects of H_2O_2 in plant-microbe relationships:

a. **Defense Mechanism:** When plants are subjected to biotic stress, such as pathogen attack or herbivore feeding, they often produce hydrogen peroxide as a part of their defense mechanism (Dikilitas et al., 2020; Kaur et al., 2022). Hydrogen peroxide acts as a signaling molecule, triggering defense responses like the expression of defense-related genes and the synthesis of secondary metabolites (Jwa et al., 2006; Akbar et al., 2023). Some microbes, particularly beneficial ones like mycorrhizal fungi, can also produce hydrogen peroxide as a defense mechanism against pathogens (Li et al., 2013; Puppo et al., 2013). This helps in protecting both the microbe and the plant.
b. **Cell Signaling:** Hydrogen peroxide can act as a signaling molecule in plant-microbe interactions (Ryals et al., 1995). It is involved in the activation of defense-related genes and the induction of systemic acquired resistance (SAR), a type of immune response that enhances the plant's resistance to subsequent infections (Wang et al., 2014; Tian et al., 2016).
c. **Root Exudates:** Plants release a variety of compounds, including hydrogen peroxide, into the rhizosphere through root exudates (Bais et al., 2006). These compounds influence the microbial community around the roots and can attract or repel specific microbes (Schrey et al., 2014; Chai and Schachtman, 2022). This can shape the composition of the rhizosphere microbiome and affect plant health (O'Neal and Alexandre, 2020).
d. **Symbiotic Relationships:** Some symbiotic relationships, such as the interaction between legumes and nitrogen-fixing rhizobia bacteria, involve the production of hydrogen peroxide (Hérouart et al., 2002; Puppo et al., 2013). In these cases, hydrogen peroxide can act as a signaling molecule to establish and maintain symbiotic relationships (Mhadhbi et al., 2009).
e. **Role in Nutrient Uptake:** Hydrogen peroxide can influence the availability and uptake of nutrients by plants (Ślesak et al., 2007; Nazir et al., 2019b). For example, it may play a role in the mobilization of iron in the rhizosphere, making it more accessible to plants (Khan et al., 2016; Silva et al., 2023).

It's important to note that the concentration and balance of hydrogen peroxide are crucial. Too much hydrogen peroxide can be harmful to plants, while moderate levels are often part of normal physiological processes. The interplay between plants and microbes involving hydrogen peroxide is complex and varies depending on the specific microorganisms and environmental conditions.

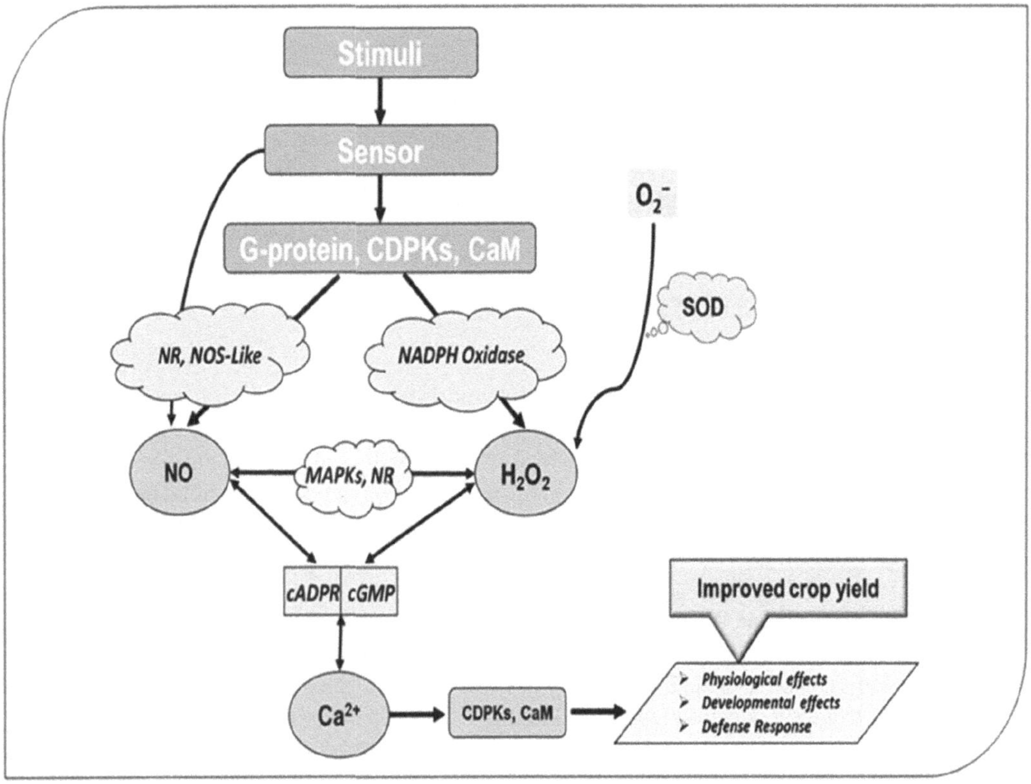

FIGURE 15.3 A diagrammatic representation illustrating the interplay among H_2O_2, NO, and Ca^{2+} in diverse plant physiological and defense processes. These molecules, H_2O_2, NO, and Ca^{2+}, can receive diverse stimuli through signaling sensors. Their interaction may involve cross-regulation, and they transmit signals to downstream molecules by activating phosphokinases such as MAPKs or influencing enzyme activity. This process serves to regulate plant development, growth, and responses to abiotic stress.

A diagrammatic representation illustrating the interplay among H_2O_2, NO, and Ca^{2+} in diverse plant physiological and defense processes is shown in Figure 15.3.

ROLE OF HYDROGEN PEROXIDE IN CROP YIELD

Various crop plants are crucial for global food security and economic well-being. Main crop plants include wheat, rice, maize, soyabean, potato, tomato, cotton, barley, sugarbeet and sugarcane. These crop plants play a vital role in sustaining human populations and supporting various industries worldwide. Advances in agricultural science and biotechnology continue to contribute to the improvement of crop varieties, enhancing their resilience, productivity, and nutritional value. Hydrogen peroxide plays key role in enhancing crop yield through various mechanisms (Orabi et al., 2018). It is often used in agriculture as a disinfectant, promoting plant health by controlling pathogens in soil and on plants (Shalaby et al., 2021). Additionally, hydrogen peroxide can stimulate plant growth by enhancing root development and nutrient absorption (Orabi et al., 2015; Habib et al., 2021). However, itis crucial to use it carefully, as excessive concentrations may harm plants (Smirnoff and Arnaud, 2019). So its use in agricultural requires precise application guidelines based on crop type and soil conditions. The effects of hydrogen peroxide on crop plants are concentration-dependent, with concentration and duration being critical factors. While moderate levels of hydrogen peroxide can play a positive role in signaling, stress responses, and growth processes, excessive levels can lead to oxidative damage and negatively impact plant health. Researchers aim to understand and optimize the beneficial effects of H_2O_2 for crop improvement while minimizing potential harm. The role of H_2O_2 in yield of crop plants is summarized in Table 15.1.

The role of H_2O_2 in wheat production is multifaceted, encompassing both beneficial and regulatory functions. H_2O_2, as an ROS, participates in signaling pathways, stress responses, and defense mechanisms, influencing various physiological processes in wheat plants. It acts as a signaling

TABLE 15.1
Showing the Role of Hydrogen Peroxide in Yield

Plant	Concentration of H_2O_2 Used	Mode of H_2O_2 Treatment	Response	References
Brassica napus	1.0 mM H_2O_2	Pot experiment Foliar spray	Improved the antioxidative defense system in plants. As a result, the application of low concentrations of H_2O_2 was associated with the highest growth and yield	Orabi et al. (2018)
Solanum lycopersicum	0.5 and 1.0 mM	Pot experiment Foliar spray	Enhanced the growth, physiological parameters, and fruit quality of the tomato plants under low-temperature conditions. The H_2O_2 treatment, especially at 1.0 mM, appeared more effective treatment in promoting these beneficial effects	Orabi et al. (2015)
Brassica napus	80 and 15 mM as	Seed priming Foliar spray	Improved performance of canola under drought stress, following exogenous application of H_2O_2, is associated with higher accumulation of osmolytes, regulation of ABA levels, and enhanced enzymatic antioxidant activities	Iqbal et al. (2018)
Pisum sativum	0%, 2% and 4%	Field experiment Seed soaking	Enhanced the growth, yield, and seed quality	Farouk and Qados (2018)
Glycine max	1 mM	Foliar spray	Promoted water retention through the biosynthesis of oligosaccharides, enhanced photosynthetic activity which ultimately contributes to improved plant resilience	Ishibashi et al. (2011)
Oryza sativa	1, 5, and 15 mmol/L	Seed priming or foliar application	Positive impact on agronomic traits like number of unfilled grains and unfilled grain weight were significantly reduced, indicating a more favorable impact on yield. The optimal concentration for achieving these positive effects is identified as 5 mmol/L	Jira-anunkul and Pattanagul (2021)
Oryza sativa	5 mmol/L	Hydroponic and soil systems	Increased germination indices, seedling growth, photosynthetic pigments, and water content in plants. These effects can be attributed to the role as a signaling molecule that activates stress tolerance mechanisms and promotes various physiological responses, ultimately contributing to improved plant performance under different conditions, including stress	Pongprayoon et al. (2013)

(Continued)

TABLE 15.1 (*Continued*)
Showing the Role of Hydrogen Peroxide in Yield

Plant	Concentration of H_2O_2 Used	Mode of H_2O_2 Treatment	Response	References
Oryza sativa	50 µM H_2O_2	Foliar application	Stimulation of the activity of key antioxidant enzymes, including superoxide dismutase (SOD), ascorbate peroxidase (APX), and glutathione reductase (GR). It also increased the levels of reduced glutathione (GSH) and facilitated proline metabolism. The expression of SOD, APX, PSBA, and PSBB was induced	Asgher et al. (2021)
Triticum aestivum	0.1 and 1 mM	Seed priming	Substantial enhancement in the growth and grain yield which was linked to heightened synthesis of photosynthetic pigments and the enhancement of leaf water relations through cellular osmotic adjustment facilitated by the accumulation of osmolytes	Habib et al. (2020)
Triticum aestivum	5 mM	Hydroponic culture Seed priming	Accumulation of osmolytes, including soluble sugars and free amino acids. These accumulated osmolytes, in turn, bolstered the photosystem and antioxidant system, thereby alleviating oxidative stress	Si et al. (2018)
Saccharum spp.	3 and 80 mmol/L	Root dipping	Increased water transport and leaf gas exchange	Silva et al. (2015)
Phaseolus vulgaris	2% for 4 hours	Seed priming	Enhanced growth parameters, photosynthetic pigments, total carbohydrates, and phytohormones	Abass and Mohamed (2011)
Phaseolus vulgaris	0.1 and 0.15 mM	Petri-dish and pot experiment Seed priming	Increased germination percentage, germination index, seed vigor index, shoot and root length, leaf water status, and photosynthetic pigments	Tania et al. (2022)
Solanum lycoperscium	0.1 or 0.5 mM for 4 hour	Pot experiment Root dipping	Improved growth, photosynthesis, and metabolic state in the plants	Nazir et al. (2019a)
Zea mays	40 mg/L for 24 hours 20 mg/L foliage applied	Pot and field experiment Seed priming and foliar spray	Enhanced morphological, yield related attributes and grain yield of spring maize	Ahmad et al. (2014)
Cucumis sativus	1.5 mM	Foliar spray	Significantly enhanced biomass, leaf relative water content (RWC), chlorophyll content, and photosynthetic rate	Sun et al. (2016)

molecule in response to abiotic stresses, such as drought, salinity, and oxidative stress. H_2O_2 plays a crucial role in the activation of stress-responsive genes and pathways, enabling wheat plants to adapt to adverse environmental conditions (Baxter et al., 2014). H_2O_2 is involved in the induction of defense responses against pathogens. It participates in the hypersensitive response (HR) and systemic acquired resistance (SAR), contributing to the plant's ability to resist fungal and bacterial infections (Apel and Hirt 2004). H_2O_2 functions as a secondary messenger in cell growth and development processes. It is implicated in the regulation of cell expansion, differentiation, and programmed cell death, influencing overall wheat plant development (Mittler, 2017). H_2O_2 plays a role in seed germination and early seedling growth. It is involved in the regulation of redox status during these critical stages, influencing the efficiency of germination and establishment of young wheat plants (Bailly, 2004). H_2O_2 is implicated in the regulation of photosynthesis, influencing both light-dependent and light-independent reactions. It participates in redox signaling pathways that modulate the efficiency of photosynthetic processes in wheat (Foyer and Noctor, 2005). In summary, H_2O_2 plays a pivotal role in wheat production by serving as a signaling molecule in stress responses, defense mechanisms, and developmental processes. Understanding the regulatory functions of hydrogen peroxide in wheat plants is essential for devising strategies to enhance stress tolerance, improve yield, and ensure sustainable wheat production in diverse environmental conditions. Hydrogen peroxide can be used in wheat and rice cultivation to enhance yield by promoting root development, improving nutrient absorption, and acting as an antifungal agent (Shetty et al., 2007; Song et al., 2013; Jira-anunkul and Pattanagul, 2021). It helps create an aerobic soil environment, fostering better conditions for root growth and nutrient uptake, ultimately contributing to increased rice productivity (Agarwal et al., 2005; Pongprayoon et al., 2013; Sohag et al., 2020).

H_2O_2 is reported to play a dual role in maize yield (Prasad et al., 1994; Sang et al., 2008). In appropriate concentrations, it acts as a signaling molecule, triggering various stress responses that can enhance plant defense mechanisms and ultimately contribute to improved yield (Ahmad et al., 2014; Ashraf et al., 2015). However, excessive levels of hydrogen peroxide can be detrimental, causing oxidative stress and damage to plant cells, negatively impacting maize growth and yield (Gondim et al., 2012; Terzi et al., 2014). Similarly, H_2O_2 can play a dual role in barley plants. In low concentrations, it serves as a signaling molecule involved in various physiological processes, including stress responses, growth, and development (Çavusoglu and Kabar, 2010; Baldus et al., 2021). However, high concentrations of hydrogen peroxide can be harmful, causing oxidative stress and damage to plant cells (Hafez and Király, 2003; Kottapalli et al., 2005). H_2O_2 plays a role in soybean yield through its involvement in various physiological processes (Puntarulo et al., 1988; Ishibashi, et al., 2011; Guler and Pehlivan, 2016). It can act as a signaling molecule, influencing plant growth and development (Dalton et al., 1986; Cheeseman, 2006). Additionally, hydrogen peroxide may help plants to respond stress by participating in defense mechanisms against pathogens (Basal and Szabó, 2020). H_2O_2 plays key role in promoting potato and tomato yield by acting as a growth stimulant and disease suppressant (Moussa et al., 2012; Deng et al., 2012; Orabi et al., 2015). When applied in appropriate concentrations, it may enhance root development, improve nutrient uptake, and protect against certain pathogens (Borden and Higgins, 2002; Nazir et al., 2019b; Abd Elhady et al., 2021). Recent study of Shalaby et al. (2023) on broccoli plants reported that foliar spraying of hydrogen peroxide and hydrogen sulfide has promising impact on physiological processes to mitigate abiotic stress and enhance plant growth and yield.

H_2O_2 plays a pivotal role in modulating various physiological processes within leguminous plants, profoundly influencing their yield (Glyanko et al., 2005; Puppo et al., 2013). As a signaling molecule, hydrogen peroxide participates in redox homeostasis, triggering defense mechanisms against environmental stressors (Hasan et al., 2016; Černý et al., 2018; León-López et al., 2020). It acts as a secondary messenger in pathways regulating root nodulation, a symbiotic interaction between legumes and nitrogen-fixing rhizobia, crucial for enhanced nutrient uptake and nitrogen assimilation (Rubio et al., 2004; Andrio et al., 2013; Orabi et al., 2015; Redanz et al., 2018; Al-Hayany, 2019). Additionally, hydrogen peroxide contributes to the regulation of antioxidant defense systems, mitigating oxidative stress induced by biotic and abiotic factors (Liu et al., 2020; Mariyam et al., 2023). This fine-tuned balance is integral for optimal photosynthesis, promoting efficient carbon assimilation and ultimately bolstering overall plant growth and reproductive success (Rahman et al., 2021; Zhang et al., 2021; Feng et al., 2023). Understanding the intricate interplay of hydrogen peroxide in legume biology provides insights for sustainable agriculture practices, offering avenues to enhance crop yield through targeted manipulation of redox signaling pathways.

H_2O_2 plays a pivotal role in modulating cereals' yield by influencing various physiological and biochemical processes (Habib et al., 2020). Acting as a signaling molecule, hydrogen peroxide participates in the regulation of plant growth and development (Jira-anunkul and Pattanagul, 2021). Its presence is intricately linked to stress responses, wherein it serves as a secondary messenger in signaling cascades triggered by environmental stimuli (Ragaey et al., 2022). In cereals, hydrogen peroxide contributes to defense mechanisms against biotic and abiotic stressors (Poudel et al., 2019; Habib et al., 2021). It functions as a key component in the plant's antioxidant defense system, mitigating oxidative damage induced by ROS (Mostafa, 2021; Arican and Demirbaş, 2022). ROS, including H_2O_2, are generated during stress conditions and act as crucial mediators in cellular responses (Bhardwaj et al., 2021; Sehar et al., 2021). Moreover, H_2O_2 is implicated in cell wall modification, influencing cell elongation and strengthening (Dunand et al., 2007; Tenhaken, 2015; Xiong et al., 2015). This, in turn, impacts overall plant architecture and grain development (Gadjev et al., 2008; Yu et al., 2017). Additionally, H_2O_2 is involved in the regulation of enzymatic activities related to photosynthesis and nutrient assimilation, thereby influencing the overall productivity of cereals (Quan et al., 2008; Smirnoff and Arnaud, 2019). Understanding the nuanced interplay of H_2O_2 in crop provides insights into potential strategies for optimizing crop yield under varying environmental conditions, offering avenues for targeted agricultural interventions and improved food security.

CONCLUSION

In conclusion, this chapter highlights H_2O_2 as a multifaceted and sustainable tool for advancing agriculture in response to the complex challenges facing global food production. Its pivotal role in regulating plant growth, development, and stress responses underscores its potential to enhance crop yield and contribute to the principles of sustainable agriculture. As the agricultural sector increasingly seeks eco-friendly alternatives, H_2O_2 emerges as a promising candidate due to its environmental friendliness and rapid decomposition into benign by-products. The synthesis of H_2O_2 within plant cells, involving intricate enzymatic and non-enzymatic processes, is tightly regulated to maintain redox balance. Moreover, the exploration of H_2O_2 in plant-microbe interactions opens avenues for understanding and harnessing its impact on soil health, nutrient acquisition, and sustainable agricultural practices. The chapter concludes by examining the context-dependent effects of H_2O_2 on major crops, emphasizing the need for precise application guidelines to maximize its benefits while minimizing potential harm. Overall, the future of agricultural practices may be shaped by the strategic incorporation of hydrogen peroxide, offering a sustainable and innovative approach toward ensuring global food security and environmental stewardship.

ACKNOWLEDGMENTS

M. Saleem gratefully acknowledges the financial support in the form of Senior Research Fellowship rendered by CSIR UGC, New Delhi Ref. NO. 677/ (CSIR-UGC NET JUNE 2018) at Aligarh Muslim University, Aligarh, India.

REFERENCES

Abass, S. M., & Mohamed, H. I. (2011). Alleviation of adverse effects of drought stress on common bean (Phaseolus vulgaris L.) by exogenous application of hydrogen peroxide. *Bangladesh Journal of Botany*, *40*(1), 75–83.

Abd Elhady, S. A., El-Gawad, H. G. A., Ibrahim, M. F., Mukherjee, S., Elkelish, A., Azab, E., ... & El-Azm, N. A. (2021). Hydrogen peroxide supplementation in irrigation water alleviates drought stress and boosts growth and productivity of potato plants. *Sustainability*, *13*(2), 899.

Agarwal, S., Sairam, R. K., Srivastava, G. C., Tyagi, A., & Meena, R. C. (2005). Role of ABA, salicylic acid, calcium and hydrogen peroxide on antioxidant enzymes induction in wheat seedlings. *Plant Science*, *169*(3), 559–570.

Ahmad, I., Basra, S. M. A., & Wahid, A. (2014). Exogenous application of ascorbic acid, salicylic acid and hydrogen peroxide improves the productivity of hybrid maize at low temperature stress. *International Journal of Agriculture and Biology*, *16*(4), 825–830.

Akbar, M. U., Aqeel, M., Shah, M. S., Jeelani, G., Iqbal, N., Latif, A., … Noman, A. (2023). Molecular regulation of antioxidants and secondary metabolites act in conjunction to defend plants against pathogenic infection. *South African Journal of Botany*, *161*, 247–257.

Al-Hayany, E. H. (2019). Effect of glutathione and hydrogen peroxide and their Interaction in the yield and its components of vigna Radiata L. Plant. *Plant Archives*, *19*(1), 1029–1035.

Andrio, E., Marino, D., Marmeys, A., de Segonzac, M. D., Damiani, I., Genre, A., … Pauly, N. (2013). Hydrogen peroxide-regulated genes in the Medicago truncatula–Sinorhizobium meliloti symbiosis. *New Phytologist*, *198*(1), 179–189.

Anjum, N. A., Gill, S. S., Corpas, F. J., Ortega-Villasante, C., Hernandez, L. E., Tuteja, N., … Fujita, M. (2022). Recent insights into the double role of hydrogen peroxide in plants. *Frontiers in Plant Science*, *13*, 843274.

Apel, K., & Hirt, H. (2004). Reactive oxygen species: Metabolism, oxidative stress, and signal transduction. *Annual Review of Plant Biology*, *55*, 373–99.

Arican, E., & Demirbaş, S. (2022). Effects of sequential hydrogen peroxide applications on salt stress tolerance in bread wheat varieties. *Journal of Agricultural Sciences*, *28*, 43.

Asgher, M., Ahmed, S., Sehar, Z., Gautam, H., Gandhi, S. G., & Khan, N. A. (2021). Hydrogen peroxide modulates activity and expression of antioxidant enzymes and protects photosynthetic activity from arsenic damage in rice (Oryza sativa L.). *Journal of Hazardous Materials*, *401*, 123365.

Ashraf, M. A., Rasheed, R., Hussain, I., Iqbal, M., Haider, M. Z., Parveen, S., & Sajid, M. A. (2015). Hydrogen peroxide modulates antioxidant system and nutrient relation in maize (Zea mays L.) under water-deficit conditions. *Archives of Agronomy and Soil Science*, *61*(4), 507–523.

Bailly, C. (2004). Active oxygen species and antioxidants in seed biology. *Seed Science Research*, 14, 93-107.

Bais, H. P., Weir, T. L., Perry, L. G., Gilroy, S., & Vivanco, J. M. (2006). The role of root exudates in rhizosphere interactions with plants and other organisms. *Annual Review of Plant Biology*, *57*, 233–266.

Balazadeh, S., Kwasniewski, M., Caldana, C., Mehrnia, M., Zanor, M. I., Xue, G. P., & Mueller-Roeber, B. (2011). ORS1, an H_2O_2-responsive NAC transcription factor, controls senescence in Arabidopsis thaliana. *Molecular Plant*, *4*(2), 346–360.

Baldus, M., Heukäufer, F., Großpietsch, C., & Methner, F. J. (2021). Accumulation of hydrogen peroxide in barley seeds–a key factor for malt quality? *Journal of the American Society of Brewing Chemists*, *79*(4), 384–393.

Barba-Espin, G., Diaz-Vivancos, P., Clemente-Moreno, M. J., Albacete, A., Faize, L., Faize, M., ... & Hernández, J. A. (2010). Interaction between hydrogen peroxide and plant hormones during germination and the early growth of pea seedlings. *Plant, Cell & Environment*, *33*(6), 981–994.

Basal, O., & Szabó, A. (2020). Ameliorating drought stress effects on soybean physiology and yield by hydrogen peroxide. *Agriculturae Conspectus Scientificus*, *85*(3), 211–218.

Baxter, A., Mittler, R., & Suzuki, N. (2014) ROS as key players in plant stress signalling. *Journal of Experimental Botany*, 65(5), 1229–1240.

Bhardwaj, R. D., Singh, N., Sharma, A., Joshi, R., & Srivastava, P. (2021). Hydrogen peroxide regulates antioxidant responses and redox related proteins in drought stressed wheat seedlings. *Physiology and Molecular Biology of Plants*, *27*, 151–163.

Bailly, C. (2004). Active oxygen species and antioxidants in seed biology. *Seed science research*, *14*(2), 93–107.

Borden, S., & Higgins, V. J. (2002). Hydrogen peroxide plays a critical role in the defence response of tomato to Cladosporium fulvum. *Physiological and Molecular Plant Pathology*, *61*(4), 227–236.

Campos-Martin, J. M., Blanco-Brieva, G., & Fierro, J. L. (2006). Hydrogen peroxide synthesis: an outlook beyond the anthraquinone process. *Angewandte Chemie International Edition*, *45*(42), 6962–6984.

Caverzan, A., Casassola, A., & Brammer, S. P. (2016). Reactive oxygen species and antioxidant enzymes involved in plant tolerance to stress. *Abiotic and Biotic Stress in Plants-Recent Advances and Future Perspectives*, *17*, 463–480.

Çavusoglu, K., & Kabar, K. (2010). Effects of hydrogen peroxide on the germination and early seedling growth of barley under NaCl and high temperature stresses. *EurAsian Journal of BioSciences*, *4*, 70.

Černý, M., Habánová, H., Berka, M., Luklová, M., & Brzobohatý, B. (2018). Hydrogen peroxide: Its role in plant biology and crosstalk with signalling networks. *International Journal of Molecular Sciences*, *19*(9), 2812.

Chai, Y. N., & Schachtman, D. P. (2022). Root exudates impact plant performance under abiotic stress. *Trends in Plant Science*, *27*(1), 80–91.

Chattopadhyay, S. N., Pan, N. C., Roy, A. N., Samanta, K. K., & Khan, A. (2021). Hybrid bleaching of jute yarn using hydrogen peroxide and peracetic acid. *Indian Journal of Fibre & Textile Research*, *46*, 78–82.

Cheeseman, J. M. (2006). Hydrogen peroxide concentrations in leaves under natural conditions. *Journal of Experimental Botany*, *57*(10), 2435–2444.

Cheeseman, J. M. (2007). Hydrogen peroxide and plant stress: A challenging relationship. *Plant Stress*, *1*(1), 4–15.

Chen, Q. M., Bartholomew, J. C., Campisi, J., Acosta, M., Reagan, J. D., & AMES, B. N. (1998). Molecular analysis of H_2O_2-induced senescent-like growth arrest in normal human fibroblasts: p53 and Rb control G1 arrest but not cell replication. *Biochemical Journal*, *332*(1), 43–50.

Cho, U. H., & Seo, N. H. (2005). Oxidative stress in Arabidopsis thaliana exposed to cadmium is due to hydrogen peroxide accumulation. *Plant Science*, *168*(1), 113–120.

Cui, J. X., Zhou, Y. H., Ding, J. G., Xia, X. J., Shi, K. A. I., Chen, S. C., … Yu, J. Q. (2011). Role of nitric oxide in hydrogen peroxide-dependent induction of abiotic stress tolerance by brassinosteroids in cucumber. *Plant, Cell & Environment*, *34*(2), 347–358.

Dalton, D. A., Russell, S. A., Hanus, F. J., Pascoe, G. A., & Evans, H. J. (1986). Enzymatic reactions of ascorbate and glutathione that prevent peroxide damage in soybean root nodules. *Proceedings of the National Academy of Sciences*, *83*(11), 3811–3815.

Deb, S., Malukani, K. K., & Patel, H. K. (2021). Understanding and manipulation of plant–microbe interaction signals for yield enhancement. *Omics Technologies for Sustainable Agriculture and Global Food Security (Vol II)*, 267–291. https://doi.org/10.1007/978-981-16-2956-3_10.

Deng, X. P., Cheng, Y. J., Wu, X. B., Kwak, S. S., Chen, W., & Eneji, A. E. (2012). Exogenous hydrogen peroxide positively influences root growth and exogenous hydrogen peroxide positively influences root growth and metabolism in leaves of sweet potato seedlings. *Australian Journal of Crop Science*, *6*(11), 1572.

Dietz, K. J. (2003). Redox control, redox signaling, and redox homeostasis in plant cells. *International Review of Cytology*, *228*, 141–193.

Dikilitas, M., Simsek, E., & Roychoudhury, A. (2020). Modulation of abiotic stress tolerance through hydrogen peroxide. *Protective Chemical Agents in the Amelioration of Plant Abiotic Stress: Biochemical and Molecular Perspectives*, 147–173. https://doi.org/10.1002/9781119552154.ch7.

Dunand, C., Crèvecoeur, M., & Penel, C. (2007). Distribution of superoxide and hydrogen peroxide in Arabidopsis root and their influence on root development: possible interaction with peroxidases. *New Phytologist*, *174*(2), 332–341.

Ellingson, B. A., Theis, D. P., Tishchenko, O., Zheng, J., & Truhlar, D. G. (2007). Reactions of hydrogen atom with hydrogen peroxide. *The Journal of Physical Chemistry A*, *111*(51), 13554–13566.

Exposito-Rodriguez, M., Laissue, P. P., Yvon-Durocher, G., Smirnoff, N., & Mullineaux, P. M. (2017). Photosynthesis-dependent H_2O_2 transfer from chloroplasts to nuclei provides a high-light signalling mechanism. *Nature Communications*, *8*(1), 49.

Fariduddin, Q., Khan, T. A., & Yusuf, M. (2014). Hydrogen peroxide mediated tolerance to copper stress in the presence of 28-homobrassinolide in Vigna radiata. *Acta Physiologiae Plantarum*, *36*, 2767–2778.

Farouk, S., & Qados, A. M. A. (2018). Enhancing seed quality and productivity as well as physio-anatomical responses of pea plants by folic acid and/or hydrogen peroxide application. *Scientia Horticulturae*, *240*, 29–37.

Feng, L., Wei, L., Liu, Y., Hu, D., Gong, W., & Liao, W. (2023). Hydrogen peroxide is involved in methyl jasmonate-induced adventitious rooting in cucumber under cadmium stress. *Scientia Horticulturae*, *309*, 111666.

Foyer, C.H., & Noctor, G. (2005) Redox homeostasis and antioxidant signaling: A metabolic interface between stress perception and physiological responses. *Plant Cell*, *17*(7), 1866–1875.

Fu, Z. Q., & Dong, X. (2013). Systemic acquired resistance: Turning local infection into global defense. *Annual Review of Plant Biology*, *64*, 839–863.

Gadjev, I., Stone, J. M., & Gechev, T. S. (2008). Programmed cell death in plants: New insights into redox regulation and the role of hydrogen peroxide. *International Review of Cell and Molecular Biology*, *270*, 87–144.

Garcia-Munoz, P., Valenzuela, L., Wegstein, D., Schanz, T., Lopez, G. E., Ruppert, A. M., ... & Keller, N. (2023). Photocatalytic synthesis of hydrogen peroxide from molecular oxygen and water. *Topics in Current Chemistry*, *381*(4), 1–73.

Gill, S. S., & Tuteja, N. (2010). Reactive oxygen species and antioxidant machinery in abiotic stress tolerance in crop plants. *Plant Physiology and Biochemistry*, *48*(12), 909–930. https://doi.org/10.1016/j.plaphy.2010.08.016.

Glyanko, A. K., Makarova, L. E., Vasil'eva, G. G., & Mironova, N. V. (2005). Possible involvement of hydrogen peroxide and salicylic acid in the legume-rhizobium symbiosis. *Biology Bulletin*, *32*, 245–249.

Gondim, F. A., Gomes-Filho, E., Costa, J. H., Alencar, N. L. M., & Prisco, J. T. (2012). Catalase plays a key role in salt stress acclimation induced by hydrogen peroxide pretreatment in maize. *Plant Physiology and Biochemistry*, *56*, 62–71.

Guler, N. S., & Pehlivan, N. (2016). Exogenous low-dose hydrogen peroxide enhances drought tolerance of soybean (Glycine max L.) through inducing antioxidant system. *Acta Biologica Hungarica*, *67*, 169–183.

Gupta, K., Sengupta, A., Chakraborty, M., & Gupta, B. (2016). Hydrogen peroxide and polyamines act as double edged swords in plant abiotic stress responses. *Frontiers in Plant Science*, *7*, 1343.

Habib, N., Ali, Q., Ali, S., Haider, M. Z., Javed, M. T., Khalid, M., … Alyemeni, M. N. (2021). seed priming with sodium nitroprusside and H_2O_2 confers better yield in wheat under salinity: ,Water relations, antioxidative defense mechanism and ion homeostasis. *Journal of Plant Growth Regulation*, *40*(6), 2433–2453.

Habib, N., Ali, Q., Ali, S., Javed, M. T., Zulqurnain Haider, M., Perveen, R., ... & Bin-Jumah, M. (2020). Use of nitric oxide and hydrogen peroxide for better yield of wheat (Triticum aestivum L.) under water deficit conditions: growth, osmoregulation, and antioxidative defense mechanism. *Plants*, *9*(2), 285.

Habibi, G. (2014). Hydrogen peroxide (H_2O_2) generation, scavenging and signaling in plants. *Oxidative Damage to Plants*, pp. 557–584. Academic Press. https://doi.org/10.1016/B978-0-12-799963-0.00019-8.

Hafez, Y. M., & Király, Z. (2003). Role of hydrogen peroxide in symptom expression of barley susceptible and resistant to powdery mildew. *Acta Phytopathologica et Entomologica Hungarica*, *38*(3–4), 227–236.

Hammond-Kosack, K. E., & Jones, J. D. G. (1996). Resistance gene-dependent plant defense responses. *The Plant Cell*, *8*(10), 1773.

Hasan, S. A., Irfan, M., Masrahi, Y. S., Khalaf, M. A., & Hayat, S. (2016). Growth, photosynthesis, and antioxidant responses of Vigna unguiculata L. treated with hydrogen peroxide. *Cogent Food & Agriculture*, *2*(1), 1155331.

Hérouart, D., Baudouin, E., Frendo, P., Harrison, J., Santos, R., Jamet, A., … Puppo, A. (2002). Reactive oxygen species, nitric oxide and glutathione: A key role in the establishment of the legume–Rhizobium symbiosis? *Plant Physiology and Biochemistry*, *40*(6–8), 619–624.

Hossain, M. A., Bhattacharjee, S., Armin, S. M., Qian, P., Xin, W., Li, H. Y., … Tran, L. S. P. (2015). Hydrogen peroxide priming modulates abiotic oxidative stress tolerance: Insights from ROS detoxification and scavenging. *Frontiers in Plant Science*, *6*, 420.

Iqbal, H., Yaning, C., Waqas, M., Rehman, H., Shareef, M., & Iqbal, S. (2018). Hydrogen peroxide application improves quinoa performance by affecting physiological and biochemical mechanisms under water-deficit conditions. *Journal of Agronomy and Crop Science*, *204*(6), 541–553.

Ishibashi, Y., Yamaguchi, H., Yuasa, T., Iwaya-Inoue, M., Arima, S., & Zheng, S. H. (2011). Hydrogen peroxide spraying alleviates drought stress in soybean plants. *Journal of Plant Physiology*, *168*(13), 1562–1567.

Jira-anunkul, W., & Pattanagul, W. (2021). Effects of hydrogen peroxide application on agronomic traits of rice (Oryza sativa L.) under drought stress. *Plant, Soil and Environment*, *67*(4), 221–229.

Juszczuk, I. M., Szal, B., & Rychter, A. M. (2012). Oxidation–reduction and reactive oxygen species homeostasis in mutant plants with respiratory chain complex I dysfunction. *Plant, Cell & Environment*, *35*(2), 296–307.

Jwa, N. S., Agrawal, G. K., Tamogami, S., Yonekura, M., Han, O., Iwahashi, H., & Rakwal, R. (2006). Role of defense/stress-related marker genes, proteins and secondary metabolites in defining rice self-defense mechanisms. *Plant Physiology and Biochemistry*, *44*(5–6), 261–273.

Kamle, M., Borah, R., Bora, H., Jaiswal, A. K., Singh, R. K., & Kumar, P. (2020). Systemic acquired resistance (SAR) and induced systemic resistance (ISR): Role and mechanism of action against phytopathogens. In *Fungal Biotechnology and Bioengineering*, 457–470. https://doi.org/10.1007/978-3-030-41870-0_20.

Kaur, S., Samota, M. K., Choudhary, M., Choudhary, M., Pandey, A. K., Sharma, A., & Thakur, J. (2022). How do plants defend themselves against pathogens-Biochemical mechanisms and genetic interventions. *Physiology and Molecular Biology of Plants*, *28*(2), 485–504.

Khan, T. A., Saleem, M., & Fariduddin, Q. (2023). Differential drought stress tolerance in five varieties of tomato (Solanum lycopersicum L.): An evaluation of photosynthesis and antioxidant system. *Journal of Soil Science and Plant Nutrition*, *23*(2), 2810–2831.

Khan, T. A., Saleem, M., & Fariduddin, Q. (2022). Recent advances and mechanistic insights on melatonin-mediated salt stress signaling in plants. *Plant Physiology and Biochemistry*, *188*, 97–107.

Khan, M. I. R., Khan, N. A., Masood, A., Per, T. S., & Asgher, M. (2016). Hydrogen peroxide alleviates nickel-inhibited photosynthetic responses through increase in use-efficiency of nitrogen and sulfur, and glutathione production in mustard. *Frontiers in Plant Science*, *7*, 44.

Kleinman, P. J. A., Spiegal, S., Rigby, J. R., Goslee, S. C., Baker, J. M., Bestelmeyer, B. T., … Walthall, C. L. (2018). Advancing the sustainability of US agriculture through long-term research. *Journal of Environmental Quality*, *47*(6), 1412–1425.

Kottapalli, B., Wolf-Hall, C. E., & Schwarz, P. (2005). Evaluation of gaseous ozone and hydrogen peroxide treatments for reducing Fusarium survival in malting barley. *Journal of Food Protection*, *68*(6), 1236–1240.

León-López, L., Escobar-Zúñiga, Y., Salazar-Salas, N. Y., Mora Rochín, S., Cuevas-Rodríguez, E. O., Reyes-Moreno, C., & Milán-Carrillo, J. (2020). Improving polyphenolic compounds: Antioxidant activity in chickpea sprouts through elicitation with hydrogen peroxide. *Foods*, *9*(12), 1791.

Li, Y., Liu, Z., Hou, H., Lei, H., Zhu, X., Li, X., … Tian, C. (2013). Arbuscular mycorrhizal fungi-enhanced resistance against Phytophthora sojae infection on soybean leaves is mediated by a network involving hydrogen peroxide, jasmonic acid, and the metabolism of carbon and nitrogen. *Acta Physiologiae Plantarum*, *35*, 3465–3475.

Litorja, M., & Ruscic, B. (1998). A photoionization study of the hydroperoxyl radical, HO_2, and hydrogen peroxide, H_2O_2. *Journal of Electron Spectroscopy and Related Phenomena*, *97*(1–2), 131–146.

Liu, L., Huang, L., Lin, X., & Sun, C. (2020). Hydrogen peroxide alleviates salinity-induced damage through enhancing proline accumulation in wheat seedlings. *Plant Cell Reports*, *39*, 567–575.

Liu, L., Huang, L., Sun, C., Wang, L., Jin, C., & Lin, X. (2021). Cross-talk between hydrogen peroxide and nitric oxide during plant development and responses to stress. *Journal of Agricultural and Food Chemistry*, *69*(33), 9485–9497.

Liu, T., Pan, Z., Vequizo, J. J. M., Kato, K., Wu, B., Yamakata, A., … Domen, K. (2022). Overall photosynthesis of H_2O_2 by an inorganic semiconductor. *Nature Communications*, *13*(1), 1034.

Liu, Y., Ye, N., Liu, R., Chen, M., & Zhang, J. (2010). H_2O_2 mediates the regulation of ABA catabolism and GA biosynthesis in Arabidopsis seed dormancy and germination. *Journal of Experimental Botany*, *61*(11), 2979–2990.

Maithani, D., Singh, H., & Sharma, A. (2021). Stress alleviation in plants using SAR and ISR: Current views on stress signaling network. In *Microbes and Signaling Biomolecules Against Plant Stress: Strategies of Plant-Microbe Relationships for Better Survival*, 7–36. https://doi.org/10.1007/978-981-15-7094-0_2.

Malik, B., Pirzadah, T. B., Tahir, I., Rehman, R. U., Hakeem, K. R., & Abdin, M. Z. (2014). Plant signaling: response to reactive oxygen species. In *Plant Signaling: Understanding the Molecular Crosstalk*, 1–38. https://doi.org/10.1007/978-81-322-1542-4_1.

Marino, D., Dunand, C., Puppo, A., & Pauly, N. (2012). A burst of plant NADPH oxidases. *Trends in Plant Science*, *17*(1), 9–15.

Mariyam, S., Bhardwaj, R., Khan, N. A., Sahi, S. V., & Seth, C. S. (2023). Review on nitric oxide at the forefront of rapid systemic signaling in mitigation of salinity stress in plants: Crosstalk with calcium and hydrogen peroxide. *Plant Science*, *336*, 111835.

Melchionna, M., Fornasiero, P., & Prato, M. (2019). The rise of hydrogen peroxide as the main product by metal-free catalysis in oxygen reductions. *Advanced Materials*, *31*(13), 1802920.

Mhadhbi, H., Mhamdi, R., Jebara, M., Limam, F., & Aouani, M. E. (2009). Legume–rhizobia symbiotic interaction under salt and drought constraints: generation of reactive oxygen species and protective role of antioxidant enzymes within nodules. *Advances in Plant Physiology*, *11*, 1–21.

Mishra, S., Singh, A., Keswani, C., Saxena, A., Sarma, B. K., & Singh, H. B. (2015). Harnessing plant-microbe interactions for enhanced protection against phytopathogens. In *Plant Microbes Symbiosis: Applied Facets*, 111–125. https://doi.org/10.1007/978-81-322-2068-8_5.

Mittler, R. (2017). ROS are good. *Trends in Plant Science*, *22*(1), 11–19. https://doi.org/10.1016/j.tplants.2016.08.002.

Møller, I.M., Jensen, P.E., & Hansson, A. (2007). Oxidative modifications to cellular components in plants. *Annual Review of Plant Biology*, *58*, 459–81.

Moschou, P.N., Sanmartin, M., Andriopoulou, A.H., Rojo, E., Sanchez-Serrano, J.J., Roubelakis-Angelakis, K.A. (2008) Bridging the gap between plant and mammalian polyamine catabolism: A novel peroxisomal polyamine oxidase responsible for a full back-conversion pathway in Arabidopsis. *Plant Physiology*, *147*(4), 1845–57.

Mostafa, L. (2021). Effect of hydrogen peroxide on the growth, fruit set, yield andquality of ewais mango trees. *Egyptian Journal of Horticulture*, *48*(1), 71–81.

Motamedi, M., Zahedi, M., Karimmojeni, H., Motamedi, H., & Mastinu, A. (2022). Effect of rhizosphere bacteria on antioxidant enzymes and some biochemical characteristics of Medicago sativa L. subjected to herbicide stress. *Acta Physiologiae Plantarum*, *44*(8), 84.

Moussa, S. A. M., El-Fadl, N. I. A., & Agamy, N. F. (2012). Role of hydrogen peroxide in improving potato tuber quality. *Alexandria Science Exchange Journal*, *33*, 73–88.

Mukherjee, S., & Corpas, F. J. (2023). H_2O_2, NO, and H_2S networks during root development and signalling under physiological and challenging environments: Beneficial or toxic? *Plant, Cell & Environment*, *46*(3), 688–717.

Nazir, F., Fariduddin, Q., & Khan, T. A. (2020). Hydrogen peroxide as a signalling molecule in plants and its crosstalk with other plant growth regulators under heavy metal stress. *Chemosphere*, *252*, 126486.

Nazir, F., Hussain, A., & Fariduddin, Q. (2019a). Hydrogen peroxide modulate photosynthesis and antioxidant systems in tomato (Solanum lycopersicum L.) plants under copper stress. *Chemosphere*, *230*, 544–558.

Nazir, F., Hussain, A., & Fariduddin, Q. (2019b). Interactive role of epibrassinolide and hydrogen peroxide in regulating stomatal physiology, root morphology, photosynthetic and growth traits in Solanum lycopersicum L. under nickel stress. *Environmental and Experimental Botany*, *162*, 479–495.

Niu, L., & Liao, W. (2016). Hydrogen peroxide signaling in plant development and abiotic responses: crosstalk with nitric oxide and calcium. *Frontiers in Plant Science*, *7*, 230.

O'Neal, L., Vo, L., & Alexandre, G. (2020). Specific root exudate compounds sensed by dedicated chemoreceptors shape Azospirillum brasilense chemotaxis in the rhizosphere. *Applied and Environmental Microbiology*, *86*(15), e01026.

Orabi, S. A., & El-Noemani, A. S. A. (2015). Role of proline in improving drought tolerance of faba bean plants via antioxidant responses to enhanced generation of superoxide anion radical and hydrogen peroxide. *American-Eurasian Journal of Sustainable Agriculture*, *7*, 31–43.

Orabi, S. A., Dawood, M. G., & Salman, S. R. (2015). Comparative study between the physiological role of hydrogen peroxide and salicylic acid in alleviating the harmful effect of low temperature on tomato plants grown under sand-ponic culture. *Scientia Agricola*, *9*(1), 49–59.

Orabi, S. A., Hussein, M. M., Zaki, S. S., & Sharara, F. A. (2018). Influence of hydrogen peroxide on growth, yield and biochemical constituents of canola plants grown under different irrigation intervals. *Current Science International*, *7*(3), 407–418.

Petrov, V. D., & Van Breusegem, F. (2012). Hydrogen peroxide—a central hub for information flow in plant cells. *AoB Plants*, 2012, pls014.

Pongprayoon, W., Roytrakul, S., Pichayangkura, R., & Chadchawan, S. (2013). The role of hydrogen peroxide in chitosan-induced resistance to osmotic stress in rice (Oryza sativa L.). *Plant Growth Regulation*, *70*, 159–173.

Poudel, A., Navathe, S., Chand, R., Mishra, V. K., Singh, P. K., & Joshi, A. K. (2019). Hydrogen peroxide prompted lignification affects pathogenicity of hemi-biotrophic pathogen Bipolaris sorokiniana to wheat. *The Plant Pathology Journal*, *35*(4), 287.

Prasad, T. K., Anderson, M. D., Martin, B. A., & Stewart, C. R. (1994). Evidence for chilling-induced oxidative stress in maize seedlings and a regulatory role for hydrogen peroxide. *The Plant Cell*, *6*(1), 65–74.

Puntarulo, S., Sánchez, R. A., & Boveris, A. (1988). Hydrogen peroxide metabolism in soybean embryonic axes at the onset of germination. *Plant Physiology*, *86*(2), 626–630.

Puppo, A., Pauly, N., Boscari, A., Mandon, K., & Brouquisse, R. (2013). Hydrogen peroxide and nitric oxide: Key regulators of the legume—rhizobium and mycorrhizal symbioses. *Antioxidants & Redox Signaling*, *18*(16), 2202–2219.

Quan, L. J., Zhang, B., Shi, W. W., & Li, H. Y. (2008). Hydrogen peroxide in plants: A versatile molecule of the reactive oxygen species network. *Journal of Integrative Plant Biology*, *50*(1), 2–18.

Ragaey, M. M., Sadak, M. S., Dawood, M. F., Mousa, N. H., Hanafy, R. S., & Latef, A. A. H. A. (2022). Role of signaling molecules sodium nitroprusside and arginine in alleviating salt-Induced oxidative stress in wheat. *Plants*, *11*(14), 1786.

Rahman, M. A., Alam, I., Sharmin, S. A., Kabir, A. H., Kim, Y. G., Liu, G., & Lee, B. H. (2021). Physiological and proteomic analyses reveal the protective roles of exogenous hydrogen peroxide in alleviating drought stress in soybean plants. *Plant Biotechnology Reports*, *15*(6), 805–818.

Raza, A., Salehi, H., Rahman, M. A., Zahid, Z., Madadkar Haghjou, M., Najafi-Kakavand, S., … Zhuang, W. (2022). Plant hormones and neurotransmitter interactions mediate antioxidant defenses under induced oxidative stress in plants. *Frontiers in Plant Science*, *13*, 961872.

Redanz, S., Cheng, X., Giacaman, R. A., Pfeifer, C. S., Merritt, J., & Kreth, J. (2018). Live and let die: Hydrogen peroxide production by the commensal flora and its role in maintaining a symbiotic microbiome. *Molecular Oral Microbiology*, *33*(5), 337–352.

Rhee, S. G. (2006). H_2O_2, a necessary evil for cell signaling. *Science*, *312*(5782), 1882–1883.

Rubio, M. C., James, E. K., Clemente, M. R., Bucciarelli, B., Fedorova, M., Vance, C. P., & Becana, M. (2004). Localization of superoxide dismutases and hydrogen peroxide in legume root nodules. *Molecular Plant-Microbe Interactions*, *17*(12), 1294–1305.

Ryals, J., Lawton, K. A., Delaney, T. P., Friedrich, L., Kessmann, H., Neuenschwander, U., … Weymann, K. (1995). Signal transduction in systemic acquired resistance. *Proceedings of the National Academy of Sciences*, *92*(10), 4202–4205.

Saleem, M., & Fariduddin, Q. (2022). Novel mechanistic insights of selenium induced microscopic, histochemical and physio-biochemical changes in tomato (Solanum lycopersicum L.) plant. An account of beneficiality or toxicity. *Journal of Hazardous Materials*, *434*, 128830.

Saleem, M., Fariduddin, Q., & Castroverde, C. D. M. (2021a). Salicylic acid: A key regulator of redox signalling and plant immunity. *Plant Physiology and Biochemistry*, *168*, 381–397.

Saleem, M., Fariduddin, Q., & Janda, T. (2021b). Multifaceted role of salicylic acid in combating cold stress in plants: A review. *Journal of Plant Growth Regulation*, *40*, 464–485.

Sang, J., Jiang, M., Lin, F., Xu, S., Zhang, A., & Tan, M. (2008). Nitric oxide reduces hydrogen peroxide accumulation involved in water stress-induced subcellular anti-oxidant defense in maize plants. *Journal of Integrative Plant Biology*, *50*(2), 231–243.

Saxena, I., Srikanth, S., & Chen, Z. (2016). Cross talk between H_2O_2 and interacting signal molecules under plant stress response. *Frontiers in Plant Science*, *7*, 570.

Schieber, M., & Chandel, N. S. (2014). ROS function in redox signaling and oxidative stress. *Current Biology*, *24*(10), R453–R462.

Schrey, S. D., Hartmann, A., & Hampp, R. (2014). Rhizosphere interactions. In *Ecological Biochemistry: Environmental and Interspecies Interactions*, 292–311. https://doi.org/10.1002/9783527686063.ch15.

Sehar, Z., Jahan, B., Masood, A., Anjum, N. A., & Khan, N. A. (2021). Hydrogen peroxide potentiates defense system in presence of sulfur to protect chloroplast damage and photosynthesis of wheat under drought stress. *Physiologia Plantarum*, *172*(2), 922–934.

Shalaby, O. A. E. S., Farag, R., & Ibrahim, M. F. (2023). Effect of hydrogen sulfide and hydrogen peroxide on growth, yield and nutrient content of broccoli plants grown under saline conditions. *Scientia Horticulturae*, *316*, 112035.

Shalaby, T. A., Abd-Alkarim, E., El-Aidy, F., Hamed, E. S., Sharaf-Eldin, M., Taha, N., ... & Dos Reis, A. R. (2021). Nano-selenium, silicon and H_2O_2 boost growth and productivity of cucumber under combined salinity and heat stress. *Ecotoxicology and Environmental Safety*, *212*, 111962.

Shetty, N. P., Mehrabi, R., Lütken, H., Haldrup, A., Kema, G. H., Collinge, D. B., & Jørgensen, H. J. L. (2007). Role of hydrogen peroxide during the interaction between the hemibiotrophic fungal pathogen Septoria tritici and wheat. *New Phytologist*, *174*(3), 637–647.

Shirasu, K., & Schulze-Lefert, P. (2000). Regulators of cell death in disease resistance. In *Programmed Cell Death in Higher Plants*, 127–141. https://doi.org/10.1007/978-94-010-0934-8_10.

Si, T., Wang, X., Zhao, C., Huang, M., Cai, J., Zhou, Q., … Jiang, D. (2018). The role of hydrogen peroxide in mediating the mechanical wounding-induced freezing tolerance in wheat. *Frontiers in Plant Science*, *9*, 327.

Sies, H. (2014). Role of metabolic H_2O_2 generation: redox signaling and oxidative stress. *Journal of Biological Chemistry*, *289*(13), 8735–8741.

Sies, H. (2017). Hydrogen peroxide as a central redox signaling molecule in physiological oxidative stress: Oxidative eustress. *Redox Biology*, *11*, 613–619.

Sies, H., & Jones, D. P. (2020). Reactive oxygen species (ROS) as pleiotropic physiological signalling agents. *Nature Reviews Molecular Cell Biology*, *21*(7), 363–383.

Silva, K. I., Sales, C. R., Marchiori, P. E., Silveira, N. M., Machado, E. C., & Ribeiro, R. V. (2015). Short-term physiological changes in roots and leaves of sugarcane varieties exposed to H_2O_2 in root medium. *Journal of Plant Physiology*, *177*, 93–99.

Silva, P. F. D., Santos, B. D. B. D., Dantas Neto, J., Melo, A. S. D., Matos, R. M. D., Bonou, S. I., … Duarte, T. F. (2023). Effect of electrical conductivity levels and hydrogen peroxide priming on nutrient solution uptake by chives in a hydroponic system. *Agriculture*, *13*(7), 1346.

Singh, A. K., Kumar, S., & Sinha, T. (2021). *Antioxidants in Plant–Microbe Interaction* (pp. 3–20). Springer Singapore, Singapore.

Singh, A., Jain, A., Sarma, B. K., Upadhyay, R. S., & Singh, H. B. (2013). Rhizosphere microbes facilitate redox homeostasis in Cicer arietinum against biotic stress. *Annals of Applied Biology*, *163*(1), 33–46.

Ślesak, I., Libik, M., Karpinska, B., Karpinski, S., & Miszalski, Z. (2007). The role of hydrogen peroxide in regulation of plant metabolism and cellular signalling in response to environmental stresses. *Acta Biochimica Polonica*, *54*(1), 39–50.

Smirnoff, N., & Arnaud, D. (2019). Hydrogen peroxide metabolism and functions in plants. *New Phytologist*, *221*(3), 1197–1214.

Sohag, A. A. M., Tahjib-Ul-Arif, M., Brestic, M., Afrin, S., Sakil, M. A., Hossain, M. T., ... & Hossain, M. A. (2020). Exogenous salicylic acid and hydrogen peroxide attenuate drought stress in rice. *Plant, Soil and Environment*, *66*(1), 7–13.

Song, Z. L., Yag, G. H., Feng, Y. Z., Ren, G. X., & Han, X. H. (2013). Pretreatment of rice straw by hydrogen peroxide for enhanced methane yield. *Journal of Integrative Agriculture*, *12*(7), 1258–1266.

Su, C., Liu, L., Liu, H., Ferguson, B. J., Zou, Y., Zhao, Y., ... Li, X. (2016). H_2O_2 regulates root system architecture by modulating the polar transport and redistribution of auxin. *Journal of Plant Biology*, *59*, 260–270.

Sun, Y., Wang, H., Liu, S., & Peng, X. (2016). Exogenous application of hydrogen peroxide alleviates drought stress in cucumber seedlings. *South African Journal of Botany*, *106*, 23–28.

Talaat, N. B. (2019). Role of reactive oxygen species signaling in plant growth and development. In *Reactive Oxygen, Nitrogen and Sulfur Species in Plants: Production, Metabolism, Signaling and Defense Mechanisms*, 225–266. https://doi.org/10.1002/9781119468677.ch10.

Tania, S. S., Rhaman, M. S., Rauf, F., Rahaman, M. M., Kabir, M. H., Hoque, M. A., & Murata, Y. (2022). Alleviation of Salt-Inhibited germination and seedling growth of kidney bean by seed priming and exogenous application of salicylic acid (SA) and hydrogen peroxide (H_2O_2). *Seeds*, *1*(2), 87–98.

Tenhaken, R. (2015). Cell wall remodeling under abiotic stress. *Frontiers in Plant Science*, *5*, 771.

Terzi, R., Kadioglu, A., Kalaycioglu, E., & Saglam, A. (2014). Hydrogen peroxide pretreatment induces osmotic stress tolerance by influencing osmolyte and abscisic acid levels in maize leaves. *Journal of Plant Interactions*, *9*(1), 559–565.

Thornton, P. K., & Herrero, M. (2014). Climate change adaptation in mixed crop–livestock systems in developing countries. *Global Food Security*, *3*(2), 99–107.

Tian, S., Wang, X., Li, P., Wang, H., Ji, H., Xie, J., ... & Dong, H. (2016). Plant aquaporin AtPIP1; 4 links apoplastic H_2O_2 induction to disease immunity pathways. *Plant physiology*, *171*(3), 1635–1650.

Tiwari, B. S., Belenghi, B., & Levine, A. (2002). Oxidative stress increased respiration and generation of reactive oxygen species, resulting in ATP depletion, opening of mitochondrial permeability transition, and programmed cell death. *Plant Physiology*, *128*(4), 1271–1281.

Torres, M. A., Dangl, J. L., & Jones, J. D. (2002). Arabidopsis gp91phox homologues AtrbohD and AtrbohF are required for accumulation of reactive oxygen intermediates in the plant defense response. *Proceedings of the National Academy of Sciences of the United States of America*, *99*(1), 517–522.

Torres, M. A., & Dangl, J. L. (2005). Functions of the respiratory burst oxidase in biotic interactions, abiotic stress and development. *Current Opinion in Plant Biology*, *8*(4), 397–403. https://doi.org/10.1016/j.pbi.2005.05.014.

Wang, C., El-Shetehy, M., Shine, M. B., Yu, K., Navarre, D., Wendehenne, D., ... Kachroo, P. (2014). Free radicals mediate systemic acquired resistance. *Cell Reports*, *7*(2), 348–355.

Wang, W., Chen, D., Zhang, X., Liu, D., Cheng, Y., & Shen, F. (2018). Role of plant respiratory burst oxidase homologs in stress responses. *Free Radical Research*, *52*(8), 826–839.

Wang, Y., Ye, X., Yang, K., Shi, Z., Wang, N., Yang, L., & Chen, J. (2019). Characterization, expression, and functional analysis of polyamine oxidases and their role in selenium-induced hydrogen peroxide production in Brassica rapa. *Journal of the Science of Food and Agriculture*, *99*(8), 4082–4093.

Winterbourn, C. C. (2013). The biological chemistry of hydrogen peroxide. *Methods in Enzymology*, *528*, 3–25.

Winterbourn, C. C. (2018). Biological production, detection, and fate of hydrogen peroxide. *Antioxidants & Redox Signaling*, *29*, 541–551.

Wojtyla, Ł., Lechowska, K., Kubala, S., & Garnczarska, M. (2016). Different modes of hydrogen peroxide action during seed germination. *Frontiers in Plant Science*, *7*, 66.

Xiong, J., Yang, Y., Fu, G., & Tao, L. (2015). Novel roles of hydrogen peroxide (H_2O_2) in regulating pectin synthesis and demethylesterification in the cell wall of rice (Oryza sativa) root tips. *New Phytologist*, *206*(1), 118–126.

Yan, J., Jiang, J., He, L., & Chen, L. (2020). Mitochondrial superoxide/hydrogen peroxide: An emerging therapeutic target for metabolic diseases. *Free Radical Biology and Medicine*, *152*, 33–42.

Yu, H., You, Y., Lei, F., Liu, Z., Zhang, W., & Jiang, J. (2015). Comparative study of alkaline hydrogen peroxide and organosolv pretreatments of sugarcane bagasse to improve the overall sugar yield. *Bioresource Technology*, *187*, 161–166.

Yu, Y., Li, Q. F., Zhang, J. P., Zhang, F., Zhou, Y. F., Feng, Y. Z., … Zhang, Y. C. (2017). Laccase-13 regulates seed setting rate by affecting hydrogen peroxide dynamics and mitochondrial integrity in rice. *Frontiers in Plant Science, 8*, 1324.

Zhang, Y., Wang, Y., Wen, W., Shi, Z., Gu, Q., Ahammed, G. J., … Guo, S. (2021). Hydrogen peroxide mediates spermidine-induced autophagy to alleviate salt stress in cucumber. *Autophagy*, *17*(10), 2876–2890.

Zimmermann, P., & Zentgraf, U. (2005). The correlation between oxidative stress and leaf senescence during plant development. *Cellular and Molecular Biology Letters*, *10*(3), 515.

Zipfel, C., & Oldroyd, G. E. (2017). Plant signalling in symbiosis and immunity. *Nature*, *543*(7645), 328–336.

16 The Interactive Prospects of H_2O_2 with Climate Extremes and Concomitant Microbial Colonization

Adil Dilawar, Iqra Shahzadi, Nasim Ahmad Yasin, Faheem Adil, Shafaq Sahar, Mohd Tanveer Alam Khan, Waheed Akram, and Aqeel Ahmad

INTRODUCTION

The interaction of hydrogen peroxide (H_2O_2) with the environment, particularly under the influence of climate extremes, is a critical yet underexplored area of study [1]. Hydrogen peroxide, a simple molecule with potent oxidative properties, is ubiquitously present in natural ecosystems. It is produced through various biochemical and photochemical processes and is a significant player in atmospheric and aquatic chemistry. Despite its simple structure, H_2O_2 exhibits a complex behavior in the environment, influencing and being influenced by a multitude of factors, including climate variability and microbial activity [2]. Climate extremes, such as heat waves, droughts, floods, and intense storms, are becoming more frequent and severe due to global climate change [3]. These extreme weather events can drastically alter environmental conditions, impacting the distribution and concentration of reactive species like H_2O_2. The stability and reactivity of H_2O_2 are highly sensitive to temperature and moisture levels, making it a dynamic component in the context of climate extremes [4]. Concomitantly, microbial colonization are profoundly affected by changes in climate. Microorganisms, the unseen majority of life forms on Earth, play essential roles in biogeochemical cycles and ecosystem functioning [5]. They exhibit remarkable adaptability to extreme conditions, often resulting in shifts in microbial community composition and metabolic activity [6]. The interaction between H_2O_2 and microbial communities under these extreme conditions adds another layer of complexity to our understanding of ecosystem dynamics. This chapter aims to explore the interactive prospects of H_2O_2 with climate extremes and the resulting impacts on microbial colonization. By examining the environmental chemistry of H_2O_2, its interactions with various climate extremes, and the responses of microbial communities, we seek to provide a comprehensive overview of the intricate relationships that define these interactions [7]. Through case studies, empirical evidence, and mechanistic insights, this chapter will illuminate the dual role of H_2O_2 as both a stressor and a regulatory agent in microbial ecology under extreme climatic conditions [7,8]. Ultimately, this exploration will underscore the importance of integrating chemical, climatic, and biological perspectives to better predict and manage ecosystem responses to an increasingly volatile climate.

DOI: 10.1201/9781032647692-16

OVERVIEW OF HYDROGEN PEROXIDE (H_2O_2)

IMPORTANCE OF STUDYING H_2O_2 IN ENVIRONMENTAL CONTEXTS

Hydrogen peroxide (H_2O_2) is a molecule of significant environmental relevance due to its widespread presence and its role as a reactive oxygen species (ROS) [9]. Understanding the behavior and impact of H_2O_2 in various environmental contexts is crucial for several reasons:

ECOLOGICAL IMPACT AND BIOGEOCHEMICAL CYCLES

H_2O_2 plays a vital role in numerous biogeochemical cycles, including the carbon, sulfur, and nitrogen cycles [10]. As a reactive molecule, it influences the oxidation and reduction processes that are fundamental to these cycles. For example, H_2O_2 can oxidize sulfur compounds, converting them into sulfate, a process that is essential for the sulfur cycle. Similarly, it can affect the nitrogen cycle by participating in the breakdown of nitrogenous compounds. Understanding these processes is essential for predicting how ecosystems respond to environmental changes and human activities.

ENVIRONMENTAL HEALTH AND TOXICOLOGY

H_2O_2 is a powerful oxidant and can be toxic to living organisms at high concentrations. It can cause oxidative stress in cells, leading to damage to proteins, lipids, and DNA. This makes it a critical factor in environmental toxicology studies [11]. Monitoring and understanding the levels of H_2O_2 in various environments can help assess the health risks to organisms, including humans, and inform strategies to mitigate these risks. This is particularly important in areas impacted by industrial pollution or agricultural runoff, where H_2O_2 levels can be artificially elevated [12].

ATMOSPHERIC CHEMISTRY AND CLIMATE REGULATION

In the atmosphere, H_2O_2 is involved in several key chemical reactions that influence air quality and climate. It participates in the formation and breakdown of ozone, a crucial component of the Earth's stratosphere that protects living organisms from harmful ultraviolet radiation. H_2O_2 also plays a role in the formation of secondary organic aerosols (SOAs), which affect cloud formation and climate regulation [13]. Studying H_2O_2 in the atmosphere helps scientists understand and predict changes in air quality and climate patterns.

MICROBIAL INTERACTIONS AND ECOSYSTEM FUNCTIONING

Microorganisms are highly sensitive to changes in H_2O_2 concentrations. While some microbes produce H_2O_2 as a defense mechanism or metabolic byproduct, others must develop strategies to neutralize its effects [14–17]. This dynamic interaction influences microbial community structure and function, affecting processes such as decomposition, nutrient cycling, and primary production [18–21]. Studying these interactions provides insights into ecosystem health and resilience, particularly under the stress of climate extremes.

CLIMATE CHANGE AND EXTREME WEATHER EVENTS

The frequency and intensity of climate extremes, such as heatwaves, droughts, and floods, are increasing due to climate change. These events can significantly alter the production, distribution, and degradation of H_2O_2 in the environment. For instance, higher temperatures can accelerate the decomposition of H_2O_2, while increased precipitation can enhance its transport and dispersion. Understanding these dynamics is essential for predicting how ecosystems will respond to future climate scenarios and for developing strategies to mitigate adverse impacts.

Environmental Monitoring and Management

H_2O_2 can serve as an indicator of oxidative stress and environmental pollution. Monitoring its levels in water bodies, soils, and the atmosphere provides valuable information about the health and status of ecosystems [22]. This information can be used to develop environmental management practices aimed at reducing pollution and protecting ecosystems. For example, managing agricultural practices to minimize runoff that leads to elevated H_2O_2 levels can help protect aquatic ecosystems.

Biotechnological and Industrial Applications

H_2O_2 is widely used in various industrial processes, including wastewater treatment, paper bleaching, and as a disinfectant. Understanding its environmental fate and impacts is crucial for developing sustainable industrial practices [23,24]. Additionally, H_2O_2-based technologies are being explored for bioremediation, where it is used to break down pollutants in contaminated environments. Studying H_2O_2 in environmental contexts helps optimize these applications, making them more efficient and environmentally friendly. In conclusion, studying H_2O_2 in environmental contexts is of paramount importance due to its multifaceted roles in ecological processes, atmospheric chemistry, and microbial interactions. It is a key component in understanding and managing the impacts of climate change, pollution, and other environmental stressors [25]. As our environment continues to face unprecedented challenges, a thorough understanding of H_2O_2 dynamics will be essential for developing strategies to preserve ecosystem health and resilience.

Hydrogen Peroxide in the Environment

Hydrogen peroxide (H_2O_2) is a naturally occurring molecule that is both produced and consumed through various processes in the environment. Understanding the natural sources and sinks of H_2O_2 is crucial for comprehending its environmental dynamics and its role in ecological and atmospheric chemistry [25]. This section delves into the various natural sources that generate H_2O_2 and the sinks that remove it from the environment.

NATURAL SOURCES OF H_2O_2

Photochemical Reactions in the Atmosphere

Photolysis of Water Vapor: One of the primary natural sources of H_2O_2 is the photolysis of water vapor in the atmosphere. Under the influence of ultraviolet (UV) radiation, water molecules (H_2O) can dissociate to form hydroxyl radicals (OH), which subsequently react to form H_2O_2 [26]. Ozone and Hydroxyl Radicals: Another significant pathway involves the reaction between ozone (O_3) and hydroxyl radicals. This reaction produces H_2O_2 as a byproduct, contributing to its atmospheric concentration.

Biological Production

Many microorganisms produce H_2O_2 as a part of their metabolic processes. For instance, during aerobic respiration, certain bacteria and fungi generate H_2O_2 as a byproduct of oxygen metabolism. Additionally, some microbes produce H_2O_2 as a defense mechanism against competing microorganisms [1]. Besides microbes, plants also produce H_2O_2 through various metabolic pathways, particularly during photosynthesis and photorespiration. It plays a role in cellular signaling and defense against pathogens.

Chemical Reactions in Water Bodies

In aquatic environments, the interaction of dissolved organic matter with sunlight can lead to the formation of H_2O_2 through photochemical reactions. This process is significant in surface waters where sunlight penetration is sufficient to drive these reactions. Furthermore, the oxidation of reduced species such as ferrous iron (Fe^{2+}) and sulfur compounds in natural waters can also produce H_2O_2. These reactions are often catalyzed by the presence of certain minerals and are influenced by the water's pH and temperature.

Atmospheric Deposition

Rain and Snow: H_2O_2 can be deposited onto the Earth's surface through precipitation. Rain and snow capture atmospheric H_2O_2 and deliver it to terrestrial and aquatic ecosystems, influencing local H_2O_2 levels.

NATURAL SINKS OF H_2O_2

Photodecomposition

UV Radiation: H_2O_2 is susceptible to photodecomposition under UV radiation. This process breaks down H_2O_2 into water and oxygen, effectively removing it from the environment [10]. This sink is particularly active in the atmosphere and surface waters exposed to sunlight.

Biological Consumption

Many microorganisms possess enzymes such as catalase and peroxidase that break down H_2O_2 into water and oxygen. This biological consumption is a significant sink for H_2O_2 in soils, sediments, and water bodies, helping to regulate its concentration. Besides microbial degradation, plants also utilize H_2O_2 for various biochemical processes. Enzymes within plant cells can degrade H_2O_2, thereby preventing its accumulation to toxic levels.

Chemical Reactions

For example, it can react with ferrous iron (Fe^{2+}) to produce ferric iron (Fe^{3+}) and hydroxyl radicals. These reactions are important in both terrestrial and aquatic environments.

Reaction with Organic Compounds: H_2O_2 can oxidize organic compounds, leading to its degradation. This process is influenced by Redox Reactions: H_2O_2 participates in various redox reactions that lead to its consumption. the presence of natural organic matter and the chemical composition of the environment.

Surface Interactions

Soil and Sediment Adsorption: H_2O_2 can be adsorbed onto soil particles and sediments, where it can undergo further chemical or biological degradation. This interaction with solid surfaces acts as a sink, particularly in terrestrial and freshwater ecosystems.

Balance and Implications

The balance between the natural sources and sinks of H_2O_2 determines its concentration and distribution in the environment. This balance is dynamic and can be influenced by various factors, including climatic conditions, biological activity, and the presence of pollutants. Understanding these natural processes is essential for predicting how H_2O_2 levels might change in response to environmental perturbations, such as climate change and human activities.

In summary, H_2O_2 is generated through a variety of natural processes, including photochemical reactions, microbial and plant metabolism, and chemical reactions in water bodies. Its removal is facilitated by photodecomposition, biological consumption, redox reactions, and interactions with soil and sediment. These natural sources and sinks are integral to maintaining the ecological balance and determining the role of H_2O_2 in environmental processes.

ANTHROPOGENIC CONTRIBUTIONS

Human activities have significantly altered the natural balance of H_2O_2 in the environment. These anthropogenic contributions can influence both the sources and sinks of H_2O_2, leading to changes in its concentration and distribution. This section examines the various ways in which human activities contribute to the production and removal of H_2O_2 in the environment.

Anthropogenic Sources of H_2O_2

Industrial Emissions

Industries involved in the production of chemicals, particularly those using processes that generate ROS, can release H_2O_2 into the environment. Factories producing hydrogen peroxide for bleaching, disinfecting, and other purposes are notable contributors. Moreover, the burning of fossil fuels in power plants, vehicles, and other combustion engines generates various pollutants, including those that can lead to the formation of H_2O_2. Nitrogen oxides (NOx) and volatile organic compounds (VOCs) emitted during combustion can react in the atmosphere to produce H_2O_2.

Agricultural Practices

The use of certain pesticides and fertilizers can result in the formation of H_2O_2. Some agrochemicals undergo photochemical reactions or degrade into compounds that produce H_2O_2 when exposed to sunlight or soil microbes. Agricultural runoff containing organic matter and pollutants can enhance the production of H_2O_2 in water bodies through photochemical reactions and microbial activity.

Urbanization and Transportation

Automobiles and other forms of transportation emit NOx and VOCs, which can react in the atmosphere to form H_2O_2. Urban areas with high traffic density often have elevated levels of these precursors, leading to increased H_2O_2 formation. Treatment plants often use H_2O_2 for disinfection purposes. Inefficient or excess use can lead to the release of H_2O_2 into water bodies, affecting local ecosystems.

Waste Management

Decomposition of organic waste in landfills can produce various gases, including those that contribute to the formation of H_2O_2. Methane and other organic compounds can undergo photochemical reactions in the atmosphere to produce H_2O_2. The burning of waste materials can release precursors that lead to the formation of H_2O_2 in the atmosphere.

Anthropogenic Sinks of H_2O_2

Environmental Remediation

H_2O_2 is commonly used in environmental cleanup efforts, particularly for the bioremediation of contaminated soils and groundwater [27,28]. It acts as an oxidizing agent to break down pollutants. While this use reduces local H_2O_2 levels, it can alter natural microbial communities and biogeochemical processes [29–31]. Moreover, in water and wastewater treatment, advanced oxidation processes (AOPs) utilize H_2O_2 to degrade organic pollutants. These processes are designed to remove contaminants but can also affect the overall balance of H_2O_2 in the treated water.

Atmospheric Deposition Control

Policies and regulations aimed at reducing NOx and VOC emissions from industrial and transportation sources indirectly control the atmospheric concentrations of H_2O_2. By limiting the precursors, these regulations help reduce the formation of H_2O_2 in the atmosphere [32].

Urban and Agricultural Management

Implementing BMPs in agriculture and urban planning can minimize runoff and reduce the input of organic matter and pollutants that contribute to H_2O_2 formation [33,34]. Practices such as proper fertilizer application, erosion control, and sustainable land use help maintain lower H_2O_2 levels. The development of green infrastructure, such as green roofs, permeable pavements, and constructed wetlands, can help mitigate urban runoff and reduce the environmental impact of pollutants that lead to H_2O_2 formation.

Environmental Chemistry of H_2O_2

H_2O_2 is an ROS with significant implications for environmental chemistry due to its oxidative properties and involvement in various biochemical and photochemical processes. Understanding the environmental chemistry of H_2O_2 involves examining its formation, degradation, reactivity, and interactions with other environmental components. This section provides a detailed overview of these aspects.

FORMATION OF H_2O_2

Photochemical Processes

In the atmosphere, H_2O_2 is primarily formed through the photolysis of water vapor and subsequent reactions involving hydroxyl radicals (OH). UV radiation splits water molecules (H_2O) into hydroxyl radicals, which can recombine to form H_2O_2. Moreover, H_2O_2 is also produced when ozone (O_3) reacts with hydroxyl radicals in the presence of sunlight. This reaction is a critical pathway for H_2O_2 formation in the troposphere.

Biological Production

Many microorganisms produce H_2O_2 as a byproduct of metabolic processes, particularly during aerobic respiration. Enzymes such as glucose oxidase and superoxide dismutase contribute to the generation of H_2O_2 in microbial cells. Plants also produce H_2O_2 during photosynthesis and photorespiration. It serves as a signaling molecule in various physiological processes, including responses to stress and pathogen attack.

Chemical Reactions in Aquatic Systems

In natural waters, the Fenton reaction, involving the reaction of ferrous iron (Fe^{2+}) with H_2O_2, can generate additional H_2O_2. This reaction is significant in environments rich in iron. H_2O_2 can also be formed through the interaction of dissolved organic matter (DOM) with sunlight, leading to photochemical production in surface waters.

DEGRADATION AND REMOVAL OF H_2O_2

Photodecomposition

H_2O_2 undergoes photodecomposition when exposed to UV radiation, breaking down into water (H_2O) and oxygen (O_2). This process is a significant sink for H_2O_2 in the atmosphere and surface waters.

Biological Degradation

Enzymes such as catalase and peroxidase, found in many microorganisms and plants, catalyze the decomposition of H_2O_2 into water and oxygen. This biological degradation is crucial for maintaining cellular redox balance and protecting organisms from oxidative stress.

Chemical Reactions

H_2O_2 participates in various redox reactions, such as the reaction with ferrous iron in the Fenton reaction, producing hydroxyl radicals and ferric iron (Fe^{3+}). These reactions are important in both terrestrial and aquatic environments. H_2O_2 reacts with reduced sulfur compounds, such as sulfide (S^{2-}), producing sulfate (SO_4^{2-}). These oxidation reactions help regulate the levels of both H_2O_2 and reduced compounds in the environment.

REACTIVITY AND ENVIRONMENTAL INTERACTIONS

Oxidative Capacity

Due to its strong oxidative properties, H_2O_2 can cause oxidative stress in living organisms, leading to damage to cellular components such as lipids, proteins, and DNA. Understanding the oxidative capacity of H_2O_2 is essential for assessing its impact on biological systems. H_2O_2 is used in environmental remediation to degrade organic pollutants through oxidation. Its effectiveness in breaking down contaminants makes it valuable for wastewater treatment and soil remediation efforts.

Role in Biogeochemical Cycles

H_2O_2 influences the carbon cycle by affecting the decomposition of organic matter. Its oxidative properties can enhance the breakdown of complex organic molecules, facilitating their conversion into simpler forms. H_2O_2 plays a role in the oxidation of sulfur and nitrogen compounds, contributing to the cycling of these essential elements. For instance, it oxidizes sulfide to sulfate and ammonium (NH^{4+}) to nitrate (NO^{3-}), impacting nutrient availability in ecosystems.

Interactions with Environmental Contaminants

H_2O_2 can mobilize heavy metals from sediments and soils by oxidizing them into more soluble forms. This process can increase the bioavailability and toxicity of metals such as mercury (Hg) and lead (Pb). In the atmosphere, reactions involving H_2O_2 can lead to the formation of secondary pollutants such as ozone and particulate matter. These secondary pollutants have significant implications for air quality and human health.

Implications for Environmental Monitoring and Management

Understanding the environmental chemistry of H_2O_2 is critical for developing effective strategies for monitoring and managing its levels in various ecosystems. By studying its sources, sinks, reactivity, and interactions, scientists can better predict the behavior of H_2O_2 under different environmental conditions and assess its impact on ecosystem health and stability.

- Regular monitoring of H_2O_2 levels in air, water, and soil is essential for assessing its environmental impact. Advanced analytical techniques, such as chemiluminescence and electrochemical sensors, are used to measure H_2O_2 concentrations accurately.

- Strategies to reduce anthropogenic sources of H_2O_2, such as emission regulations and best management practices in agriculture and industry, can help mitigate its environmental impact.
  - Utilizing H_2O_2 in environmental remediation requires careful consideration of its reactivity and potential side effects. Optimizing its use in bioremediation and advanced oxidation processes can enhance the effectiveness of cleanup efforts while minimizing adverse impacts.

CLIMATE EXTREMES: DEFINITIONS AND TYPES

Climate extremes refer to unusual, severe, or unseasonal weather events that significantly deviate from the average weather patterns for a specific region. These events can have profound impacts on ecosystems, human health, infrastructure, and economies. Understanding the definitions and types of climate extremes is crucial for predicting their occurrence, mitigating their effects, and developing strategies for resilience and adaptation. This section explores the various definitions and types of climate extremes.

Meteorological Extremes

Meteorological extremes are of two types. (i) Weather Extremes; include short-term events such as heatwaves, heavy rainfall, thunderstorms, and hurricanes. These events are characterized by their intensity and sudden onset, often lasting from hours to days. (ii) Climate Extremes; encompass longer-term variations and anomalies, such as prolonged droughts or persistent abnormal temperatures. These are often assessed relative to long-term climatic averages and trends.

Hydrological Extremes

There are also two types of hydrological extremes. (i) Floods; occur when water bodies overflow, inundating surrounding areas. This can be due to excessive rainfall, rapid snowmelt, or storm surges (Figure 16.1). Floods are classified based on their origin, such as riverine, flash, or coastal floods. (ii) Droughts; are prolonged periods of deficient precipitation, resulting in water scarcity. They can impact agriculture, water supply, and ecosystems. Droughts are classified based on their severity and duration, including meteorological, agricultural, and hydrological droughts.

Temperature Extremes

There are two subtypes of temperature extremes. (i) Heatwaves; are extended periods of excessively high temperatures, often accompanied by high humidity. They pose serious health risks and can cause significant ecological and economic damage. (ii) Cold Waves; are periods of unusually low temperatures that can lead to frost, snowstorms, and ice formation. These events can disrupt transportation, agriculture, and energy supply systems (Figure 16.2).

Wind Extremes

Wind extremes are also classified into two different types. (i) Tornadoes; are violently rotating columns of air that extend from thunderstorms to the ground. They are characterized by their high wind speeds and destructive potential. (ii) Hurricanes and Typhoons; are large-scale tropical cyclones with sustained winds exceeding 74 mph. Hurricanes occur in the Atlantic and Northeast Pacific, while typhoons occur in the Northwest Pacific.

FIGURE 16.1 A tree under stress conditions due to water logging after extreme high precipitation in Florida, USA.

FIGURE 16.2 Plants facing abrupt temperature drop in Wuhan, Central China.

WILDFIRES

Wildfires are uncontrolled fires that spread rapidly through vegetation. They can be triggered by natural causes, such as lightning, or human activities, such as unattended campfires or arson. Wildfires destroy forests, homes, and infrastructure, release significant amounts of CO_2, and pose severe health risks due to smoke inhalation. They also lead to loss of biodiversity and degradation of ecosystems.

CLIMATE EXTREMES: A STEP FORWARD FROM CLIMATE CHANGE

Extreme climates refer to regions or areas that experience severe weather conditions, abrupt climatic changes, or environmental characteristics that deviate suddenly and significantly from the norm. These climates are typically characterized by intense heat, extreme cold, heavy precipitation, strong winds, or other severe weather events. Extreme climates are not necessarily a direct result of steady climate change, but they can be influenced by long-term climate patterns and changes. Climate change refers to the long-term alteration of temperature, precipitation, wind patterns, and other aspects of the Earth's climate system. While climate change can lead to an increase in the frequency or intensity of certain extreme weather events, such as heatwaves, droughts, storms, or heavy rainfall, extreme climates can also occur naturally in specific regions due to geographical factors, atmospheric circulation patterns, or other local conditions. Extreme climates are not exclusively caused by steady climate change but can be influenced by it. They can occur naturally in specific regions due to a combination of geographical factors, atmospheric conditions, and climate patterns. Climate change, however, can exacerbate certain extreme weather events and potentially lead to more frequent or severe occurrences in some areas. Here we will discuss some important aspects and issues faced by researchers working in extreme climates.

A comprehensive classification and characterization of extreme environmental events (EEEs) have been performed by NOAA (National Oceanic and Atmospheric Administration) (https://www.noaa.gov/). According to the NOAA classification system, EEEs are classified into six distinct categories, (i) extreme high temperature, (ii) extreme low temperature, (iii) extreme high precipitation, (iv) extreme low precipitation, (v) wind, and (vi) drought. As a researcher in the relevant field, I know several of my colleagues are having trouble fully understanding extreme low precipitation and drought. Most people think they are the two names of a single phenomenon. But it's not in reality. The problem lies in its close relatedness in most of the geographic areas situated globally. Mostly, rain brings water and eliminates drought and its impacts on life. However, studying the impact of drought on crops and living organisms assumes the drought is a duration of the period of NO RAIN. Suppose an area faces >600 mm of rain in less than 48 hours and has no rain for the next six months. It will be the area affected by two EEEs (facing Extreme Precipitation and Drought as well). A somewhat similar situation has happened in Beihai city of Guangxi region, China.

> The city of Yulin in the Guangxi region experienced 35 hours of non-stop rain on 8–9 June, while the nearby city of Beihai was flooded after 614.7mm of rainfall over 24 hours in the same period. This is approximately a third of the city's average yearly precipitation, and a June record for the Guangxi region. It is in stark contrast to May, when Guangxi experienced its lowest rainfall in 60 years.
>
> *https://www.theguardian.com/environment/2023/jun/16/weather-tracker-extreme-rainfall-heat-china-asian-heatwave*

Another problem arises about the arithmetic calculation of the climatic thresholds beyond which the climate can be classified as an 'Extreme Climate'. Of course, the threshold levels of EEEs will be different in each region according to its normal environmental conditions. For example, certain regions near the Earth's poles, such as the Arctic or Antarctica, experience extreme cold climates due to their high latitudes and polar ice caps. These extreme cold climates have existed for thousands of years and are not solely a consequence of recent climate change. Similarly, areas prone to hurricanes or typhoons in tropical regions have a naturally higher risk of extreme weather events. To classify weather as an EEE, it should be much above (below) normal or extreme conditions as those falling in the upper (lower) tenth percentile of the local, period of record (Climate Extreme Index (CEI) by NOAA) (https://www.ncei.noaa.gov/access/monitoring/cei/definition). Therefore, for each class of climate, the researchers should focus on the highest tenth percentile and the lowest tenth percentile of the values to decide the threshold level of EEEs.

INTERACTIONS BETWEEN H_2O_2 AND CLIMATE EXTREMES

H_2O_2 is an ROS that plays a significant role in environmental chemistry and biology. Its interactions with climate extremes, such as heatwaves, droughts, floods, and storms, are complex and can influence its concentration, distribution, and reactivity. Understanding these interactions is crucial for predicting the behavior of H_2O_2 under changing climatic conditions and assessing its impact on ecosystems. This section explores the various ways in which H_2O_2 interacts with different climate extremes.

Influence of Temperature Extremes

Heatwaves

High temperatures during heatwaves can enhance the photochemical production of H_2O_2 in the atmosphere. UV radiation increases, leading to more intense photolysis of water vapor and ozone, which generates hydroxyl radicals that subsequently form H_2O_2. Elevated temperatures also accelerate the decomposition of H_2O_2. The balance between its production and decomposition rates determines its overall concentration during heatwaves. Heatwaves can stress microbial communities, affecting their ability to produce and degrade H_2O_2. Some microbes may increase H_2O_2 production as a stress response, while others may experience reduced enzymatic activity needed to decompose H_2O_2.

Cold Waves

Lower temperatures and reduced sunlight during cold waves can decrease the photochemical production of H_2O_2. Colder temperatures may increase the stability of H_2O_2, slowing down its decomposition and prolonging its presence in the environment. Cold waves can inhibit microbial metabolism, reducing both the production and degradation of H_2O_2 by microorganisms.

Effects of Hydrological Extremes

Reduced Water Availability

Droughts reduce the availability of water, concentrating solutes and affecting the chemistry of soils and water bodies. This can influence the reactivity of H_2O_2, as its interactions with various ions and organic compounds become more pronounced. Reduced water flow and lower water levels can increase the exposure of dissolved organic matter to sunlight, potentially enhancing the photochemical production of H_2O_2 in remaining water bodies. Drought stress in plants can lead to increased production of H_2O_2 as part of the oxidative stress response. This can influence soil chemistry as plants release H_2O_2 into the rhizosphere. Ambient soil moisture is critical for microbial activity. Drought conditions can suppress microbial metabolism, reducing the production and decomposition of H_2O_2. Some drought-resistant microbes might produce more H_2O_2 as a defense mechanism against desiccation.

Impacts of Flooding and Excess Moisture

Flooding increases water volumes, diluting solutes and potentially reducing the concentration of H_2O_2. However, increased water movement can also enhance the distribution and mixing of reactive species, influencing H_2O_2 dynamics. Floodwaters can carry organic and inorganic materials that participate in redox reactions, influencing the production and degradation of H_2O_2. Excessive and prolonged flooding can engender water logging and create anaerobic conditions in soils and sediments, altering microbial communities and reducing the activity of H_2O_2-degrading enzymes such as catalase and peroxidase. Flooded conditions can also stress plants, potentially leading to increased H_2O_2 production as part of the plant's oxidative stress response.

Extreme Storms and Atmospheric Dynamics

Hurricanes and Typhoons

Strong winds and turbulence associated with hurricanes and typhoons can enhance the mixing of atmospheric layers, distributing H_2O_2 and its precursors more widely. Heavy rainfall can remove H_2O_2 from the atmosphere through wet deposition, delivering it to terrestrial and aquatic ecosystems.

Thunderstorms and Lightning

Lightning generates hydroxyl radicals through the ionization of air, which can lead to increased production of H_2O_2 in storm clouds. Thunderstorms can also facilitate the scavenging of H_2O_2 from the atmosphere, impacting its distribution and concentration.

NATURE OF CLIMATE INTERACTIONS WITH H_2O_2

Synergistic and Antagonistic Interactions

The combination of climate extremes and elevated H_2O_2 levels can exacerbate oxidative stress in plants and microorganisms, leading to more severe damage and disruption of ecological processes. However, in some cases, certain climate extremes might reduce H_2O_2 levels, potentially alleviating some oxidative stress. For example, heavy rainfall can dilute H_2O_2 concentrations in water bodies, reducing its reactivity.

Positive and Negative Feedback Mechanisms

Climate extremes can create conditions that enhance H_2O_2 production, which in turn can influence climate processes. For example, increased H_2O_2 levels can affect cloud formation and precipitation patterns. Conversely, some interactions might act as negative feedback mechanisms, stabilizing H_2O_2 levels despite changing environmental conditions. For example, increased microbial activity during moderate warming might enhance H_2O_2 degradation, counteracting its increased production. The interactions between H_2O_2 and climate extremes are multifaceted and can significantly influence environmental chemistry and ecosystem dynamics. By understanding these interactions, scientists can better predict the impacts of climate change on H_2O_2 levels and develop strategies to mitigate its effects. This knowledge is crucial for managing ecosystems and protecting biodiversity in the face of increasing climate variability and extreme weather events.

MICROBIAL COLONIZATION IN EXTREME CLIMATES

Microbial colonization in extreme climates involve the adaptation and survival of microorganisms in environments characterized by harsh conditions such as high temperatures, severe cold, prolonged drought, and flooding. These extreme environments pose significant challenges to microbial life, yet many microbes have developed unique mechanisms to thrive under such conditions. Understanding microbial colonization in these extreme climates is crucial for elucidating their roles in ecosystem functioning, biogeochemical cycles, and their potential applications in biotechnology and environmental management.

Types of Microbes Involved

Thermophiles

Thermophiles are microorganisms that thrive at high temperatures, typically between 45°C and 80°C. Hyperthermophiles can survive at even higher temperatures, often exceeding 80°C. Common habitats for thermophiles include hot springs, geothermal vents, and compost heaps. They are also found in industrial processes that involve high temperatures. Thermophiles possess heat-stable enzymes and proteins, unique membrane lipids that remain functional at high temperatures, and efficient DNA repair mechanisms to withstand thermal stress (Figure 16.3).

Psychrophiles

Psychrophiles are microorganisms that thrive at low temperatures, typically below 15°C, with optimal growth occurring at around 0°C–10°C. Psychrophiles are commonly found in polar regions, deep ocean waters, glaciers, and permafrost. They are also present in refrigerated environments and cold storage facilities. Psychrophiles produce cold-active enzymes, possess membrane lipids that maintain fluidity at low temperatures, and have antifreeze proteins to prevent ice crystal formation within cells.

Halophiles

Halophiles are microorganisms that thrive in high-salt environments, such as salt lakes, saline soils, and salt mines. They are found in environments with salt concentrations ranging from 3% to 30% or higher, including the Dead Sea and salt flats. Halophiles have specialized ion pumps to expel excess salt, compatible solutes to balance osmotic pressure, and proteins that remain functional in high-salt conditions.

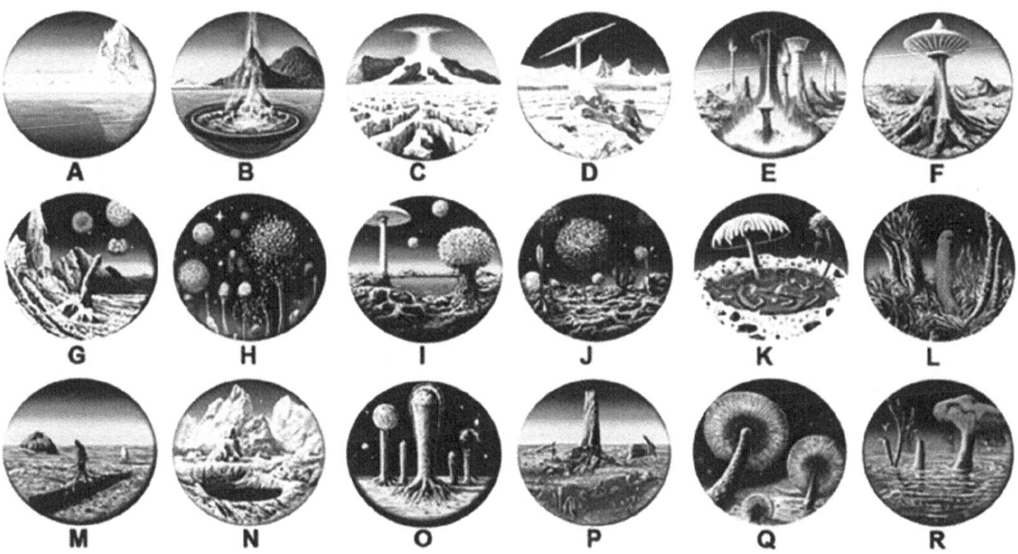

FIGURE 16.3 Graphical illustration of extremophiles and their classification.

Acidophiles and Alkaliphiles

Acidophiles thrive in acidic environments with pH levels below 5, while alkaliphiles thrive in alkaline environments with pH levels above 9. These microbes are found in acid mine drainage, sulfuric hot springs, and acidic soils, whereas alkaliphiles are found in soda lakes and alkaline soils. Acidophiles have mechanisms to pump out excess protons and maintain internal pH, while alkaliphiles possess adaptations to maintain cellular functions at high pH levels.

Xerophiles

Xerophiles are microorganisms that can survive in extremely dry conditions, such as deserts and arid environments. They are found in desert soils, rock surfaces, and other environments with minimal water availability. Xerophiles produce protective extracellular matrices, accumulate compatible solutes to retain water, and have efficient DNA repair mechanisms to counteract desiccation-induced damage.

ROLE OF H_2O_2 IN MICROBIAL ECOLOGY

H_2O_2 plays a multifaceted role in microbial ecology, influencing microbial community structure, interactions, and metabolic processes. As an ROS, H_2O_2 can act both as a signaling molecule and as a stress factor, impacting the dynamics of microbial ecosystems. This section explores the various ways H_2O_2 affects microbial ecology, including its antimicrobial properties, effects on microbial growth and survival, and interactions with microbial metabolism.

Oxidative Stress and Microbial Defense

H_2O_2 is a potent oxidizing agent that can cause oxidative stress in microorganisms. It can react with cellular components such as lipids, proteins, and nucleic acids, leading to cell damage and death. This oxidative stress is a crucial factor in controlling microbial populations and preventing the overgrowth of pathogenic species. Microorganisms have evolved various defense mechanisms to mitigate the damaging effects of H_2O_2. These include the production of antioxidant enzymes such as catalase, peroxidase, and superoxide dismutase, which neutralize H_2O_2 and other ROS. Additionally, some microbes can produce small molecules like glutathione that scavenge ROS and protect cellular components from oxidative damage.

Ecological Interactions

H_2O_2 can influence microbial competition and predation within ecosystems. Some microbes produce H_2O_2 to inhibit the growth of competitors or as a defense mechanism against predatory microorganisms. For example, certain bacteria and fungi secrete H_2O_2 as part of their competitive strategy to dominate specific ecological niches. In some symbiotic relationships, H_2O_2 plays a role in signaling and defense. For instance, in lichens, which are symbiotic associations between fungi and algae or cyanobacteria, the production of H_2O_2 can help protect the symbiotic partners from environmental stress and microbial attack.

Microbial Growth and Survival

The effects of H_2O_2 on microbial growth and survival are concentration dependent. At low concentrations, H_2O_2 can act as a signaling molecule, while at higher concentrations, it can inhibit growth and induce cell death. This dual role makes H_2O_2 a critical factor in regulating microbial community dynamics. High levels of H_2O_2 can cause significant cellular damage, leading to lipid peroxidation, protein denaturation, and DNA damage. These effects can inhibit microbial growth and survival, particularly in sensitive species or under conditions where antioxidant defenses are overwhelmed.

Adaptation and Resistance

Microbes have developed stress response pathways to adapt to fluctuating H_2O_2 levels. These pathways involve the upregulation of genes encoding antioxidant enzymes and proteins that repair damaged cellular components. The ability to mount an effective stress response can determine the resilience of microbial species to oxidative stress. Some microbes form biofilms as a protective strategy against H_2O_2. Biofilms are complex communities of microorganisms embedded in an extracellular matrix, which can provide a physical barrier to H_2O_2 penetration. This biofilm lifestyle enhances microbial survival under oxidative stress conditions and contributes to the persistence of microbial populations in harsh environments.

INTERACTIONS WITH MICROBIAL METABOLISM

Metabolic Pathways

H_2O_2 is produced as a byproduct of aerobic respiration in many microorganisms. The electron transport chain generates superoxide radicals, which are then converted to H_2O_2 by superoxide dismutase. Microbes must manage this intracellular H_2O_2 to prevent oxidative damage. In anaerobic conditions, certain fermentation pathways can also produce H_2O_2, influencing microbial metabolism and community structure. H_2O_2 participates in various redox reactions within microbial cells. These reactions are crucial for energy production, detoxification, and metabolic regulation. For example, the Fenton reaction, involving H_2O_2 and iron, produces highly reactive hydroxyl radicals, which can further impact microbial metabolic processes.

Quorum Sensing and Communication

At low concentrations, H_2O_2 can function as a signaling molecule in microbial communities. It plays a role in quorum sensing, a process by which microbes communicate and coordinate their behavior based on population density. H_2O_2-mediated signaling can influence biofilm formation, virulence factor production, and other collective behaviors. H_2O_2 can mediate interactions between different microbial species. For instance, some bacteria produce H_2O_2 to inhibit the growth of competing species, while others may use it to signal beneficial interactions, such as cooperation in biofilm formation or metabolic cross-feeding.

ECOLOGICAL AND ENVIRONMENTAL IMPLICATIONS

Nutrient Cycling

H_2O_2 plays a role in the decomposition of organic matter and the mineralization of nutrients. Its oxidative properties can break down complex organic molecules, making nutrients more accessible to microbes and plants [35–37]. This process is essential for nutrient cycling in ecosystems, particularly in soils and aquatic environments [38]. H_2O_2 influences the cycling of elements such as carbon, nitrogen, sulfur, and iron. For example, it can oxidize reduced sulfur compounds to sulfate or convert ferrous iron (Fe^{2+}) to ferric iron (Fe^{3+}), affecting the availability and mobility of these elements in the environment.

Environmental Stress and Adaptation

The production and effects of H_2O_2 are influenced by climate extremes such as heatwaves, droughts, and floods. These conditions can alter microbial activity and community composition, impacting the ecological roles of H_2O_2. For instance, heat stress may increase H_2O_2 production in some microbes, while drought conditions can suppress microbial activity and H_2O_2 degradation. H_2O_2 is used in environmental remediation to degrade pollutants through oxidative processes. Understanding its role in microbial ecology helps optimize bioremediation strategies, as microbes can both produce and degrade H_2O_2, influencing its effectiveness in contaminant degradation.

Biodiversity and Community Structure

Climate extremes and H_2O_2 levels can act as selective pressures, shaping microbial community structure. Species with robust antioxidant defenses and stress response mechanisms are more likely to survive and proliferate. This selection process can lead to changes in microbial diversity and ecosystem function. The presence of multiple microbial species with similar ecological functions can enhance ecosystem resilience to climate extremes. Functional redundancy ensures that critical processes, such as nutrient cycling and organic matter decomposition, continue despite shifts in community composition due to H_2O_2 stress.

MECHANISMS OF H_2O_2-MICROBE INTERACTIONS UNDER CLIMATE EXTREMES

H_2O_2 interacts with microbial communities in various ways, and these interactions can be significantly affected by climate extremes such as heatwaves, droughts, floods, and storms. Understanding the mechanisms of these interactions under such conditions is crucial for predicting microbial responses to climate change and managing ecosystem health. This section explores the biochemical pathways, microbial defense mechanisms, and the synergistic and antagonistic effects of H_2O_2 on microbes under extreme climatic conditions.

Oxidative Stress and Microbial Defense Mechanisms

Production of Reactive Oxygen Species (ROS)

Elevated temperatures during heatwaves can increase the metabolic activity of microbes, leading to enhanced production of ROS, including H_2O_2. High temperatures can also cause protein denaturation and membrane fluidity changes, contributing to oxidative stress. Cold stress can disrupt cellular homeostasis and induce oxidative stress. Reduced metabolic rates may decrease ROS production, but cold-induced damage to cellular structures can still lead to increased H_2O_2 levels.

Antioxidant Enzyme Systems

Microorganisms produce antioxidant enzymes such as catalase and peroxidase to detoxify H_2O_2. These enzymes catalyze the decomposition of H_2O_2 into water and oxygen, protecting cells from oxidative damage [39,40]. Under climate extremes, the expression and activity of these enzymes can be upregulated as part of the stress response. Superoxide Dismutase (SOD) converts superoxide radicals into H_2O_2, which is then degraded by catalase and peroxidase [41–43]. The coordinated activity of these enzymes is crucial for managing oxidative stress under extreme temperatures, drought, and flooding.

Non-Enzymatic Antioxidants

Microbes produce small-molecule antioxidants such as glutathione and thioredoxin to scavenge ROS. These compounds help maintain redox balance and protect cellular components from oxidative damage. During climate extremes, the synthesis and recycling of these antioxidants are critical for microbial survival. Some microorganisms synthesize carotenoids and polyphenols, which have antioxidant properties. These molecules can quench ROS and stabilize cell membranes, providing additional protection under stress conditions.

FUTURE RESEARCH DIRECTIONS

Gaps in Current Knowledge

Despite significant advancements in understanding H_2O_2 and its interactions with microbial ecology, several critical gaps remain in our knowledge. Addressing these gaps is essential for a comprehensive understanding of H_2O_2 dynamics, especially under the influence of climate extremes. This section highlights key areas where further research is needed to enhance our understanding of H_2O_2-microbe interactions and their implications for ecosystem health and resilience.

While photochemical reactions are known to produce H_2O_2, the exact mechanisms and rates of these reactions under varying environmental conditions are not fully understood. More research is needed to quantify the contributions of different sources to H_2O_2 levels in diverse ecosystems. The specific pathways through which different microbial species produce and decompose H_2O_2 remain inadequately characterized. Understanding these pathways in greater detail, including the role of microbial communities in modulating H_2O_2 concentrations, is crucial.

The effects of temperature fluctuations and light intensity on H_2O_2 production and stability need further investigation. Current models may not accurately predict H_2O_2 dynamics in response to extreme weather events and seasonal variations. The interactions of H_2O_2 with various environmental chemicals, including pollutants and natural organic matter, are not fully understood. These interactions can significantly impact H_2O_2 stability and reactivity, influencing its ecological effects. While some microbial stress response mechanisms to H_2O_2 are known, the genetic regulation and molecular details of these pathways are not comprehensively understood. Research is needed to identify key regulatory genes and proteins involved in H_2O_2 detoxification. The evolutionary adaptations of microbial species to high H_2O_2 environments, particularly under climate extremes, are not well-documented. Studies on the genetic and phenotypic changes that confer resistance to oxidative stress can provide valuable insights.

The impact of H_2O_2 on microbial community interactions, including competition, cooperation, and predation, requires further exploration. Understanding how H_2O_2 influences community structure and function is essential for predicting ecological outcomes. The role of H_2O_2 in biofilm development and maintenance is not fully elucidated. Biofilms provide a protective environment against oxidative stress, and understanding the dynamics of H_2O_2 within biofilms is critical for managing microbial populations in natural and artificial settings.

The combined effects of heatwaves, droughts, and elevated H_2O_2 levels on microbial activity and ecosystem processes are not well understood. Research should focus on how these stressors interact to influence microbial physiology and ecosystem functions. The temporal dynamics of H_2O_2 production and decomposition during prolonged heatwaves and droughts need to be characterized. Understanding how these dynamics change over time can help predict long-term ecological impacts.

CONCLUSION

The interactive prospects of H_2O_2 with climate extremes and concomitant microbial colonization reveal a complex web of biochemical, ecological, and environmental dynamics. H_2O_2, a reactive oxygen species, plays a multifaceted role in both environmental chemistry and microbial ecology, acting as both a signaling molecule and a stressor. Its interactions with climate extremes such as heatwaves, droughts, floods, and storms significantly influence its concentration, distribution, and reactivity, thereby impacting microbial communities and broader ecosystem processes.

Climate extremes alter the production and stability of H_2O_2. Heatwaves and high temperatures enhance photochemical production but also accelerate decomposition, creating a dynamic balance that affects microbial activity. Cold waves increase H_2O_2 stability but reduce its production. Droughts concentrate solutes and organic matter, potentially increasing photochemical production of H_2O_2, while floods dilute H_2O_2 but can enhance its distribution and interactions with other chemicals.

Microorganisms exhibit a range of adaptations to manage oxidative stress induced by H_2O_2. These include the production of antioxidant enzymes such as catalase and peroxidase, synthesis of non-enzymatic antioxidants like glutathione, and the formation of biofilms for collective protection. Microbial responses to H_2O_2 are crucial for maintaining ecosystem stability, especially under the stress of climate extremes. H_2O_2 influences key biogeochemical cycles, including the cycling of carbon, nitrogen, sulfur, and iron. It affects nutrient availability and organic matter decomposition, playing a pivotal role in ecosystem functioning. Climate extremes can exacerbate the oxidative stress on microbial communities, potentially altering species composition and ecosystem processes.

REFERENCES

1. M.A. Hossain, S. Bhattacharjee, S.-M.M. Armin, P. Qian, W. Xin, H.-Y.Y. Li, D.J. Burritt, M. Fujita, L.-S.S.P. Tran, Hydrogen peroxide priming modulates abiotic oxidative stress tolerance: insights from ROS detoxification and scavenging, *Front Plant Sci* 6 (2015). https://doi.org/10.3389/fpls.2015.00420.
2. T.A. Khan, M. Yusuf, Q. Fariduddin, Hydrogen peroxide in regulation of plant metabolism: Signalling and its effect under abiotic stress, *Photosynthetica* 56 (2018) 1237–1248. https://doi.org/10.1007/s11099-018-0830-8.
3. T. Zhu, X. Deng, X. Zhou, L. Zhu, L. Zou, P. Li, D. Zhang, H. Lin, Ethylene and hydrogen peroxide are involved in brassinosteroid-induced salt tolerance in tomato, *Sci Rep* 6 (2016) 1–15. https://doi.org/10.1038/srep35392.
4. X. Zhang, L. Zhang, Y. Sun, S. Zheng, J. Wang, T. Zhang, Hydrogen peroxide is involved in strigolactone induced low temperature stress tolerance in rape seedlings (Brassica rapa L.), *Plant Physiol Biochem* 157 (2020) 402–415. https://doi.org/10.1016/j.plaphy.2020.11.006.
5. Q. Fariduddin, T.A. Khan, M. Yusuf, Hydrogen peroxide mediated tolerance to copper stress in the presence of 28-homobrassinolide in Vigna radiata, *Acta Physiol Plant* 36 (2014) 2767–2778. https://doi.org/10.1007/s11738-014-1647-0.
6. Y.P. Jiang, F. Cheng, Y.H. Zhou, X.J. Xia, W.H. Mao, K. Shi, Z.X. Chen, J.Q. Yu, Hydrogen peroxide functions as a secondary messenger for brassinosteroids-induced CO_2 assimilation and carbohydrate metabolism in Cucumis sativus, *J Zhejiang Univ Sci B* 13 (2012) 811–823.
7. F. Nazir, Q. Fariduddin, T.A. Khan, Hydrogen peroxide as a signalling molecule in plants and its crosstalk with other plant growth regulators under heavy metal stress, *Chemosphere* 252 (2020) 126486. https://doi.org/10.1016/j.chemosphere.2020.126486.
8. C. Shi, C. Qi, H. Ren, A. Huang, S. Hei, X. She, Ethylene mediates brassinosteroid-induced stomatal closure via Gα protein-activated hydrogen peroxide and nitric oxide production in Arabidopsis, *Plant J* 82 (2015) 280–301. https://doi.org/10.1111/tpj.12815.
9. R. Konieczny, A.K. Banaś, E. Surówka, Ż. Michalec, Z. Miszalski, M. Libik-Konieczny, Pattern of antioxidant enzyme activities and hydrogen peroxide content during developmental stages of rhizogenesis from hypocotyl explants of Mesembryanthemum crystallinum L., *Plant Cell Rep* 33 (2014) 165–177. https://doi.org/10.1007/s00299-013-1520-4.
10. T.A. Khan, M. Yusuf, A. Ahmad, Z. Bashir, T. Saeed, Q. Fariduddin, S. Hayat, H.-P.P. Mock, T. Wu, Proteomic and physiological assessment of stress sensitive and tolerant variety of tomato treated with brassinosteroids and hydrogen peroxide under low-temperature stress, *Food Chem* 289 (2019) 500–511. https://doi.org/10.1016/j.foodchem.2019.03.029.
11. F. Nazir, Q. Fariduddin, A. Hussain, T.A. Khan, Brassinosteroid and hydrogen peroxide improve photosynthetic machinery, stomatal movement, root morphology and cell viability and reduce Cu-triggered oxidative burst in tomato, *Ecotoxicol Environ Saf* 207 (2021) 111081. https://doi.org/10.1016/j.ecoenv.2020.111081.
12. M.I.R. Khan, N.A. Khan, A. Masood, T.S. Per, M. Asgher, Hydrogen peroxide alleviates nickel-inhibited photosynthetic responses through increase in use-efficiency of nitrogen and sulfur, and glutathione production in mustard, *Front Plant Sci* 7 (2016). https://doi.org/10.3389/fpls.2016.00044.
13. T.A. Khan, Q. Fariduddin, M. Yusuf, Effect of exogenously sourced hydrogen peroxide treatments on growth, photosynthesis and antioxidant traits in two contrasting cultivars of tomato: A mode and concentration dependent study, *Acad J Agric Res* 6 (2018) 19–29. https://doi.org/10.15413/ajar.2018.1201.
14. A. Ahmad, W. Akram, R. Sardar, N.A. Yasin, Editorial: Interactive effects of plant growth-promoting microbes and nanoparticles on the physiology, growth, and yield of crops, *Front Plant Sci* 15 (2024). https://doi.org/10.3389/fpls.2024.1338470.

15. S. Shafique, S. Shafique, A. Ahmad, Biochemical and molecular screening of varieties of chili plants that are resistant against Fusarium wilt infection, *Eur J Microbiol Immunol (Bp)* 8 (2018) 12–19. https://doi.org/10.1556/1886.2017.00031.
16. J. Khan, Z. Bashir, A. Ahmad, W. Tariq, A. Yousaf, M. Gohar, Mathematical modeling of cotton leaf curl virus with respect to environmental factors, *Eur J Microbiol Immunol (Bp)* 5 (2015) 172–176. https://doi.org/10.1556/1886.2015.00012.
17. Z. Bashir, A. Ahmad, S. Shafique, T. Anjum, S. Shafique, W. Akram, Hypersensitive response — A biophysical phenomenon of producers, *Eur J Microbiol Immunol (Bp)* 3 (2013) 105–110. https://doi.org/10.1556/EuJMI.3.2013.2.3.
18. A. Yousaf, A. Qadir, T. Anjum, A. Ahmad, Identification of microbial metabolites elevating vitamin contents in barley seeds, *J Agric Food Chem* 63 (2015) 7304–7310. https://doi.org/10.1021/acs.jafc.5b01817.
19. A. Noman, M. Aqeel, N. Khalid, W. Islam, T. Sanaullah, M. Anwar, S. Khan, W. Ye, Y. Lou, Zinc finger protein transcription factors: Integrated line of action for plant antimicrobial activity, *Microb Pathog* 132 (2019) 141–149. https://doi.org/10.1016/j.micpath.2019.04.042.
20. A. Ahmad, Y. Ashraf, In Vitro and In Vivo Management of Alternaria Leaf Spot of Brassica campestris L., *J Plant Pathol Microbiol* 7 (2016) 1000365. https://doi.org/10.4172/2157-7471.1000365.
21. S. Ahmed, H. Liu, A. Ahmad, W. Akram, E.K.N. Abdelrahman, F. Ran, W. Ou, S. Dong, Q. Cai, Q. Zhang, X. Li, S. Hu, X. Hu, Characterization of anti-bacterial compounds from the seed coat of chinese windmill palm tree (trachycarpus fortune), *Front Microbiol* 8 (2017) 1894. https://doi.org/10.3389/fmicb.2017.01894.
22. K.J. Dietz, L. Vogelsang, H_2O_2 sensing in immunity, *Nature Plants* 8 (2022) 1140–1141. https://doi.org/10.1038/s41477-022-01256-1.
23. U. Chakraborty, B. Pradhan, Oxidative stress in five wheat varieties (Triticum aestivum L.) exposed to water stress and study of their antioxidant enzyme defense system, water stress responsive metabolites and H_2O_2 accumulation, *Braz J Plant Physiol* 24 (2012) 117–130. https://doi.org/10.1590/S1677-04202012000200005.
24. G. Bi, M. Hu, L. Fu, X. Zhang, J. Zuo, J. Li, J. Yang, J.-M. Zhou, The cytosolic thiol peroxidase PRXIIB is an intracellular sensor for H_2O_2 that regulates plant immunity through a redox relay, *Nature Plants* 8 (2022) 1160–1175. https://doi.org/10.1038/s41477-022-01252-5.
25. X. Xia, C. Gao, L. Song, Y. Zhou, K. Shi, J. Yu, Role of H_2O_2 dynamics in brassinosteroid-induced stomatal closure and opening in S olanum lycopersicum, *Plant Cell Environ* 37 (2014) 2036–2050.
26. C. Cloix, E. Kaiserli, M. Heilmann, K.J. Baxter, B.A. Brown, A. O'Hara, B.O. Smith, J.M. Christie, G.I. Jenkins, C-terminal region of the UV-B photoreceptor UVR8 initiates signaling through interaction with the COP1 protein, *Proc Natl Acad Sci USA* 109 (2012) 16366–16370. https://doi.org/10.1073/pnas.1210898109.
27. A. Ahmad, Y. Liu, Phenology forcing model to estimate phenology shifting ability of extreme environmental events, *Front Plant Sci* 13 (2022) 3237. https://doi.org/10.3389/FPLS.2022.961335.
28. A. Ahmad, Y. Liu, Q. Ge, Assessing environmental thresholds in relation to plant structure and nutritional value for improved maize calendar ensuring food security, *Sci Total Environ* 834 (2022) 155120. https://doi.org/10.1016/j.scitotenv.2022.155120.
29. N.A. Yasin, W.U. Khan, S.R. Ahmad, A. Ali, A. Ahmad, W. Akram, Imperative roles of halotolerant plant growth-promoting rhizobacteria and kinetin in improving salt tolerance and growth of black gram (Phaseolus mungo), *Environ Sci Pollut Res* 25 (2018) 4491–4505. https://doi.org/10.1007/s11356-017-0761-0.
30. N.A. Yasin, W. Akram, W.U. Khan, S.R. Ahmad, A. Ahmad, A. Ali, Halotolerant plant-growth promoting rhizobacteria modulate gene expression and osmolyte production to improve salinity tolerance and growth in Capsicum annum L., *Environ Sci Pollut Res* 25 (2018) 23236–23250. https://doi.org/10.1007/s11356-018-2381-8.
31. N.A. Yasin, W.U. Khan, S.R. Ahmad, A. Ali, A. Ahmad, W. Akram, Effect of enterobacter sp. CS2 and EDTA on the phytoremediation of ni-contaminated soil by Impatiens balsamina, *Pol J Environ Stud* 28 (2019) 425–433. https://doi.org/10.15244/pjoes/76179.
32. A. Hussain, Amna, M.A. Kamran, M.T. Javed, K. Hayat, M.A. Farooq, N. Ali, M. Ali, H. Manghwar, F. Jan, H.J. Chaudhary, Individual and combinatorial application of Kocuria rhizophila and citric acid on phytoextraction of multi-metal contaminated soils by Glycine max L, *Environ Exp Bot* 159 (2019) 23–33. https://doi.org/10.1016/j.envexpbot.2018.12.006.
33. A. Ahmad, I. Shahzadi, S. Mubeen, N.A. Yasin, W. Akram, W.U. Khan, T. Wu, Karrikinolide alleviates BDE-28, heat and Cd stressors in Brassica alboglabra by correlating and modulating biochemical attributes, antioxidative machinery and osmoregulators, *Ecotoxicol Environ Saf* 213 (2021) 112047. https://doi.org/10.1016/j.ecoenv.2021.112047.

34. N.A. Yasin, W.U. Khan, S.R. Ahmad, A. Ahmad, W. Akram, M. Ijaz, Role of acinetobacter sp. CS9 in improving growth and phytoremediation potential of catharanthus longifolius under cadmium stress, *Pol J Environ Stud* 28 (2018) 435–443. https://doi.org/10.15244/pjoes/80806.
35. M. Akbar, S.A. Chohan, N.A. Yasin, A. Ahmad, W. Akram, A. Nazir, Mycorrhizal inoculation enhanced tillering in field grown wheat, nutritional enrichment and soil properties, *Peer J* 11 (2023) e15686. https://doi.org/10.7717/peerj.15686.
36. A. Ahmad, T.A. Khan, S. Mubeen, I. Shahzadi, W. Akram, T. Saeed, Z. Bashir, R. Wang, M. Alam, S. Ahmed, D. Hu, G. Li, T. Wu, Metabolic and proteomic perspectives of augmentation of nutritional contents and plant defense in vigna unguiculata, *Biomolecules* 10 (2020) 224. https://doi.org/10.3390/biom10020224.
37. A. Ahmad, W. Akram, I. Shahzadi, R. Wang, D. Hu, Z. Bashir, W. Jaleel, S. Ahmed, W. Tariq, G. Li, T. Wu, N. Ahmad Yasin, S. Shafique, Benzenedicarboxylic acid upregulates O48814 and Q9FJQ8 for improved nutritional contents of tomato and low risk of fungal attack, *J Sci Food Agric* 99 (2019) 6139–6154. https://doi.org/10.1002/jsfa.9836.
38. M.M. Zaheer, N.A. Yasin, S.R. Ahmad, W.U. Khan, A. Ahmad, A. Ali, S.U. Rehman, Amelioration of cadmium stress in gladiolus (*Gladiolus grandiflora* L.) by application of potassium and silicon, *J Plant Nutr* 41 (2018) 461–476. https://doi.org/10.1080/01904167.2017.1385808.
39. H.Z. Raza, A.A. Shah, Z. Noreen, S. Usman, S. Zafar, N.A. Yasin, S.R.M. Sayed, F.A. Al-Mana, H.O. Elansary, A. Ahmad, Farzana habib, M. Aslam, Calcium oxide nanoparticles mitigate lead stress in Abelmoschus esculentus though improving the key antioxidative enzymes, nutritional content and modulation of stress markers, *Plant Physiol Biochem* 206 (2024) 108171. https://doi.org/10.1016/j.plaphy.2023.108171.
40. S. Faiz, N.A. Yasin, W.U. Khan, A.A. Shah, W. Akram, A. Ahmad, A. Ali, N.H. Naveed, L. Riaz, Role of magnesium oxide nanoparticles in the mitigation of lead-induced stress in *Daucus carota* : modulation in polyamines and antioxidant enzymes, *Int J Phytoremed* 24 (2022) 364–372. https://doi.org/10.1080/15226514.2021.1949263.
41. A. Ahmad, W.U. Khan, A. Ali Shah, N.A. Yasin, S. Naz, A. Ali, A. Tahir, A. Iram Batool, A.A. Shah, N.A. Yasin, S. Naz, A. Ali, A. Tahir, Synergistic effects of nitric oxide and silicon on promoting plant growth, oxidative stress tolerance and reduction of arsenic uptake in Brassica juncea, *Chemosphere* 262 (2021) 128384. https://doi.org/10.1016/j.chemosphere.2020.128384.
42. A.A. Shah, S. Aslam, M. Akbar, A. Ahmad, W.U. Khan, N.A. Yasin, B. Ali, M. Rizwan, S. Ali, Combined effect of Bacillus fortis IAGS 223 and zinc oxide nanoparticles to alleviate cadmium phytotoxicity in Cucumis melo, *Plant Physiol Biochem* 158 (2021) 1–12. https://doi.org/10.1016/j.plaphy.2020.11.011.
43. N.A. Yasin, W.U. Khan, S.R. Ahmad, A. Ali, S. Ahmed, A. Ahmad, Effect of Bacillus fortis 162 on growth, oxidative stress tolerance and phytoremediation potential of Catharanthus roseus under chromium stress, *Int J Agric Biol* 20 (2018) 1513–1522. https://doi.org/10.17957/IJAB/15.0655.

Index

Note: **Bold** page numbers refer to tables and *italic* page numbers refer to figures.